Conversion of SI Units to English Units

To Convert from	To	MULTIPLY BY: ACCURATE	COMMON
Joule (J)	Foot-pound-force (ft-lbf)	$7.375\ 620 \times 10^{-1}$	0.737
Kilogram (kg)	Slug	$6.852\ 178 \times 10^{-2}$	0.068 5
Kilogram (kg)	Pound-mass (lbm avoirdupois)	$2.204\ 622 \times 10^{0}$	2.20
Kilogram (kg)	Ton (short 2000 lbm)	$1.102\ 311 \times 10^{-3}$	0.001 10
Kilometer (km)	Mile (mi U.S. Statute)	$6.213\ 713 \times 10^{-1}$	0.621
Kilowatt (kw)	Horsepower (hp)	$1.341\ 022 \times 10^{0}$	1.34
Meter (m)	Foot (ft)	$3.280\ 840 \times 10^{0}$	3.28
Meter (m)	Inch (in.)	$3.937\ 008 \times 10^{1}$	39.4
Meter^3 (m^3)	Gallon (gal U.S.)	$2.641\ 720 \times 10^{2}$	264.
Newton (N)	Pound-force (lbf avoirdupois)	$2.248\ 089 \times 10^{-1}$	0.225
Pascal (Pa)	Pound-force/foot^2 (lbf/ft^2)	$2.088\ 543 \times 10^{-2}$	0.0209
Pascal (Pa)	Pound-force/inch^2(psi)	$1.450\ 370 \times 10^{-4}$	0.000 145
Watt (W)	Foot-pound-force/second (ft-1bf/s)	$7.375\ 620 \times 10^{-1}$	0.737

Engineering Mechanics of Deformable Bodies

Fourth edition

Edward F. Byars
Professor of Mechanical Engineering and Mechanics
Clemson University

Robert D. Snyder
Dean of Engineering
University of North Carolina at Charlotte

Helen L. Plants
Professor of Mechanical Engineering
West Virginia University

HarperCollins*Publishers*

Sponsoring Editor: Cliff Robichaud
Project Editor: Bob Greiner
Designer: T. R. Funderburk
Production: Delia Tedoff
Compositor: Syntax International Pte. Ltd.
Printer and Binder: Vail-Ballou Press Inc.
Art Studio: Vantage Art, Inc.
Cover: Ben Kann

Engineering Mechanics of Deformable Bodies, Fourth Edition
Copyright © 1983 by Edward F. Byars, Robert D. Snyder, and Helen L. Plants

Library of Congress Cataloging in Publication Data

Byars, Edward Ford, 1925–
  Engineering mechanics of deformable bodies.

  Includes index.
  1. Strength of materials. 2. Mechanics, Applied.
I. Snyder, Robert D.   II. Plants, Helen Lester.
III. Title.
TA405.B92   1983   620.1′12   82-15699
ISBN 0–06–041109–0

Contents

Chapter 3 / Experimental Mechanical Properties of Engineering Materials

Chapter 4 / Mechanical Response of Materials

Chapter 5 / Members in Simple Tension, Compression, and Shear

Chapter 6 / Torsion

Chapter 7 / Beam Stresses

Chapter 8 / Beam Deflections I

Chapter 13 / Work and Energy Methods

Chapter 14 / Nonstatic Loads, Strain Concentrations, and Time-Dependent Properties

Chapter 15 / Experimental Mechanics

Preface

Although the purpose and philosophy of this book have not changed, this fourth edition contains some very significant revisions aimed primarily at improving teaching and learning the subject matter. We have reworked and reorganized much of the content. However, the book, which presupposes a working knowledge of statics and integral calculus, continues to be a text for a one- or two-semester sophomore-junior course for all engineering students—as well as a resource for practicing engineers. The writing style is rather informal, the presentation a blend of rigorous proofs and physically intuitive arguments imbued with a strong flavor of the real world of engineering.

Specifically, each chapter begins with a list of objectives giving students learning targets. Some individual sections have been shortened and some new ones added to sharpen the focus of the presentation. More illustrative and exercise problems have been added to provide better gradation from the more simple to the more complex applications of the theory. Also, we have increased the use of SI units throughout the book and appendixes.

As for changes within chapters, the first ten have been substantially revised and reorganized, with new material added. The chapters dealing with mechanical properties of materials have been reorganized in response to users' suggestions. Material has been added on joints, statically indeterminate members, section modulus, and columns. Design has been emphasized throughout. The basic discussion of beams has been expanded from two chapters to three. Of particular note is the treatment of shear and moment diagrams in which we have reinforced their importance and facilitated learning by introducing practice in the use of the diagrams before we teach the use of shear and moment equations and singularity functions. To do this, we have judiciously repeated some ideas and applications at points throughout the three chapters.

The two senior authors wish to acknowledge the contributions and efforts of our junior author, Helen Plants, without whose participation this fourth edition would be nothing more than a fond hope. In turn she wishes to recognize the help given by Kenneth Plants and Virginia Kiddy in the preparation of the manuscript and by Dean Dubbe, Porter Arbogast, Richard Yih, Robert Minehart, and Steve Chang in the solution of the problems. Last, she would like to acknowledge her intellectual debt to Glenn Murphy, whose excellent text profoundly influenced her approach to teaching mechanics, and upon whose problems many of hers are modelled.

Edward F. Byars
Robert D. Snyder
Helen L. Plants

To the student

This book will introduce you to the analysis of nonrigid or deformable bodies. What are "deformable" bodies? They are simply engineering members (or pieces, structures, parts, components, or whatever you want to call them) that change geometrically and mechanically when loaded. We present relatively simple members in this book. The more complex members and the more complex loadings will come in later design courses.

It has been only during the twentieth century that the study of deformable bodies has gained its rightful recognition as a basic engineering science. Before this time much design was done on a trial-and-error basis where rules of thumb were used extensively and the reliability of the results depended on whose thumb was used. In today's competitive engineering world, where economy and reliability are paramount, and with the present use of high-speed machines operating under great ranges of temperature and other conditions, the obsolete rules of thumb no longer suffice, and the engineers must have a basic knowledge of deformable bodies upon which to build sound design theories.

In this book we present and discuss some of the fundamental principles concerning the behavior of deformable bodies. These principles are some of the tools of the trade with which you can undertake the big job of determining the load-carrying ability of various types of engineering components.

As we analyze these various simple components, we must continually ask ourselves three basic questions:

1 How is the body loaded?

2 What is its shape (geometry)?

3 What is it made of (material)?

We discuss these questions many times throughout this book.

Even though we are considering the subject matter of this book as a basic engineering science, we must be realistic and practical. Real engineering conditions and responses often do not fit our idealizations. Some examples of idealizations of conditions are: considering pin joints to be smooth, considering a load to be applied at a point, neglecting the weight of a member, neglecting small temperature gradients in the bodies, and neglecting the dimension changes of loaded bodies when writing equilibrium equations.

Some examples of idealizations of responses are: considering the materials to be homogeneous and isotropic, assuming that the behavior of a body's surface is indicative of its internal behavior, and assuming in many cases that the material behaves similarly under dissimilar loading conditions.

Some of these idealizations do not appreciably affect the validity of the solution, but in other cases idealizations can drastically limit the usefulness of a particular solution. Watch out for these as we go along, and be aware of their significance.

Good luck!

Chapter 1
Stress

1-1 / Objectives

Upon completion of this chapter, you will be able to:

1 Define stress.
2 Differentiate between normal stress and shearing stress.
3 Compute simple internal loads.
4 Compute normal and shearing stress caused by simple loads.
5 Compute the state of stress on any plane when given the states of stress on orthogonal planes.
6 Identify principal stresses and planes.
7 Draw and interpret Mohr's circle for stress.
8 Compute maximum shearing stress in three-dimensional problems.

Figure 1-1

1-2 / Introduction

The primary purpose of most machines and structures, such as those in Fig. 1-1, is to transmit or support loads. A load-carrying device must be properly designed and fabricated to withstand the loads that will be applied to the device while it is performing its intended function. If the device is improperly designed or fabricated, or if the actual applied loads exceed the design specifications, the device will probably fail to perform its intended function with possible serious consequences. A well-engineered structure or machine greatly minimizes the possibility of costly failures.

In this book we wish to develop the fundamental concepts needed to determine the load-carrying capacity of some basic structural and machine members such as beams, columns, and shafts. While the structural engineer and the machine design engineer are usually the experts in such matters, any engineer (electrical, industrial, chemical, and so on) might be called upon to determine or estimate the load-carrying capacity of some type of device encountered at work or leisure. This book should provide you with the tools of the trade for handling such problems.

We assume that you have had a course in statics in which you dealt with equilibrium and resultants of forces and force systems. In all probability the bodies on which the forces were acting were considered to be *rigid* and no consideration was given to the *deformation* or dimension changes caused by the applied loads. Indeed, you probably did not even worry about whether the body was strong enough to support the loads. In this textbook the *strength* of a body and the *deformation* of the body are the two central considerations. However, free-body diagrams and equilibrium equations remain of utmost importance throughout this book.

1-3 / Internal reactions

In statics much of your time was spent in determining the *external* reactions on a loaded body or structure necessary to keep the body in statical equilibrium. In Fig. 1-2, the idealized foundation supports at *a* and *b* exert external reactions on the loaded A-frame as indicated in the free-body diagram in Fig. 1-2(b). On the other hand, if we were to consider the horizontal member *c-d-e*, the external reactions would be those at *c* and *d* exerted by the pin joints connecting this member with the other members of the structure, as shown in Fig. 1-2(c). Thus, external reactions are those forces and/or couples that one body exerts on another usually through some sort of connecting or restraining device.

Assuming the free-body diagram is correct and complete, we can usually determine the external reactions by applying the requirements of statical equilibrium, namely, that the resultant force and resultant moment must be zero; that is,

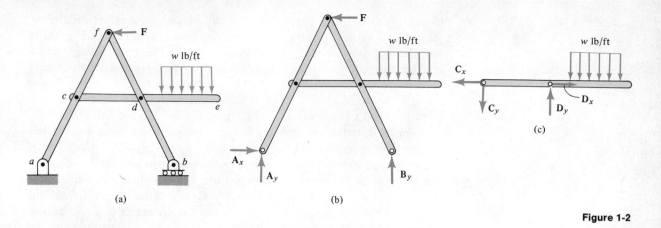

(a)

(b)

(c)

Figure 1-2

$$\sum \mathbf{F} = 0 \qquad \sum \mathbf{M} = 0 \qquad\qquad (1\text{-}1)$$

In some cases these requirements will not be sufficient to determine all the reactions and the problem is *statically indeterminate*. We will encounter such situations later in this book, but for the time being we assume the situation to be statically determinate.

Consider now the beam in Fig. 1-3(a) and its corresponding FBD

Figure 1-3

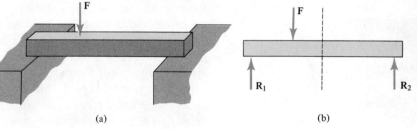

(a)

(b)

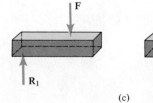

(c)

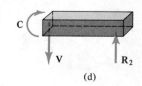

(d)

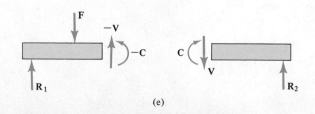

(e)

(free-body diagram) in Fig. 1-3(b). Instead of considering the entire beam, suppose we visually cut the beam into two halves as shown in Fig. 1-3(c) with the appropriate external reactions on each. It is immediately apparent that the right half cannot be in equilibrium as it is shown in this figure since $\mathbf{R}_2 \neq \mathbf{0}$. In order for this half of the beam to be in equilibrium, the left half of the beam has to exert some forces and/or couples upon the right half and vice versa. In this particular case, the left half must exert a downward vertical force $\mathbf{V}$ for forcewise equilibrium and a clockwise couple $\mathbf{C}$ for momentwise equilibrium as shown in Fig. 1-3(d). Conversely, the right half exerts "equal but opposite" reactions $-\mathbf{V}$ and $-\mathbf{C}$ upon the left half of the beam (Newton's third law) as shown in Fig. 1-3(e). The reactions $\mathbf{V}$ and $\mathbf{C}$ (or $-\mathbf{V}$ and $-\mathbf{C}$) are called *internal reactions* because they are reactions that one *part* of a body exerts on another *part* of the same body to hold the body together in statical equilibrium. If the material of the body is not strong enough to exert the required internal reactions throughout the body, the body will break or deform severely. Thus, the analysis of internal reactions throughout a body is a major step in determining the load-carrying capacity of that body.

Internal reactions, like the external reactions you learned to compute in statics, are either forces, couples, or forces and couples. There is no way that a force can be replaced with a couple although both may exist on a single plane. Each force and each couple may be considered to act on the plane independently of all other loads.

An internal force can have three possible results on a body. (1) It may produce tension in the body, causing adjacent elements to move farther apart, lengthening the body. (2) It may compress the body, causing adjacent elements to move closer together, shortening the body. (3) It may shear the body, causing adjacent planes to slip on one another, changing the angular relationships within the body. Figure 1-4(a), (b), and (c) illustrate these three possible effects, respectively.

Tension and compression are easy to visualize and understand for most people, but shear often presents a difficulty. It may be helpful to note that scissors are called shears because one blade slides past the other. Another helpful analogy is to consider the loaded body as being composed

Figure 1-4

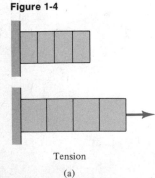

Tension

(a)

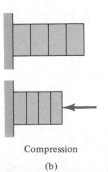

Compression

(b)

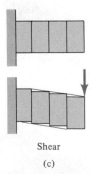

Shear

(c)

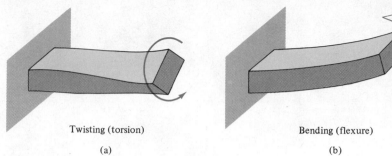

Twisting (torsion)

(a)

Bending (flexure)

(b)

Figure 1-5

of an infinite number of thin plates like a deck of cards. Under the in-fluence of a shearing force the rectangular deck slides over and becomes a parallelepiped. All three sorts of force have some secondary effects on size and shape but these will be considered later.

Internal couples have two possible effects on a loaded body. A couple may cause twisting or bending as illustrated in Fig. 1-5(a) and (b), respectively. It should be noted that in Fig. 1-4 the forces causing tension and compression are normal to the plane on which they act, while the force causing shear is parallel to the plane. If a force acting on a plane is neither normal nor parallel to the plane, it must be resolved into two components, one normal to the plane and one parallel to the plane. Couples must be handled in the same manner. A couple oblique to the plane on which it acts must be resolved into a couple causing bending and another couple causing twisting. The breaking down of internal loads into the components that are normal and parallel to the plane on which they act is a vital step in the solution of problems.

Let us now apply these ideas to a body of some general shape loaded in an arbitrary manner as shown in Fig. 1-6(a). If we pass a cutting plane through the body and draw the FBD of the left section, the internal reaction might be a single force, a single couple, or a force-couple system such as shown in Fig. 1-6(b) where a double-arrowed vector is used to

Figure 1-6

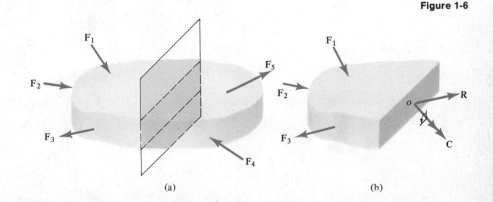

(a)

(b)

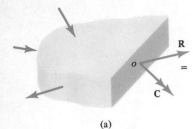
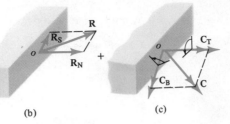

Figure 1-7

indicate the couple **C** by means of the right-hand rule, as indicated by the circular arrow. Note that the value of **C** will depend upon the location of the force **R**. In most instances we will take point o to be the centroid of the cross section.

Although the force **R** and the couple **C** can be resolved into rectangular "horizontal" and "vertical" components, it is much more meaningful from a "strength" viewpoint to decompose these reactions in the following manner:

1 The internal force **R** is resolved into a force $\mathbf{R_N}$ normal (perpendicular) to the cut surface and a force $\mathbf{R_S}$ tangent (parallel) to the cut surface as shown in Fig. 1-7(b). The normal force exerts a tension or compression on the material whereas the tangent force tends to shear the material (thus the subscript S).

2 The internal couple **C** is resolved into a couple $\mathbf{C_T}$ perpendicular to the surface and a couple $\mathbf{C_B}$ parallel to the surface as shown in Fig. 1-7(c). Physical interpretations of these couples show that $\mathbf{C_T}$ tends to twist the body whereas $\mathbf{C_B}$ tends to bend the body, thus the subscripts T and B.

The following example problems illustrate these ideas of internal reactions.

EXAMPLE 1-1

Find the internal reactions at sections a-a and b-b in the bracket shown in Fig. 1-8.

Solution:

A FBD of the entire bracket is shown in Fig. 1-9(a). Noting that BC is a two-force member so that $\mathbf{F}_{BC}$ must have the line of action shown, we take

$$\sum M\hat{A} = 0$$

$$(\tfrac{4}{5}F_{BC} \times 0.6) - (1200 \times 1.6) = 0$$

$$F_{BC} = \tfrac{5}{4} \times 3200 = 4000 \text{ N}$$

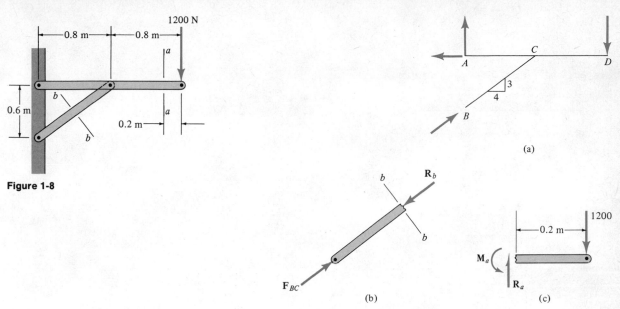

Figure 1-8

(a)

(b)

(c)

Figure 1-9

Figure 1-9(b) shows a FBD of the portion of BC below section b-b. We can see that

$$R_b = F_{BC} = 4000 \text{ N compression}$$

Figure 1-9(c) shows the portion of the horizontal member to the right of section a-a. Taking moments about the left edge, we find

$$M_a - (1200 \times 0.2) = 0$$

$$M_a = 1200 \times 0.2 = 240 \text{ N·m bending}$$

Considering $\sum \mathbf{F}_y = 0$ gives

$$R_a = 1200 \text{ N shear}$$

EXAMPLE 1-2

For the crankshaft shown in Fig. 1-10(a) determine the internal reactions at section a-a and section b-b.

Solution:

A FBD of that portion of the crankshaft to the right of section a-a is shown in Fig. 1-10(b) with the internal reactions labeled $\mathbf{R}_N$, $\mathbf{R}_S$, $\mathbf{C}_T$, and $\mathbf{C}_B$. Let $\mathbf{i}$, $\mathbf{j}$, $\mathbf{k}$ be orthogonal unit vectors. Then, summing forces

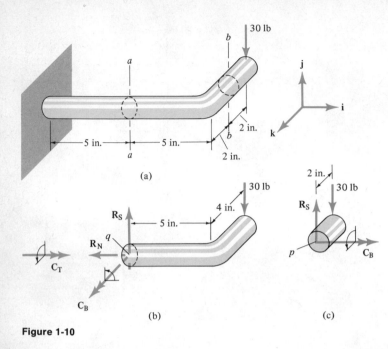

(a)

(b)   (c)

Figure 1-10

$$\sum \mathbf{F} = 0$$

$$-30\mathbf{j} - R_\mathrm{N}\mathbf{i} + R_\mathrm{S}\mathbf{j} = 0$$

$$R_\mathrm{N} = 0 \text{ lb} \qquad R_\mathrm{S} = 30 \text{ lb as shown}$$

Summing moments with respect to point q yields

$$\sum \mathbf{M}_q = 0$$

$$-30(4)\mathbf{i} - 30(5)\mathbf{k} + C_\mathrm{T}\mathbf{i} + C_\mathrm{B}\mathbf{k} = 0$$

$$C_\mathrm{T} = 120 \text{ in.-lb} \qquad C_\mathrm{B} = 150 \text{ in.-lb as shown}$$

Hence, at section a-a the crankshaft is being sheared, twisted, and bent simultaneously.

A FBD for section b-b is shown in Fig. 1-10(c) with internal reactions $\mathbf{R}_\mathrm{S}$ and $\mathbf{C}_\mathrm{B}$. (It should be apparent to you that there is no normal force or twisting couple on this section.) Summing forces

$$-30\mathbf{j} + R_\mathrm{S}\mathbf{j} = 0$$

$$R_\mathrm{S} = 30 \text{ lb as shown}$$

Summing moments with respect to point p gives

$$-30(2)\mathbf{i} + C_\mathrm{B}\mathbf{i} = 0$$

$$C_\mathrm{B} = 60 \text{ in.-lb as shown in Fig. 1-10(c).}$$

Hence, section b-b is being sheared and bent but *not* twisted.

PROBLEMS

1-1 to 1-22 For each of the devices shown, determine the internal reactions at *a-a*, *b-b*, and *c-c*. Resolve each reaction into components normal and parallel to the indicated section. Identify each component as tension, compression, shear, twisting, or bending.

Problem 1-1

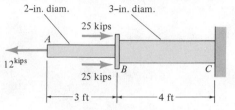

Problem 1-2

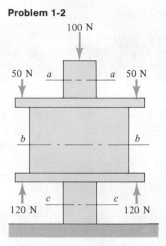

Problem 1-3

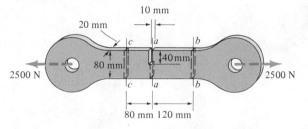

Problem 1-4

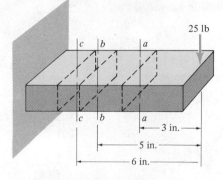

Problem 1-5

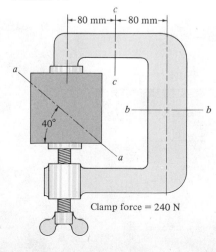

Problem 1-6

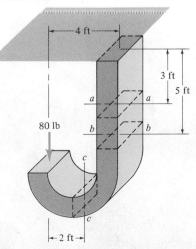

Problem 1-7

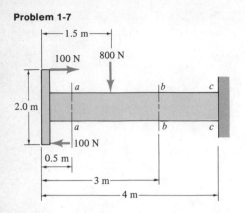

Problem 1-8

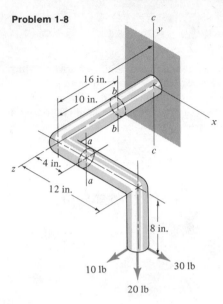

Problem 1-9

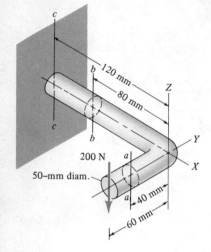

Problem 1-10

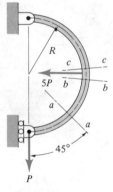

Problem 1-11

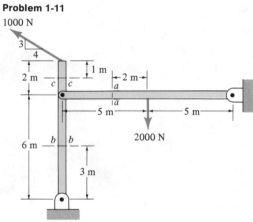

Problem 1-12

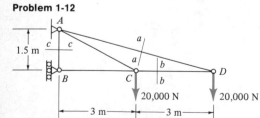

Problem 1-13

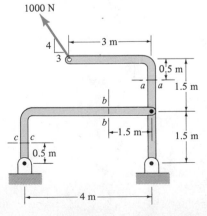

Problem 1-14

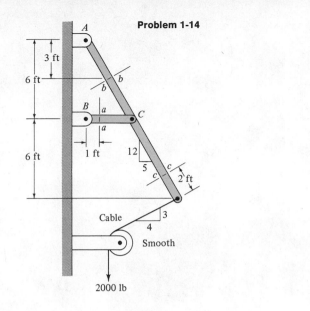

A

3 ft

6 ft

b

b

B

a

C

a

1 ft

12

5

c

c

2 ft

Cable

3

4

Smooth

2000 lb

Problem 1-15

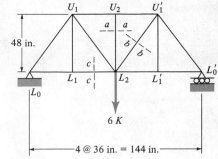

U_1 U_2 U_1'

48 in.

a a

b

b

c

L_1 c

L_2 L_1' L_0'

L_0

6 K

4 @ 36 in. = 144 in.

Problem 1-16

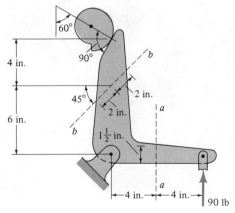

60°

90°

b

4 in.

45°

2 in.

2 in.

b

$1\frac{1}{2}$ in.

a

6 in.

a

4 in. 4 in.

90 lb

Problem 1-17

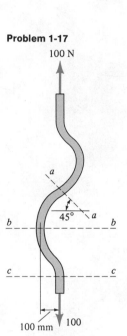

100 N

a

45° a

b b

c c

100 mm 100

Problem 1-18

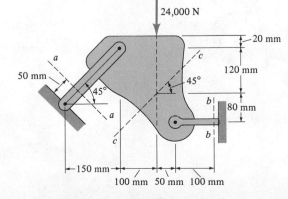

24,000 N

20 mm

a

c

50 mm

120 mm

45°

45°

a

b

80 mm

c

b

150 mm

100 mm 50 mm 100 mm

Problem 1-19

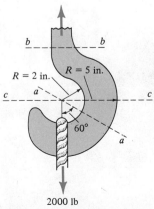

b b

R = 2 in. R = 5 in.

c a c

60°

a

2000 lb

Problem 1-20

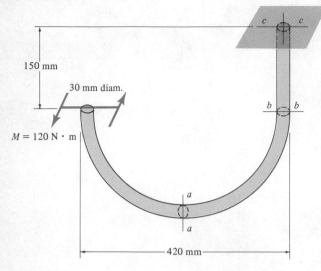

Problem 1-21

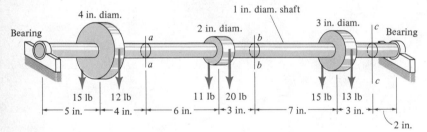

Problem 1-22

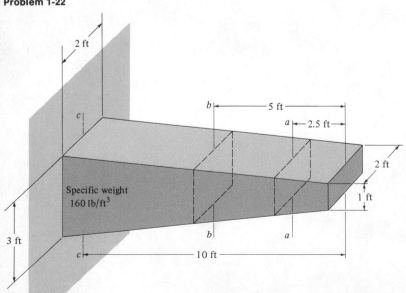

1-4 / The stress concept

The idea of internal reactions is a useful scheme for determining how strong a structural member must be in terms of the magnitude and type of forces or couples the member must transmit. But the question as to the actual strength of a structural member is inherently tied to such factors as the size and shape of the member and the material of which it is made. For example, our experience leads us to surmise that a $\frac{1}{2}$-in.-diameter bolt is considerably stronger than a similar $\frac{1}{4}$-in.-diameter bolt, and we attribute this simply to the fact that the $\frac{1}{2}$-in. bolt is bigger and has more material than the $\frac{1}{4}$-in. bolt. Hence, we anticipate that the size and possibly the shape will enter into the evaluation of the strength of a structural member.

As a second example, our experience leads us to believe that steel is stronger than, say, wood. Probably, what we mean by stronger is that, given a piece of steel and a piece of wood of the *same* size and shape, the steel one can withstand a correspondingly greater load of any type than can the wood one before breaking. Intuitively, then, when comparing the relative strength of two bodies of material, we should make the comparison on the basis of the same geometric and loading configurations. However, it is quite apparent that all structural members do not have the same geometry and loading conditions, and therefore, to compare the relative strengths of two different structural members, we must devise some common denominator that incorporates the size, shape, and loading. Accordingly, we are led to introduce the concept of *stress*.

Thus far we have been dealing with the net internal reactions manifested by a force **R** and a couple **C**. However, these resultants are actually the cumulative effects of the many, many minute cohesive forces that are exerted by the individual material particles across the cut section of the body. In Fig. 1-11 is illustrated the idea of forces on individual particles. Here each particle having a cross-sectional area Δa resists two component loads, namely $\Delta \mathbf{R}_N$ normal to Δa and $\Delta \mathbf{R}_S$ parallel to Δa. The magnitude and direction of the force $\Delta \mathbf{R}$ may vary from particle to particle

Figure 1-11

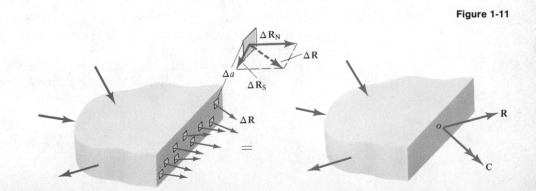

and, in general, are dependent upon the size of the area Δa. If we wish, we could define an average force intensity on the particle by dividing the magnitude of the force $\Delta \mathbf{R}$ by the corresponding area Δa.

Although these particles can be quite small, they are nevertheless finite. Furthermore, in most materials on a microscopic level there are voids or cracks between the particles. It is convenient, however, to think of the material as being "smeared out" in a fashion such that the cut section is a smooth continuous surface of voidless material. In this way the areas Δa can be made as small as we like; we indicate this by using the symbol da. Likewise, the corresponding forces $\Delta \mathbf{R}$ become $d\mathbf{R}$ with

$$d\mathbf{R} = d\mathbf{R_N} + d\mathbf{R_S} \tag{1-2}$$

and we write

$$dR_N = \sigma \, da \tag{1-3a}$$

$$dR_S = \tau \, da \tag{1-3b}$$

The intensities σ and τ are called the normal and shear stress respectively, and are usually thought of as

$$\sigma = \lim_{\Delta a \to 0} \frac{\Delta R_N}{\Delta a} = \frac{dR_N}{da} \tag{1-4}$$

$$\tau = \lim_{\Delta a \to 0} \frac{\Delta R_S}{\Delta a} = \frac{dR_S}{da} \tag{1-5}$$

since Δa can be made as small as we like on our "smeared out" or voidless material surface.

In simplistic terms, stress is the internal force per unit area and is used to measure the strength of a member. It is usually assumed that if the stress in a member is kept below a certain level, the member will perform properly, while if the stress exceeds a certain level, the member will fail in some manner. A substantial portion of this book is devoted to determining stresses caused by loads acting on real structural and machine members.

There are two general types of normal stresses: tensile and compressive. Tensile normal stress is produced by a force acting away from the plane area—a pulling force. Compressive normal stress is produced by a force acting toward the plane area—a pushing force. Shear stresses, however, are produced by forces parallel to the plane area, and there is no need for further classification.

It should be noted that stress, either normal or shear, must be associated with a plane area and is not simply a force vector.* Normal stress is derived from, and should always be associated with, a force perpendic-

---

* In a more sophisticated mathematical treatment, stress is a quantity of higher order called a tensor.

Table 1-1   Systems of Units

System	Force	Length	Stress	Energy
Engineering English	pound (lb)	foot (ft) inch (in.)	lb/ft^2 $lb/in.^2$ (psi)	ft-lb in.-lb
Absolute metric (cgs)	dyne (dyn)	centimeter (cm)	dyn/cm^2	erg = dyn-cm
International	newton (N)	meter (m)	pascal (Pa) = N/m^2	joule (J) = $N \cdot m$

ular to a particular plane area. Shear stress is derived from, and should be associated with, a force parallel to a plane area. Care must be taken to avoid the use of *stress* as a synonym for *force*.

The fundamental dimensions of stress are force divided by length squared. Traditionally, the most commonly used units in engineering in the United States have been pounds per square inch ($lb/in.^2$ or psi). However, the International System (SI) units are gaining in usage in the United States and are used extensively throughout the world. Table 1-1 gives the units of force, length, stress, and energy for three systems and Table A-13 in the Appendix gives the various conversion factors.

You, as an engineer, should develop a feel for these units in relation to engineering materials; for example, as you progress in this course you should acquire a first-hand knowledge of the order of magnitude of the significant properties of the common engineering materials as listed in Table A-1 of the Appendix. In addition to the units given in Table 1-1, there are two units of convenience that are frequently used in engineering work. These are the kip, or kilopound, used with the English system, and the megapascal used with SI units. One kip is 1000 lb, one megapascal (MPa) is 10^6 Pa, and one ksi (kips per square inch), is 1000 psi ($lb/in.^2$).

1-5 / State of stress

Let us assume for the time being that stress is a physically significant quantity that can be used to measure the strength of a load-carrying member. As engineers, we now face the question: Given a structural member of some size, shape, and material loaded in some manner, how do we determine the stress in the member? First of all, you should realize that stress as defined in the previous section is an *internal effect* and is *not* physically measurable in the manner that we measure an external force (such as weight) or a surface traction (such as pressure). An attempt to implant a "stress-measuring device" in a solid body would involve actually cutting the body, implanting the device, and then patching the body back together, all of which would unquestionably alter and damage the body, making our measurements virtually worthless insofar as the original unaltered body is concerned. Also, since the internal reactions are apt to vary throughout the entire body, to determine the stress everywhere in the body would require an innumerable number of such experimental measurements throughout the body. Faced with this dilemma, the engineer tries to develop some rational theory relating the external applied loads and external geometry of the member to the internal stresses. Sometimes these theories are quite sophisticated, other times rather crude, with the choice often dictated by the peculiar requirements, economics, or complexities of a given problem. The engineer's job is to do the best he or she can within the imposed limitations and difficulties.

To begin our analysis of stress let us consider a rather simple combination of geometry and loading, namely, a slender rod with a cross-sectional area A under a tensile load $\mathbf{P}$ acting through the centroid of the cross section as shown in Fig. 1-12(a). From the stress concept we can visualize that the pulling load produces tensile (pulling) stresses σ on the internal cross-sectional area A perpendicular to the axis of the load as indicated in Fig. 1-12(b). If the cross section is small compared with the length of the member, it is reasonable to conjecture that for practical purposes the value of σ is the same over the entire cross section, say σ_{av}. Then, since the net cumulative effect of all the internal forces is equal to the load $\mathbf{P}$, we have

$$P = \int_{\substack{\text{cross} \\ \text{section}}} \sigma \, da = \sigma_{av} \int_A da = \sigma_{av} A$$

or

$$\sigma_{av} = \frac{P}{A} \tag{1-6}$$

Thus we have obtained our first load-stress relation between an applied load $\mathbf{P}$ and an average tensile stress σ_{av}. This is an extremely important result and one that is infinitely useful in a variety of engineering applications. In order to use it, one needs only to make sure that the load $\mathbf{P}$ has been correctly identified and is indeed acting at the centroid of the area A where the stress is to be determined.

Figure 1-12

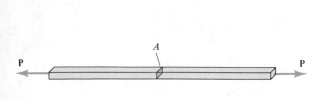

(a)

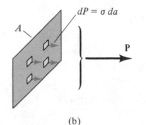

(b)

Figure 1-13

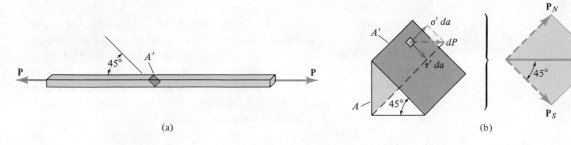

(a)   (b)

However, before we bask in the simplicity of this result, suppose we had considered a cross section A' cut at a $45°$ angle to the axis of the member rather than the perpendicular one of our previous consideration. This situation is depicted in Fig. 1-13.

For this cross section the resultant internal reaction will be composed of both a normal force and a shear force with associated normal and shear stress, σ' and τ', respectively. As before, if we assume that the value of σ' (and τ') is the same over the entire cross section A' we would obtain

$$P_N = \int_{A'} \sigma' \, da = \sigma'_{av} A' \tag{1-7a}$$

$$P_S = \int_{A'} \tau' \, da = \tau'_{av} A' \tag{1-7b}$$

Now,

$$P_N = P \sin 45°$$

$$P_S = P \cos 45°$$

and

$$A = A' \sin 45°$$

Thus

$$\sigma'_{av} = \frac{P_N}{A'} = \frac{P \sin 45°}{A/\sin 45°}$$

$$= \frac{P}{A} (\sin 45°)^2 = \frac{P}{A} \frac{1}{2} \tag{1-8a}$$

and

$$\tau'_{av} = \frac{P_S}{A'} = \frac{P \cos 45°}{A/\sin 45°}$$

$$= \frac{P}{A} \cos 45° \sin 45° = \frac{P}{A} \frac{1}{2} \tag{1-8b}$$

This is a most significant result in that it says that the normal stress σ'_{av} on this $45°$ section is half the value of the normal stress σ_{av} on the perpendicular section. Also, while there appeared to be no shear stress on the perpendicular section, a shear stress τ'_{av} does exist on the 45° section.

It should be reasonably apparent that if we considered yet another cross-sectional area, say at 60° to the axis of the member, we would obtain yet another combination of normal and shear stresses, σ'' and τ'', say. Furthermore, we could consider any number of other possible cross sections each resulting in a possibly different combination of normal and shear stresses. Thus we begin to see that there is no such thing as "the stress" in this body but rather a multitude of possible combinations of stress. We refer to this multitude of possible combinations as a *state of stress*. This particular example of the simple slender rod is called a *uniaxial* state of stress.

In the next section we shall develop this idea of the state of stress much more fully. For now, the important thing is to begin to understand the interrelation between both force and area in producing a stress. The following example is another illustration of the need to identify both the proper force and the area upon which it acts in determining a stress.

EXAMPLE 1-3

The eye-bar in Fig. 1-14 is fusion welded across a 30° seam as shown. Find the average normal and shear stresses on the seam. Determine the values of the maximum normal and shear stresses in the main body of the eye-bar.

Solution:

A FBD of the eye-bar cut along the weld is shown in Fig. 1-14(b) in which **R** has been resolved into normal and shear components. The transverse cross-sectional area of the bar is

$$A = (30)(10) = 300 \text{ mm}^2$$

Figure 1-14

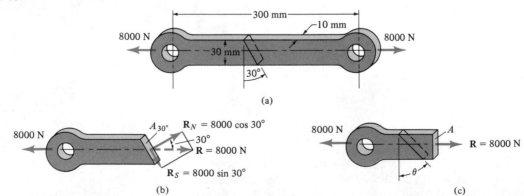

(a)

(b)

(c)

so that the weld area is

$$A_{30°} = \frac{A}{\cos 30°} = \frac{300}{\cos 30°} \, mm^2$$

Thus, the average normal and shear stresses on the weld seam are

$$\sigma_{30°} = \frac{R_N}{A_{30°}} = \frac{8000 \cos 30°}{300/\cos 30°}$$

$$= \frac{8000}{300} (\cos 30°)^2 = 20 \, N/mm^2$$

$$= 20 \times 10^6 \, Pa = 20 \, MPa \tag{a}$$

$$\tau_{30°} = \frac{R_S}{A_{30°}} = \frac{8000 \sin 30°}{300/\cos 30°}$$

$$= \frac{8000}{300} \sin 30° \cos 30° = 11.55 \, N/mm^2$$

$$= 11.5 \times 10^6 \, Pa = 11.5 \, MPa \tag{b}$$

To determine the maximum normal and shear stresses in the main body of the bar, consider an area oriented at some arbitrary angle θ as shown in Fig. 1-14(c). The procedure for determining the normal and shear stresses on this area would be the same as for the stresses on the welded seam except that the angle θ would replace the 30° in equations (a) and (b) above. Thus

$$\sigma_\theta = \frac{8000}{300} (\cos \theta)^2 \tag{c}$$

$$\tau_\theta = \frac{8000}{300} \sin \theta \cos \theta \tag{d}$$

To find the value of θ that maximizes the values of (c) and (d) we differentiate and set equal to zero.

$$\frac{d\sigma_\theta}{d\theta} = \frac{8000}{300} 2 \cos \theta \, (-\sin \theta) = 0$$

$$\therefore \theta = 0°$$

$$\frac{d\tau_\theta}{d\theta} = \frac{8000}{300} (-\sin^2 \theta + \cos^2 \theta) = 0$$

$$\therefore \sin \theta = \cos \theta$$

$$\theta = 45°$$

Therefore, the maximum normal stress occurs on the area perpendicular to the axis of the bar.

$$\sigma_{max} = \frac{8000}{300} (\cos \theta)^2 = 26.66 \text{ N/mm}^2 = 26.66 \times 10^6 \text{ N/mm}^2$$

$$= 26.7 \times 10^6 \text{ Pa} = 26.7 \text{ MPa}$$

Similarly, the maximum shear stress occurs on an area at 45° with the axis of the bar.

$$\tau_{max} = \frac{8000}{300} \sin 45° \cos 45°$$

$$= 13.33 \text{ N/mm}^2 = 13.3 \times 10^6 \text{ Pa} = 13.3 \text{ MPa}$$

Important Remark. A comment about numerical accuracy and significant figures is in order. Hand calculators and large computers are capable of providing numerical results to six, eight, ten, or more figures. However, the output answer cannot be any more accurate than the input data that the calculator used to obtain the answer. Thus if the input data have three accurate figures the calculator cannot provide an answer that is accurate to eight figures even though it may display those eight figures.

Sometimes an engineer does not know how accurate the available data are. In the above example the dimensions were given as 30 mm and 10 mm and the load as 8000 N. How accurate are these numbers? It is doubtful that the bar would be *exactly* 10.000 ... by 30.000 ... or that the load would be *exactly* 8000.0000 .... Therefore, rather than presuming an accuracy of one or four digits or presuming that the data are exact, we have given the answers in scientific notation with three significant figures. We shall continue this policy throughout the text except in those instances when circumstances or common sense dictate otherwise.

PROBLEMS

1-23 In Prob. 1-2 the cross section at sections *a-a* and *c-c* is 40 × 20 mm. Compute the average stress on each cross section.

1-24 Compute the average stress on each of the cross sections indicated in Prob. 1-1.

1-25 The block in Prob. 1-5 is 100 × 100 × 60 mm. (The 60 mm dimension is perpendicular to the page.) Find the state of stress on plane *a-a*.

1-26 The truss in Prob. 1-12 is composed of flat steel bars 20 × 80 mm. Find the stress at section *a-a*.

1-27 The shaded portions of the frame in Prob. 1-14 are composed of flat bars 6 × 2 in. Find the stress at section *a-a*.

1-28 Find the stress at each section indicated in the illustration.

Problem 1-28

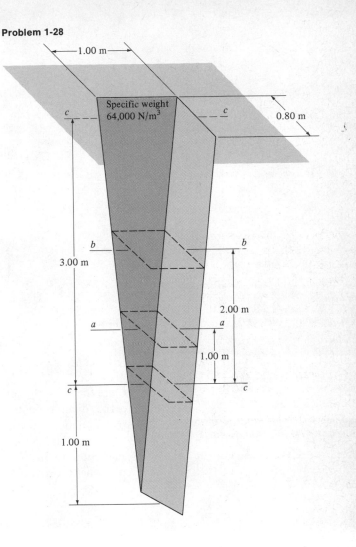

1-6 / General state of stress

Stress at a point in a material body has been defined as a force per unit
area. But, as we have seen, this definition is somewhat ambiguous since
it depends upon what area we consider at the point. To see this, consider
a point q in the interior of the body shown in Fig. 1-15(a). Let x, y, z
be the usual rectangular coordinate axes and let us pass a cutting plane
through point q perpendicular to the x axis as shown in Fig. 1-15(b). If
da_x is the "area surrounding point q in the cut plane perpendicular to the
x axis," then by the definitions of Eqs. 1-3a and 1-3b, the force at point q
can be written in terms of rectangular components as

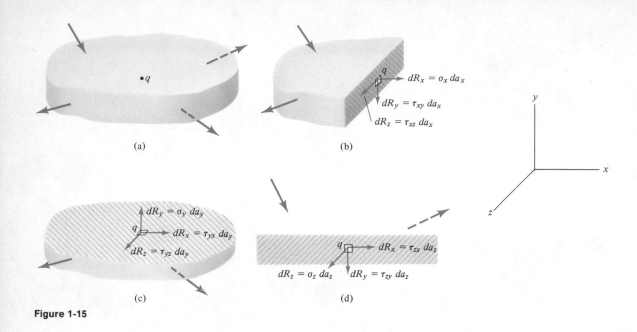

Figure 1-15

$$dR_x = \sigma_x \, da_x \qquad dR_y = \tau_{xy} \, da_x \qquad dR_z = \tau_{xz} \, da_x \qquad (1\text{-}9a)$$

where the subscript x denotes that the area under consideration is normal (perpendicular) to the x axis, while the second subscript on the shear stresses denotes the direction of the shear force vector.

In a similar fashion, let us pass a cutting plane through point q perpendicular to the y axis as shown in Fig. 1-15(c). The corresponding component forces can be written as

$$dR_y = \sigma_y \, da_y \qquad dR_x = \tau_{yx} \, da_y \qquad dR_z = \tau_{yz} \, da_y \qquad (1\text{-}9b)$$

where once again the first subscript y indicates the plane of the area and the second denotes the direction of the force.

Finally, by passing a cutting plane perpendicular to the z axis as in Fig. 1-15(d) we have

$$dR_z = \sigma_z \, da_z \qquad dR_x = \tau_{zx} \, da_z \qquad dR_y = \tau_{zy} \, da_z \qquad (1\text{-}9c)$$

Now it is quite likely that each of the nine dR's in Figs. 1-15(b)–(d) will have a different value, and consequently each of the nine stresses σ_x, σ_y, σ_z, τ_{xy}, τ_{xz}, τ_{yx}, τ_{yz}, τ_{zx}, τ_{zy} will also likely have a different value. Therefore, what is the stress at point q? We seem to have at least nine choices! There is no such thing as *the stress* at point q, but rather there is a combination or *state of stress* at point q. It is convenient to depict a state of stress by the scheme in Fig. 1-16 in which the stresses on three mutually perpendicular planes are labeled in the manner described above. In this figure the directions of the arrows (up-down, left-right) have been chosen somewhat arbitrarily merely for purposes of illustration. As we shall see presently, these directions cannot be chosen quite so arbitrarily.

The state of stress depicted in Fig. 1-16 is called the *general* or *triaxial state of stress* that can exist at any interior point of a loaded body. In dealing with states of stress the engineer is confronted with two problems.

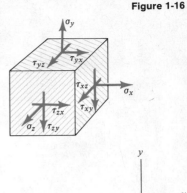

Figure 1-16

1 How does he or she determine the state of stress of a point; that is, how does one calculate values for σ_x, σ_y, τ_{xy}, and so on?

2 How does one determine the *maximum* value of the stress (normal or shear) at a point? After all, the x, y, z axes were chosen rather arbitrarily and the corresponding stresses σ_x, σ_y, τ_{xy}, and so on, may not be the maximum possible values for the stresses at the point. Furthermore, the material of the body is totally unaware of our choice of axes but it is keenly aware of all the existing stresses that are tending to break it.

We shall deal with the first question in subsequent chapters. The second question is primarily a question in the theory of linear algebra and we shall not pursue it in its fullest generality but will deal with a more simple but quite practical situation in the next section. Until then we summarize our results thus far by saying that the general state of stress at any point in a loaded body is described by nine component stresses, three normal stresses and six shear stresses, with the physical interpretations of Fig. 1-16.

1-7 / Plane stress

While it is entirely possible that the state of stress at an interior point of a loaded body may be the general triaxial one depicted in Fig. 1-16, what about a point at the outer surface of the body as, for example, point p in Fig. 1-17(a)? If no external loads are applied directly at point p,

Figure 1-17

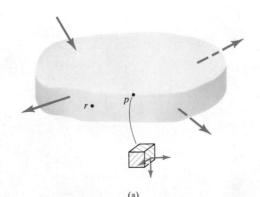

(a)

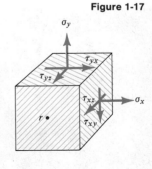

(b)

the top surface and front surface would be stressfree and we might anticipate that the state of stress at p would be as shown with stresses existing
only on the x plane. Point p is rather special in that being on the edge
of the body in Fig. 1-17(a) it has two exterior free surfaces. Generally a
point on the surface of a body has one rather than two free surfaces as,
for example, point r. For point r then, we might expect the state of stress
to be that shown in Fig. 1-17(b). This situation typifies the state of stress
that is likely to exist at most surface points in real load-carrying bodies.
Indeed, as we shall see in later chapters, this situation is also likely to
exist at most interior points of a loaded body and we shall now examine
this situation in greater detail.

On the basis of our knowledge of stress thus far, let us draw a free-body
diagram of a small chunk (dx, dy, dz) of material as shown in Fig. 1-18(a).
In this figure, the z surfaces (front and back) are assumed to be stressfree.
The forces on the x and y surfaces have been labeled using the stress
notation introduced earlier with the primes merely indicating, for example,
that the stress σ'_x on the left face of the chunk need not be the same value
as the corresponding stress σ_x on the right face. For such a chunk of
material, equilibrium requires that the sums of forces and moments must
be zero. Considering moments about the x axis and expressing the da's
in terms of the lengths dx, dy, and dz yields

$$\tau_{yz}(dx\,dz)\frac{dy}{2} + \tau'_{yz}(dx\,dz)\frac{dy}{2} = 0$$

$$\tau'_{yz} = -\tau_{yz} \tag{1-10a}$$

Similarly, for moments about the y axis

$$\tau'_{xz} = -\tau_{xz} \tag{1-10b}$$

Summing forces in the z direction gives

$$\tau_{xz}(dy\,dz) + \tau_{yz}(dx\,dz) - \tau'_{xz}(dy\,dz) - \tau'_{yz}(dx\,dz) = 0$$

Figure 1-18

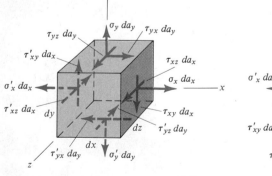

(a)

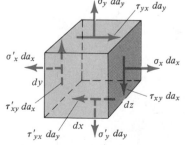

(b)

which, when combined with Eqs. 1-10a and 1-10b, yields

$$2\tau_{xz}\,dy + 2\tau_{yz}\,dx = 0 \qquad (1\text{-}10\text{c})$$

Now dx and dy are independent dimensions of the chunk so that Eq. 1-10c leads to the result

$$\tau_{xz} = \tau_{yz} = 0 \qquad (1\text{-}10\text{d})$$

Hence, this result leads to the conclusion that the stress situation in Fig. 1-18(a) can exist on an equilibrated body only if in fact it reduces to that shown in Fig. 1-18(b) in which there are no z forces (or z stresses). All the existing forces (and stresses) lie in the xy plane, and this situation is called a *plane state of stress* or *biaxial state of stress*. This is the situation that exists in many types of load-carrying members and its analysis is of utmost importance in their design.

Although we have conjectured that the state of stress on a real chunk of material might be that of Fig. 1-18(b), something is still not quite right because the shear forces as shown on this chunk would tend to rotate the chunk in a clockwise direction. Considering equilibrium of moments about the z axis,

$$\tau_{xy}(dy\,dz)\frac{dx}{2} + \tau'_{xy}(dy\,dz)\frac{dx}{2} + \tau_{yx}(dx\,dz)\frac{dy}{2} + \tau'_{yx}(dxdz)\frac{dy}{2} = 0$$

$$\tau_{xy} + \tau'_{xy} + \tau_{yx} + \tau'_{yx} = 0$$

Since the dimensions dx and dy can be made arbitrarily small, the difference between primed and unprimed values can be made arbitrarily small so that the above equation becomes

$$2\tau_{xy} + 2\tau_{yx} = 0$$

$$\tau_{xy} = -\tau_{yx}$$

The minus sign indicates that the directions of the shear stresses τ_{xy} (or τ_{yx}) in Fig. 1-18(b) are incorrect if equilibrium is to exist. Aside from this, however, the magnitude of the shear stress τ_{xy} must be the same as the magnitude of τ_{yx}. In physical terms, the vertical shear forces and the horizontal shear forces must produce "equal but opposite" *couples*. Figures 1-19(a) and (b) illustrate permissible states of plane stress whereas that of Fig. 1-18(b) is *not* permissible. Note carefully the arrows on all the shear force vectors. For either case in Fig. 1-19, we must have equality in magnitude of the shear stresses

$$\tau_{xy} = \tau_{yx} \qquad (1\text{-}11)$$

In summary, a plane state of stress is described by three values of stress, namely, two normal stresses σ_x and σ_y, and the value of the corresponding shear stress τ_{xy} or τ_{yx}.

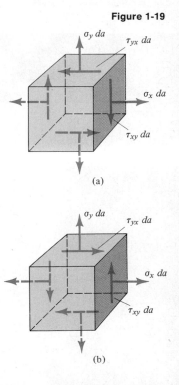

Figure 1-19

(a)

(b)

1-8 / Analysis of plane stress

Consider a point q in some sort of structural member like that shown in Fig. 1-20(a) and assume that in some manner we are able to describe the state of plane stress existing at point q; that is, by some method we are able to measure or calculate the stresses σ_x, σ_y, and τ_{xy}. These stresses are indicated on the two-dimensional diagram of Fig. 1-20(b) which is a common way of representing a state of plane stress. Thus, we assume that we know the stresses at point q that are acting on the vertical x plane and the horizontal y plane. (Remember, x and y plane mean the planes *perpendicular* to the x and y axes, respectively.) But what about the infinite number of other planes passing through point q such as the θ plane in Fig. 1-20(c)? What are the stresses on this plane?

We must realize that the material is unaware of what we have called the x and y axes. That is, the material has to resist the loads regardless of how we wish to name them or whether they are horizontal, vertical, or otherwise. Furthermore, the material will break when the intensity of the forces becomes too great regardless of how we look at the body. Thus a fundamental problem in engineering design is to determine the maximum normal stress or maximum shear stress at any particular point in the body. Assuming that at the point in question we can determine σ_x, σ_y, and τ_{xy} in Fig. 1-20(b), there is no reason to believe a priori that these are the maximum normal stress and shear stress occurring at that point. The maximum stresses are probably associated with some other plane. It is, therefore, important to be able to obtain from the assumed known stresses σ_x, σ_y, and τ_{xy}, the stresses σ_θ and τ_θ associated with other planes at this point.

For example, suppose that we wish to find the stresses for some other plane at the same point, such as the θ plane shown in Fig. 1-20(c), where θ is an arbitrary angle. Consider an elemental chunk of material such as that shown in Fig. 1-20(d). For convenience, we assume this element (chunk) to be one unit thick ($dz = 1$) and proportioned such that

Figure 1-20

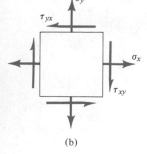

(a)

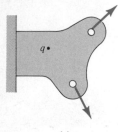

(b)

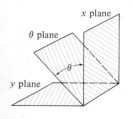

(c)

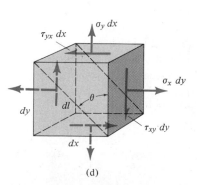

(d)

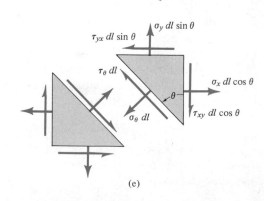

(e)

$$dx = dl \sin \theta \qquad dy = dl \cos \theta \qquad (1\text{-}12)$$

where dl is the length of the diagonal of the front face. We now split this chunk into two triangular-shaped wedges with stresses σ_θ and τ_θ acting on the diagonal face as indicated in Fig. 1-20(e). In this figure we have used Eq. 1-12 and the fact that the chunk is one unit thick in labeling the force vectors on the upper wedgelike element. Since this wedge is in equilibrium, summing forces parallel to the σ_θ stress gives

$$\overset{+\nearrow}{\sum} F = 0$$

$$-\sigma_\theta dl + (\sigma_x \, dl \cos \theta) \cos \theta - (\tau_{xy} \, dl \cos \theta) \sin \theta$$
$$-(\tau_{yx} \, dl \sin \theta) \cos \theta + (\sigma_y \, dl \sin \theta) \sin \theta = 0$$

Dividing out the dl and rearranging yields

$$\sigma_\theta = \sigma_x \cos^2 \theta + \sigma_y \sin^2 \theta - 2\tau_{xy} \sin \theta \cos \theta \qquad (1\text{-}13)$$

Must use top angle [handwritten annotation]

where we have used Eq. 1-11, namely $\tau_{xy} = \tau_{yx}$. Similarly, by summing forces parallel to τ_θ will yield

$$\tau_\theta = (\sigma_x - \sigma_y) \sin \theta \cos \theta + \tau_{xy}(\cos^2 \theta - \sin^2 \theta) \qquad (1\text{-}14)$$

Equations 1-13 and 1-14 are called the *stress transformation equations for plane stress*. If the state of stress at a point q can be described in terms of values for σ_x, σ_y, and τ_{xy}, then these equations can be used to determine the normal and shear stress acting on any other plane through point q specified by the angle θ.

In using these equations we must adopt an algebraic sign convention. Notice that for $\theta = 0°$

$$\sigma_{0°} = \sigma_x \qquad \tau_{0°} = \tau_{xy}$$

while for $\theta = 90°$

$$\sigma_{90°} = \sigma_y \qquad \tau_{90°} = -\tau_{xy}$$

Then, if we consider the shear stress on the vertical plane to be algebraically positive (negative), the shear stress on the horizontal plane will be algebraically negative (positive). Accordingly, it is convenient to adopt the following sign convention for using Eqs. 1-13 and 1-14.

1 Tensile stresses will be considered positive, compressive stresses negative.

2 Shear stresses corresponding to clockwise couples (for example, τ_{xy} in Fig. 1-20(b)) will be considered positive; shear stresses corresponding to counterclockwise couples (τ_{yx} in Fig. 1-20(b)) will be considered negative. Thus, shear stresses on mutually perpendicular planes will *always* have *opposite signs* insofar as the stress transformation equations are concerned.

Equations 1-13 and 1-14 are often written in the alternative forms

$$\sigma_\theta = \frac{\sigma_x + \sigma_y}{2} + \left(\frac{\sigma_x - \sigma_y}{2}\right) \cos 2\theta - \tau_{xy} \sin 2\theta \qquad (1\text{-}15)$$

$$\tau_\theta = \left(\frac{\sigma_x - \sigma_y}{2}\right) \sin 2\theta + \tau_{xy} \cos 2\theta \qquad (1\text{-}16)$$

where we have used the double-angle identities

$$\sin^2 \theta = \frac{1 - \cos 2\theta}{2}$$

$$\cos^2 \theta = \frac{1 + \cos 2\theta}{2} \qquad (1\text{-}17)$$

$$2 \sin \theta \cos \theta = \sin 2\theta$$

Equations 1-13 to 1-16 express the desired unknown stresses σ_θ and τ_θ in terms of the known stresses σ_x, σ_y, and τ_{xy} and the angle θ. For any particular known state of stress, the quantities σ_x, σ_y, and τ_{xy} are unique and single valued, and they are therefore constants in these equations. So the dependent variables σ_θ and τ_θ are actually functions of only one independent variable θ. This means that for any known state of plane stress, we can find the value of θ for which the normal stress σ_θ is a maximum or minimum by differentiating Eq. 1-15 with respect to θ and setting the derivative equal to zero. Thus,

$$\frac{d\sigma_\theta}{d\theta} = \left(\frac{\sigma_x - \sigma_y}{2}\right)(-\sin 2\theta)(2) - \tau_{xy}(\cos 2\theta)(2) \qquad [1\text{-}15a]$$

Setting the derivative equal to zero and solving for θ_{max}, we obtain

$$\left(\frac{\sigma_x - \sigma_y}{2}\right) \sin 2\theta_{max} = -\tau_{xy} \cos 2\theta_{max}$$

$$\tan 2\theta_{max} = \frac{-\tau_{xy}}{\left(\dfrac{\sigma_x - \sigma_y}{2}\right)} \qquad (1\text{-}18)$$

Likewise the value of θ'_{max} that will maximize τ_θ can be obtained by setting the derivative of Eq. 1-16 equal to zero. The result is

$$\tan 2\theta'_{max} = \frac{\left(\dfrac{\sigma_x - \sigma_y}{2}\right)}{\tau_{xy}} \qquad (1\text{-}19)$$

Equations 1-18 and 1-19 will each yield two values of θ. One value indicates the orientation of the plane of the maximum stress, and the other indicates the orientation for the minimum stress.

Closer examination of Eqs. 1-18 and 1-19 shows that $\tan 2\theta_{max}$ from Eq. 1-18 is the negative reciprocal of the value from Eq. 1-19. This means that the angle 2θ for which σ_θ is a maximum or minimum will differ by 90° from the angle 2θ for which τ_θ is a maximum. Therefore, the orientation of the plane for maximum normal stress will differ from that for maximum shear stress by 45°. Also note from Eq. 1-15a that setting the derivative of σ_θ equal to zero is equivalent to setting τ_θ from Eq. 1-16 equal to zero. Thus, no shear stress is present on the plane for which the normal stress is a maximum or minimum.

PROBLEMS

1-29 to 1-36 Using the "wedge" method of analysis, determine the normal and shear stresses on the indicated planes shown in the accompanying illustrations, without appealing to the transformation equations.

Problem 1-29 **Problem 1-30**

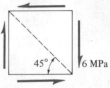

Problem 1-31 **Problem 1-32** **Problem 1-33** **Problem 1-34**

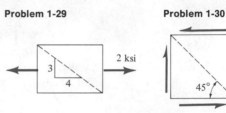

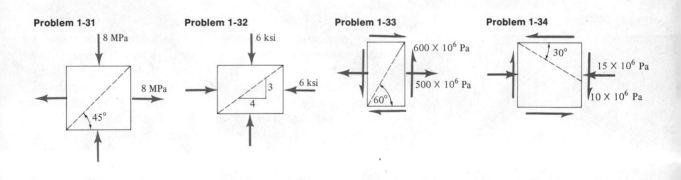

Problem 1-35 **Problem 1-36**

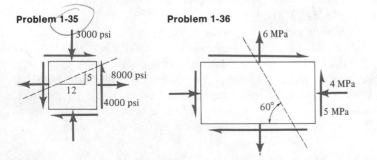

1-9 / Mohr's circle for plane stress

Construction of Mohr's circle Although the equations in section 1-8 are useful, a graphical representation of them can be used more easily and better illustrates their physical significance. If we square Eqs. 1-15 and 1-16, we get the following relationships:

$$\left[\sigma_\theta - \left(\frac{\sigma_x + \sigma_y}{2}\right)\right]^2 = \left[\left(\frac{\sigma_x - \sigma_y}{2}\right)\cos 2\theta - \tau_{xy}\sin 2\theta\right]^2$$

$$\tau_\theta^2 = \left[\left(\frac{\sigma_x - \sigma_y}{2}\right)\sin 2\theta + \tau_{xy}\cos 2\theta\right]^2$$

Now by adding the above two equations, expanding, and rearranging, we obtain

$$\left[\sigma_\theta - \left(\frac{\sigma_x + \sigma_y}{2}\right)\right]^2 + \tau_\theta^2 = \left(\frac{\sigma_x - \sigma_y}{2}\right)^2 + \tau_{xy}^2 \qquad (1\text{-}20)$$

By slight rearrangement, Eq. 1-20 can be put into the form of an equation of a circle. Thus,

$$\left[\sigma_\theta - \left(\frac{\sigma_x + \sigma_y}{2}\right)\right]^2 + [\tau_\theta - 0]^2 = \left[\sqrt{\left(\frac{\sigma_x - \sigma_y}{2}\right)^2 + \tau_{xy}^2}\right]^2 \qquad (1\text{-}20a)$$

$$[x - \quad (a) \quad]^2 + [y - b]^2 = [\quad\quad c \quad\quad]^2 \qquad (1\text{-}21)$$

The equation of a circle as presented in most mathematics books is written directly below the combined transformation equation (Eq. 1-20a) above where a is the x coordinate of the center of the circle, b is the y coordinate of the center, and c is the radius. Equation 1-20a represents the equation of a circle for which the x coordinate of the center is $(\sigma_x + \sigma_y)/2$, the y coordinate of the center is 0, and the radius is the radical term. As before, σ_θ and τ_θ are the unknown variables and correspond to x and y, respectively, in the circle equation. Now, by adopting the proper sign convention and assuming that $\sigma_x > \sigma_y > 0$, we can plot Eq. 1-20a as a circle based on rectangular axes and the variables σ_θ and τ_θ. Figure 1-21(a) shows Mohr's circle of stress, which is so called because it was developed by Otto Mohr (1835–1918).*

Each of the infinite number of points on the circle in Fig. 1-21(a) represents one of the infinite number of possible orientations of Plane ③ in Fig. 1-21(b). Each point on the circle has two coordinates, σ_θ and τ_θ, which represent magnitudes and signs of the stresses for the corresponding orientation of the plane. Tensile stresses are assumed to be positive, compressive stresses are negative, and shear stresses that produce clockwise couples are positive. One Mohr's circle completely represents a

* An outstanding professor of structural mechanics in Germany.

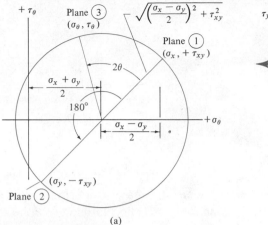

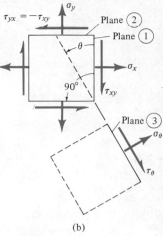

(a)

(b)

Figure 1-21

state of *biaxial* stress at one particular point in a loaded body. The point on the circle labeled Plane ① gives the stresses on the vertical plane marked Plane ① on the element; and the point on the circle labeled Plane ② gives the stresses on the horizontal plane marked Plane ② on the element. These two points are 180° apart on the circle, while the corresponding planes on the element are 90° apart. In general, an angular measurement of 2θ on the circle corresponds to an angular measurement of θ on the element. *Remember: θ on the element, 2θ on the circle.*

Comparison of Eqs. 1-20a and 1-21 shows that b must equal zero. Therefore, the center of the circle must *always* be on the horizontal σ_θ axis, and the circle can be plotted if the coordinates of any two diametrically opposite points are known. This means that Mohr's circle for plane stress at a point in a loaded body can be plotted if the stresses are known on any two perpendicular planes at the point, that is, on two sides of an element such as the one in Fig. 1-21(b).

EXAMPLE 1-4

Assume that the stresses at a critical point in some engineering member are shown in Fig. 1-22(a). Sketch Mohr's circle.

Solution:

We first locate the points that represent the stresses on Plane ① and Plane ② of the element. Since these points are at opposite ends of a diameter of the circle, a line connecting the two points is a diameter, and the intersection of the horizontal axis with the diameter is the center. We now sketch the circle.

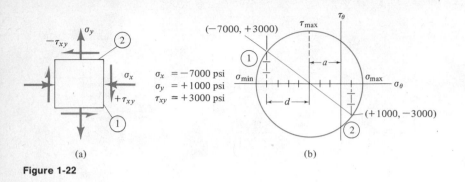

$\sigma_x = -7000$ psi
$\sigma_y = +1000$ psi
$\tau_{xy} = +3000$ psi

(a)   (b)

Figure 1-22

A complete analysis of the circle can be made from this rough sketch by using geometry and trigonometry. An accurate graphical construction is usually unnecessary. The distance a in Fig. 1-22(b) from the origin to the center can be found as follows:

$$a + d = 7000 \qquad \text{(magnitude of } \sigma_\theta \text{ for Plane } ①)$$

$$d = a + 1000 \qquad \text{(from similar triangles)}$$

Therefore, $a = 3000$ and $d = 4000$. The coordinates of the center of the circle are -3000 and 0, and its radius is $\sqrt{4000^2 + 3000^2} = 5000$.

Analysis of Mohr's circle Once Mohr's circle is drawn for known values of σ_x, σ_y, and τ_{xy} at a point, the maximum stresses at the point can be found. The coordinate of the right-hand extremity of that circle is the algebraic maximum normal stress; the coordinate of the left-hand extremity is the minimum normal stress; and the radius is the maximum shear stress. It is important to note that the points representing the maximum and minimum normal stresses are on the horizontal σ_θ axis, as shown in Fig. 1-23, and this condition means that there is no shear stress on the planes of maximum and minimum normal stress at the point in the body. These normal stresses, which act on planes of no shear, are called *principal stresses*. They occur on planes that are perpendicular to one another (180° apart on the circle) and are called *principal planes*.

The minimum normal stress is just as important as the maximum. The algebraic minimum stress could have a magnitude greater than that of the maximum principal stress if the state of stress were such that the center of the circle is to the left of the origin, as in Fig. 1-22. The magnitude of the maximum or minimum principal stress is the horizontal coordinate of the center of the circle plus or minus the radius. In Fig. 1-22(b), the maximum normal stress is $(-3000 + 5000) = 2000$ psi; the minimum is $(-3000 - 5000) = -8000$ psi; and $\tau_{max} = 5000$ psi.

The angle of orientation of the principal planes with respect to known

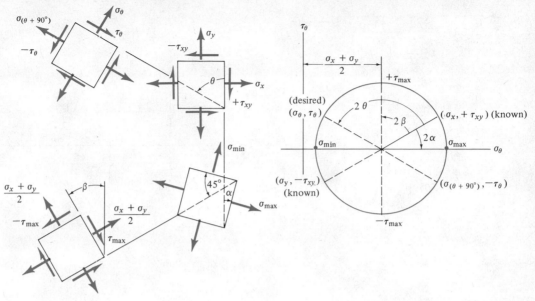

Figure 1-23

planes can be found from the circle. *Remember: θ on the element corre-sponds to 2θ on the circle.* In Fig. 1-23, the angle 2θ on the circle from a point representing an orientation for which the stresses are known to any other point representing an orientation for which the stresses are desired is referenced *from* the known point and measured in the same direction as the angle θ on the element between the corresponding known plane and the desired plane. Satisfy yourself about this condition by going from a known plane to a principal plane in Fig. 1-23. Also try going from a known plane to any other desired plane. Then try going from a known plane to a plane of maximum shear stress.

You should see from the circle that a point for τ_{max} is 90° away from a point for a principal stress. Thus a plane of maximum shear will *always* make an angle of 45° with a principal plane. You should also note that since stresses on perpendicular faces of any element are given by the coordinates of diametrically opposite points on the circle, the sum of the two normal stresses for any and all orientations of the element is a constant. This sum is an invariant for any particular state of stress.

EXAMPLE 1-5

While in operation, the torque and thrust on a propeller shaft produce the state of stress indicated in Fig. 1-24(a). Determine (a) the values and directions of the principal stresses and (b) the value and direction of the maximum shear stress.

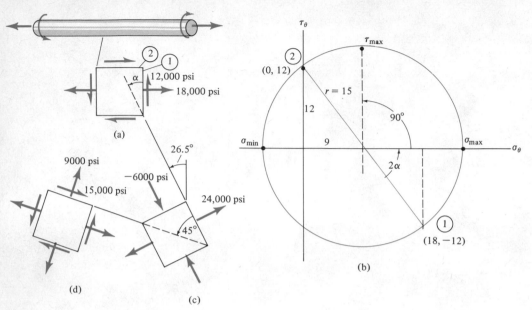

Figure 1-24

Solution:

In our computations, the stresses will be expressed in kips per square inch (ksi), a kip being 1000 lb. Thus, with our adopted sign convention we have

$$\sigma = +18 \text{ ksi} \qquad\qquad \sigma = 0$$

Plane ① 　　　　　　　　 Plane ②

$$\tau = -12 \text{ ksi} \qquad\qquad \tau = +12 \text{ ksi}$$

These points are plotted and the Mohr circle drawn in Fig. 1-24(b). The center of the circle is at

$$\frac{\sigma_x + \sigma_y}{2} = \frac{18 + 0}{2} = 9$$

and its radius is

$$r = \sqrt{9^2 + 12^2} = 15$$

(a) The maximum stress is at the right extremity of the circle:

$$\sigma_{max} = \text{center plus radius}$$
$$= 9 + 15 = 24 \text{ ksi}$$

Similarly, the minimum stress is at the left extremity:

$$\sigma_{min} = \text{center minus radius}$$
$$= 9 - 15 = -6 \text{ ksi}$$

Hence the principal stresses are 24,000 psi (tensile) and 6000 psi (compressive). As for their directions, note that the point representing σ_{max} on the circle is at an angle of 2α counterclockwise from the point representing Plane ①, that is, the vertical plane of the shaft. Thus, the plane of maximum normal stress is oriented at an angle α counterclockwise from Plane ①. This is indicated in Fig. 1-24(c) which shows an element oriented parallel to the principal planes. From the circle we see

$$\tan 2\alpha = \frac{12}{9} \quad \text{therefore} \quad 2\alpha = 53° \quad \alpha = 26.5°$$

(b) The value of maximum shear stress is given by the upper extremity of the circle and is simply the radius.

$$\tau_{max} = 15,000 \text{ psi}$$

As for its direction, notice that the point τ_{max} on the circle is 90° counterclockwise from the point σ_{max}. Hence, the plane of τ_{max} is 45° counterclockwise from the plane of σ_{max}. This plane is indicated in Fig. 1-24(d), which shows an element oriented parallel to the planes having the largest shear stresses. As a final observation, note that on this element there are normal stresses as well as shear stresses, while on the principal element in Fig. 1-24(c) there are no shear stresses.

EXAMPLE 1-6

A timber beam is to be loaded in such a manner that the resulting state of stress near the load will be that shown in Fig. 1-25(a). It is feared that the timber will split if the shear stress parallel to the grain reaches or exceeds 4×10^6 Pa. Is the beam safe?

Figure 1-25

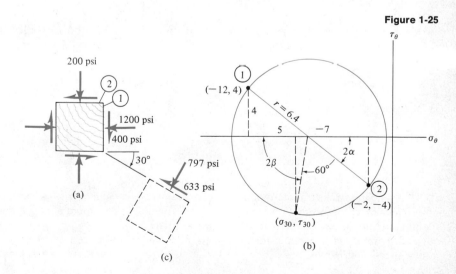

Solution:

The circle is constructed in the usual manner in Fig. 1-25(b), where we have omitted the hundreds on the values.

$$\text{center} = \frac{-12 - 2}{2} = -7$$

$$\text{radius} = \sqrt{4^2 + 5^2} = 6.4$$

The plane of the grain lies at an angle of 30° clockwise from the horizontal ② plane. Hence we will measure clockwise an angle of 2 × 30° = 60° from the ② point on the circle, thus locating the point on the circle that represents the stresses on the plane of grain. To determine the value of τ_{30}, first observe that

$$\tan 2\alpha = \frac{4}{5} \qquad 2\alpha = 38.7°$$

then

$$2\beta + 2\alpha + 60° = 180°$$

$$2\beta = 81.3°$$

Hence, by trigonometry

$$\tau_{30} = -r \sin 2\beta$$

$$= -(6.4) \sin 81.3° = -6.33$$

so that the shear stress parallel to the grain is 633 psi in the direction shown in Fig. 1-25(c).

The given critical value for this shear stress is 4 × 10^6 Pa or, by Table A-13 in the Appendix, 4 × 145 = 580 psi which is smaller than the value of 633. Hence, on this basis, the timber is unsafe. For completeness, we have shown the compressive stress σ_{30} acting perpendicular to the grain.

PROBLEMS

1-37 Using the stress transformation equations, show that the sum of the normal stresses on any two perpendicular planes is a constant for a state of plane stress. What does this result mean in regard to the Mohr circle of stress?

1-38 to 1-49 For each of the states of plane stress shown, (a) plot the Mohr circle, (b) determine the principal stresses and planes, and (c) determine the maximum shear stress and its direction.

1-50 to 1-55 In Probs. 1-31 to 1-36, determine the stresses on the indicated planes by use of Mohr's circle.

Problem 1-38

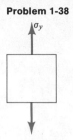

Problem 1-39

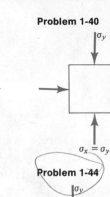

$\sigma_x = \sigma_y$

Problem 1-40

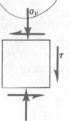

$\sigma_x = \sigma_y$

Problem 1-41

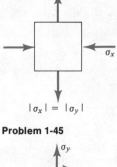

$|\sigma_x| = |\sigma_y|$

Problem 1-42

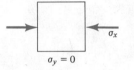

$\sigma_y = 0$

Problem 1-43

τ

Problem 1-44

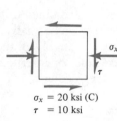

τ

$|\sigma_y| = 2|\tau|$

Problem 1-45

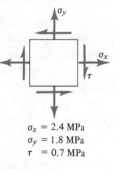

τ

$|\sigma_x| = |\sigma_y| = 3\,|\tau|$

Problem 1-46

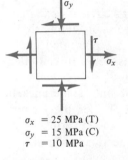

$\sigma_x = 25$ MPa (T)
$\sigma_y = 15$ MPa (C)
$\tau = 10$ MPa

Problem 1-47

σ_x

τ

$\sigma_x = 20$ ksi (C)
$\tau = 10$ ksi

Problem 1-48

σ_y

σ_x

τ

$\sigma_x = 2.4$ MPa
$\sigma_y = 1.8$ MPa
$\tau = 0.7$ MPa

Problem 1-49

σ_y

σ_x

τ

$\sigma_x = 150 \times 10^6$ Pa (C)
$\sigma_y = 150 \times 10^6$ Pa (T)
$|\tau| = 150 \times 10^6$ Pa

Note: In the following problems, determine the unknown quantities by use of Mohr's circle.

1-56 Determine the stress on the indicated planes.

Problem 1-56

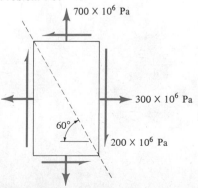

700×10^6 Pa

300×10^6 Pa

200×10^6 Pa

$60°$

1-57 Given: $\sigma_x = +6$ MPa, $\sigma_y = -2$ MPa, $\tau_{xy} = +2$ MPa. If $\sigma_\theta = +3$ MPa, find θ and τ_θ.

Problem 1-57

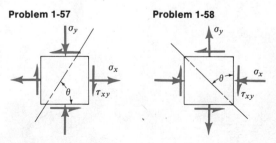

σ_y

σ_x

τ_{xy}

θ

Problem 1-58

σ_y

σ_x

τ_{xy}

θ

1-58 Given: $\sigma_x = -5$ ksi, $\sigma_y = +1$ ksi, $\tau_{xy} = +4$ ksi. If $\tau_\theta = -2$ ksi, find σ_θ and θ.

1-59 A block made of heavy-grained wood will crack along the grain if the shear stress along the grain ever exceeds 5×10^6 Pa. If $\sigma_y = 8 \times 10^6$ Pa, what range of values may σ_x assume without the block cracking?

Problems 1-59 and 1-61

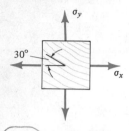

1-60 Two wedges of material are held together by some adhesive compound to form a cube. The joint will break if the tensile stress on the joint exceeds 15 MPa. What is the maximum value σ_y may have, if $\sigma_x = 8$ MPa and $\tau_{xy} = 5$ MPa?

Problem 1-60

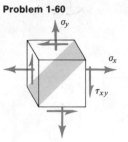

1-61 A block made of heavy-grained wood will crack along the grain if the tensile stress normal to the grain ever exceeds 500 psi. If $\sigma_y = 700$ psi, what range of values may σ_x assume without the block cracking?

1-10 / Absolute maximum shear stress

The discussion in section 1-9 was confined to Mohr's circle for *one* plane of stress. Triaxial states of stress can be analyzed by use of Mohr's circles, but the technique is much more involved and is not presented in this book. However, we shall consider one particular aspect of three-dimensional stress analysis. In order to determine the *absolute maximum* shear stress on an element subjected to *plane* stress, it may sometimes be necessary to analyze a triaxial state of stress. Consider the element shown in Fig. 1-26(a), which is subjected to a biaxial state of stress.

We can rotate the element about an axis perpendicular to the *xy* plane of stress so that its sides are parallel to the principal planes, as shown in Fig. 1-26(b); in (c) the element is shown in three dimensions,

Figure 1-26

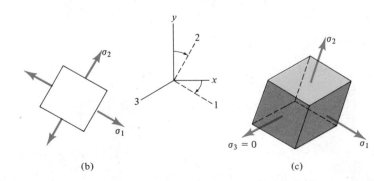

(a)　　　　　　(b)　　　　　　(c)

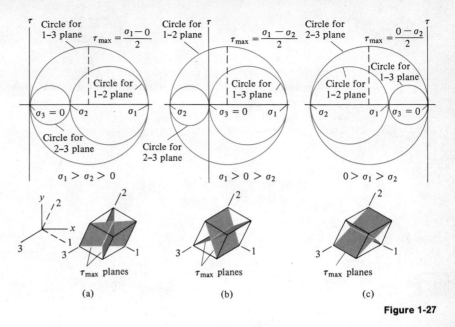

Figure 1-27

and the principal stresses are labeled σ_1, σ_2, and σ_3. Remember that for plane stress σ_3 is zero.

Mohr's circle can be plotted for *each* biaxial state of stress, that is, for the 1-2 plane, the 1-3 plane, and the 2-3 plane. Figure 1-27(a) shows Mohr's circles for the case where σ_1 is greater than σ_2 and both are tensile. Figure 1-27(b) is for the case where σ_1 is a tensile stress and σ_2 is a compressive stress of different magnitude. Figure 1-27(c) is for the case where both σ_1 and σ_2 are compressive and σ_2 has the greater magnitude. The absolute maximum shear stress is equal to one-half of the algebraic difference between the maximum and minimum principal stresses. In Fig. 1-27(a), the maximum principal stress is σ_1 and the minimum is σ_3 or zero. In (b), the maximum principal stress is σ_1 and the minimum is σ_2. In (c), the maximum principal stress is σ_3 or zero and the minimum is σ_2.

Although we have considered only cases where the third principal stress σ_3 is zero, it can be proved that for *any* triaxial state of stress where σ_3 is not zero, the maximum shear stress is always one-half of the difference between the maximum and minimum principal stresses. Thus,

$$\tau_{max} = \frac{1}{2} (\sigma_{max} - \sigma_{min}) \qquad (1\text{-}22)$$

Note: In each of the following problems, determine the principal stresses and the principal planes in which they act. Also, assuming that $\sigma_z = 0$ in each case, determine the maximum shear stress for each case.

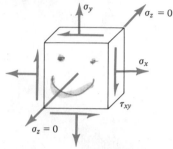

Problems 1-62 to 1-71

1-62	$\sigma_x = +20 \times 10^6$ Pa	$\sigma_y = 0$	$\tau_{xy} = +5 \times 10^6$ Pa
1-63	$\sigma_x = -2000$ psi	$\sigma_y = +6000$ psi	$\tau_{xy} = +4000$ psi
1-64	$\sigma_x = -2$ MPa	$\sigma_y = -8$ MPa	$\tau_{xy} = -4$ MPa
1-65	$\sigma_x = -10$ ksi	$\sigma_y = -20$ ksi	$\tau_{xy} = -10$ ksi
1-66	$\sigma_x = +500$ psi	$\sigma_y = +500$ psi	$\tau_{xy} = 0$
1-67	$\sigma_x = 0$	$\sigma_y = 0$	$\tau_{xy} = -8$ ksi
1-68	$\sigma_x = -5$ ksi	$\sigma_y = +5$ ksi	$\tau_{xy} = +3$ ksi
1-69	$\sigma_x = -20$ MPa	$\sigma_y = +10$ MPa	$\tau_{xy} = +10$ MPa
1-70	$\sigma_x = +6000$ psi	$\sigma_y = +15,000$ psi	$\tau_{xy} = +10,000$ psi
1-71	$\sigma_x = +600$ psi	$\sigma_y = +1000$ psi	$\tau_{xy} = -500$ psi

1-11 / Calculator solution

If it is necessary to do a large number of problems involving the analysis of plane stress it becomes reasonable to consider the use of a computer or programmable calculator.

Considering the number of calculator models available and the rapidity with which manufacturers are improving and changing current models, it is not practical to give you specific instructions for programming your calculator. It is, however, practical to present the underlying logic necessary for all programmed solutions. Figure 1-28 presents that logic by means of a flowchart. By using the chart you should find it relatively easy to write a program that suits your own needs.

Appendix I presents a Fortran program derived from the flowchart.

1-72 to 1-92 Solve Probs. 1-38 to 1-58 by means of a programmable calculator.

Figure 1-28

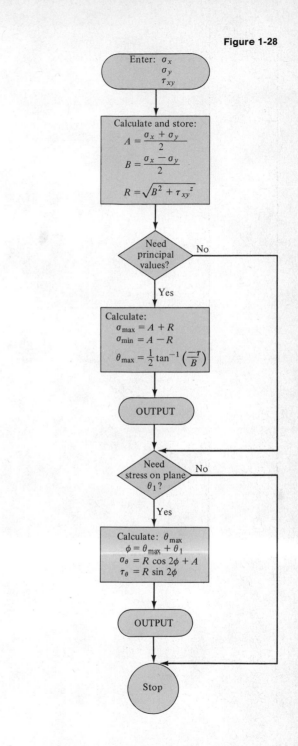

Chapter 2
Strain

2-1 / Objectives

Upon completion of this chapter, you will be able to:

1 Define strain.
2 Identify normal strain and shearing strain.
3 Compute deformation from geometric descriptions.
4 Compute normal and shearing strains from deformation and from temperature change.
5 Compute the strains associated with any axis when given the strains associated with a pair of orthogonal axes.
6 Identify principal strains.

2-2 / Deformations

As stated earlier, mechanics of deformable bodies concerns the relationships between external loads and the resulting internal force intensities and dimension changes. Chapter 1 dealt with the internal force intensities, which were defined as stresses. This chapter will be a similar treatment of dimension changes.

The primary function of an engineering member is to resist loads. When loads of any type are applied to a member, the member will always undergo dimension changes. In other words, the loads alter the size and/or the shape of the body. Such dimension changes may or may not be visible to the naked eye depending on the degree to which the loads alter the body.

A dimension change is called a deformation and will be denoted by the letter *e*. A deformation that causes an increase of a length is commonly called an elongation, or extension; and a deformation that causes a decrease of a length will be called a contraction, or compression. A deformation that results in an angular distortion is called a shear deformation.

It is possible for an element in any solid body to undergo deformations of various types and in many directions when loads are applied. Hence, the general analysis of deformation is a three-dimensional problem. Although the discussions in this chapter will be confined to the deformations occurring in one plane, you should not lose sight of the fact that deformations may be occurring simultaneously in directions perpendicular to the plane under consideration.

Figure 2-1

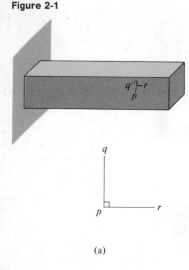

(a)

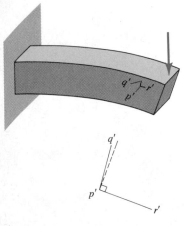

(b)

2-3 / Displacement

In order to visualize the deformations occurring locally within some small region of a deformable body, consider the cantilever beam shown in Fig. 2-1(a). At some point *p* on the surface we have scribed two very short lines *pq* and *pr*. Then, if this beam is loaded as shown in Fig. 2-1(b), the points *p*, *q*, and *r* will be *displaced* to new positions *p'*, *q'*, *r'* as indicated in the figure. Also the lines *pq* and *pr* have been rotated. (Although it is quite possible that the lines *p'q'* and *p'r'* will no longer be straight lines, if they are "short enough" they can be considered to be straight.) We also observe, however, that the length of *p'r'* is longer than *pr* and *p'q'* is shorter than *pq*. Also, what was originally a right angle between lines *pr* and *pq* is no longer a right angle between *p'r'* and *p'q'*. Hence we see that the local displacements produced by the force can be classified into four general types:

1 Translation of points

2 Rotation of lines

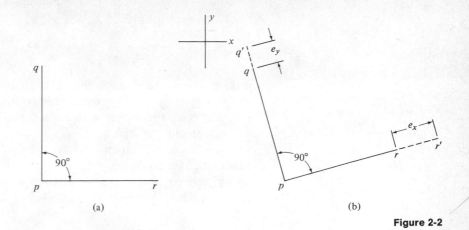

Figure 2-2

3 Change of length; that is, elongation and contraction

4 Distortion; that is, angle changes between lines

The first two types of displacement occur because of movement of the entire body, while the latter two are associated with the *local deformation* within the body. In what follows we are concerned primarily with displacements associated with local deformations.

2-4 / Definition of strain

Axial strain Consider two very short lines emanating from some point p of a deformable body (Fig. 2-2). These two lines suffice to define the plane to which we are restricting our analysis of deformation. Assume that the body is deformed in such a manner that no distortion occurs at point p; that is, the right angle at p remains a right angle although the lengths pr and pq become $p'r'$ and $p'q'$, respectively. For convenience, we have chosen the x and y axes to be parallel to pr and pq, respectively. Then, pr has undergone an elongation e_x, the notation here being for the elongation e of a line segment originally parallel to the x axis. Similarly pq has undergone an elongation e_y.

By definition, the average axial strain ε_{av} is the axial deformation (elongation or contraction) per unit length. Hence

$$\varepsilon_{av} = \frac{e}{L}$$

and, in particular

$$(\varepsilon_{av})_x = \frac{e_x}{L_{pr}} \qquad (\varepsilon_{av})_y = \frac{e_y}{L_{pq}}$$

The average strain

$$\varepsilon_{av} = \frac{e}{L}$$

is the strain most commonly used in engineering computations. However, strain frequently varies considerably over the length of a body so that it is necessary to develop a more accurate expression in order to take account of such variation.

To do this, as in Chapter 1, we assume that the material is "smeared out" to smooth over all the voids and irregularities. Then the original material lines pr and pq can be made as small as we please and we define the axial strains at p as

$$\varepsilon_x = \lim_{L_{pr} \to 0} \frac{e_{pr}}{L_{pr}} = \frac{de_x}{dx} \qquad \varepsilon_y = \lim_{L_{pq} \to 0} \frac{e_{pq}}{L_{pq}} = \frac{de_y}{dy} \qquad (2\text{-}1)$$

Notice that ε_x is not necessarily equal to ε_y any more than e_x was equal to e_y. Axial strain resulting from an elongation is commonly called a tensile strain, while that resulting from a contraction is called a compressive strain.

Shear strain   Consider now two orthogonal very short material lines emanating from a point p of a deformable body as before. Assume now that the body deforms, in such a manner that the lengths of pr and pq remain unchanged but that the right angle has been increased by an amount ϕ, as shown in Fig. 2-3. Then

$$\sin \phi = \frac{e_s}{L_{pq'}} = \frac{e_s}{L_{pq}}$$

where e_s is the shear deformation. In most engineering applications ϕ will be quite small so that

$$\phi \cong \sin \phi = \frac{e_s}{L_{pq}}$$

where ϕ is expressed in radians and is called the average shear strain.

Figure 2-3

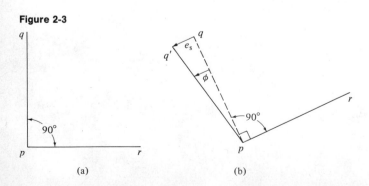

(a) (b)

Once again we assume the material to be voidless or smeared out so that pq and pr can be as small as we wish and we define the shear strain, denoted by γ, at point p as

$$\gamma = \lim_{L_{pq} \to 0} \frac{e_s}{L_{pq}} = \frac{de_s}{dL} \qquad (2\text{-}2)$$

But from above we see that γ at p will correspond to ϕ, the angle of distortion. Thus the shear strain is usually interpreted physically as an angle change between two originally orthogonal very short line elements. It is very important that you understand the distinction between deformation and strain. Deformation is a change of linear dimension and has units of length, such as inches, feet, or meters. Strain is deformation per unit length and is therefore a *dimensionless* quantity. However, axial strain is usually expressed in terms of inches per inch (or millimeters per millimeter) and microinches (millionths of an inch) per inch (or microns per meter). Axial strain is also commonly expressed as percent deformation. The axial strain for engineering materials in ordinary use seldom exceeds 0.002 in./in. (mm/mm), which is equivalent to 2000 microinches/in. or 0.2 percent. Shear strain, since it is an angle change, is often expressed in radians. A convenience unit often used by experimental workers is microstrain. One microstrain is one microinch/ inch or one micrometer/meter and is written $1\ \mu\varepsilon$. Both normal and shearing strain may be given in microstrains.

EXAMPLE 2-1

A manila rope is used in hoisting a heavy load. When unloaded the rope is exactly 10 m long. However, when the hoisting process begins the upper end of the rope is 10.33 m above the load when the rope finally begins to "take the strain" and raise the load. What is the average axial strain in the rope at that instant?

Solution:

The average axial strain will be:

$$\varepsilon_{av} = \frac{e}{L} = \frac{\text{deformation}}{\text{original length}}$$

The deformation is the difference between the original length and the final length so that

$$e = 10.33 - 10.00 = 0.33\ \text{m}$$

and

$$\varepsilon_{av} = 0.33/10 = 0.033\ \text{m/m}$$

This answer may also be expressed as a 3.3 percent strain.

EXAMPLE 2-2

Figure 2-4 shows a block before and after a horizontal force is applied midway between A and A_1. Calculate the average shearing strain. The deformation is exaggerated for clarity.

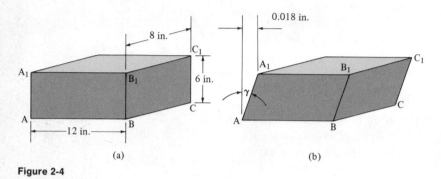

(a)   (b)

Figure 2-4

Solution:

We may determine shearing strain by observing angular changes within the straining body. We look for the angle which indicates the amount of distortion as a result of the load. In this problem the angle is γ in Fig. 2-4(b). For a small angle

$$\gamma \simeq \sin \gamma \simeq \tan \gamma = \frac{0.018}{6} = 0.003 \ \frac{\text{in.}}{\text{in.}} = 3000 \ \mu\varepsilon$$

This answer may also be interpreted as

$$\gamma = 0.003 \ \text{rad}$$

EXAMPLE 2-3

A relatively rigid bar is pivoted at a and held in position by a wire attached at b as shown in Fig. 2-5(a). A vertical load applied to the bar causes it to rotate 10° from its initial horizontal position. What is the average axial strain induced in the wire due to this rotation?

Solution:

A schematic diagram of the displacements is shown in Fig. 2-5(b) in which the rotation has been exaggerated for clarity. By definition the average axial strain in the wire will be

$$\varepsilon_{\text{av}} = \frac{\text{final length} - \text{initial length}}{\text{initial length}}$$

Now the final length is

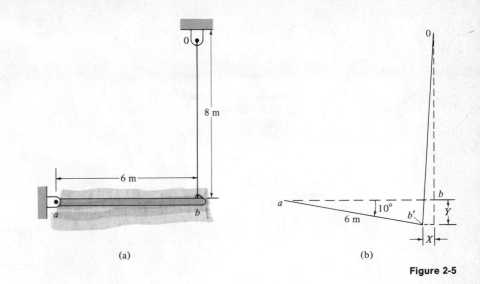

Figure 2-5

$$L_f = \sqrt{(L_i + Y)^2 + (X)^2}$$

where X is the horizontal displacement and Y is the vertical displacement of point b. By trigonometry

$$X = 6 - 6 \cos 10° = 6 - 5.91 = 0.09 \text{ m}$$

$$Y = 6 \sin 10° = 1.04 \text{ m}$$

Thus

$$L_f = \sqrt{(8 + 1.04)^2 + (0.09)^2} = 9.04 \text{ m}$$

and

$$\varepsilon_{av} = \frac{9.04 - 8}{8} = 0.13 \text{ m/m}$$

EXAMPLE 2-4

The axial strain in any direction due to thermal expansion is given by the relationship $\varepsilon = \alpha \Delta T$, where α is the coefficient of thermal expansion and ΔT is the change in temperature. What are the final size and shape of the thin homogeneous plate shown in Fig. 2-6(a) if (a) the entire plate undergoes a temperature increase of M degrees, (b) the plate is subjected to a temperature gradient given by $\Delta T = Kx$? (c) What is the shear strain γ_{xz} in the plate for case (a)? (d) What is the shear strain for case (b) assuming the z edge to be fixed?

Solution:

(a) If the entire plate undergoes a uniform temperature change of M degrees, for free expansion the axial strain in any direction will be uniform and of value

$$\varepsilon = \alpha M$$

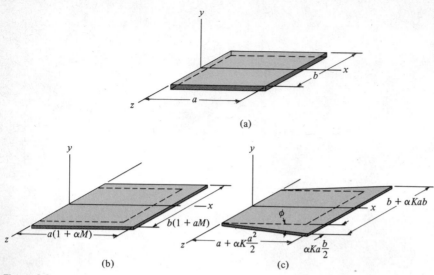

Figure 2-6

Hence, the elogation in the x and z directions will be

$$e_x = \varepsilon a \qquad e_z = \varepsilon b$$

respectively, and the plate will still be rectangular with dimensions

$$L_x = a(1 + \alpha M) \qquad L_z = b(1 + \alpha M)$$

shown in Fig. 2-6(b).

(b) If the plate undergoes a temperature gradient given by $\Delta T = Kx$, then at any particular point $p(x, z)$ the axial strain will be given by $\varepsilon = \alpha Kx$. From Eq. 2-1

$$de_x = \varepsilon \, dx \qquad de_z = \varepsilon \, dz$$

Integration yields

$$e_x = \alpha K \frac{x^2}{2} \qquad e_z = \alpha Kxz$$

Since α is usually many orders of magnitude less than unity, the shape of the plate is shown in Fig. 2-6(c), where the deformations are exaggerated.

(c) As remarked in (a), under a uniform free expansion the plate remains rectangular everywhere and hence no distortion occurs. Thus

$$\gamma_{xz} = 0 \text{ everywhere}$$

(d) For case (b) recall

$$e_x = \alpha K \frac{x^2}{2} \qquad e_z = \alpha K x z$$

and from Fig. 2-6(c) we see clearly that distortion occurs. Also note that the maximum distortion occurs along the $z = \pm b/2$ edges of the plate and the angle of distortion is

$$\tan \phi = \frac{\alpha K a (b/2)}{a + (\alpha K a^2/2)} = \frac{\alpha K (b/2)}{1 + (\alpha K a/2)}$$

Now, since $\alpha \ll 1$ so that the deformations and distortions are small, we can say that

$$\tan \phi \cong \phi = \gamma_{max} = \alpha K \frac{b}{2}$$

and in general the shear strain for pairs of lines parallel to the x-z axes is given by

$$\gamma_{xz} = \frac{de_z}{dx} = \alpha K z$$

PROBLEMS

2-1 A 3-in.-long rubber band is stretched until its length is 4 in. What is the induced average axial strain?

2-2 A 100-mm rubber band (200 mm perimeter) is slipped over a solid roll of paper 100 mm in diameter. What average axial strain is induced in the band?

2-3 A spherical balloon initially contains 1 m³ of gas. What will be the induced circumferential strain in the balloon if enough additional gas is pumped into the balloon so as to double its original volume?

2-4 An unloaded straight vine is 10 m long. It stretches 80 mm when Tarzan is at the bottom of his swing. What axial strain is imposed?

2-5 A cylindrical gas tank having an inside radius of 8 in. undergoes an internal pressure change such that the radius becomes 8.05 in. Compute the average change in surface (membrane) strain in the tank.

2-6 Consider the flexible cable shown wrapped around the cone in the illustration. What is the *change* in the average strain of the cable if it is forced down to a depth of $h/2$?

Problem 2-6

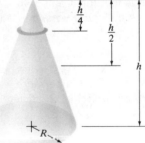

2-7 A very stiff bar (assumed to be rigid) is suspended from two wires of unequal length, as shown in the illustration. Initially, point B is 10 mm higher than point A. The bar is then loaded in such a manner that the bar eventually becomes horizontal. (a) What is the relationship between the deformation of the wire at A and that of the wire at B? (b) What is the relationship between the strain in the wire at A and that in the wire at B? (c) What is the vertical displacement of point C in terms of the deformation of the wire at A?

Problem 2-7

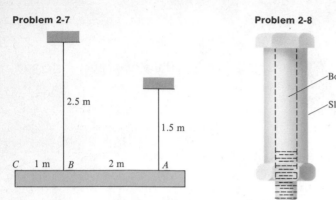

Problem 2-8

Problem 2-12

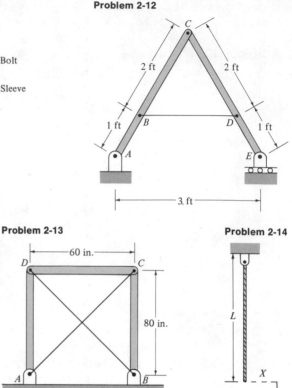

2-8 The bolt-and-sleeve combination shown in the illustration is adjusted until the nut is just in contact with the sleeve. The bolt has a pitch of 24 threads per inch. The nut is then tightened two full turns. If any deformations of the bolt head and nut are neglected, what is the relationship between the deformation of the bolt stem and the deformation of the sleeve? (b) What is the relationship between the strain in the bolt stem and the strain in the sleeve?

2-9 A turnbuckle has a $\frac{1}{4}$-in.-diameter screw, 20 threads per inch (each end) and is used to tighten a steel cable that is fastened between two rigid walls 80 in. apart. If the cable is taut (unstrained) to start, how much axial strain will the cable be subjected to if the turnbuckle is tightened two full turns?

2-10 In Prob. 2-9, what would the axial strain in the cable be if (a) after tightening two turns, the temperature were lowered 30°F? (b) if the temperature were raised 30°F? See the Appendix for coefficient of thermal expansion.

2-11 A thin metal hoop has an inside diameter of 1400 mm at room temperature. It is then heated until it easily slips on a mandrel 1406.6 mm in diameter. If the mandrel is assumed to be rigid, what is the circumferential strain in the hoop when it cools down to room temperature?

2-12 The very stiff rods AC and CE are held in the positions shown in the illustration by the pins at A and C and the taut flexible wire BD. If point E is then given a horizontal displacement of 1.2 in. to the right, what is the change in the average strain in the wire?

2-13 The pin-connected rods BC, CD, and DA, which can be assumed to be rigid, are held in the positions shown in the illustration by the taut flexible wires AC and BD. If point D is then given a horizontal dis-

Problem 2-13

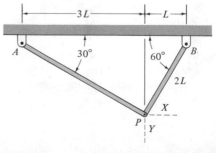

Problem 2-14

placement of 8 in. to the right, what is the change in the average strain in each of the wires?

2-14 A cable of length L ft hangs in the vertical position shown. If the free end is displaced X ft to the right and Y ft downward, what strain will be induced in the wire? If both X and Y are small compared with L, what is a good approximate value for the strain?

2-15 Two rods AP and BP are pinned together as shown. What will be the average strain induced in

Problems 2-15 and 2-16

each rod if point P is displaced X ft to the right? What are approximate values for these strains if X is small compared with L?

2-16 Same as Prob. 2-15 except point P is displaced downward an amount Y ft.

2-17 The rigid bar of length L shown in the illustration is supported by two vertical deformable wires. The bar is rotated about the vertical axis a-a through an angle θ measured in the horizontal plane, but the bar is also constrained so that its elevation does not change. What will be the average axial strain in each wire due to this rotation? If θ is small, what will be an approximate value for this strain?

Problem 2-17

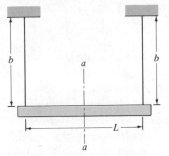

2-18 Two rigid straight plates of length L are separated by a rubber block whose dimensions are L and h as shown in the illustration. The upper plate AB is then moved a small distance e_0 to the right, but

Problem 2-18

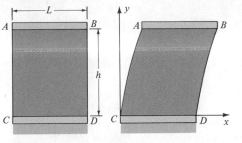

remains horizontal and straight, while the plate CD remains stationary. What is the shear strain γ_{xy} at the points $(L/2, h/3)$ and $(0, 0)$ if (a) the original vertical sides remain straight; (b) the original vertical sides deform into parabolas for which the equation is $x = ky^2$?

2-19 A square rubber plate has two diagonal lines scribed on its surface as shown. The plate is then deformed as indicated by the dashed lines. What is the axial strain in the diagonal lines? What is the shear strain between these lines at point O?

Problem 2-19

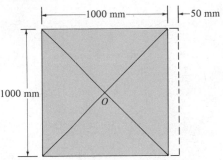

2-20 A torsional spring composed of an outer metal ring, a disk of hard rubber, and an inner ring is shown in the illustration. The outer ring is rotated through a small angle ϕ relative to the inner ring. Assume that the original radial lines remain straight during the deformation. Find an expression for the shear strain in the rubber.

Problem 2-20

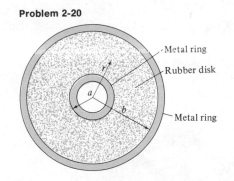

2-5 / Analysis of plane strain

In section 2-4, axial strain and shear strain were analyzed individually. Generally, a point p in a body will undergo axial strain and shear strain simultaneously as indicated in Fig. 2-7. Thus, in general, a state of plane

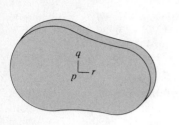

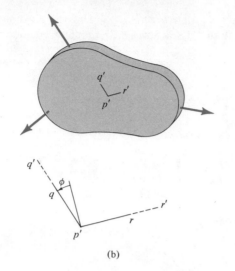

(a)   (b)

Figure 2-7

strain at a point is characterized by two axial strains and one shear strain. Since we are interested only in the relation between the initial and final configurations, we can assume that the final configuration was achieved by an axial straining process followed by a shear straining process or vice versa. *So long as the deformations are small, the order of the straining processes is immaterial,* so for convenience we shall assume the former. Also, since we are considering the strains at a point, we shall label the original lengths dx and dy, as in Fig. 2-8.

A pair of intersecting infinitesimal line elements may be considered to define a point p in a body. The line elements are made parallel to the axes of a specific two-dimensional coordinate system, such as the x-y system. Therefore, any two orthogonal line elements can be used to show only the strains associated with that particular set of axes. In Fig. 2-8 the shear strain γ_{xy} and the axial strains ε_x and ε_y are associated only with line elements initially parallel to the x and y axes.

Figure 2-8

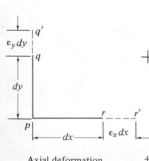

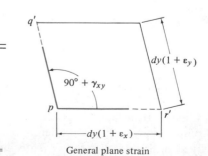

Axial deformation   +   Shear deformation   =   General plane strain

Transformation of axial strain Many engineering problems require the determination of the strains or deformations associated with a particular coordinate system. As a result, it is often necessary to relate the strains associated with one coordinate system to the strains associated with another coordinate system. For example, suppose we know the strains ε_x, ε_y, and γ_{xy} associated with the x-y coordinate system, and we wish to know the strains associated with the n-s coordinate system, which is oriented at some angle θ with the x-y system, as shown in Fig. 2-9(a). In order to determine the strains in the n-s system, we must relate the deformations parallel to the n and s axes to the deformations of line elements parallel to the x and y axes. Figure 2-9(b) shows undeformed line elements associated with the x-y system and having such proportions that

$$dx = dn \cos \theta \qquad dy = dn \sin \theta$$

or

$$\cos \theta = \frac{dx}{dn} \qquad \sin \theta = \frac{dy}{dn}$$

Figure 2-9(c) shows these elements in a state of plane strain.

The axial strain in the n direction, denoted by ε_θ, would be the axial deformation de_θ in that direction divided by dn. This deformation can be found by appropriately adding or subtracting the component deformations in the n direction, which are represented by the sides of triangles ①, ②, and ③ parallel to the n axis. That side of triangle ① parallel to dn is $\varepsilon_y \, dy \sin \theta$; that of triangle ② is $\varepsilon_x \, dx \cos \theta$; and that of triangle ③ is $dy(1 + \varepsilon_y)\gamma_{xy} \cos \theta$. Hence,

$$\varepsilon_\theta = \frac{de_\theta}{dn} = \frac{\varepsilon_y \, dy \sin \theta + \varepsilon_x \, dx \cos \theta - dy(1 + \varepsilon_y)\gamma_{xy} \cos \theta}{dn} \qquad (2\text{-}3)$$

By substituting $\cos \theta$ for dx/dn and $\sin \theta$ for dy/dn and neglecting the product $\gamma_{xy}\varepsilon_y$, which is extremely small compared with γ_{xy} or ε_y, we obtain

$$\varepsilon_\theta = \varepsilon_x \cos^2 \theta + \varepsilon_y \sin^2 \theta - \gamma_{xy} \cos \theta \sin \theta \qquad (2\text{-}4)$$

From Eq. 2-4 we can obtain the axial strain along any axis lying in the plane of analysis. This equation is called a strain transformation equation. Note that to use this equation we must employ a sign convention consistent with that assumed in the derivation. The angle θ is positive when measured counterclockwise from the x axis; tensile axial strains are positive, and compressive axial strains are negative; *a positive shear strain is one that causes an increase in the angle at p when the x axis is taken as the "horizontal" reference line.*

Transformation of shear strain Figure 2-9(c) also shows that the line

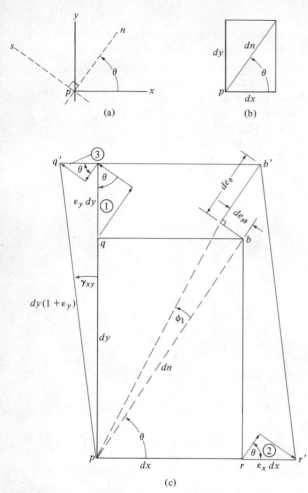

(a)

(b)

(c)

Figure 2-9

dn parallel to the n axis has undergone a small change of orientation, denoted by ϕ_1. The angle ϕ_1, being very small, can be expressed in radians by dividing the deformation $de_{s\theta}$ by the length dn. The deformation $de_{s\theta}$ can be obtained by adding vectorially the component deformations *perpendicular* to the n axis. Thus, referring to the triangles of Fig. 2-9(c),

$$\phi_1 = \frac{de_{s\theta}}{dn} = \frac{\varepsilon_y \, dy \cos \theta - \varepsilon_x \, dx \sin \theta + dy(1 + \varepsilon_y)\gamma_{xy} \sin \theta}{dn} \tag{2-5}$$

$$= \varepsilon_y \sin \theta \cos \theta - \varepsilon_x \cos \theta \sin \theta + \gamma_{xy} \sin^2 \theta \tag{2-6}$$

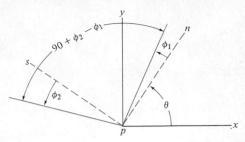

Figure 2-10

However, ϕ_1 does *not* represent the shear strain associated with the *n-s* coordinate system. Shear strain is the *change* in the 90° angle between two lines. In order to determine the shear strain associated with the *n-s* coordinate system, we must determine the angle change at p between the *n* and *s* axes.

Figure 2-10 shows that the line element parallel to the *n* axis will undergo a change in orientation denoted by ϕ_1, and the line parallel to the *s* axis will undergo a change in orientation denoted by ϕ_2. The angle at p before distortion was 90°. The angle at p after distortion will be $90° + \phi_2 - \phi_1$. Therefore, the change in the angle at p will be $\phi_2 - \phi_1$, and this quantity will be the shear strain γ_θ associated with the *n-s* system.

To determine ϕ_2, we observe that ϕ_2 will be given by Eq. 2-6 if we substitute $\theta + 90°$ for θ since the *s* axis was originally at an angle of $\theta + 90°$ from the *x* axis. Thus

$$\phi_2 = \varepsilon_y \sin(\theta + 90°) \cos(\theta + 90) - \varepsilon_x \cos(\theta + 90) \sin(\theta + 90)$$
$$+ \gamma_{xy} \sin^2(\theta + 90) \tag{2-7}$$
$$= -\varepsilon_y \cos\theta \sin\theta + \varepsilon_x \sin\theta \cos\theta + \gamma_{xy} \cos^2\theta \tag{2-8}$$

The shear strain γ_θ is, therefore, from Eqs. 2-6 and 2-8

$$\gamma_\theta = \phi_2 - \phi_1 = 2(\varepsilon_x - \varepsilon_y)\sin\theta\cos\theta + \gamma_{xy}(\cos^2\theta - \sin^2\theta) \tag{2-9}$$

From Eq. 2-9, which is another strain transformation equation, we can determine the shear strain associated with any set of axes lying in the plane of analysis. The sign convention must be the same as that for Eq. 2-4.

It should be noted that while the foregoing derivations involve fairly intricate geometry and trigonometry, the resulting transformation equations are quite easy to use and the Mohr's circle solutions for strain are very simple.

2-6 / Mohr's circle for plane strain

By use of the double-angle identities of trigonometry, the strain transformation equations (Eqs. 2-4 and 2-9) can be written as

$$\varepsilon_\theta = \frac{\varepsilon_x + \varepsilon_y}{2} + \left(\frac{\varepsilon_x - \varepsilon_y}{2}\right) \cos 2\theta - \frac{\gamma_{xy}}{2} \sin 2\theta \qquad (2\text{-}10)$$

$$\frac{\gamma_\theta}{2} = \left(\frac{\varepsilon_x - \varepsilon_y}{2}\right) \sin 2\theta + \frac{\gamma_{xy}}{2} \cos 2\theta \qquad (2\text{-}11)$$

You should note the similarity between the stress transformation equations (Eqs. 1-15 and 1-16) and the strain transformation equations (Eqs. 2-10 and 2-11).

Since we were able to express the stress transformation equations in the form of an equation representing a circle (Eq. 1-21), we can do the same thing with the strain transformation equations. By transposing squaring, adding, and simplifying, we obtain the following result:

$$\left[\varepsilon_\theta - \left(\frac{\varepsilon_x + \varepsilon_y}{2}\right)\right]^2 + \left[\frac{\gamma_\theta}{2} - 0\right]^2 = \left[\sqrt{\left(\frac{\varepsilon_x - \varepsilon_y}{2}\right)^2 + \left(\frac{\gamma_{xy}}{2}\right)^2}\right]^2 \qquad (2\text{-}12)$$

We can plot Eq. 2-12 as a circle on a rectangular coordinate system in which the abscissa is ε_θ and the ordinate $\gamma_\theta/2$. Such a plot is called Mohr's circle for plane strain, and its properties are very similar to those of Mohr's circle for plane stress.

As an example, suppose that at some point in a loaded body we know the strains associated with a particular set of axes which we will call the x and y axes. For convenience, we shall assume that all the strains are positive and that ε_x is the larger axial strain. Figure 2-11(a) shows Mohr's circle for strain plotted for the assumed conditions. Figure 2-11(b) shows the orientations of the n and s axes and axes labeled 1 and 2 with respect to the original x and y axes. The line elements associated with the various axes are shown in Fig. 2-11(b) merely to illustrate the distortion resulting from the respective shear strains.

You should study Fig. 2-11 very carefully. The following important features can be observed from this figure:

Figure 2-11

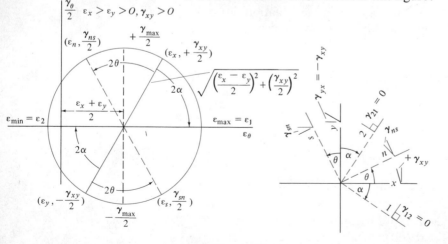

1   The center of the circle is always on the ε_θ axis.

2   Each point on the circle represents two quantities for the point in the loaded body, namely, an axial strain along a particular axis and one-half of the value of the shear strain associated with that same axis. In every case *a positive shear strain produces an increase in the right angle and a negative shear strain produces a decrease in that angle.* Satisfy yourself about this fact for each situation in Fig. 2-11(b) by rotating the book so as to make the axis in question a "horizontal" reference line. Note that the first letter of the subscript for γ corresponds to this horizontal reference line.

3   Any angle in the x-y coordinate system corresponds to twice that angle on the circle; that is, θ *on the body,* 2θ *on the circle.* Also recall that in order to use Eqs. 2-4 and 2-9 correctly, θ had to be measured from the x axis with the counterclockwise direction being positive. However, in using the circle, we can refer any desired axis or coordinate system to either the x axis or the y axis, as long as we correctly reference 2θ on the circle.

4   Two diametrically opposite points on the circle represent the *axial strains* and the *shear strain* associated with a *set* of perpendicular axes, such as the x and y axes, the n and s axes, and the 1 and 2 axes.

5   The algebraic sum of the axial strains at a point is the same for all sets of axes; that is, $\varepsilon_x + \varepsilon_y = \varepsilon_n + \varepsilon_s = \varepsilon_1 + \varepsilon_2$.

6   The maximum and minimum axial strains, denoted by ε_1 and ε_2, occur along the axes 1 and 2. These strains are called the principal strains, and the axes 1 and 2 are called the principal axes. No shear strain is associated with the principal axes. Line elements parallel to the principal axes will have no angular distortion and no shear strain.

7   The magnitude of the maximum shear strain is twice the radius of the circle; that is, the diameter of the circle represents the magnitude of the maximum shear strain

---

EXAMPLE 2-5

Some point on the surface of a loaded body was found to have a tensile strain of 0.0001 in./in. in the x direction and a compressive strain of 0.0007 in./in. in the y direction, and the angular distortion γ_{xy} was -0.0006 rad. (a) Determine the values and directions of the principal strains. (b) Determine the magnitude of the maximum shear strain. (c) Determine the axial and shear strains associated with the n axis which is 30° counterclockwise from the x axis, as indicated in Fig. 2-12(a).

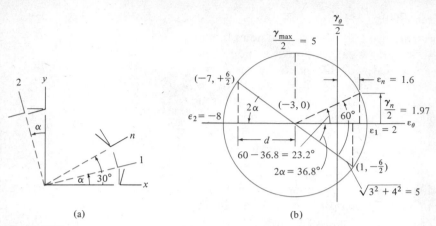

(a)   (b)

Figure 2-12

Solution:

First, the points representing the known strains are plotted and joined by a line, which is a diameter of the Mohr circle. The center of the circle is midway between these points, and the circle is shown in Fig. 2-12(b). In this circle, actual strains are multiplied by 10^4 for use as distances. Thus,

$$\varepsilon_x = 0.0001 \text{ in./in.} = 1 \text{ unit}$$

$$\varepsilon_y = -0.0007 \text{ in./in.} = -7 \text{ units}$$

$$\gamma_{xy} = -0.0006 \text{ rad} = -6 \text{ units}$$

$$d = \frac{1+7}{2} = 4$$

The coordinates of the center of the circle are -3 and 0, and its radius is $\sqrt{4^2 + 3^2} = 5$.

(a) The principal strains are equal to the center coordinate plus and minus the radius. They are

$$\varepsilon_1 = -3 + 5 = +2 \qquad \varepsilon_2 = -3 - 5 = -8$$

So $\varepsilon_1 = 0.0002$ in./in., tensile, and $\varepsilon_2 = -0.0008$ in./in., compressive. The direction of ε_1 is at an angle α counterclockwise from the x axis, and ε_2 is at an angle α counterclockwise from the y axis. From the circle,

$$\tan 2\alpha = \tfrac{3}{4} \qquad 2\alpha = 36.8° \qquad \alpha = 18.4°$$

(b) The magnitude of the maximum shear strain is equal to twice the radius of the circle. So

$$\gamma_{max} = 2(\gamma/2)_{max} = 2(5) = 10 \text{ units}$$

or

$$\gamma_{max} = 0.0010 \text{ rad}$$

(c) The point representing the n axis on the circle is located 60° counterclockwise from the point representing the x axis. A radial line from this point to the center is then drawn. The angle between this radial line and the ε_0 axis is

$$60° - 2\alpha = 60° - 36.8° = 23.2°$$

We now see that ε_n is equal to the coordinate of the center of the circle plus the base of the 23.2° triangle, and $\gamma_n/2$ is the vertical side of this triangle. Thus,

$$\varepsilon_n = -3 + 5 \cos 23.2° = -3 + 4.6 = 1.6 \text{ units}$$

or

$$\varepsilon_n = 0.00016 \text{ in./in., tensile}$$

$$\gamma_n/2 = 5 \sin 23.2° = 1.97 \text{ units}$$

and

$$\gamma_n = 0.000394 \text{ rad, positive}$$

PROBLEMS

2-21 Using the strain transformation equations, show that the sum of the axial strains in two perpendicular directions is a constant. What does this imply in regard to the Mohr circle for strain?

2-22 Using the strain transformation equations, show that the angle which maximizes the shear strain differs by 45° from that which maximizes the axial strain.

Note: For the given state of plane strain in each of the problems from Prob. 2-23 to Prob. 2-29, determine the values of the principal strains and their directions. The x and y axes are horizontal and vertical, respectively, and all strains are in (in./in.) $\times$ 10^6, or micro-strain.

2-23 $\varepsilon_x = 400$ $\varepsilon_y = -2000$ $\gamma_{xy} = 1000$

2-24 $\varepsilon_x = -600$ $\varepsilon_y = 800$ $\gamma_{xy} = -700$

2-25 $\varepsilon_x = 800$ $\varepsilon_y = 2000$ $\gamma_{xy} = -1600$

2-26 $\varepsilon_x = 500$ $\varepsilon_y = -1000$ $\gamma_{xy} = -2000$

2-27 $\varepsilon_x = 1200$ $\varepsilon_y = -1200$ $\gamma_{xy} = 2400$

2-28 $\varepsilon_x = 3000$ $\varepsilon_y = -1800$ $\gamma_{xy} = 1400$

2-29 $\varepsilon_x = 730$ $\varepsilon_y = 1480$ $\gamma_{xy} = 400$

2-30 Rework Prob. 2-19 by use of Mohr's circle for strain. Why do these answers differ slightly from those obtained by the earlier method?

2-31 Two diagonal lines are scribed on a block of rubber. The block is then deformed as indicated by the dashed lines. What is the resulting strain of the lines? What is the final length of each of these lines? What is the angle change at O?

Problem 2-31

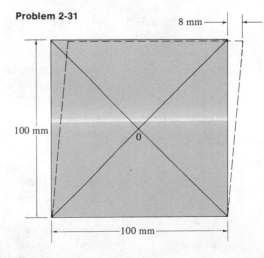

2-32 Two perpendicular axial strain gages A and B are bonded to the surface of a gas tank. When the tank is pressurized, it is known that the ratio of the principal strains will be $\varepsilon_{max}/\varepsilon_{min} = 3.0$. Determine ε_{max} and ε_{min} in terms of measured strains ε_A and ε_B.

2-7 / Calculator solution

As was true in the case of plane stress, it is possible to program your calculator to solve for principal strains or strain in a specified direction given two orthogonal strains and the associated shearing strain. It is not efficient to do so for only one or two problems but it is very efficient if several similar problems must be solved.

The programs in section 1-11 may be modified for use in problems involving strain transformations by means of the following steps:

1 Enter ε_x instead of σ_x.
5 Read ε_{min} instead of σ_{min}.

2 Enter ε_y instead of σ_y.
6 Read $\gamma_{max}/2$ instead of τ_{max}.

3 Enter $\gamma_{xy}/2$ instead of τ_{xy}.
7 Read ε_θ instead of σ_θ.

4 Read ε_{max} instead of σ_{max}.
8 Read $\gamma_\theta/2$ instead of τ_θ.

PROBLEMS

2-33 to 2-39 Solve Probs. 2-23 to 2-29 by means of a programmed calculator.

2-8 / Strain measurement and rosette analysis

The definitions of stress and strain "at a point" are analytical conveniences. For instance, we have defined stress as a force intensity on an infinitesimal area. As such, it would be physically impossible for us to see or measure stress. A finite load, however, can be measured, and a large portion of this book is concerned with relating measurable loads to the theoretical stresses existing in engineering structures.

In this chapter we have defined strain as a deformation per unit length, and we have analyzed plane strain at a point. Although it is physically possible to measure *finite* deformations on the surface of a body, a *finite* deformation can only be measured over a *finite* length. So the idea of strain at a *point* in a body is an analytical convenience. However, if a deformation is measured over a relatively small length, an average deformation per unit length can be evaluated and interpreted as the approximate value of a strain at the "finite point" of measurement. On this basis, an axial strain can be approximated by first measuring the elongation or contraction over a small length and then dividing this deformation by the length over which the deformation occurred.

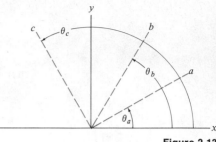

Figure 2-13

It is much more difficult to approximate a shear strain since it would be necessary to measure the angular distortion occurring at a point on the surface of a body. Generally, the various mechanical, optical, and electrical methods of measuring *axial* deformations are far simpler and more reliable than are the methods for measuring angular distortions. In Chapter 15 some of the more common methods of strain measurement are discussed.

In most problems involving strain analysis, the engineer is interested in the values and directions of the principal strains. In cases where the directions of the principal strains are known, their values can be approximated by measuring the deformations along the principal axes. However, there are a great many situations in which neither the directions nor the values of the principal strains are known. The principal strains at a point can be determined in such cases from any three known axial strains at that point. This is indeed fortunate, since it eliminates the somewhat difficult task of measuring shear strains.

In Fig. 2-13, suppose that the axial strains along the axes *a*, *b*, and *c* are known, as well as the angles θ_a, θ_b, and θ_c. By using these known values and applying Eq. 2-4, we obtain three simultaneous equations involving the three unknowns ε_x, ε_y, and γ_{xy}. These equations follow:

$$\varepsilon_a = \varepsilon_x \cos^2 \theta_a + \varepsilon_y \sin^2 \theta_a - \gamma_{xy} \cos \theta_a \sin \theta_a$$

$$\varepsilon_b = \varepsilon_x \cos^2 \theta_b + \varepsilon_y \sin^2 \theta_b - \gamma_{xy} \cos \theta_b \sin \theta_b$$

$$\varepsilon_c = \varepsilon_x \cos^2 \theta_c + \varepsilon_y \sin^2 \theta_c - \gamma_{xy} \cos \theta_c \sin \theta_c$$

The solution of these equations will yield ε_x, ε_y, and γ_{xy}, and we will then be able to plot Mohr's circle, from which the principal strains can be determined.

A device for measuring three or more simultaneous axial strains is called a rosette strain gage. Generally the axes of strain measurement are arranged in a definite pattern, so as to make the solution of the above equations numerically simple. For example, in a rectangular rosette the axial strains are measured along three axes that are 45° apart, while in a delta rosette the three axes are 60° apart as shown in Fig. 2-14. There are various graphical and semigraphical methods for solving strain rosettes

Figure 2-14

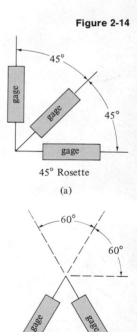

45° Rosette

(a)

Delta rosette

(b)

which eliminate the need for the solution of simultaneous equations.*

For the more commonly used rosettes, it is possible and even reasonable to reduce the equations to a somewhat easier form. For the 45° rosette, choose your axes so that $\theta_a = 0$, $\theta_b = 45°$, and $\theta_c = 90°$. The equations then become:

$$\varepsilon_a = \varepsilon_x$$

$$\varepsilon_b = \varepsilon_x\left(\frac{\sqrt{2}}{2}\right)\left(\frac{\sqrt{2}}{2}\right) + \varepsilon_y\left(\frac{\sqrt{2}}{2}\right)\left(\frac{\sqrt{2}}{2}\right) - \gamma_{xy}\left(\frac{\sqrt{2}}{2}\right)\left(\frac{\sqrt{2}}{2}\right)$$

$$= \tfrac{1}{2}(\varepsilon_x + \varepsilon_y - \gamma_{xy})$$

$$\varepsilon_c = \varepsilon_y$$

Thus

$$\varepsilon_x = \varepsilon_a$$

$$\varepsilon_y = \varepsilon_c$$

$$\gamma_{xy} = \varepsilon_a + \varepsilon_c - 2\varepsilon_b$$

For the delta rosette, choose your axes so that $\theta_a = 0$, $\theta_b = 60°$, and $\theta_c = 120°$. The equations then become

$$\varepsilon_a = \varepsilon_x$$

$$\varepsilon_b = \varepsilon_x\left(\frac{1}{2}\right)^2 + \varepsilon_y\left(\frac{\sqrt{3}}{2}\right)^2 - \gamma_{xy}\left(\frac{\sqrt{3}}{2}\right)\left(\frac{1}{2}\right)$$

$$= \frac{1}{4}\varepsilon_x + \frac{3}{4}\varepsilon_y - \frac{\sqrt{3}}{4}\gamma_{xy}$$

$$\varepsilon_c = \varepsilon_x\left(-\frac{1}{2}\right)^2 + \varepsilon_y\left(\frac{\sqrt{3}}{2}\right)^2 - \gamma_{xy}\left(\frac{\sqrt{3}}{2}\right)\left(-\frac{1}{2}\right)$$

$$= \frac{1}{4}\varepsilon_x + \frac{3}{4}\varepsilon_y + \frac{\sqrt{3}}{4}\gamma_{xy}$$

Simultaneous solution of the last two equations produces the following results:

$$\varepsilon_x = \varepsilon_a$$

$$\varepsilon_y = \tfrac{1}{3}(2\varepsilon_b + 2\varepsilon_c - \varepsilon_a)$$

$$\gamma_{xy} = \frac{2\sqrt{3}}{3}(\varepsilon_c - \varepsilon_b) = 1.154(\varepsilon_c - \varepsilon_b)$$

---

* A good presentation of a graphical solution of strain rosettes may be found in Chapters 6 and 7 of *The Strain Gage Primer*, 2d ed., by C. C. Perry and H. R. Lissner (New York: McGraw-Hill, 1962).

When you have determined ε_x, ε_y, and γ_{xy} from these equations, it will still be necessary to plot Mohr's circle to find the principal strains and the maximum shearing strain. It should be noted that the use of these equations will always cause your direction angles to be measured from the axis along which strain gage a is oriented.

EXAMPLE 2-6

A rectangular rosette strain gage was applied to the surface of a beam, as shown in Fig. 2-15(a). The strains along the axes a, b, and c, respectively, were $\varepsilon_a = +0.0008$ in./in., $\varepsilon_b = -0.0006$ in./in., and $\varepsilon_c = -0.0004$ in./in. Determine the directions and values of the principal strains at point O. Express your answer in microns per meter.

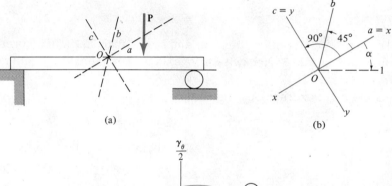

(a)   (b)

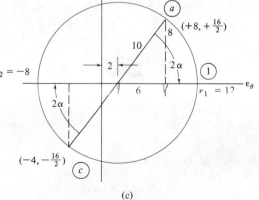

(c)

Figure 2-15

Solution:

Let the a axis of the gage be coincident with the x axis of an arbitrary coordinate system, as illustrated in Fig. 2-15(b). Then $\theta_a = 0°$, $\theta_b = 45°$, and $\theta_c = 90°$. Substituting these values and the measured strains in Eq. 2-4, we obtain the following results:

$$\varepsilon_a = \varepsilon_x(1) + \varepsilon_y(0) - \gamma_{xy}(1)(0) = \varepsilon_x$$

$$\varepsilon_x = 0.0008 \text{ in./in.}$$

$$\varepsilon_c = \varepsilon_x(0) + \varepsilon_y(1) - \gamma_{xy}(0)(1) = \varepsilon_y$$

$$\varepsilon_y = -0.0004 \text{ in./in.}$$

$$\varepsilon_b = \varepsilon_x \tfrac{1}{2} + \varepsilon_y \tfrac{1}{2} - \gamma_{xy}\sqrt{\tfrac{1}{2}}\sqrt{\tfrac{1}{2}}$$

$$\gamma_{xy} = -2\varepsilon_b + (\varepsilon_x + \varepsilon_y)$$
$$= -2(-0.0006) + (0.0008 - 0.0004) = +0.0016 \text{ rad}$$

We can now plot Mohr's circle and determine the principal strains, as shown in Fig. 2-15(c). In this circle, values of strain are multiplied by 10^4 for use as distances. The computations follow:

$$\tan 2\alpha = \tfrac{8}{6} \qquad 2\alpha = 53.2° \qquad \alpha = 26.6°$$

$$\varepsilon_1 = 2 + 10 = 12 \qquad \varepsilon_2 = 2 - 10 = -8$$

Thus, $\varepsilon_1 = 0.0012$ in./in. (tensile) and $\varepsilon_2 = 0.0008$ in./in. (compressive). The maximum strain occurs along an axis 26.6° clockwise from axis a. The minimum strain occurs along an axis 26.6° clockwise from axis c.

Finally, to express our answer in micrometers per meter,

$$\varepsilon_1 = 0.0012 \text{ in./in.} = 1200 \text{ } \mu\text{in./in.} = 1200 \text{ } \mu\text{m/m}$$

$$\varepsilon_2 = -0.0008 \text{ in./in.} = -800 \text{ } \mu\text{m/m}$$

You must realize that a strain measured by means of a rosette represents only an average value of the strain over the measuring length. It is therefore physically impossible to determine the *exact* value of a strain at a *point*. The results obtained in Example 2-6 represent only a good approximation of the principal strains at point O.

PROBLEMS

The data in each of the problems from Probs. 2-40 to 2-43 were obtained from 45° rosettes oriented so that $\theta_a = 0°$. Find the values and directions of the principal strains. The strains are in (mm/mm) $\times 10^6$.

2-40 $\varepsilon_a = 450$ $\qquad \varepsilon_b = -420$ $\qquad \varepsilon_c = -1100$

2-41 $\varepsilon_a = 200$ $\qquad \varepsilon_b = 900$ $\qquad \varepsilon_c = 600$

2-42 $\varepsilon_a = 2000$ $\qquad \varepsilon_b = 450$ $\qquad \varepsilon_c = 300$

2-43 $\varepsilon_a = 400$ $\qquad \varepsilon_b = 200$ $\qquad \varepsilon_c = 200$

Problems 2-40 to 2-43

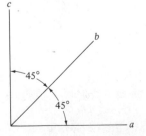

The data in each of the problems from Probs. 2-44 to 2-47 were obtained from delta rosettes oriented so that $\theta_a = 0$. Find the values and directions of the principal strains. The strains are given in microstrain.

2-44 $\varepsilon_a = 2000$ $\varepsilon_b = 1000$ $\varepsilon_c = -500$

2-45 $\varepsilon_a = 1100$ $\varepsilon_b = 400$ $\varepsilon_c = -550$

2-46 $\varepsilon_a = 200$ $\varepsilon_b = 900$ $\varepsilon_c = 600$

2-47 $\varepsilon_a = 1100$ $\varepsilon_b = 200$ $\varepsilon_c = -600$

The data in Probs. 2-48 and 2-49 are in microstrain. Determine the values and directions of the principal strains.

2-48 $\varepsilon_c = -2010$ $\varepsilon_e = -430$ $\varepsilon_g = +270$

2-49 $\varepsilon_b = 1100$ $\varepsilon_d = 200$ $\varepsilon_f = -600$

2-50 The strain readings from a delta rosette strain gage applied to a submarine hull were recorded at one instant during a severe maneuver with the following results: ε_A was 2360 μm/m compressive; ε_B was 780 μm/m compressive; and ε_C was 2410 μm/m tensile. Did the maximum strain at the point exceed the design strain of 3000 μm/m compressive?

Problems 2-44 to 2-47

Problems 2-48 and 2-49

Problem 2-50

Chapter 3
Experimental mechanical properties of engineering materials

3-1 / Objectives

Upon completion of this chapter, you will be able to:

1   Given load-deformation data, plot a stress-strain curve.
2   Identify the modulus of elasticity.
3   Identify the proportional limit.
4   Define Poisson's ratio and use it to determine transverse strain.
5   Describe the elastic limit.
6   Define ductility and brittleness.
7   Sketch a stress-strain curve for a brittle material.
8   Sketch a stress-strain curve for a ductile material.
9   Find the yield strength of a given material for a given offset.
10   Find the modulus of resilience of a given material from its stress-strain curve.
11   Find the modulus of toughness of a given material from its stress-strain curve.
12   Describe the most prevalent modes of failure.
13   Determine the most probable mode of failure for a given device.
14   Select an appropriate material for a given application by the comparison of stress-strain curves.

3-2 / The stress-strain curve

The simplest and most common experiment for measuring the mechanical response of engineering structural materials is the uniaxial tensile or compression test. In this test a slowly increasing (quasi-static) load is applied to a specimen of material and the resulting load-deformation (elongation or contraction) data are recorded. Figure 3-1(a) shows a typical screw-type testing machine, while Fig. 3-1(b) shows a close-up of the specimen with an extensometer attached for measuring the axial deformation. Many such machines are equipped to record the continuous load-deformation data automatically throughout the test. Figure 3-2(a) shows a typical tensile specimen with the gripping region thicker than the test region so as to reduce the possibility of failure due to the gripping of the specimen in the test machine. The initial gage length over which the deformation is measured is usually 2 to 8 in. and the test region diameter is in the neighborhood of $\frac{1}{2}$ to 1 in.

Figure 3-1

(a)

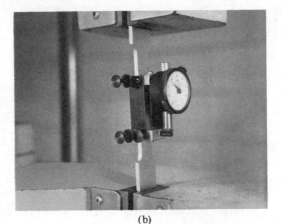

(b)

Figure 3-2

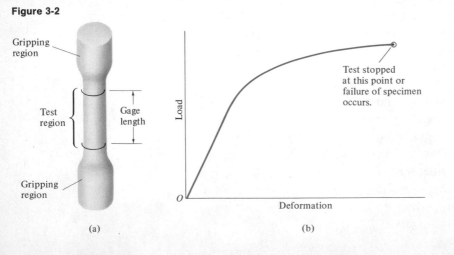

(a)

(b)

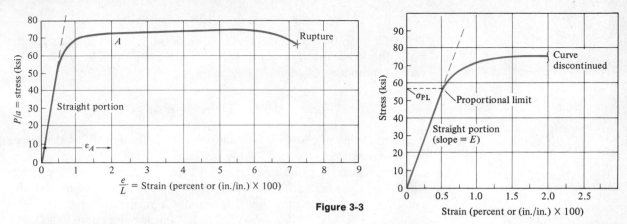

Figure 3-3

Figure 3-4

Figure 3-2(b) shows a typical load-deformation curve for some specimen of material. No numbers are shown in this curve for a very important reason: *The values of load and deformation would depend greatly upon the size of the test specimen.* Intuitively we would expect a thick specimen to require a larger load than a thinner specimen to produce the same amount of deformation. Similarly, a specimen whose gage length was 8 in. would deform more than a similar specimen whose gage length was only 2 in. Thus, to characterize the response of the *material*, the load-deformation data must be quantified in terms of quantities that are independent of the specimen size. Accordingly, the load is divided by the cross-sectional area to obtain values of axial stress (load per unit area), while the deformation is divided by the initial gage length so as to obtain values of axial strain (deformation per unit length). The resulting scaled plot is called a *uniaxial engineering stress-strain curve.* Such a curve is considered to be a reliable and reproducible measure of the uniaxial mechanical response of engineering structural materials.

In Fig. 3-3 is shown a plot of stress versus strain for a tensile test on an extruded rod of a common aluminum alloy. As is customary in engineering, stress is plotted as the ordinate. This curve extends from zero strain to rupture. From an engineering-design viewpoint, the first part of the curve, or that up to about point A, is the most important portion since a material of this kind is usually unusable if the strain is much greater than ε_A. The more important initial portion of the curve is shown to a larger scale in Fig. 3-4.

It is generally true that for most metallic engineering materials the stress is proportional to strain for the first portion of the curve, that is, a linear response. The stress corresponding to the first point at which the curve deviates from the initial straight, or linear, portion is called the *proportional limit.* Up to the proportional limit the stress change $\Delta\sigma$ is equal to the corresponding strain change $\Delta\varepsilon$ multiplied by some constant. Thus,

$$\Delta\sigma \propto \Delta\varepsilon \qquad \Delta\sigma = (\Delta\varepsilon) \times \text{constant}$$

This relationship is a simple manifestation of the familiar Hooke's law in which this constant of proportionality is called the *modulus of elasticity* and is denoted by E.

$$E = \frac{\Delta\sigma}{\Delta\varepsilon} \quad \text{for} \quad \sigma \leq \sigma_{PL} \quad \text{and} \quad \varepsilon \leq \varepsilon_{PL} \tag{3-1}$$

The modulus of elasticity can, and should, be thought of as the slope of the initial linear portion of the uniaxial stress-strain curve. It is a constant, which is unique for a particular material, and is a measure of the *stiffness* of the material. Since $\Delta\varepsilon$ is usually many orders of magnitude smaller than $\Delta\sigma$, the modulus E is usually a relatively large number; for example, it is 30,000,000 psi (207×10^9 Pa) for steel and 10,000,000 psi (70×10^9 Pa) for aluminum (see Table A-1 in the Appendix). For most metallic engineering materials the initial portion of the stress-strain curve for uniaxial compressive stress has the same slope as that for tensile stress. Hence, the tensile modulus of elasticity usually has the same value as the compressive modulus although this need not be true for all materials.

You must realize that the results obtained from any one uniaxial stress-strain test would represent only the response of that one particular sample of one particular material loaded in one particular direction. Thus, by themselves, the data for *one* test are of little use. However, if the material is homogeneous, then one test of one sample should be indicative of all such similar material loaded in the manner of the sample. Also, if the response of the material is independent of the orientation of the load axis of the sample, we say the material is *isotropic*. On the other hand, if the response is orientation *dependent*, the material is *anisotropic*. Most engineering materials are homogeneous and isotropic to a considerable degree. In this text we shall assume the materials under consideration to be homogeneous and isotropic unless specifically stated otherwise.

Before getting to a more detailed analysis of uniaxial stress-strain curves we might ask if there are shear stress-shear strain curves. Such a curve can be obtained from a plot of torque applied to a circular cylindrical shaft versus the resulting angle of twist. The conversion of the torque-twist curve to a shear stress-shear strain curve is more involved than is the conversion in the previous axial case. This conversion can be found in various references.*

As in the case of uniaxial stress, the shear stress generally is proportional to the shear strain for the first portion of the curve. Thus,

$$\Delta\tau \propto \Delta\gamma \qquad \Delta\tau = (\Delta\gamma) \times \text{constant}$$

---

* A. Nadai, *Theory of Flow and Fracture of Solids*, Chap. 21 (New York: McGraw-Hill, 1950).

This constant is called the *modulus of elasticity in shear* and is denoted by G. It is also commonly called the *modulus of rigidity* or simply the *shear modulus.* For a shear stress not greater than the proportional limit in shear,

$$G = \frac{\Delta\tau}{\Delta\gamma} \qquad \tau \le \tau_{PL} \qquad \gamma \le \gamma_{PL} \qquad (3\text{-}2)$$

The stiffness of a material, that is, its ability to resist deformation, is judged by the size of its elastic modulus, and the elastic strength of the material is sometimes judged by the magnitude of the stress at the proportional limit. These are determined from the straight-line portion of the stress-strain curve. However, materials have many other properties, many of which may also be determined from such a curve and which prove to be of great practical use. These will be discussed in later sections.

Caution: Our entire discussion of modulus of elasticity is confined to the initial linear portion of the stress-strain curve, that is, the part below the proportional limit. The words *elasticity* and *elastic* are often used ambiguously in engineering. A more detailed discussion of these words is presented in a later section.

3-3 / Poisson's ratio

You have no doubt noticed that when you stretch a rubber band, the cross section gets smaller; and if you squeeze a solid rubber ball, the diameter in a transverse plane (one at right angles to the axial squeezing forces) gets larger. Engineering materials also behave in this manner, but to a lesser degree and the changes in dimensions are not easily discernible to the naked eye. It can be shown experimentally that in a homogeneous and isotropic material subjected to a *uniaxial* state of stress, the resulting transverse strain is directly proportional to the axial strain in the direction of the load for stresses below the proportional limit. Thus,

$$\varepsilon_{\text{transverse}} \propto \varepsilon_{\text{axial}} \quad \text{or} \quad \varepsilon_{\text{transverse}} = \varepsilon_{\text{axial}} \times \text{constant}$$

The ratio of the magnitude of the transverse *strain* to the magnitude of the axial *strain* is called Poisson's ratio.* It will here be denoted by μ. Therefore,

$$\mu = \frac{|\varepsilon_{\text{transverse}}|}{|\varepsilon_{\text{axial}}|} \qquad (3\text{-}3)$$

Poisson concluded that the theoretical value for the ratio μ was $\frac{1}{4}$ for any isotropic body. For most metallic engineering materials, the actual

---

* The French mathematician and scientist S. D. Poisson (1781–1840) first defined this ratio as an elastic constant.

value is in the vicinity of $\frac{1}{3}$ for strains less than the proportional limit (see Table A-1 in the Appendix). Poisson's ratio can be measured experimentally rather easily by the simultaneous use of two strain gages on a test specimen subjected to uniaxial tensile or compressive stress; one gage is parallel to the load, and the other is perpendicular to it.

This discussion of Poisson's ratio is confined to strains in the elastic range below the proportional limit. For strains beyond the proportional limit, the ratio increases and in some cases approaches the limiting value of $\frac{1}{2}$.*

PROBLEMS

3-1 Following are data from the first part of a tensile test of a 0.505-in.-diam. specimen of ZK60A magnesium alloy extrusion. A 2-in. gage length extensometer was used. Plot the stress-strain curve. Determine the elastic modulus and the proportional-limit stress.

Load, lb	Extensometer reading, in.	Load, lb	Extensometer reading, in.
0	0	7000	0.0120
700	0.0010	7500	0.0130
1300	0.0020	7700	0.0140
1960	0.0030	8070	0.0150
2560	0.0040	8260	0.0160
3300	0.0050	8370	0.0170
3780	0.0060	8400	0.0190
4440	0.0070	8420	0.0210
5000	0.0080	8440	0.0230
5540	0.0090	8490	0.0250
6080	0.0100	8500	0.0270
6560	0.0110	Data stopped—No rupture	

3-2 Following are data taken from a compressive test of a 25-mm-diam. specimen of nuclear grade graphite. The strain readings are an average of two electrical resistance strain gages that were on opposite sides. The gage axes were parallel to the load. Note that these gages read directly in microns per meter. Plot the stress-strain curve. Do you notice anything unique about the elastic modulus and the proportional-limit stress for this material?

Load, N	Strain, μm/m	Load, N	Strain, μm/m
0	0	6000	3,100
1000	350	7000	4,300
2000	750	8000	6,100
3000	1,160	9000	8,600
4000	1,700	9700	11,000 rupture
5000	2,400		

3-3 Following are results from tensile test data of 2024-T4 aluminum alloy (extrusion) at room temperature. Plot the stress-strain curve. Use a straightedge for the linear portion and an irregular curve for the nonlinear portion. Take great care in the transition near the proportional-limit stress.

Stress, psi	Strain, in./in.	Stress, psi	Strain, in./in.
0	0	30,000	0.00288
2,000	0.00020	32,000	0.00307
4,000	0.00035	34,000	0.00326
6,000	0.00055	36,000	0.00345
8,000	0.00074	38,000	0.00364
10,000	0.00092	40,000	0.00382
12,000	0.00114	42,000	0.00401
14,000	0.00135	44,000	0.00420
16,000	0.00152	46,000	0.00442
18,000	0.00175	48,000	0.00460
20,000	0.00193	50,000	0.00486
22,000	0.00211	52,000	0.00548
24,000	0.00230	53,000	0.00740
26,000	0.00249	54,000	0.00940
28,000	0.00268		

* For a discussion of the Poisson effect beyond the proportional limit, see F. R. Shanley, *Strength of Materials*, pp. 171–173 (New York: McGraw-Hill, 1957).

Problem 3-8

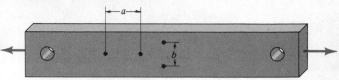

3-4 Find the proportional-limit stress and the elastic modulus for the material in Prob. 3-3.

3-5 Following are data for a compression test of a 1.75 × 1.75-in. clear birch specimen 8 in. long. The gage length of the compressometer was 6.00 in. Plot the stress-strain curve.

Stress, psi	Strain, in./in.	Stress, psi	Strain, in./in.
0	0	7,840	0.00606
654	0.00154	8,500	0.00654
1,308	0.00206	9,150	0.00717
1,961	0.00249	9,810	0.00801
2,615	0.00288	10,110	0.00873
3,270	0.00332	10,460	0.00948
3,925	0.00373	10,790	0.01065
4,580	0.00409	10,950	0.01149
5,230	0.00456	11,110	0.01256
5,890	0.00486	11,280	0.01385
6,540	0.00518	11,430	0.01591 (ultimate)
7,190	0.00563		

3-6 Find the proportional-limit stress and the elastic modulus for the material in Prob. 3-5.

3-7 The following are data from a quasi-static tensile test of compact bone from the human femur. The specimen had a 5 × 15 mm cross section and a 20 mm gage length. Plot the stress-strain curve and find the proportional limit.

Load, N	Elongation, mm	Load, N	Elongation, mm
0	0	9,500	0.1600
1,200	0.0200	10,200	0.1800
2,500	0.0400	10,650	0.2000
3,600	0.0600	11,100	0.2400
5,000	0.0800	10,900	0.2800
6,200	0.1000	10,350	0.3200
7,100	0.1200	10,275	0.3250
8,350	0.1400	(rupture)	

3-8 At some point in the elastic range of the specimen shown in the illustration, the elongation of a is 2500 × 10^{-6} in. and the contraction of b is 375 × 10^{-6} in. The original lengths were a = 2.50 in. and b = 1.25 in. What is Poisson's ratio for the material of the specimen?

3-9 The original 2-in. longitudinal gage length of a cylindrical compression specimen decreases 0.0024 in. During the same time the circumference, which was originally 7.16 in., increases 0.0021 in. Assuming that the action was elastic, find Poisson's ratio for the material.

3-10 A steel bar is 12 × 1 × $\frac{1}{4}$ in. before it is loaded. A tensile load is applied parallel to the 12-in. dimension and an increase in that dimension of 0.006 in. is noted. Assuming Poisson's ratio to be $\frac{1}{4}$, determine the change in the other two dimensions.

3-11 A cube initially 100 mm on each side is loaded in uniaxial compression. The loaded dimensions are 99.940 × 100.024 × 100.024 mm. Find Poisson's ratio for the material from which the cube was made. Assume elastic action.

3-12 A tensile specimen has an initial cross section 40 × 10 mm and a gage length of 100 mm. When loaded the gage length is found to be 100.12 mm long. Assume elastic action and determine the loaded dimensions of the cross section if $\mu = \frac{1}{3}$.

3-4 / Elastic and plastic deformations

Materials are classified or grouped in an almost limitless variety of ways. First, they are classified as solids, liquids, fluids, gases, vapors, and so on. We shall concern ourselves only with solids, although many of our definitions will apply to materials of the other types. Solid materials are further classified according to their molecular or crystalline structure, physical and chemical properties, thermal properties, electrical properties, mechanical properties, and so on. In this chapter, we shall discuss some of the mechanical and thermal properties that are of particular importance to the engineer and designer. For the most part, the mechanical properties will be those obtained experimentally from a *uniaxial* tension or compression test as described earlier in this chapter.

Before we discuss in detail specific experimental and design properties, it will be helpful to define some common general terms relating to material behavior. Several of these terms, such as elastic, plastic, stiffness, and brittle, are often used too loosely, incorrectly, and ambiguously.

Elastic action   *Elastic* is an adjective meaning "capable of recovering size and shape after deformation." If a material is subjected to load, deformation will result. If, upon release of the load, the material returns to its original size and shape, it has undergone *elastic* action or *elastic* deformation. The stress was an *elastic* stress within the *elastic* range. *Elastic limit* is the maximum *uniaxial* stress that can be applied to a material without causing any permanent deformation. *Elastic range* is the range of stress below the elastic limit.

Nowhere in this discussion of the word elastic have we said or implied that the load and the deformation must be proportional or linearly related in the elastic range. They usually are, as we saw from the linear first portions of the stress-strain curves discussed earlier in this chapter, but linearity is not a necessary condition for a material to be elastic. Many engineering materials behave as indicated in Fig. 3-5(a); however, some behave as in (b) or (c) while in the elastic range. When a material behaves as in (c), the stress-strain relationship is *not* single valued since the strain corresponding to any particular stress will depend on the loading history.

In engineering literature the terms elastic action or elastic range are used quite commonly to mean a linear stress-strain behavior. It is common for an engineer to say elastic limit when he means proportional limit. This practice has come about naturally, because for most engineering materials the proportional-limit stress and the elastic-limit stress are approximately the same. It should be noted that, by definition, the proportional limit must be obtained from a load-deformation or stress-strain curve, but the elastic limit *cannot* be obtained from such a single curve. If desired, the elastic limit may be obtained from a laboratory test in which increasing loads are applied and released, and then a check for

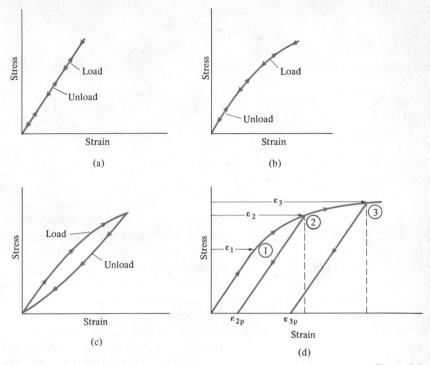

Figure 3-5

residual deformation is made. The value of the elastic limit determined experimentally in this way is dependent on the sensitivity of the measuring instrument used and on other variables such as time effect. Consequently, the elastic limit is seldom used and the easily determined and approximately equal proportional limit is substituted for it. You should recognize the difference between the terms and exercise reasonable caution in their use.

Plastic action   *Plastic* deformation, or *permanent set*, is any deformation that remains in the material *after* the load has been removed. All deformation is composed of plastic and elastic deformation. However, when the plastic deformation is negligible compared with the elastic deformation, the material is said to be elastic, and vice versa.

Figure 3-5(d) illustrates the idea of elastic (recoverable) and plastic (permanent) strain. If the material is stressed to level ① and then released, for all practical purposes the strain will return to zero. If the material is stressed to level ② and then released, the material will recover the amount $\varepsilon_2 - \varepsilon_{2_p}$, where ε_{2_p} will be the plastic strain remaining after the load is released. Similarly, for level ③ the permanent strain will be ε_{3_p}. Thus, for stress levels above the elastic limit, the total strain is composed of an elastic portion and a plastic portion.

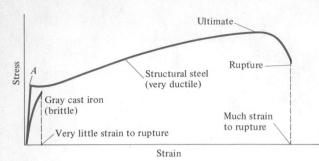

Figure 3-6

A material that undergoes very little plastic deformation before rupture is said to be *brittle*. A material that undergoes a great amount of plastic deformation before rupture is said to be *ductile*. Since elastic deformations are generally small, the usual measure of ductility (or brittleness) is the total percent elongation up to rupture of a 2-in. gage length tensile specimen. Sometimes the percent reduction of cross-sectional area of a tensile specimen after rupture is also used as a measure of ductility. A very ductile material, such as structural steel, may have an elongation of 30 percent at rupture, whereas a brittle material, such as gray cast iron or glass, will have relatively little elongation at rupture. Figure 3-6 illustrates the difference in stress-strain curves for a ductile material and a brittle material. Compare the ductilities of several engineering materials in Table A-1 of the Appendix.

You should not confuse the term stiffness with the term brittleness. Stiffness is the ability of a material to resist deformation, and the modulus of elasticity and the shear modulus are measures of stiffness. Thus, while all steels have approximately the same elastic stiffness (30×10^6 psi), there are brittle steels (hardened tool steels) and there are ductile steels (mild structural steel). Similarly, a material with a low stiffness might be comparatively brittle (glass).

3-5 / Ultimate strength and rupture stress

The ultimate tensile strength of a material is the *maximum* tensile stress the material can withstand before rupture in a tensile test in which the load is applied *slowly*. It is obtained by dividing the ultimate load by the original cross-sectional area of the test specimen (measured before loading). The ultimate tensile strength is the stress corresponding to the uppermost point on the stress-strain curve. It is an important design property, particularly for a brittle material.

The ultimate compressive strength of a material is obtained in a similar manner from a compressive test. Many ductile materials do not exhibit a clearly defined rupture in a compression test. For such materials, the load becomes indefinitely large as the compression specimen "flattens out," and the term ultimate compressive strength is therefore meaningless. The usefulness of a material of this kind in compression is measured by some other property, such as the compressive yield strength, or by its resistance to buckling.

The rupture stress is, as the name implies, the stress *at the time of rupture*. It is obtained by dividing the load at rupture by the original cross-sectional area of the test specimen. For a brittle material, the ultimate stress and the rupture stress are likely to have the same value. If you look at several of the stress-strain diagrams in this chapter for ductile materials, however, you will see that the rupture stress is not necessarily the same as the ultimate stress.

In a tensile test of most ductile engineering materials there is a very pronounced reduction in cross-sectional area of the specimen in the vicinity where rupture occurs. This phenomenon, which occurs shortly before rupture, is called "necking." In the stress-strain curve for the ductile material in Fig. 3-6, necking occurs after the load reaches the ultimate strength. During this necking process, the test specimen becomes considerably weakened. Hence, the applied load drops rapidly as rupture is approached. Rupture stress is not usually an important quantity from a design standpoint, and it is seldom given in tables of mechanical properties for materials. However, for a ductile material the difference between the ultimate stress and the rupture stress is indicative of the degree of deformation that occurs during the necking process.

3-6 / True stress and true strain

During a uniaxial tensile or compressive test, the cross section of the test member changes because of the Poisson effect. This lateral change is greater in the plastic range since Poisson's ratio increases in that range. For a ductile material in tension, the change in cross section is quite drastic just before rupture, because of the necking effect. For the ordinary engineering stress-strain curve discussed so far in this text, the stress at any point is based on the *original* cross-sectional area.

The *true stress* in a tension or compression test is found by dividing the load by the *actual* cross-sectional area. To obtain data for plotting true stress versus strain, the actual cross section would have to be measured for each plotted point. The additional precise measurements required make it more difficult to perform this type of test. As seen from Fig. 3-7, there is no appreciable difference between true stress and ordinary

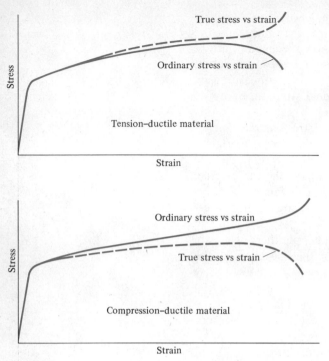

Figure 3-7

stress in the range of stress usually encountered in engineering members. Ordinary stress is much more commonly used by the engineer since in the design of a member the engineer is concerned with finding the *original* cross-sectional area required to support specific forces.

Similarly, one can define other measures of strain rather than the engineering strain based on *original* gage length. For example, the *true* or *logarithmic* strain ε_t is defined by

$$d\varepsilon_t = \frac{dL}{L} \tag{3-4}$$

where L is the instantaneous length rather than the original fixed gage length. Integration of Eq. 3-4 leads to

$$\varepsilon_t = \ln L_f - \ln L_o = \ln \frac{L_f}{L_o} \tag{3-5}$$

where the subscripts indicate final and original lengths.

The values of the true strain will differ very little from those of engineering strain for the magnitudes usually encountered in structural materials. True strain, however, can be a useful measure in dealing with highly deformable materials.

3-7 / Yield point and yield strength

Mild steel (for example, SAE 1020 or ASTM A-36), which is one of the most widely used engineering materials, has an unusual feature in its static stress-strain relationship. At the end of its linear elastic range there is a sudden and complete loss of stiffness, which is characterized by a sharp "knee" in the curve, as at point *A* in Fig. 3-6. At this point the material undergoes a rather rapid and extensive *plastic* deformation with *no* accompanying *increase* in load. Such a deformation process is called *yielding.*

The yield-point stress is specifically defined as the stress corresponding to the first point on the stress-strain curve at which there is a large increase in strain for no increase in stress. Although the yield point is easily obtained from the curve (the first place where the slope is zero), it is also commonly obtained simply by watching the load-indicator dial on the testing machine during a uniaxial tensile or compressive test. If the material has a yield point, and if the specimen is being loaded at a uniformly slow rate, the dial will halt temporarily at the yield point, indicating that the load remains constant even though the specimen is still being stretched (strained). This point is referred to as the yield point by "halt of dial" (or "drop of beam" for an old lever-type testing machine). The flaking off of the brittle mill scale on the surface of a specimen of ductile *hot rolled* mild steel is also a simple, but effective, indication that the yield point has been reached.

Although few materials exhibit a yield point, it is of great interest because it occurs in mild steel, which is such an important engineering material. For most purposes, mild steel is used only in the elastic range of stress. Since the yield point marks the end of the elastic range, the yield-point stress is commonly used as a measure of the "usable strength" of the material.

For a material with no yield point, there is no simple measure of usable strength. Hence, a somewhat arbitrary measure has been developed. From experience, it was decided that a measure of the usable strength would be the stress causing a certain permissible permanent strain (set) after release of the load in a uniaxial test. For most metallic materials, a strain of 0.2 percent (or 0.002 in./in.) is commonly used for the permissible set, although 0.1 percent is sometimes used for ferrous metals.

As we now know, the stiffness of a material decreases when it is loaded above the proportional limit into the inelastic range. If a material with no yield point is loaded into the inelastic range and is then unloaded, the unloading curve will be essentially parallel to the initial linear portion of the loading curve.* See Fig. 3-8(a). The usual method of determining

* There will be a slight curving off toward the origin at the bottom of the unloading curve. This residual action is known as the Bauschinger effect. See *Strength of Materials*, by F. R. Shanley, pp. 118, 157 (New York: McGraw-Hill, 1957).

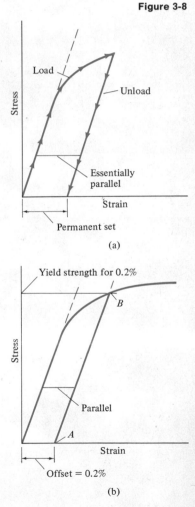

Figure 3-8

(a)

(b)

the yield strength corresponding to a specific permanent set is as follows: First, offset the stress-strain curve by the desired amount, such as 0.2 percent in Fig. 3-8(b), to locate point A. Then through point A draw a straight line AB parallel to the original straight portion of the stress-strain curve. The stress level at B, at which this construction line crosses the stress-strain curve, is the *yield strength.*

The yield strength determined in the manner just outlined is approximately the stress that will cause the specified permissible set. This stress is commonly used as a measure, or at least as an index, of the usable strength of a metallic engineering material that does not exhibit a yield point. The yield strength should always be associated with a specific offset or permanent set. Engineers and technical writers are often careless about this requirement. If no offset is given, 0.2 percent is usually implied, since this is by far the most common offset.

Yield point and yield strength are very important design parameters for the classification of engineering materials because they are easily reproducible. Properties such as proportional limit and elastic limit cannot be reproduced so easily. For example, suppose that a group of 20 students (on second thought, let us say 20 experienced engineers) was given a set of numerical values for plotting a stress-strain curve, and each one was asked to determine the proportional limit and the yield strength for a 0.2-percent offset. It would not be surprising to find that their values of the proportional limit would vary by 15 to 25 percent, while the variation in their values of the yield strength would probably be less than 5 percent. This illustration is indicative of the difference in reproducibility.

3-8   Resistance to energy loads

Work is produced by a force actihg through a distance. The loads acting on an engineering material move through certain distances as the material undergoes deformations. Therefore, when loads are applied to a body, work is being done on the body, and energy is absorbed by the body. Energy is the capacity or ability to do work. The energy that a body absorbs as the result cf its deformation under load is called *strain energy.* Strain energy that can be recovered as the loading is released is called *elastic strain energy.* In fact, a more sophisticated definition of an elastic body is as one for which the entire strain energy is recoverable. In an elastic body, no energy is dissipated during the deformation process. No perfectly elastic body exists, but practically speaking, an elastic body is one for which the dissipated energy is negligible.

Consider a load-deformation curve for a uniaxial tensile test, such as that shown in Fig. 3-9(a). By definition, the total work done on the body by loading up to some value of deformation e is

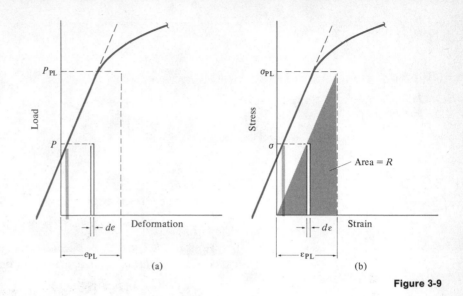

Figure 3-9

$$W = \int_0^e P\,de$$

which in turn equals the area under the P versus e curve up to the point corresponding to the particular deformation e. The percentage of the strain energy that is recoverable elastic energy will depend on the shape and path of the unloading curve (see Fig. 3-5). The amount of energy dissipated will be the difference between the total energy (the area under the loading curve) and the recovered energy (the area under the unloading curve). For most common engineering materials, the loading and un-loading curves will be very nearly the same as long as the load-deformation relationship is linear.

The total strain energy determined from a load-deformation curve is not really indicative of the material since the results will depend on the size of the test specimen. In order to eliminate the specimen size as a factor, we consider the strain energy per unit volume (sometimes called the strain energy density). This quantity can be obtained from a uniaxial test simply by dividing the total strain energy by the volume of the specimen; however, since both the stress and the strain are uniformly distributed in the uniaxial test, the strain energy per unit volume becomes

$$U = \frac{W}{V} = \int_0^e \frac{P\,de}{AL} = \int_0^\varepsilon \sigma\,d\varepsilon$$

which in turn equals the area under the stress-strain curve; see Fig. 3-9(b).

The units of strain energy per unit volume usually are inch-pounds (Newton-meters) of energy per cubic inch (cubic meters) of volume:

$$U = \frac{\text{in.-lb}}{\text{in.}^3} \quad \text{or} \quad \frac{\text{N·m}}{\text{m}^3}$$

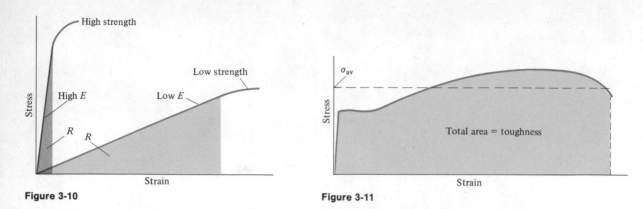

Figure 3-10

Figure 3-11

The area under the linear portion of the uniaxial stress-strain curve is a measure of a material's ability to store elastic energy. This measure is called the *modulus of resilience* and will be denoted by R. Thus,

$$R = \int_0^{\varepsilon_{PL}} \sigma \, d\varepsilon = \frac{1}{2} \sigma_{PL} \varepsilon_{PL} = \frac{\sigma_{PL}^2}{2E}$$

As illustrated in Fig. 3-10, it is possible for a material with low strength and a low elastic modulus to have a larger modulus of resilience than a material with high strength and a high elastic modulus.

The area under the entire stress-strain curve is a measure of the strain energy per unit volume required to rupture a material. This measure is called the *modulus of toughness* or just *toughness* of the material. Toughness is an important engineering property, since it indicates a material's resistance to energy loads before rupture. For example, a high degree of toughness would be important for the material in an automobile bumper, a highway guard rail, or an airplane landing gear.

The general expression for toughness is

$$T = \int_0^{\varepsilon_{rup}} \sigma \, d\varepsilon$$

However, there is seldom any simple analytical way of expressing σ as a function of ε. This integral should be thought of as the area under the entire stress-strain curve (see Fig. 3-11). The area can usually be obtained with sufficient accuracy by visually approximating an overall average stress from the diagram and multiplying this average stress by the strain at rupture from the diagram.

3-9 / Effects of other variables on mechanical properties

In this elementary text we are primarily concerned with deformable bodies subjected to static or slowly applied (quasi-static) loads at approximately room or ambient temperatures. Disregarding time, temperature, and type of load is in many cases a great simplification because these items can be major factors influencing the stress-strain relationship.

In our usual room-temperature tests of materials, we apply the load slowly and assume that the time rates of change of stress and strain are very small, a slow strain rate being of the order of 0.0001 in./in./sec. For dead loading (constant load), it is usually assumed that the strain rate is zero (unless yielding is occurring). That is

$$\frac{d\sigma}{dt} = 0 \qquad \frac{d\varepsilon}{dt} = 0$$

This assumption is also made for most engineering materials in use at moderate temperatures. A high rate of loading during a laboratory test usually results in an apparent increase in the tensile proportional limit and ultimate strength of the material.

There are, of course, materials that do exhibit significant viscoelastic (time-dependent) characteristics so that even under constant load (constant stress) the strain may change with time. This behavior is called *creep*. That is,

$$\frac{d\varepsilon}{dt} \neq 0 \qquad (\sigma = \text{const})$$

Creep is usually more pronounced, and often becomes a critical problem, at elevated temperatures. Some materials, such as metals with low melting points (for example, lead) and many plastics, exhibit appreciable creep at room temperatures. Engineering members designed and stressed "elastically" on the basis of an ordinary static stress-strain relationship may rupture by creep after a long period, perhaps years, at a constant stress level.

Sometimes the strain is kept constant, but the stress changes (usually decreases) with time. This behavior is called *relaxation*. That is,

$$\frac{d\sigma}{dt} \neq 0 \qquad (\varepsilon = \text{const})$$

Relaxation is also more prevalent at higher temperatures. This phenomenon often occurs in a bolted connection where the stress relaxes with time while the strain remains relatively constant. Creep and relaxation are discussed further in Chapters 4 and 14.

The stress in an engineering member often changes continuously with time because the loading is applied and removed continually. This action is referred to as repeated loading, or fatigue loading, and is discussed in more detail in Chapter 14. Suffice it to say here that repeated loading of a ductile material can cause the material to lose its ductility and result in a brittlelike fracture of the material. (Try breaking a paper clip by repeatedly bending it back and forth.) This loss of ductility, called *cold working*, can be a serious design problem in situations involving vibrating machines or structures.

Temperature has a considerable influence on the stress-strain relationship and properties of practically every engineering material. In the case of an engineering metal the ultimate strength, yield strength, and stiffness decrease appreciably with increasing temperatures, while the percent elongation (ductility) increases as the temperature rises. The reverse is generally true as the temperature is lowered. The loss of ductility in steels at very low temperatures became a major engineering problem in World War II, and as a result a great research effort was undertaken in this area. Development of materials to withstand the cryogenic temperatures of liquid oxygen and hydrogen was necessary for the space program.

In recent years the behavior of materials at extremely high temperatures has remained one of engineering's most vexing problems. In the past operating temperatures in gas turbines, rocket motors, hypersonic aircraft, space structures, and nuclear devices, for example, have been limited to lower, less thermally efficient values because of the unavailability of materials that will maintain their strength properties at extremely high temperatures. Today, sustained repeated loading at well above 2000°F is possible with some materials. Many new temperature-resistant materials such as graphite and tungsten composites are under development for use at higher temperatures.

PROBLEMS

3-13 Determine the yield strength for 0.2-percent offset and the moduli of resilience and toughness for the aluminum alloy of Figs. 3-3 and 3-4.

3-14 For the magnesium alloy of Prob. 3-1, evaluate
(a) the yield strength for 0.2-percent offset,
(b) the modulus of resilience, and
(c) the strain energy up to the point at which the data were discontinued.

3-15 For the graphite material of Prob. 3-2, evaluate
(a) the yield strength based on 0.2-percent offset,
(b) the modulus of toughness, and
(c) the ultimate strength.

Express your answer in both the International and American systems.

3-16 For the aluminum alloy of Prob. 3-3, evaluate
(a) the yield strength for a 0.2-percent offset,
(b) the modulus of resilience, and

(c) the strain energy up to the point where the data were discontinued.

3-17 For the birch timber of Prob. 3-5, evaluate
(a) the yield strength for a 0.5-percent offset,
(b) the modulus of resilience, and
(c) the modulus of toughness.

3-18 For the human bone of Prob. 3-7, evaluate
(a) the yield strength for a 0.2-percent offset,
(b) the modulus of resilience, and
(c) the modulus of toughness.

Express your answer in both the International and American systems.

3-19 to 3-22 For the stress-strain curve in the illustrations, determine (a) the elastic modulus, (b) the proportional-limit stress, (c) the yield strength for an offset of 0.2 percent, and (d) the modulus of resilience.

Problem 3-19

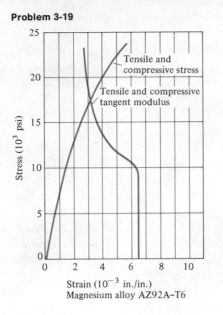

Stress (10^3 psi) vs Strain (10^{-3} in./in.)

Tensile and compressive stress

Tensile and compressive tangent modulus

Magnesium alloy AZ92A–T6

Problem 3-20

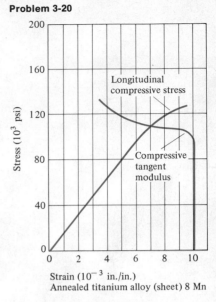

Stress (10^3 psi) vs Strain (10^{-3} in./in.)

Longitudinal compressive stress

Compressive tangent modulus

Annealed titanium alloy (sheet) 8 Mn

Problem 3-21

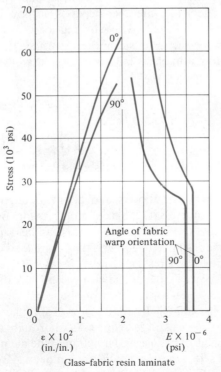

Stress (10^3 psi)

$0°$

$90°$

Angle of fabric warp orientation

$90°$ $0°$

$\varepsilon \times 10^2$ (in./in.)

$E \times 10^{-6}$ (psi)

Glass–fabric resin laminate

Problem 3-22

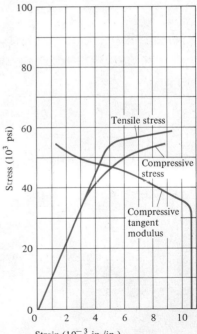

Stress (10^3 psi) vs Strain (10^{-3} in./in.)

Tensile stress

Compressive stress

Compressive tangent modulus

Compressive tangent modulus (10^6 psi)
Thickness, 0.250 in. to 1.499 in.
Aluminum alloy 2024–T4

3-10 / Failure and safety

Failure occurs when a member or structure ceases to perform the function for which it was designed. A member can cease to perform its function, or fail, for any one or more of a number of reasons. These *modes of failure* will be discussed in the next section. The word *failure* is often mistakenly used to mean fracture or break (separation). Fracture is a common and important type of failure, but every failure is not due to a fracture. Some failures can occur even before inelastic action or permanent deformation of the member. It is possible for a member or structure to cease to perform its function because of excessive *elastic* deformation. As an example, a machine lathe that is too flimsy (is not stiff enough) will not hold the close tolerances necessary for performing its function properly. This excessive elastic deformation may occur at a very low stress level. Thus you must remember that failure of a member is defined with reference to the function of the member, and not necessarily to its degree of destruction.

Since the primary function of an engineering member is to resist loads, we must relate failure to load; that is, failure will occur when the load reaches a value, called the failure load, at which the member ceases to perform its function. The failure load will be denoted by P_f. The margin of safety of a member depends on how close the working load P_w is to the failure load. The working load, or design load, is the load that the member actually supports in normal use. The *factor of safety* is defined as the ratio of the failure load to the working load. If N denotes the factor of safety

$$N = \frac{P_f}{P_w} \tag{3-6}$$

$$P_f = NP_w \tag{3-6a}$$

The factor of safety should be greater than unity. If it is unity, the working load equals the failure load, and there is no margin of safety.

Many handbooks and design books define the factor of safety as a ratio of stresses. This definition is valid only if the stresses can be directly related to the failure load; this can be done in many design problems. In some instances, however, the failure load cannot be directly related to any particular stress, and in such cases it is incorrect to apply the factor of safety to a stress. The best way to stay out of trouble in this respect is to base the factor of safety on the load.

The intelligent choice of a proper factor of safety is often difficult, to say the least. It requires a thorough and intimate knowledge of all aspects of the use and function of the member. Textbooks on design cover the selection of safety factors in detail. By measurements and theoretical calculations, an engineer can generally predict with good accuracy at what load a member will fail, but the decision as to how close to this failure load the member should operate must be based on many tangible, intangible, and interrelated factors. Often the factor of safety is increased because of uncertainties or ignorance. A few of the questions to be answered are:

1   What would an unpredictable failure cost in lives, in dollars, and in time?

2   How reliable are the material properties used in the calculations of the design?

3   How reliable is the material itself?

4   To what extent are the assumptions used in the design relationships valid?

5   Is the fabrication or manufacture exactly as specified in the design?

6   Can the extra weight of the structure resulting from a high factor of safety be tolerated?

7   Who will use the product or member—an experienced expert or an unskilled, careless person who is likely to misuse (overload) the member?

3-11 / Modes of failure and design criteria

A few of the more common modes of failure will now be discussed briefly.

Fracture by static load   When a fracture occurs in a *brittle* material, it is usually sudden and complete in nature and is likely to begin with a crack in an area of high stress concentration. Failure by fracture under static loads, however, is not so common in ductile materials as in brittle materials. In a ductile member, failure usually occurs as a result of excessive inelastic action which leads to very large overall deformations long before fracture.

Breaking a piece of chalk by bending or twisting it between the fingers is an example of fracture of a brittle material. A ductile steel leader (the wire between the hook and the line) in a deep-sea fishing rig will fail by fracture when subjected to the pull imposed by a large fish if it is designed only for smaller fishes. It does not cease to perform its function until fracture occurs. The same is true of a thin, ductile nonferrous blowout diaphragm in a safety valve.

Fracture by repeated load   Fracture caused by a repeated load is commonly referred to as a *fatigue* failure. Regardless of whether the material is brittle or ductile, no appreciable inelastic deformation is associated with this mode of failure. This mode is responsible for a large number of the failures in engineering. Such failures are often catastrophic in nature. The failure load for this mode cannot be predicted easily. The nature of the fatigue failures is discussed in more detail in Chapter 14. Let it suffice here to say that this type of fracture usually starts at a microscopic imperfection in a highly stressed area, and that the resulting crack progresses as the repetitions of loading are continued. Finally the crack grows to such an extent that one more application of the load results in a sudden and complete fracture of the member. Failure of an internal moving part of an engine or turbine would most likely be due to fatigue. Similarly, many aircraft structural failures can be traced to fatigue resulting from vibration.

General yielding   When a structure or member fails by general yielding, it loses its ability to support the load. General yielding can occur only in a member or structure of a ductile material. It is necessary to distinguish general yielding from localized yielding, which often occurs in a ductile member at a point of high stress concentration but is not widespread enough to significantly affect the overall structural integrity of the member. General yielding must be widespread enough throughout one or more complete sections of the structure to allow the cumulative deformations to render the structure unfit for performing its function. These deformations usually lead to total collapse. This mode of failure is not associated with fracture, although in some cases of total collapse, localized ruptures may occur as secondary effects.

If you try to hang an extra-heavy overcoat on a common wire coat hanger, it ceases to perform its function because of general yielding. A heavy snow often causes a thin and poorly supported metal roof to collapse as the result of general yielding.

Excessive elastic deformation   At what stage elastic deformation becomes excessive depends, of course, on the function of the member. If the deformation is great enough to make the member cease to perform its function, the member can be said to have failed. For example, if you walk on a long, thin plank to get across a creek, and it deflects down into the water and lets you get your feet wet, then it has failed. It does not have to "break" to fail. In fact, in this case the deflection is elastic and the plank itself would not in any way be damaged. The flimsy lathe mentioned in the first paragraph of section 3-10 is another example of failure due to excessive elastic deformation.

Excessive inelastic deformation   Failure due to excessive inelastic deformation is similar to the mode of failure just described except the deformations required to cause failure result in stresses beyond the elastic limit. This failure is distinguished from general yielding in that the *amount of deformation* is the quantity that prevents the member from functioning properly. On the other hand, in general yielding the member or structure loses its ability to support the load.

Buckling Failure by buckling occurs when a member or structure becomes unstable. Failures resulting from instability usually occur rapidly and without warning, and such failures are generally catastrophic in nature. Secondary effects may be general yielding and/or fracture. As an example of failure by buckling, roll a sheet of paper into a tube about an inch or two in diameter. Then put one flat end on the desk, and slowly apply a vertical load downward at the other flat end with the palm of your hand. The tube is surprisingly strong, but when it collapses, it does so with little or no warning. Can you think of any specific engineering member or structure that would probably buckle if its design load were drastically exceeded?

Many other types of failure may occur in various engineering structures, but these are usually peculiar to the specific structure. For example, a pressure vessel might spring a leak as a result of any one of many possible actions, and a leak would certainly constitute failure. In most structures, however, failure usually results from one or more of the six modes previously discussed.

PROBLEMS

3-23 What is the probable mode of structural failure of each of the following:
(a) toothpick
(b) watch balance spring
(c) garden hose
(d) sailboat mast

3-24 What is the probable mode of structural failure of each of the following:
(a) beer can
(b) guitar string
(c) umbrella frame
(d) paper clip

3-25 What is the probable mode of structural failure of each of the following:
(a) vaulting pole (fiberglass, metal)
(b) golf club shaft
(c) fishing rod
(d) baseball bat

3-26 What is the probable mode of structural failure of each of the following:

(a) outboard motor drive shaft
(b) astronaut's EVA tether
(c) boiler tube in steam plant
(d) wheel bearing

3-27 What is the probable mode of structural failure of each of the following:
(a) automobile fender
(b) door key
(c) lower-limb prosthesis (artificial leg)
(d) dentures (false teeth)

3-28 What is the probable mode of structural failure of each of the following:
(a) plate glass window
(b) electric motor-generator coupling shaft
(c) aerosol spray can
(d) high-pressure O_2 tank in spacecraft

3-29 What is the probable mode of structural failure of each of the following:
(a) automobile suspension torsion bar
(b) automobile tire

(c) automobile bumper

(d) connecting rod in auto engine

3-30 What is the probable mode of structural failure of each of the following:

(a) aircraft control cable

(b) windshield in commercial aircraft

(c) aircraft wing spar

(d) turbine blade in jet engine

3-31 to 3-38 In Probs. 3-23 to 3-30, (1) decide which of the following qualities is the most important for the object in question: strength, stiffness, resilience, toughness, ductility; (2) select the best material for the object from the types of materials whose stress-strain curves are shown in the illustration. Be prepared to defend your choice. All curves are plotted to the same scale.

Problems 3-31 to 3-38

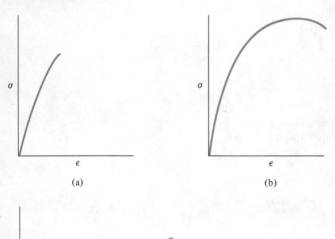

(a)

(b)

(c)

3-12 / The nature of solids

Engineering solids are made up of numerous particles of various sizes. The particles may be atoms, crystals, molecules, fibers, pieces of gravel, grains of sand, or even voids. Some of these particles are macroscopic (visible to the unaided eye); some are microscopic (for example, crystals); and some are submicroscopic (for example, atoms). The cohesive forces between the particles, and the energies associated with them, are the major factors influencing the mechanical properties of a solid that are of interest to the engineer.

Starting with the submicroscopic scale, we realize that all materials are composed of atoms. Bonds that are formed between atoms enable them to combine in very large groups to form solids. Adjacent atoms in a solid exert attractive and repulsive forces, which tend to maintain the atoms at an equilibrium distance from one another. The forces making up the bonds are balanced at this equilibrium spacing. Any tendency to change this spacing causes an unbalance that tends to pull (or push) the atoms back to the equilibrium spacing. The curves in Fig. 3-12 are plots of attractive force, repulsive force, and resultant force versus the distance between the atoms. When there is a tendency, due to some external cause (for example, tensile load), to increase the spacing of the atoms, a positive or pulling force is built up between the atoms, and this force tends to cause the atoms to return to the equilibrium spacing. If an external cause tends to decrease the spacing, there is set up between the atoms a negative or compressive force that also tends to cause a return to the equilibrium spacing. This "returning" tendency manifests itself as stored elastic energy, which was discussed in section 3-8.

The atoms are not completely stationary at the equilibrium spacing. There is continual movement in the form of vibration because of the heat present. The amplitude of these oscillations due to thermal agitation is of the order of one-tenth of the equilibrium spacing. The equilibrium spacing is of the order of 10^{-7} mm. We see then that the atoms are continually oscillating back and forth about the equilibrium spacing.

Notice in Fig. 3-12 that beyond the spacing at which the cohesive strength is developed, the resultant force decreases as the spacing increases. The strength of any engineering solid on the macroscopic level is only a fraction of the theoretical cohesive strength considered here on a submicroscopic level.

How are groups of atoms built up to form solids? Usually the atoms

Figure 3-12

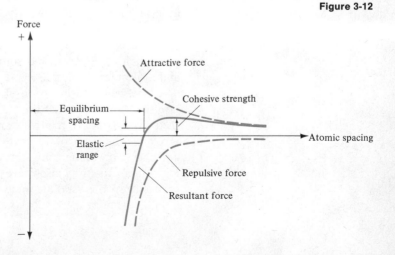

combine to form crystals or molecules. Solids are generally aggregates of these crystals or molecules. These structural "building blocks" are extremely important in determining the mechanical behavior of engineering solids.

Most metallic engineering solids are crystalline; that is, they are composed of many crystals. A crystal is made up of an *orderly* array of atoms. As a crystal is formed, the atoms arrange themselves in three-dimensional rows so that each atom is at the equilibrium distance from the adjacent atoms. This structural arrangement is called a space lattice. A small portion of a simple cubic lattice is represented in Fig. 3-13. Such a lattice can be thought of as a three-dimensional array of cubic blocks with an atom at each corner where four cubes come together.

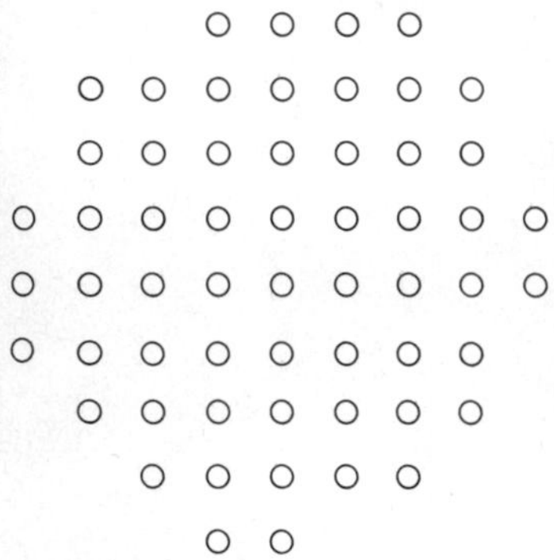

One crystal may consist of millions of atoms, the number depending on the crystal size. Crystal size (or grain size, as it is often referred to) can vary widely. Although crystals usually are of microscopic size, they can sometimes be quite large, and under special conditions they can be made to grow to several inches. Engineering tests on single crystals are possible in research laboratories.

In some materials atoms arrange themselves into chainlike or sheetlike structural units called molecules. A molecule is the smallest particle that retains the chemical properties of the original material. As molecules are united to form solids, molecular bonds are created between them. These bonds are not nearly so strong as are the atomic bonds discussed earlier. Chainlike molecules often intertwine in a solid. Such an arrangement has an appreciable effect on the mechanical behavior of the material.

Solids are usually classified as crystalline, amorphous, or a combination of these two. A crystalline material is usually made up of a number of crystals, in which case it may be called polycrystalline, whereas an amorphous material is usually made up of molecules. Crystalline materials have a much more orderly atomic arrangement, and as a result generally have greater density and strength than amorphous materials. Most engineering metals are crystalline. Some important amorphous engineering materials are wood, plastics, glass, and rubber. Other materials, such as concrete and some ceramics, are combinations of crystals and molecules.

A brief general discussion of this kind may be misleading since it may give the impression that the makeup of solids and the factors influencing their properties are relatively simple. Of course, such is not the case. It is outside the scope of this text to consider other details of atomic physics and chemistry influencing atomic bonds. It is important for the engineering student to study elsewhere such things as the major influence of imperfections on crystalline materials and the chemistry of polymerization in influencing properties of some amorphous materials.

Figure 3-13

3-13 / The nature of elastic and plastic action

Elastic action   Elastic action in a polycrystalline material, such as almost any engineering metal, consists primarily of distortions of the atomic space lattice of the crystals. In the case of tension, there are increases in the distances between the atoms in the direction of external loading. These increases develop attractive forces between the atoms to balance the external applied forces. For compression the distances are decreased, and repulsive forces are built up. Accumulations of these small increases (for tension) result in overall elastic elongations. Upon release of the external load, the atoms return to their equilibrium spacing, and essentially each atom maintains the same position relative to all other atoms. In Fig. 3-12 the part of the curve for resultant force versus distance just to either side of the equilibrium spacing can be considered approximately straight for the small changes in spacing considered here. Therefore, for most metals important in engineering, elastic action is essentially linear.

In an amorphous material or in a material that is a combination of amorphous and crystalline portions, elastic action is often approximately linear, but it could quite possibly be nonlinear. Although the same type of action as described in the preceding paragraph may occur with atoms of the molecules of an amorphous material, the configurations and arrangements of the molecules may contribute to a nonlinear elastic action. This nonlinear action is likely in such materials as rubber and thermoplastic polymers. "Combination" type materials that are important in engineering, such as concrete and plastic-fiberglass laminates, often exhibit this nonlinear characteristic.

Elastic action is essentially an atomic and/or molecular distortion caused by the external loads, wherein the atoms and molecules maintain similar relative positions while the loads are being applied or removed. When the load is entirely removed, the material on a macroscopic, microscopic, or submicroscopic scale essentially goes back to its original configuration.

Nature of yielding   When external loads reach a sufficient magnitude, the elastic range ends, and either one of two effects is possible. The material could rupture or it could yield. If a material were to rupture at the end of the elastic range, it would be said to be perfectly brittle. Although glass and a few other common engineering materials are very brittle, they yield slightly before rupture.

What is the nature of yielding and how is it different from elastic action? Yielding occurs in a crystalline solid material when the distances between atoms increase to the extent that atomic bonds are broken and place changes occur in the atomic lattice. The cumulative effect of such place changes is a greater overall elongation. When elongation with

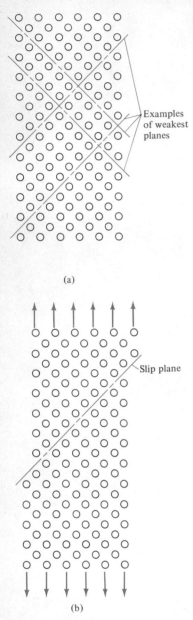

(a)

Examples of weakest planes

Slip plane

(b)

Figure 3-14

yielding is plotted against load, the curve becomes nonlinear. These place changes are changes of positions of atoms relative to each other. The behavior is therefore *not reversible* upon release of the load. Yielding or initial plastic action, as it is often called, is a *permanent* action.

The atomic place changes associated with yielding are transcrystalline; that is, they occur on planes through the crystals. Yielding is a *slipping* action, associated with a tendency to shear, and it occurs on planes of relative weakness through the crystal. The locations of these planes of weakness depend on the specific arrangement of atoms in the space lattice, and this arrangement varies for different metals. Generally the weakest planes are those in which the atoms are most closely spaced. An example of slip in a portion of a crystal subjected to uniaxial tension is illustrated in Fig. 3-14. Only a microscopic portion of a crystal is represented, and this figure is, of course, somewhat idealized. Elastic lattice distortion is not shown in the figure. Notice that place changes are shown in view (b), and that new bonds have been established between atoms along the slip plane and different neighboring atoms. If a steel engineering member were to yield when in uniaxial tension, the initial yield planes on a macroscopic level would show that the statistical accumulation of thousands of submicroscopic slips is a shear phenomenon and is associated with maximum shear stress because the overall yields would develop on planes making angles of approximately 45° with the external load.

Where and how does yielding start in a loaded body? From a macroscopic viewpoint, it is reasonable to assume that yielding will start in a region of high stress. Regardless of attempts to distribute stress uniformly in any real member of an engineering material, there will always be some regions of relatively higher stress concentration and therefore higher strain concentration. A concentration may be caused by the method of loading, the geometry of the member, or a material imperfection such as a flaw (due perhaps to inclusions of foreign matter or voids). Now consider the region of high stress submicroscopically. Although the arrangement in an atomic lattice of metallic crystals has a high degree of order, it is not absolutely perfect. There are different types of imperfections, called dislocations, which provide starting places for the slip planes.

Yielding of crystalline materials In a polycrystalline material, yielding or plastic strain is primarily a rearranging of atoms along certain planes of slip with little or no change in the final atomic distance. The new atomic bonds formed are just as strong as the old ones. Therefore, yielding in a metal is not generally associated with any weakening or loss of strength. For example, if a mild-steel bar is stressed in tension or compression somewhat beyond the yield point, is then unstressed, and is again stressed to the yield point, it will be found that the yield-point stress is just as high for the second time or possibly even higher.

A combination of elastic action and plastic action results in an overall deformation. Since elastic action involves changes of the atomic spacing, it is accompanied by a volume change, which is an increase for tension and a decrease for compression. Remember that Poisson's ratio for most materials is only about $\frac{1}{3}$ for the elastic range. Since plastic action involves primarily place changes, rather than atomic spacing changes, it is *not* accompanied by any significant volume change. Can you now give a valid physical argument to show that Poisson's ratio in the plastic range should be $\frac{1}{2}$ for metals?

To get a correct mental picture of the yielding process in an engineering metal, you must have an appreciation of the relative scales on which the slipping actions occur. The overall observable behavior is the cumulative statistical effect of a great many atomic, microscopic, and macroscopic actions. For example, in a single crystal many dislocations may cause the presence of many slip planes, and the length of each slip may be more than a thousand times the distance between atoms. Also, across a typical section of an engineering member there may be hundreds of crystals. Hence, the observable effect may be the result of a very great number of unobservable actions.

In a static tensile test of a flat specimen of mild steel, yielding will start across some section, perhaps at a grip. Once this region of the material yields initially, it is strengthened and resists further yielding; but other portions, usually adjacent, then begin to yield. This progression of yielding eventually covers the entire specimen between grips before the specimen as a whole will sustain additional load. As a result there is a rather flat region on the stress-strain curve for this material. In Fig. 3-15 the flat region is between points A and B.

Notice that the strain at B in Fig. 3-15 is approximately 20 times the limiting elastic strain at A. When point B on the curve has been reached, all the material in the specimen will have yielded initially. Further yielding or slipping in the crystals will require more stress, mainly because of the interference to further slip provided by the crystal boundaries and adjacent crystals. There are other rather complex factors that contribute to this strengthening, but a discussion of such factors is beyond the scope of this

Figure 3-15

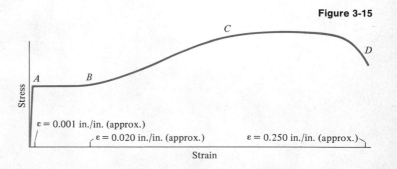

text. This strengthening phenomenon, which takes place approximately between *B* and *C* in Fig. 3-15, is called strain hardening or work hardening. Most engineering metals exhibit some degree of strain hardening.

Yielding of amorphous materials   In an amorphous material, or in a combination amorphous-crystalline material, yielding is also a manifestation of atomic and molecular bond failure. Usually, however, the number of new bonds formed is much lower than the number of bonds broken. As a result, there is a general weakening of the material, rather than a strengthening like that which may occur when a polycrystalline material yields. Other rather complex internal behaviors may take place in amorphous or "combination" type materials. Fragmentation of, and friction between, different materials may occur in a combination type of material such as concrete. In some polymer plastics and similar materials long-chain molecules often continue to slip past each other as molecular bonds are broken, even though new bonds are being formed continually. This action actually causes a kind of viscous flow.

The science of rheology is devoted to the study of materials that flow to some extent. In fact, one basic law of rheology says that all materials flow to some degree. However, the most interesting and difficult to describe are those materials that are solid but border on being fluidlike. These are referred to as viscoelastic materials. Many new load-carrying plastics are viscoelastic as are load-carrying biological materials such as bone, skin, tendon, and muscle tissue. The essential feature of a viscoelastic material is that the behavior under load is time dependent. Except for the elementary discussion of rheological models and viscoelastic response in the first part of Chapter 4, further discussion of this interesting behavior is left to more advanced courses.

Chapter 4
Mechanical response of materials

where η i
deformat

3 Altho
elastic s
combina
several t

(a) *Ma*
and
(b) *Voi*
line
(c) *Sta*
Fig

4-1 / Objectives

Upon completion of this chapter, you will be able to:

1 Derive response relationships for simple mechanical systems and idealized solids.
2 Sketch idealized curves from response relationships.
3 Find the tangent modulus and secant modulus associated with a given stress.
4 State the generalized Hooke's law and use it to determine triaxial stress and strain.
5 Relate the elastic modulus, the shear modulus, and Poisson's ratio by means of an equation.
6 Compute bulk modulus and cubical dilatation.

experimental plots usually do not pass through the origin of the load and deformation axes, since there is always a certain amount of play or looseness that must be eliminated in a real system before deformation begins, all analytical expressions discussed here are based on the assumption that the stress-strain curve passes through the origin of the co-ordinates. It is assumed that any necessary adjustment for a displaced origin is made when the load scale is converted to stress and the deformation scale to strain.

Before the mathematical models for each of the five types of curves are discussed, two other quantities sometimes used in the expressions will be defined. The first is the *tangent modulus*, denoted by E_t, which is the *slope* of the stress-strain curve at any point *beyond* the proportional limit:

$$E_t = \lim_{\Delta\varepsilon \to 0} \frac{\Delta\sigma}{\Delta\varepsilon} = \frac{d\sigma}{d\varepsilon} \quad \text{for} \quad \sigma > \sigma_{PL} \qquad \varepsilon > \varepsilon_{PL} \qquad (4\text{-}5)$$

Generally the curve beyond the proportional limit is nonlinear, and the tangent modulus is a variable as for curves of types IV and V. The tangent modulus may be a constant, as for a curve of type III; may be zero, as for a curve of type II; or may not exist, as for a curve of type I.

The second quantity is the *secant modulus*, denoted by E_s, which is the ratio of the stress to the strain at any point beyond the proportional limit:

$$E_s = \frac{\sigma}{\varepsilon} \quad \text{for} \quad \sigma > \sigma_{PL} \qquad \varepsilon > \varepsilon_{PL} \qquad (4\text{-}6)$$

Figure 4-8 shows the difference between E and E_t and E_s.

We shall now discuss the analytical expression or mathematical model and the method of obtaining it for each type of curve in Fig. 4-7. All curves are assumed to be adjusted so that they pass through the origin.

Type I: The analytical expression for the simple linear curve is $\sigma = E\varepsilon$, where the constant E is simply the measured slope $\Delta\sigma/\Delta\varepsilon$ of the experimental curve.

Type II: The expression for the initial linear portion is the same as

Figure 4-8

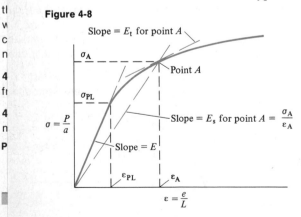

that for a curve of type I, or $\sigma = E\varepsilon$, where $\varepsilon \leq \varepsilon_{PL}$. For the "flat-topped" portion, where $\varepsilon > \varepsilon_{PL}$, the expression is $\sigma = $ constant. The constant stress is measured directly from the curve.

Type III: The expression for the initial linear portion is the same as that for a curve of type I or type II; thus, $\sigma = E\varepsilon$, where $\varepsilon \leq \varepsilon_{PL}$. For the second linear portion, where $\varepsilon > \varepsilon_{PL}$, the expression is $(\sigma - \sigma_{PL}) = E_t(\varepsilon - \varepsilon_{PL})$. The constants E_t, σ_{PL}, and ε_{PL} are measured directly from the curve, as indicated in the figure.

Types IV and V: An analytical expression for a part of a stress-strain curve that is nonlinear cannot be obtained so easily as one for a linear part. An expression for the linear part of a curve of type IV is like that for a curve of type I, II, or III. For a curve that is partially or wholly nonlinear, a commonly used approximation is the following expression given by Ramberg and Osgood.*

$$\varepsilon = \frac{\sigma}{E} + K\left(\frac{\sigma}{E}\right)^n \qquad (4\text{-}7)$$

The first term σ/E represents a straight line with slope E. For a curve of type V, which has no linear part, E is the slope of the tangent to the curve at the origin and is sometimes called the *initial tangent modulus*. The second term $K(\sigma/E)^n$ is the deviation of the stress-strain curve from the straight line with slope E. For a curve of type IV the constants K and n are such that this second term is negligible below the proportional limit. To obtain a Ramberg-Osgood expression from experimental data, it is necessary to obtain suitable values of K and n. Ramberg and Osgood show how to do this in their publication.

Shear stress-strain curves   An analytical expression for a stress-strain curve for shear can be obtained by adopting much the same procedure as that used for uniaxial stress, *provided of course that the stress-strain curve for shear can be obtained from some experimental data.* As was pointed out earlier, the procedure for obtaining a stress-strain curve for shear is more involved than that for uniaxial stress.

PROBLEMS

4-9   For the concrete shown in Fig. 5-32(a), determine the tangent modulus and the secant modulus for a stress of 3000 psi.

4-10   For the material shown in Fig. 3-4, determine the tangent modulus and the secant modulus for a stress of 70 ksi.

4-11   Following are data from the first part of a tensile test of a 0.505-in.-diam. specimen of ZK60A magnesium alloy extrusion. A 2-in. gage length extensometer was used. Plot the stress-strain curve. Determine the tangent modulus and the secant modulus for a stress of 40,000 psi.

* Walter Ramberg and William Osgood, *Description of Stress-Strain Curves by Three Parameters*, NASA-TN 902, July 1943.

Load, lb	Extensometer reading, in.	Load, lb	Extensometer reading, in.
0	0	7000	0.0120
700	0.0010	7500	0.0130
1300	0.0020	7700	0.0140
1960	0.0030	8070	0.0150
2560	0.0040	8260	0.0160
3300	0.0050	8370	0.0170
3780	0.0060	8400	0.0190
4440	0.0070	8420	0.0210
5000	0.0080	8440	0.0230
5540	0.0090	8490	0.0250
6080	0.0100	8500	0.0270
6560	0.0110	Data stopped—No rupture	

4-12 For the material of Prob. 4-11 determine the tangent modulus and the secant modulus for a stress of 42,000 psi.

4-13 For the material of Prob. 3-2, determine the E_t and E_s for a stress of 15×10^6 Pa. Also, convert your answers to psi.

4-14 For the material of Prob. 3-3, determine the E_t and E_s for a stress of 53,000 psi.

4-15 For the material of Prob. 3-5, determine the E_t and E_s for a stress of 10,000 psi.

4-16 For the material of Prob. 3-7, determine the secant modulus and tangent modulus for a stress of 14×10^7 Pa. Also, convert your answers to psi.

4-17 A material has a stress-strain relationship that can be approximated by the equation

$$\varepsilon = \frac{\sigma}{2 \times 10^7} + 200 \left(\frac{\sigma}{2 \times 10^6} \right)^3$$

Find the secant modulus and the tangent modulus for a stress level of 25,000 psi.

4-18 An aluminum alloy has a stress-strain relationship that can be approximated by the equation

$$\varepsilon = \frac{\sigma}{10^7} + \left(\frac{\sigma}{10^5} \right)^{10}$$

Find the secant modulus and the tangent modulus for a stress level of 50,000 psi.

4-19 For the stress-strain relationship of Prob. 4-17, what is the strain energy absorbed up to a stress level of 25,000 psi?

4-20 For the aluminum alloy whose stress-strain relationship is given in Prob. 4-18, find the strain energy absorbed up to a stress level of 60,000 psi.

4-5 / Generalized Hooke's law

As we have seen, for stresses not greater than the proportional limit,

$$\varepsilon = \frac{\sigma}{E} \tag{4-8}$$

and

$$\mu = \frac{|\varepsilon_{transverse}|}{|\varepsilon_{axial}|} \tag{3-3}$$

These equations express the relationship between stress and strain (Hooke's law) for a *uniaxial* state of stress only when the stress is not greater than the proportional limit. Let us now consider the general triaxial state of stress shown in Fig. 4-9(a).

In order to analyze the deformational effects produced by all the stresses, we shall consider the effects of one axial stress at a time. Since we presumably are dealing with strains of the order of one percent or less, these effects can be superposed arbitrarily. Figures 4-9(b), (c), and (d) show these effects separately. The undeformed element is represented by dashed lines, and the deformed element by solid lines. The deformations in these figures are greatly exaggerated. Figure 4-9(b) shows that the stress σ_x causes an increase in the x dimension and a decrease in the y and z dimensions because of the Poisson effect. Since this figure represents a uniaxial state of stress in the x direction, the strains in the x, y, and z directions, respectively, would be as follows:

$$\varepsilon_x = \frac{\sigma_x}{E} \qquad \varepsilon_y = -\mu\varepsilon_x \qquad \varepsilon_z = -\mu\varepsilon_x \qquad (4\text{-}9)$$

Then the resulting deformations would be

$$de_x = \varepsilon_x\,dx \qquad de_y = -\mu\varepsilon_x\,dy \qquad de_z = -\mu\varepsilon_x\,dz \qquad (4\text{-}9\text{a})$$

Here the minus signs indicate contractions.

In a similar manner, Figs. 4-9(c) and (d) illustrate the deformations produced by the stresses σ_y and σ_z, respectively. By applying the previous analysis to Fig. 4-9(c), we obtain

$$\varepsilon_y = \frac{\sigma_y}{E} \qquad \varepsilon_x = -\mu\varepsilon_y \qquad \varepsilon_z = -\mu\varepsilon_y \qquad (4\text{-}10)$$

$$de_y = \varepsilon_y\,dy \qquad de_x = -\mu\varepsilon_y\,dx \qquad de_z = -\mu\varepsilon_y\,dz \qquad (4\text{-}10\text{a})$$

For Fig. 4-9(d),

$$\varepsilon_z = \frac{\sigma_z}{E} \qquad \varepsilon_x = -\mu\varepsilon_z \qquad \varepsilon_y = -\mu\varepsilon_z \qquad (4\text{-}11)$$

$$de_z = \varepsilon_z\,dz \qquad de_x = -\mu\varepsilon_z\,dx \qquad de_y = -\mu\varepsilon_z\,dy \qquad (4\text{-}11\text{a})$$

The *total* deformation in any one direction would be the sum of the deformations in that direction resulting from all the individual stresses. Thus

$$de_x = \overbrace{\frac{\sigma_x}{E}dx}^{\text{from }\sigma_x} - \overbrace{\mu\frac{\sigma_y}{E}dx}^{\text{from }\sigma_y} - \overbrace{\mu\frac{\sigma_z}{E}dx}^{\text{from }\sigma_z} \qquad (4\text{-}12)$$

If we divide through by dx, we obtain

$$\frac{de_x}{dx} = \varepsilon_{x(\text{total})} = \frac{\sigma_x}{E} - \mu\frac{\sigma_y}{E} - \mu\frac{\sigma_z}{E} \qquad (4\text{-}12\text{a})$$

In a similar manner, the total strains in the y and z directions become

$$\varepsilon_{y(\text{total})} = \frac{\sigma_y}{E} - \mu\frac{\sigma_x}{E} - \mu\frac{\sigma_z}{E} \qquad (4\text{-}12\text{b})$$

$$\varepsilon_{z(\text{total})} = \frac{\sigma_z}{E} - \mu\frac{\sigma_x}{E} - \mu\frac{\sigma_y}{E} \qquad (4\text{-}12\text{c})$$

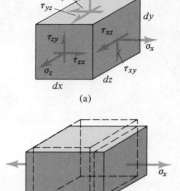

(a)

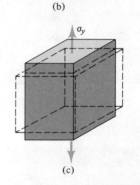

(b)

(c)

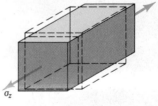

(d)

Figure 4-9

The shear stresses were not considered in the foregoing analysis. It can be shown that for *isotropic* materials a shear stress will produce only its corresponding shear strain and will not influence the axial strains. We can, however, write Hooke's law for the individual shear strains and stresses in the following manner:

$$\gamma_{xy} = \frac{\tau_{xy}}{G} \tag{4-13a}$$

$$\gamma_{xz} = \frac{\tau_{xz}}{G} \tag{4-13b}$$

$$\gamma_{yz} = \frac{\tau_{yz}}{G} \tag{4-13c}$$

Equations 4-12 and 4-13 are usually called the *generalized Hooke's law* and are the constitutive equations for *linear elastic isotropic* materials. When these equations are used as written, the strains can be completely determined from known values of the stresses. The constitutive equations can also be written explicitly in strains (see Prob. 4-24). You should remember that σ_x, σ_y, and σ_z are not necessarily principal stresses, but are merely the normal stresses in three mutually perpendicular directions.

We have remarked in Chapter 1 that the case of plane stress ($\sigma_z = \tau_{xz} = \tau_{yz} = 0$) is of particular importance to the engineer. For this special situation, Eqs. 4-12 and 4-13 reduce to

$$\varepsilon_x = \frac{\sigma_x}{E} - \mu \frac{\sigma_y}{E}$$

$$\varepsilon_y = \frac{\sigma_y}{E} - \mu \frac{\sigma_x}{E}$$

$$\varepsilon_z = -\mu \frac{\sigma_x}{E} - \mu \frac{\sigma_y}{E} \tag{4-14}$$

$$\gamma_{xy} = \frac{\tau_{xy}}{G}$$

along with their inverse relations

$$\sigma_x = \frac{E}{1 - \mu^2} (\varepsilon_x + \mu\varepsilon_y)$$

$$\sigma_y = \frac{E}{1 - \mu^2} (\varepsilon_y + \mu\varepsilon_x) \tag{4-15}$$

$$\tau_{xy} = G\gamma_{xy}$$

Hooke's law is probably the most well known and widely used constitutive equation for engineering materials. However, we do not wish to leave the impression that all engineering materials are linear elastic isotropic ones, particularly in this age when new materials are being developed every day. Indeed, many useful materials exhibit nonlinear response, are not elastic, and have a high degree of anisotropy. Unfortunately, mechanical models and constitutive equations for such materials are often extremely difficult to obtain and utilize. The modern engineer faces many interesting challenges in trying to make efficient and effective use of the structural capabilities of these new materials.

4-6 / Relationship between elastic constants E, G, and μ

The constitutive equations for a linear elastic isotropic material given by Eqs. 4-12 and 4-13 contain three parameters, namely, E, G, and μ. It can be shown by various analytical and physical arguments that such a material has at most two *independent* parameters and that any other parameters can be expressed in terms of two known independent ones. To demonstrate this we shall utilize the biaxial case of pure shear stress (which is not uncommon in engineering members). Figure 4-10(a) shows an element in a state of pure shear and labeled with the existing shear stress. Figure 4-10(b) is Mohr's stress circle for pure shear. This circle shows the following facts:

1 For pure shear the center of the circle must be at the origin.

2 The maximum principal stress, minimum principal stress, and maximum shear stress all have the same magnitude.

3 The algebraic sum of the normal stresses for any orientation of the element is always zero.

Figure 4-10(c) shows the element oriented with its faces parallel to the

Figure 4-10

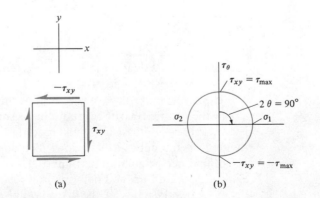

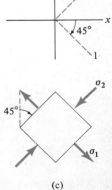

(a)   (b)   (c)

principal planes. So σ_1 and σ_2 are the maximum and minimum principal stresses, respectively. The generalized Hooke's law for this biaxial state of stress may be expressed by using Eqs. 4-12 and 4-13 and letting $\sigma_3 = 0$. The results are:

$$\varepsilon_1 = \frac{\sigma_1}{E} - \mu \frac{\sigma_2}{E} \tag{4-16a}$$

$$\varepsilon_2 = \frac{\sigma_2}{E} - \mu \frac{\sigma_1}{E} \tag{4-16b}$$

$$\gamma_{xy} = \frac{\tau_{xy}}{G} \tag{4-16c}$$

For pure shear, $\sigma_1 = +|\tau_{xy}|$ and $\sigma_2 = -|\tau_{xy}|$; so the principal strains ε_1 and ε_2 may be expressed as follows:

$$\varepsilon_1 = \frac{\tau_{xy}}{E} + \mu \frac{\tau_{xy}}{E} = \frac{\tau_{xy}}{E}(1 + \mu) \tag{4-17}$$

$$\varepsilon_2 = -\frac{\tau_{xy}}{E} - \mu \frac{\tau_{xy}}{E} = -\frac{\tau_{xy}}{E}(1 + \mu) \tag{4-17a}$$

These are principal strains, and they therefore represent the right-hand and left-hand extremities of Mohr's *strain* circle, which is shown in Fig. 4-11.

From Fig. 4-11 we see that the magnitude of the maximum shear strain is the diameter of the circle, or twice ε_1. So

$$\gamma_{\max} = \gamma_{xy} = 2\varepsilon_1 \tag{4-18}$$

Substituting values from Eqs. 4-16c and 4-17 into Eq. 4-18, we get

$$\frac{\tau_{xy}}{G} = \frac{2\tau_{xy}(1 + \mu)}{E}$$

$$\frac{1}{G} = \frac{2(1 + \mu)}{E} \tag{4-19}$$

$$G = \frac{E}{2(1 + \mu)}$$

Equation 4-19 relates the elastic constants E, G, and μ. This equation shows that these elastic constants are not independent of each other, and experimental tests have proved that this relationship is valid.

Another defined elastic constant that is sometimes useful to the engineer is the bulk modulus. The bulk modulus is the average normal stress divided by the cubical dilatation. The cubical dilatation is the change in volume per original unit volume. Hence if an element originally has sides dx, dy, and dz, and deforms into an element with sides $dx(1 + \varepsilon_x)$, $dy(1 + \varepsilon_y)$, and $dz(1 + \varepsilon_z)$, the dilatation becomes

Figure 4-11

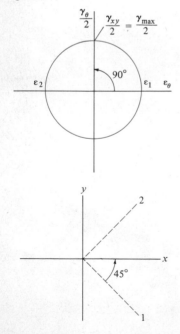

$$\text{dilatation} = \frac{V_f - V_o}{V_o}$$

$$= \frac{(1 + \varepsilon_x)(1 + \varepsilon_y)(1 + \varepsilon_z)\, dx\, dy\, dz - dx\, dy\, dz}{dx\, dy\, dz}$$

$$\approx \varepsilon_x + \varepsilon_y + \varepsilon_z = \varepsilon_V$$

if products of strain are considered negligible compared with the strains themselves. Thus, by definition the bulk modulus K becomes

$$K = \frac{\dfrac{\sigma_x + \sigma_y + \sigma_z}{3}}{\varepsilon_x + \varepsilon_y + \varepsilon_z} = \frac{E}{3(1 - 2\mu)} \qquad (4\text{-}20)$$

The intermediate details of this derivation are left as an exercise for the student (Prob. 4-25). Note that the bulk modulus is not an independent elastic constant but is dependent upon E and μ. Thus, while we have defined four elastic constants E, G, μ, and K, only two (any two) are needed to completely know the elastic properties of an isotropic elastic material.

PROBLEMS

4-21 Write Hooke's law for plane strain, explicit in stress and explicit in strain.

4-22 If Poisson's ratio for the magnesium alloy of Prob. 3-1 is 0.34, what is the modulus of rigidity?

4-23 Give an argument that the value of Poisson's ratio for a material obeying Hooke's law must lie between -1 and $+\frac{1}{2}$.

4-24 Express generalized Hooke's law for stresses in terms of the strains ε_x, ε_y, ε_z, γ_{xy}, γ_{xz}, and γ_{yz}.

4-25 Carry out the details in the derivation of Eq. 4-20.

4-26 Prove rigorously that for plane stress in an isotropic material obeying Hooke's law the principal directions of stress coincide with the principal directions of strain. *Notes 4-20*

4-27 Show that for a linearly elastic isotropic material the normal stress σ_i in any direction can be expressed as $\sigma_i = C_1 \varepsilon_i + C_2 D$, where ε_i is the axial strain in the same direction as σ_i and D is the dilatation. Express C_1 and C_2 in terms of E and μ.

4-28 Two axial strain gages A and B are bonded to *Notes*

the outside surface of a thin-walled pressure vessel in such a manner that their axes are perpendicular to each other. It is known that the principal stresses σ_1 and σ_2 in the wall of the vessel are in a ratio of two to one. Obtain a relationship between the larger principal stress σ_1 and the sum $\varepsilon_A + \varepsilon_B$ in terms of E and μ of the wall material.

Note: For Probs. 4-29 to 4-32 assume that $\sigma_z = 0$ and determine the values and directions of the principal stresses and the maximum shear stress in each case. Strains are in microinches per inch and E, K, and G in psi $\times 10^6$.

4-29 $\varepsilon_x = 400$ $\varepsilon_y = -2000$ $\gamma_{xy} = 1000$
4-21 $E = 21, \ \mu = \frac{1}{3}$

4-30 $\varepsilon_x = -600$ $\varepsilon_y = 800$ $\gamma_{xy} = -700$
$E = 30, \ \mu = 0.4$

4-31 $\varepsilon_x = 800$ $\varepsilon_y = 2000$ $\gamma_{xy} = -1600$
$E = 9.0, \ G = 3.2$

4-21
4-32 $\varepsilon_x = 500$ $\varepsilon_y = -1000$ $\gamma_{xy} = -2000$
$E = 24, \ G = 8.0$

Note: For Probs. 4-33 to 4-35 assume that $\sigma_z = 0$ and determine the values and directions of the principal stresses and the maximum shear stress in each case. Strains are in micrometers per meter and E, K, and G in gigapascals.

4-33 $\varepsilon_x = 1200$ $\varepsilon_y = -1200$ $\gamma_{xy} = 2400$
       $G = 35$, $\mu = 0.3$

4-34 $\varepsilon_x = 3000$ $\varepsilon_y = -1800$ $\gamma_{xy} = 1400$
       $G = 70$, $\mu = \frac{1}{4}$

4-35 $\varepsilon_x = 730$ $\varepsilon_y = 1480$ $\gamma_{xy} = 400$
       $E = 210$, $K = 140$

4-36 During a tensile test on a steel specimen

($E = 30 \times 10^6$ psi, $\mu = \frac{1}{4}$) the axial stress reaches 20,000 psi. What is the corresponding value of the cubical dilatation? *4-21*

4-37 A rectangular block of aluminum $1 \times 2 \times 4$ in. is placed into a pressure chamber and subjected to a pressure of 8000 psig. If $E = 9 \times 10^6$ psi and $\mu = \frac{1}{3}$, what will be the decrease in volume of the block?

4-38 An electrical strain gage is bonded to the surface of a tensile test specimen. Inadvertently the gage axis is cocked at an angle of 15° with the longitudinal axis of the specimen. Determine the relation in terms of E and μ between the axial stress σ and the strain in the gage ε_g. *4-21*

not on exam →

Chapter 5
Members in simple tension, compression, and shear

5-1 / Objectives

Upon completion of this chapter you will be able to:

1 Determine the magnitude and location of an axial load when given the stress distribution for the loaded member and the stress strain curve for the material of which it is composed. *Don't worry about this*
2 Determine the stress for members in simple tension, compression, or shear.
3 Calculate the stresses in simple connections and riveted joints.
4 Calculate the stresses in thin-walled pressure vessels.
5 Calculate the deformation caused by centroidal loads.
6 Design a simple axially loaded member for criteria based on strength and deformation.
7 Solve problems involving statically indeterminate axially loaded members.

5-2 / Introduction

The earlier chapters of this book have taught you the basic concepts and procedures of stress analysis without paying a great deal of attention to the application of those procedures. This chapter and those that follow it will take a much more detailed look at deformable bodies as they exist in the real world and will help you to develop skills in the analysis and design of such members.

The primary function of a structural or machine member is to support and/or transmit loads. In Chapter 1 we saw that the external applied loads produce internal reactions that can be resolved into normal and shear forces and bending and twisting couples (refer to Figs. 1-4 and 1-5). In this chapter we shall look at some types of load-carrying members whose net internal reaction is primarily a normal or shear force. In Chapter 6 we shall look at members that are being twisted (called shafts), and in Chapters 7, 8, and 9 we shall examine members that are being bent (called beams).

In this and the succeeding chapters we shall be concerned with solving two basic types of engineering problems. The first problem is to relate the applied external loads to the internal stresses so as to obtain a measure of the strength of the member. The second problem is to relate the applied loads to the strains so as to determine the deformation of the member. As we shall see, the load-deformation relations can be quite useful in solving heretofore statically indeterminate problems. As you study the succeeding sections, bear in mind that the primary purpose is to develop a rational *procedure* for solving the two basic problems described above.

5-3 / Axially loaded members

Consider the eye-bar shown in Fig. 5-1(a). If this member is loaded by equilibrated pulling forces at each end, we know that these forces will have to be equal in magnitude, be opposite in direction, and have the same line of action. If we pass a cutting plane perpendicular to the line of action of the load, the resulting net internal reaction will be equal to a tensile load of P. It is reasonable to conjecture that this net internal reaction is the accumulative effect of the normal tensile stress σ distributed over the entire cut cross-sectional area so that

$$P = \int_{\substack{\text{cut} \\ \text{cross section}}} \sigma \, da \qquad (5\text{-}1)$$

where the integral is interpreted as a summation of the infinitesimal forces $\sigma \, da$. Equation 5-1 is the basic *load-stress relation* for an axially loaded two-force member. If our conjecture is correct, the state of stress in the

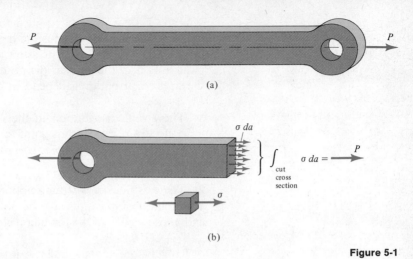

Figure 5-1

member will be uniaxial so that the maximum shear stress will occur at 45° with the axis of the member.

To generalize our discussion of axially loaded members, consider a member of *arbitrary but constant* cross section subjected to a resultant tensile force P parallel to the y axis as shown in Fig. 5-2(a). By passing a cutting plane perpendicular to the line of action of P and applying the same reasoning as before, we again arrive at the basic load-stress relation

$$P = \int_{\text{area}} \sigma \, da \qquad (5\text{-}1)$$

for an axially loaded member.*

Figure 5-2

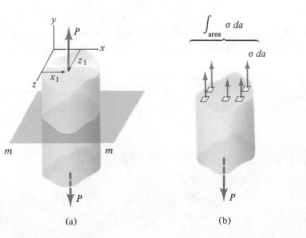

* Slender compression members will not be considered in this chapter because of their buckling mode of failure. This mode will be considered in more detail in Chapter 10.

When Eq. 5-1 is to be applied to any axially loaded member, an expression for the stress σ must be determined *before* the integral $\int_{area} \sigma\, da$ can be evaluated. The expression for stress is a mathematical function indicating how the stress is distributed throughout the body. In general, the stress distribution is dependent on two major factors.

1 The strain distribution in the member. This distribution indicates how the strain varies from point to point throughout the body. *In most cases, the strain distribution is governed by the geometry of the member and the location of the line of action of the applied load.*

2 The existing stress-strain relationship for the material of the member. This relationship can usually be obtained from the stress-strain diagram and a knowledge of the loading history of the member.

Thus, before any statement can be made about the distribution of the stress throughout a loaded member, two questions must be answered: (1) How is the strain distributed or what is the kinematic response of the member to the applied load? (2) How is the stress related to the strain? Throughout this and the succeeding chapters we shall continually be asking ourselves these two vital questions.

EXAMPLE 5-1

The member shown in Fig. 5-3(a) has a uniform longitudinal axial strain of 0.0006 mm/mm. It is made of a material with the stress-strain curve shown in Fig. 5-3(b). Find the load P.

Solution:

Since the strain is less than 0.0010, we know (by referring to the stress-strain curve) that the proportional limit is not exceeded and that the stress is 126×10^6 Pa. Since the strain was said to be uniform, the stress also will be uniform and the stress distribution is shown in Fig. 5-3(c).

Figure 5-3

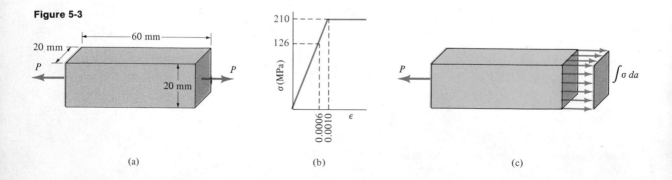

(a)   (b)   (c)

By equilibrium

$$P = \int_{area} \sigma \, da$$

since the stress is constant in this loading

$$P = \sigma A = 126 \times 10^6 \times 0.02 \times 0.02$$
$$= 50,400 \text{ N}$$

EXAMPLE 5-2

The member shown in Fig. 5-4(a) is axially loaded. Because of symmetry, the strain does not *vary* in the z and x directions. After the load P was applied, the strain varied linearly in the y direction from 0.0030 in./in. tensile along AA to 0.0010 in./in. compressive along BB, as in Fig. 5-4(b). If the material has the stress-strain curve (tensile and compressive) of Fig. 5-4(c), determine the magnitude and vertical location of P.

Figure 5-4

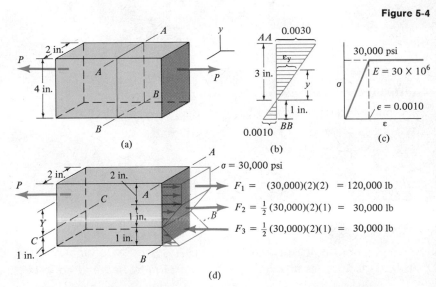

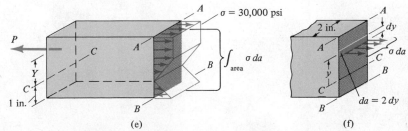

Solution:

From the stress-strain curve we see that as long as the axial strain is equal to or less than 0.0010 in./in., the stress will be directly proportional to the strain. For strains larger than 0.0010, the stress has a constant value of 30,000 psi. Hence, the stress distribution is shown in Fig. 5-4(d), where the fibers below CC are in compression and those above CC are in tension. The stress distribution of Fig. 5-4(d) may be interpreted merely as a distributed load or, more conveniently, as three distributed loads each of which acts effectively at the centroid of the "volume" of the respective distribution, as shown in Fig. 5-4(d). Remember that each load is computed as the volume of a parallelepiped in Fig. 5-4(d). Thus

$$\sum F_x = 0$$

$$P - 120{,}000 - 30{,}000 + 30{,}000 = 0$$

$$P = 120{,}000$$

Also,

$$\sum M_{CC} = 0$$

$$PY - 120{,}000(2) - 30{,}000(\tfrac{2}{3}) - 30{,}000(\tfrac{2}{3}) = 0$$

$$Y = 2.33 \text{ in.}$$

Alternative solution:

For a more formal mathematical approach, we refer to Fig. 5-4(e) and (f).

$$\sum F_x = 0$$

$$+P - \int_{area} \sigma \, da = 0$$

$$P = \int_{-1}^{+3} \sigma 2 \, dy$$

To evaluate the integral, σ must be expressed as a function of y. Because of the change in distribution that occurs at $y = +1$, the integral must be handled in two parts. Thus,

$$P = \int_{-1}^{+3} \sigma 2 \, dy = \int_{-1}^{+1} \sigma 2 \, dy + \int_{+1}^{+3} \sigma 2 \, dy$$

From Fig. 5-4(e) and (f), we have, for $-1 \le y \le +1$,

$$\frac{\sigma}{30{,}000} = \frac{y}{1} \quad \text{or} \quad \sigma = 30{,}000y$$

For $+1 \le y \le +3$,

$$\sigma = 30{,}000$$

Hence,

$$P = \int_{-1}^{+1} (30{,}000y)2\,dy \int_{+1}^{+3} (30{,}000)2\,dy$$

$$= 60{,}000 \left.\frac{y^2}{2}\right|_{-1}^{+1} + 60{,}000y\left.\right|_{+1}^{+3}$$

$$= 30{,}000 - 30{,}000 + 180{,}000 - 60{,}000 = 120{,}000 \text{ lb}$$

To determine the vertical location of the line of action of P, we must satisfy rotational equilibrium. For this purpose a moment equation with respect to CC will be convenient. If the distance from CC to the line of action of P is denoted by Y,

$$\overset{\curvearrowright}{\sum} M_{CC} = 0$$

$$-PY + \int_{-1}^{+3} y(\sigma\,da) = 0$$

$$PY = \int_{-1}^{+3} y\sigma 2\,dy$$

Again, two integrals will be necessary. Using the expressions for stress previously obtained, we have

$$PY = \int_{-1}^{+1} y(30{,}000y)2\,dy + \int_{+1}^{+3} y(30{,}000)2\,dy$$

$$= 60{,}000 \left.\frac{y^3}{3}\right|_{-1}^{+1} + 60{,}000 \left.\frac{y^2}{2}\right|_{+1}^{+3}$$

$$= 20{,}000 + 20{,}000 + 270{,}000 - 30{,}000 = 280{,}000 \text{ in.-lb}$$

Hence, the distance from CC to P is

$$Y = \frac{280{,}000}{120{,}000} = 2.33 \text{ in.}$$

and the distance from the bottom is 3.33 in.

PROBLEMS

Note: In each of the problems from 5-1 to 5-12, determine the magnitude and vertical location of the resultant load P.

5-1 The member in part (a) of the illustration is loaded parallel to its longitudinal axis in such a manner that the strain does not vary in the z direction. After the resultant load P was applied, the strain varied linearly from 0.0005 in./in. tensile along AA to 0 in./in. along BB, as indicated in part (b) of the illustration. The stress-strain curve (tensile and compressive) for the material is as shown in part (c).

Problem 5-1

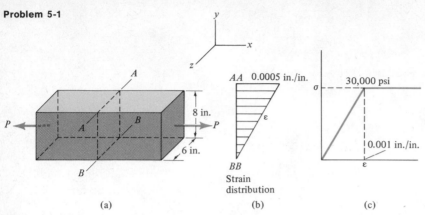

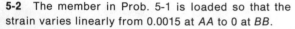

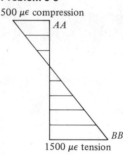

(a) (b) (c)

5-2 The member in Prob. 5-1 is loaded so that the strain varies linearly from 0.0015 at *AA* to 0 at *BB*.

5-3 The member in Prob. 5-1 is loaded in compression so that the strain does not vary in the *z* direction. After the load was applied, the strain varied linearly from 0.0015 m/m compressive along *AA* to 0.0005 m/m compressive along *BB*, as indicated in part (a) of the illustration for Prob. 5-3. The material has the stress-strain curve (tensile and compressive) shown in part (b).

Problem 5-3

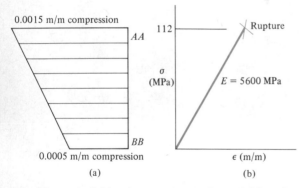

(a) (b)

5-4 The material for the member in Prob. 5-3 has the stress-strain curve shown in part (c) of the illustration for Prob. 5-1.

5-5 The member in Prob. 5-1 is loaded in such a manner that the strain does not vary in the *z* direction. After the load *P* was applied, the strain varied linearly from 500 $\mu\varepsilon$ compression at *AA* to 1500 $\mu\varepsilon$ tensile at *BB*, as indicated in the illustration for Prob. 5-5. The material has the stress-strain curve (tensile and

compressive) shown in part (c) of the illustration for Prob. 5-1.

Problem 5-5

5-6 The material for the member in Prob. 5-5 has the stress-strain curve shown in part (b) of the illustration for Prob. 5-3.

5-7 The member in Prob. 5-1 is loaded in such a manner that the strain does not vary in the *z* direction. After the load *P* was applied, the strain varied linearly from 8000 $\mu\varepsilon$ tensile at *BB* to 0 at *AA*. The stress-strain curve (tensile and compressive) for the material is shown in the illustration for Prob. 5-7.

Problem 5-7

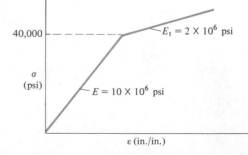

5-8 The member in Prob. 5-1 is axially loaded in such a manner that the strain varies linearly as shown in the illustration for Prob. 5-5. The material is linear elastic with a tensile elastic modulus of $E = 84 \times 10^9$ Pa and a compressive elastic modulus of 105×10^9 Pa.

5-9 The member in Prob. 5-7 is loaded so that the strain varies as shown in the illustration for Prob. 5-9.

Problem 5-9

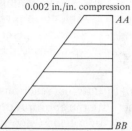

0.002 in./in. compression

AA

BB

0.006 in./in. compression

5-10 The material for the member in Prob. 5-1 has a stress-strain relationship that can be approximated by $\sigma = k\varepsilon^{1/2}$, in which $k = 3 \times 10^9$ Pa.

5-11 The material for the member in Prob. 5-5 has a stress-strain relationship that can be approximated by $\sigma = k\varepsilon^{1/3}$, in which $k = 12,000$ psi.

5-12 The material for the member in Prob. 5-5 has a stress-strain relationship that can be approximated by $\sigma = k\varepsilon^{1/2}$ where $k = 400,000$ psi for tensile strain and $k = -500,000$ psi for compressive strain.

5-13 The eccentricity e in the illustration is 0.125 in. The strain gage axes are vertical. (a) If the strain gage at B reads 933 μin./in., compressive, what should the gage at A read? Assume that there is linear strain variation from A to B and that the material is mild steel with a yield-point stress of 30,000 psi. (b) Determine the load P.

Problem 5-13

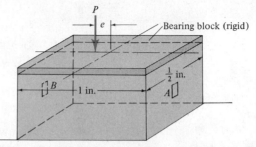

5-14 The member in the illustration for Prob. 5-13 is made of mild steel with a yield-point stress of 30,000 psi. (a) If the gage at A reads 600 μin./in., compressive, and P is 10,000 lb, what should the gage at B read? Assume linear strain variation. (b) Determine the eccentricity e.

5-15 Solve Prob. 5-13 if the gage at B reads 1200 microstrain.

5-16 Solve Prob. 5-14 if P is 13,000 lb.

5-17 What will be the readings of the gages at A and B in the illustration for Prob. 5-13 if P is 30,000 N and e is 2.5 mm? The material is mild steel with a yield-point stress of 210 MPa, and the member is 12 × 25 mm.

5-18 Solve Prob. 5-17 if P is 45,000 N.

5-4 / Centroidal internal reactions

The initial step in any problem involving the determination of stress is the calculation of the internal reactions of the plane cross-sectional area on which the stress is to be found. This is done in the following manner:

1 Draw a line through a picture of the body at the location where you wish to check the stress. The line represents the edge of a transverse cutting plane. The cutting plane should divide the body into *two* parts. (If you are working with a three-dimensional picture, you must, of course, represent the plane in three dimensions.)

2   Draw a free-body diagram of either of the two parts showing all loads acting on that part. You can ignore the other part and its loads completely. The cutting plane you drew in step 1 should form one edge of your figure, and it represents the edge of the area on which you are finding the internal reactions.

3   The internal reactions should be shown as either forces or couples. All forces should be shown acting through the centroid of the cut area, hence they are *centroidal reactions*. The couples are also shown acting on the cut plane. (Refer to section 1-3.) It is usually desirable to resolve the force reactions into components parallel to and normal to the plane cross-sectional area, that is, normal and shear forces. Similarly, the couples are usually resolved into twisting and bending couples.

4   Use the laws of statics to determine the values of the internal reactions necessary to have the free body in equilibrium. These are the appropriate values of the reactions on the cross-sectional area you identified in step 1.

The foregoing routine is presented in general terms since we shall be referring back to it from time to time. Most problems in this chapter will produce centroidal internal reactions that are only forces; that is, there will be no twisting or bending effects (they will come in later chapters). We refer to this condition as *centroidal loading*. The centroidal forces that you find will be either normal to the plane causing tension or compression stresses, or parallel to it causing shear stresses.

EXAMPLE 5-3

The turnbuckle in Fig. 5-5 is tightened until the force in each cable is 1000 N. (a) Find the internal axial force in an eye-bolt; (b) find the internal axial force in one side of the link.

Solution:

Figure 5-6(a) is a free body of the entire turnbuckle. Lines *a-a* and *b-b* indicate the location of cutting planes for parts (a) and (b) of the problem. Figure 5-6(b) shows the portion to the left of plane *a-a* and Fig. 5-6(c) shows the internal reaction necessary for equilibrium. Figures 5-6(d) and 5-6(e) show the equivalent steps for finding the forces in the link.

The foregoing has been a review of material covered in Chapter 1. It is reiterated here for the convenience of the student. If you wish to practice this procedure before going on to the calculation of stress, the following problems are appropriate.

Figure 5-5

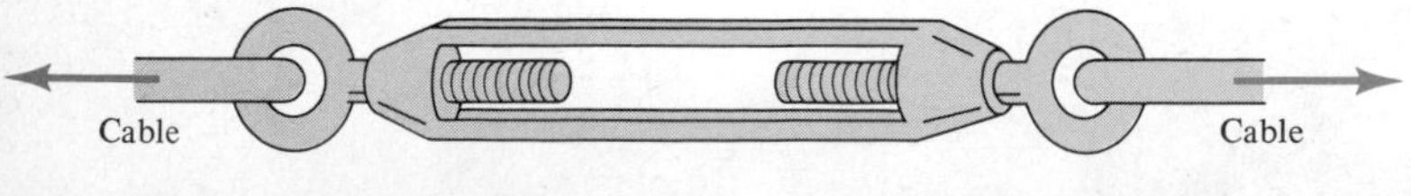

Cable   Cable

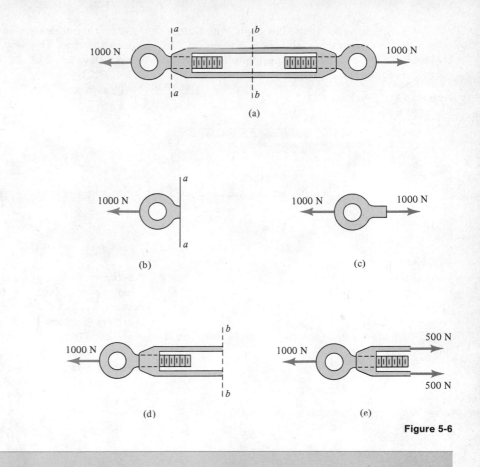

Figure 5-6

In the following problems, be sure to indicate whether your answer is tension, compression, or shear.

5-19 The catch shown in the illustration failed along the plane indicated by the broken line when the loads shown were applied. What was the force on the failure plane?

Problem 5-19

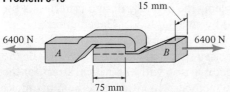

5-20 In the illustration the tension in the bolt joining the two parts is 1000 lb. What is the force acting on the contact area between parts?

Problem 5-20

5-21 A spring is added to the device in Prob. 5-20 as shown in the illustration. What is the force on the contact area if the force in the spring is 750 lb compression and the force in the bolt is unchanged?

Problem 5-21   Problem 5-22

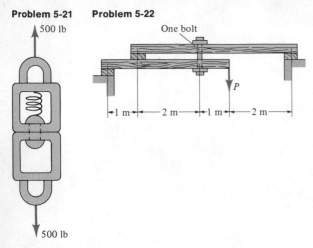

5-22 Find the force in the bolt in the illustration if the load *P* is 12 kN.

5-23 The illustration shows part of the landing-gear assembly for a glider. Determine the compressive force developed on *BC* if the reaction on the wheel is 10,000 N.

Problem 5-23

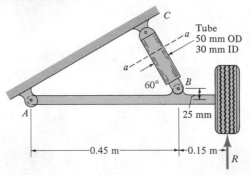

5-24 In the timber truss shown in the illustration the wood has a tendency to shear along the grain at corners *A* and *C*. Evaluate the shearing forces on the critical sections if the applied load is 15,000 lb as shown. (Treat all joints as though they were pin connected when determining the forces in the members.)

Problem 5-24

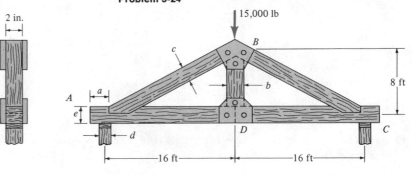

5-5 / Stresses caused by centroidal axial loading

In a great many engineering problems the primary design criterion is the maximum normal stress produced by an axial load. The engineer is often called upon to determine the proper dimensions and material of an axially loaded member that is to resist a specified load, or to investigate the stress in a particular member for comparison with the critical stress for the particular material and use of the member. In either case he must make use of the relationship between load and stress. We shall now determine a specific load-stress relationship for a particular, but common, type of situation, namely, centroidal loading.

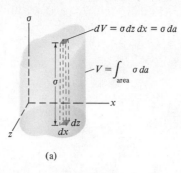

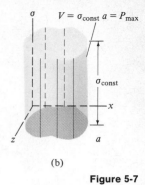

(a)   (b)

Figure 5-7

Figure 5-7(a) is a graphical representation of the integral $\int_{area} \sigma \, da$, where σ varies over the area a. In this figure we can interpret the integral $\int_{area} \sigma \, da$ as a volume having a variable height represented by σ and a cross-sectional area in the xz plane. Also, we can see that, for any desired maximum stress, the maximum obtainable volume would be one with uniform height σ_{const}, as in Fig. 5-7(b). Thus,

$$\text{maximum volume} = \sigma_{const} \times \text{area}$$

Since the volume represents the integral $\int_{area} \sigma \, da$, which in turn is equal to the axial load P (from Eq. 5-1), the *maximum* axial load for any desired maximum stress is obtained when the stress is uniform over the cross-sectional area. Hence,

$$P_{max} = \int_{area} \sigma_{const} \, da = \sigma_{const} \int_{area} da$$

$$P_{max} = \sigma_{const} A \quad \text{or} \quad \sigma = \frac{P}{A} \tag{5-2}$$

When the normal stress in an axially loaded member is *uniformly* distributed over its cross-sectional area, the member is said to be in "pure" tension or "pure" compression, the nature of the stress depending on the type of the applied load. Before we plunge ahead, however, and apply Eq. 5-2 to engineering members, let us first investigate the conditions necessary to produce a uniform stress distribution.

Let us first determine the location of the line of action of the externally applied force P that is necessary to produce a uniform stress distribution. We first write an equilibrium moment equation for the free-body diagram in Fig. 5-8(c) by taking the moments of the forces with respect to the x axis.

$$\sum M_x = 0$$

$$Pz_1 - \int_{area} (\sigma \, da)z = 0$$

$$Pz_1 = \int_{area} (\sigma \, da)z$$

But for pure tension σ is uniform over the cross-sectional area and $P = \sigma A$. So

$$(\sigma A)z_1 = \sigma \int_{\text{area}} z\,da$$

$$z_1 = \frac{\int z\,da}{A} = \bar{z} \tag{5-3a}$$

Similarly,

$$x_1 = \bar{x} \tag{5-3b}$$

Here $\bar{x}$ and $\bar{z}$ are the x and z coordinates of the *centroid* of the cross section. These results show that for uniform stress distribution *the line of action of the resultant externally applied force must be coincident with the longitudinal centroidal axis of the member*; that is, the member must undergo centroidal loading. This statement is true whether the external load is considered concentrated or distributed and is a *necessary* (but not sufficient) condition for the stress to be uniform.

It was pointed out in Chapters 3 and 4 that most engineering materials can be considered to be *homogeneous* and *isotropic*; so let us consider the material of the body in Fig. 5-8 to be homogeneous and isotropic. In Fig. 5-8(a) two parallel planes aa and bb are passed through an unloaded body so that they are perpendicular to the y axis and are separated by a distance L_{ab}. Let us assume now that the stress is uniform across section aa, section bb, and every parallel section between aa and bb. As indicated in Fig. 5-8(b), the state of stress is uniaxial, and the corresponding strain in the y direction can be obtained directly from the appropriate stress-strain curve of the material.

Since the material is homogeneous, the uniform stress across section aa implies a *uniform axial strain* ε_y across section aa. The same is true for

Figure 5-8

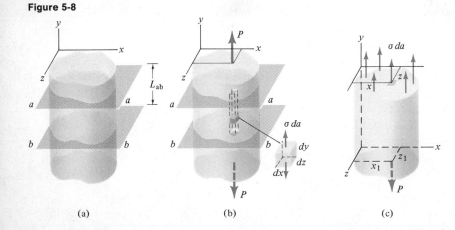

(a)   (b)   (c)

section *bb* and any section between *aa* and *bb*. Thus, the elongation of *any* longitudinal fiber of material between *aa* and *bb* is given by

$$\frac{de_y}{dy} = \varepsilon_y \qquad e_y = \int_{\substack{over \\ L_{ab}}} \varepsilon_y \, dy \qquad\qquad (5\text{-}4)$$

Since all fibers have the same strain and the same original length, all fibers will undergo the same deformation. *Thus sections aa and bb must remain parallel, if the member is to have a uniform stress distribution.* A more general statement is that *parallel cross sections must remain parallel.* This condition must be satisfied for any length, such as L_{ab}, of an axially loaded *homogeneous* member if it is to have uniform stress distribution within that length. The condition applies regardless of whether the stress is above or below the proportional limit of the material.

It should be mentioned that the requirement for uniform stress distribution in a homogeneous axially loaded straight member—*parallel sections must remain parallel*—generally will not be satisfied in the immediate vicinity of the applied external load. However, if the longitudinal dimension of the body is relatively large in comparison with the transverse dimensions, the localized effects near the applied load disappear in sections sufficiently distant from its point of application.*

Although the preceding analysis was concerned with a homogeneous member of constant cross section, Eq. 5-2 has been found to be applicable for homogeneous symmetrical members of gradually varying cross section. However, any abrupt change in cross section will produce a nonuniform strain distribution and parallel cross sections will not remain parallel. Figure 5-9 illustrates how the line of action of the load and the geometry of the member affect the strain distribution.

In Fig. 5-9(a) the member is centrally loaded, straight, and has a uniform cross section. The strain is uniform, except for localized effects near the point of application of the load. Although the member in (b) is curved, it is straight and of uniform cross section in the length *ab*. Also, the line of action of the load coincides with the longitudinal centroidal axis in this length. So the load produces uniform strain in the length *ab*. In (c) the loading on the length *ab* is eccentric, and the strain distribution is nonuniform.

The cross section of the member in Fig. 5-9(d) varies *gradually*, but the line of action of the load coincides with the longitudinal centroidal axis of the member. Although the stress varies from cross section to cross section in the length *ab*, there is essentially uniform stress across any one particular cross section. If the taper is more severe, however, the assumption of

* A French mathematician, Saint Venant, first showed that localized effects diminish rapidly as you consider sections further and further away from such locations. This is a topic usually covered in courses on the theory of elasticity.

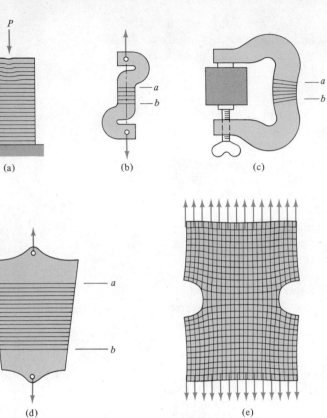

Figure 5-9

uniform and uniaxial stress across the cross section becomes quite questionable. In Fig. 5-9(e) there is an abrupt change in the cross section of the member at the notch. As a result, the strain distribution over the cross section is not uniform in the vicinity of the notch.

The strain distribution and the accompanying stresses produced by eccentric loading and bending will be studied in Chapter 7. Localized nonuniform strain distributions resulting from abrupt changes in geometry are called *strain concentrations*. Such *strain concentrations* are often very significant in determining the load-carrying capacity of structural members, particularly for certain types of materials and loading conditions. The importance of strain concentrations and their accompanying *stress concentrations* as a design consideration is discussed in Chapter 14.

It should be noted that the foregoing discussion and equation do not apply to slender compressive members. Slender members loaded in compression fail in buckling—an action akin to bending—which will be studied in Chapter 10.

To summarize:

$$\sigma = \frac{P}{A}$$

if, and only if

1 The member is subject to "simple" tension or compression with an axial load P passing through the centroid of its cross sectional area A.

2 The load results in uniform tensile stress or the load results in compressive stress and the member is short and stocky in proportions.

3 The member is composed of an isotropic homogeneous material.

4 The member is free from abrupt changes in cross section.

5-6 / Working problems involving simple tensile, compressive, or shear stresses

Many problems involving centroidal loads may be roughly divided into two classes:

1 Problems where the dimensions of the body are known and stresses or a load must be found.

2 Design problems where the load and allowable stresses are known and the dimensions of the body must be found.

These problems may be solved in a routine manner, but each step must be performed with care. The steps are as follows:

1 *Decide upon the critical section.* This is the section at which you need to determine the stress. It may be given in the statement of the problem or you may have to identify it by thinking about the probable failure of the body. Sometimes you will need to examine more than one section. If this is the case, examine only one at a time; that is, complete the following steps for each section before you move on to the next.

2 *Draw a free-body diagram to determine the internal reactions acting at the critical section.* To do this, pass an imaginary cutting plane through the body at the critical section. It must divide the body into two parts. Draw all the known loads acting on one part and the internal reaction acting on the critical section.

3 *Find the internal reactions.* Use the equations of equilibrium to determine the internal reaction acting on the cut section. Make sure every reaction is either normal or parallel to the critical area. If it does not come that way in your first solution, replace it by appropriate components.

4 *Solve.* Use the equations

$$\sigma = \frac{P}{A} \quad \text{or} \quad \tau = \frac{P}{A}$$

to determine the unknown quantities. Remember that: P is the centroidal load on the critical section; σ is the normal stress on the critical section; τ is the shear stress; and A is the appropriate area of the critical section. Make sure all parts of your calculation relate to a single section.

EXAMPLE 5-4

The timber block in Fig. 5-10 is $4 \times 4 \times 4$ in. It is subject to an axial compression load as shown. The maximum allowable shearing stress along the grain is limited to 1000 psi. Determine the maximum value that P may attain.

Solution:

Figure 5-11(a) shows a transverse cutting plane passed through the block parallel to the grain. Figure 5-11(b) shows a front view of the block and plane. Figure 5-11(c) is a free-body diagram of the upper portion of the block. Figure 5-11(d) is also a free-body diagram showing the resolution of P into components normal and parallel to the cutting plane.

Figure 5-10

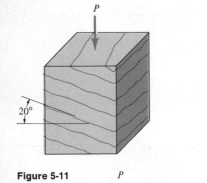

Figure 5-11

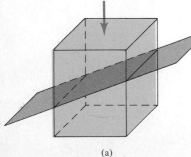

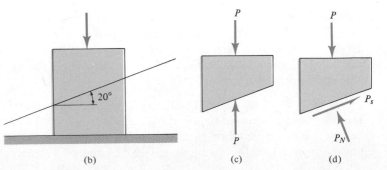

(a)   (b)   (c)   (d)

The shearing force P_s is

$$P_s = P \sin 20°$$

The area on which P_s acts is

$$A = 4 \frac{4}{\cos 20°} = 17.03 \text{ in.}^2$$

The maximum shearing stress that may occur on the area is 1000 psi. Thus

$$\tau = \frac{P_s}{A}$$

$$1000 = \frac{P \sin 20°}{17.03}$$

$$P = \frac{1000(17.03)}{\sin 20°} = 49,800 \text{ lb}$$

EXAMPLE 5-5

Determine the magnitude of the axial stress developed in the 3-mm-diameter control cable shown in Fig. 5-12 due to an applied force of 250 N.

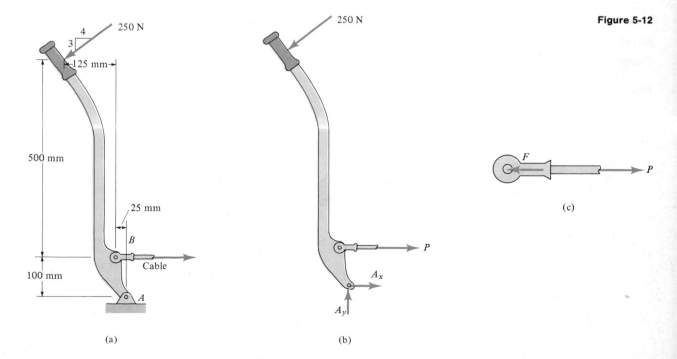

Figure 5-12

(a)   (b)

5-56 The longitudinal joint in a boiler transmits 360 kN/m. The boiler plate is 20 mm thick and each strap is 12 mm thick. The rivets are 18 mm in diameter. Find the maximum average tensile stress developed in any plate and the maximum nominal bearing stress between rivets and plate.

Problem 5-56

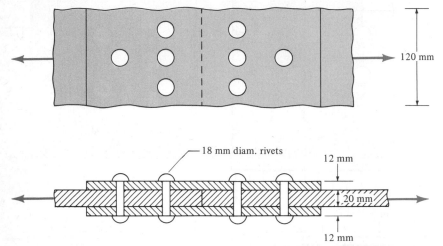

120 mm

18 mm diam. rivets

12 mm

20 mm

12 mm

5-10 / Thin-walled cylindrical pressure vessels

A further application of centroidal loading relying heavily on free-body analysis is thin-walled circular pressure vessels. Consider a section of a thin-walled pressure vessel carrying a fluid of negligible weight under a *gage* pressure p, as shown in Fig. 5-25(a). An end view of the unpressurized vessel with two radial planes OA and OD is shown in Fig. 5-25(b). Figure 5-25(c) illustrates the deformation of an infinitesimal section of the wall of the cylinder caused by the pressure p. It is assumed that the radial planes remain radial and that the thickness of the wall does not change due to the pressure. From this figure the circumferential strains in the inside and outside fibers are

$$\varepsilon_{\text{cir}} = \frac{R_i \, d\theta - r_i \, d\theta}{r_i \, d\theta} = \frac{R_i - r_i}{r_i} \quad \text{(inside)} \qquad (5\text{-}5a)$$

$$\varepsilon_{\text{cir}} = \frac{R_o \, d\theta - r_o \, d\theta}{r_o \, d\theta} = \frac{R_o - r_o}{r_o} \quad \text{(outside)} \qquad (5\text{-}5b)$$

Now if

$$R_o = R_i + t \quad \text{and} \quad r_o = r_i + t$$

Eq. 5-5b becomes

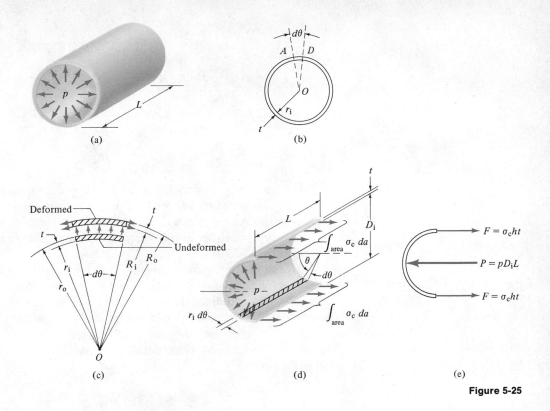

Figure 5-25

$$\varepsilon_{\text{cir}} = \frac{R_i + t - (r_i + t)}{r_i + t} = \frac{R_i - r_i}{r_i + t} \approx \frac{R_i - r_i}{r_i}$$

if t is small in comparison with r_i. Hence, for these conditions the outside and inside circumferential strains are the same, and we assume that the strain is uniform throughout the thickness of the wall.

This assumption will be reasonably good if the wall thickness is not more than about $\frac{1}{10}$ of the radius. Such vessels are classified here as "thin-walled." A skin-diver's air tank would be considered a thin-walled vessel. A rifle barrel would be thick-walled, and there would be a non-uniform circumferential strain through the wall.*

If the circumferential strain is uniform and the material is homogeneous, the circumferential stress σ_c in the wall of the pressure vessel will be uniform. An analysis of the free body in Fig. 5-25(d) will result in the desired relationship between gage pressure and circumferential stress.

---

* For an analysis of thick-walled cylinders see *Advanced Mechanics of Materials*, by Seely and Smith, Chapter 10 (New York: Wiley, 1952).

Thus,

$$\overset{+}{\underset{\leftarrow}{\sum}} F = 0$$

$$\int_{-\pi/2}^{+\pi/2} (pLr_i \, d\theta) \cos\theta - 2\int_{area} \sigma_c \, da = 0$$

$$pLr_i \sin\theta \Big|_{-\pi/2}^{+\pi/2} - 2\sigma_c Lt = 0$$

$$\sigma_c = \frac{2pLr_i}{2Lt} = \frac{pD_i}{2t} \tag{5-6}$$

The circumferential stress σ_c is produced by the internal *gage* pressure only, and this pressure will produce no shear stress along a longitudinal section of the vessel. A circumferential stress is commonly called a *hoop* stress. A simplified free-body diagram showing the forces involved is shown in Fig. 5-25(e).

If a vessel has, or could be considered to have, closed ends and contains a fluid under a gage pressure p, as in Fig. 5-26(a), the wall of the vessel will have a longitudinal stress as well as a circumferential stress. If the deformation of the vessel does not alter its basic geometry, that is, if the vessel does not "bulge," then the longitudinal stress σ_L in the wall of the vessel will be uniform, since parallel transverse sections would remain parallel. From Fig. 5-26(b)

$$\overset{+}{\underset{\rightarrow}{\sum}} F = 0$$

$$(p)\left(\frac{\pi D_i^2}{4}\right) - \int_{area} \sigma_L \, da = 0$$

$$\frac{p\pi D_i^2}{4} = \sigma_L(\pi D_{av} t)$$

Since $D_{av} \approx D_i$ when t is very small in comparison with D_i,

$$\sigma_L = \frac{pD_i}{4t} \tag{5-7}*$$

Thus, the longitudinal stress produced in the wall of a vessel by an internal pressure is equal to one-half of the circumferential stress due to the same pressure. From this analysis we see that a biaxial state of stress exists in the wall of a thin-walled pressure vessel with closed ends. Actually there also exists a third normal stress in the wall, called the radial stress, which is due to the compressive effects of the pressure on the internal surface of the vessel. However, the radial stresses are usually neglected in the analysis of a thin-walled vessel.

* If D_i were replaced by D_{av} in Eqs. 5-6 and 5-7, the results would agree more closely with those obtained from the theory of elasticity. However, for thin-walled vessels, $D_i \approx D_{av} \approx D_o$.

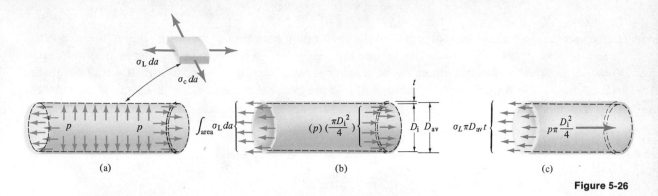

Figure 5-26

Caution: *Circumferential stresses* exist on *longitudinal sections* [see Fig. 5-25(d)]. *Longitudinal* stresses exist on *transverse sections* [see Fig. 5-26(b)]. Do not confuse the two. The best way to avoid such confusion is to always draw the appropriate free body [Fig. 5-25(e) or 5-26(c)]. This will provide a second benefit since many problems can be solved directly from free-body analysis.

EXAMPLE 5-13

A cylindrical rocket motor case represented in Fig. 5-27(a) is made from a sheet of special steel 0.080 in. thick. The case is 70 in. in diameter and 8 ft long. It is subjected to 125,000 lb of tensile longitudinal thrust, in addition to internal pressure. What is the operating pressure, if the tensile stress in the wall must not exceed 210,000 psi?

Figure 5-27

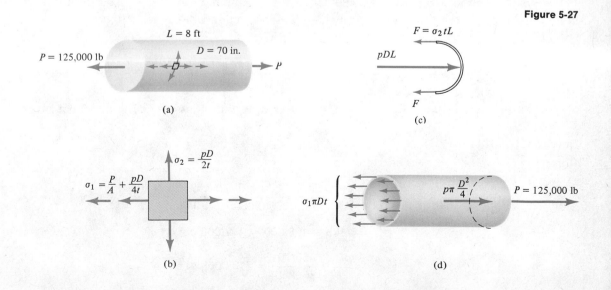

Solution:

The stresses on an element along the longitudinal and transverse planes are shown in Fig. 5-27(b). No shear acts on these planes. The longitudinal stress σ_1 is the sum of the stress due to internal pressure and that due to the thrust. Refer to the free body shown in Fig. 5-27(d). If A represents the area of the wall in a transverse section, and p is the internal pressure, the stress σ_1 in pounds per square inch is

$$\sigma_1 = +\frac{P}{A} + \frac{pD}{4t} = \frac{125,000}{\pi 70(0.080)} + \frac{p(70)}{4(0.080)}$$

$$= 7105 + 218.8p$$

The circumferential stress σ_2 is due to internal pressure alone, since the thrust does not cause any stress in this direction. Refer to Fig. 5-27(c). This stress in pounds per square inch is

$$\sigma_2 = \frac{pD}{2t} = \frac{p(70)}{2(0.080)} = 437.5p$$

It is not obvious which is the larger stress. So each will be equated to the allowable stress to determine the allowable pressure. For σ_1,

$$218.8p + 7105 = 210,000$$

$$p = 927 \text{ psi}$$

For σ_2,

$$437.5p = 210,000$$

$$p = 480 \text{ psi}$$

The *smaller* of the two results, or 480 psi, is the maximum allowable pressure. If it is exceeded, the circumferential tensile stress on any longitudinal section will exceed 210,000 psi.

EXAMPLE 5-14

Find the maximum shear in the rocket case of Example 5-13 if the pressure is 480 psi.

Solution:

Referring to Fig. 5-27(b), we find that:

$$\sigma_1 = \frac{P}{A} + \frac{pD}{4t} = \frac{125,000}{\pi 70(0.080)} + \frac{480(70)}{4(0.080)}$$

$$= 7105 + 105,000 = 112,000 \text{ psi}$$

$$\sigma_2 = \frac{pD}{2t} = \frac{480(70)}{2(0.080)} = 210,000 \text{ psi}$$

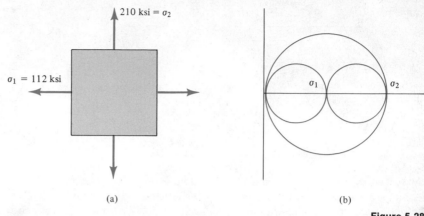

(a)

(b)

Figure 5-28

The element and its stresses are shown in Fig. 5-28(a).

The stress in the third direction is 0 at the outer face, 480 psi at the inner face. The latter value is negligible compared to the computed stresses.

Referring to Eq. 1-22, with $\sigma_{max} = 210$ ksi, $\sigma_{min} = 0$

$$\tau_{max} = \tfrac{1}{2}(\sigma_{max} - \sigma_{min}) = \tfrac{1}{2}(210 - 0) = 105 \text{ ksi}$$

Figure 5-28(b) shows the appropriate Mohr's circles.

PROBLEMS

5-57 Derive a relationship between the maximum tensile stress in the wall and the internal gage pressure for a thin-walled spherical pressure vessel.

5-58 A thin-walled circular tube carries a fluid stream at a gage pressure of 3×10^6 N/m^2. Find the maximum tensile stress and the maximum shear stress developed in the wall of the tube. The inside diameter of the tube is 200 mm and the wall thickness is 4 mm. Express your answer in both International and American units.

5-59 The diameter of a spherical pressure vessel is 1.5 m, and its wall is 2.5 mm thick. What internal pressure is possible in the vessel if the maximum shear stress may not exceed 83 MPa?

5-60 A cylindrical tank 1 m in diameter and 3 m in length is closed at both ends. It is composed of structural-steel plate 8 mm thick. Determine the principal strains on the outer surface of the boiler when the internal pressure is 1.40 MPa.

5-61 The closed tank in the illustration has a spiral seam as shown that is welded. If the tank is 36 in. in diameter, how much shearing and normal force will be carried per linear inch of weld when the tank is pressurized to 200 psi?

Problem 5-61

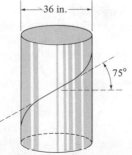

5-62 Two 10-ft-diameter hemispheres are joined to form a spherical pressure vessel. What is the total force transmitted per inch of length of joint when the internal pressure is 300 psi?

Problem 5-63

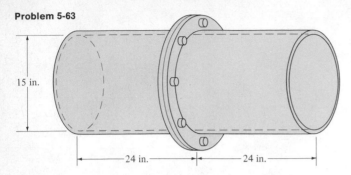

15 in.

24 in. 24 in.

5-63 Two cylindrical containers 15 in. in diameter and 24 in. long are joined by bolting their flanges together as indicated in the illustration. How many 1-in.-diameter structural-steel bolts are required to hold the cylinders together when the assembly is subjected to an internal pressure of 200 psi? The tensile stress in the bolts is not to exceed 8000 psi.

5-64 Find the minimum wall thickness required for the assembly of Prob. 5-63 if it is constructed of structural steel with a factor of safety of 2.50.

5-65 Determine the minimum diameter of structural-steel bolts required for the cylinder of Prob. 5-63 if the tensile stress in the eight bolts is not to exceed 10,000 psi for an internal pressure of 200 psi.

5-66 Determine the percent increase in volume of a thin-walled cylindrical pressure tank expressed as a function of its dimensions, the internal pressure, and the mechanical properties for the material of which it is made. Assume elastic action and neglect products of very small quantities.

5-67 A closed-ended cylindrical pressure vessel of aluminum, for which $E = 90 \times 10^3$ MPa and $\mu = \frac{1}{3}$, has an inside diameter of 1.5 m and a thickness of 6 mm. While the vessel is unpressurized, two electrical strain gages are bonded on the curved longitudinal surface at *right* angles to each other. The tank is then pressurized until the strain gages read 750 and 450 μin./in., respectively. Assume that the aluminum behaves elastically. (a) What are the normal stresses in the tank wall in the directions of the gages? (b) What are the longitudinal and circumferential stresses in the tank wall? (c) What is the pressure in the tank?

5-68 A thin hoop of a material for which $E = 270 \times 10^3$ MPa and $\alpha = 6.5 \times 10^{-6}$ m/m/°F has an inside diameter of 375 mm, a thickness of 6 mm, and a width of 25 mm. The hoop is to be sweated onto a rigid mandrel 375.25 mm in diameter. (a) What will be the hoop stress in the hoop? (b) What will be the bearing stress between the hoop and the mandrel?

5-69 A thin-walled spherical pressure vessel having a constant wall thickness t in. and an initial inside radius R in. is subjected to an internal gage pressure p psi. Calculate (a) the change in radius and (b) the change in volume of the sphere due to the pressure. Assume elastic behavior.

5-70 A cylindrical rocket motor case is made from a 0.035-in.-thick high-strength steel alloy sheet that has a uniaxial tensile yield strength of 200,000 psi and an ultimate tensile strength of 240,000 psi at the operating temperature. The case is 70 in. in diameter and 20 ft long. Eight of these motors are used in a cluster to propel a 1,000,000-lb-thrust rocket. Each motor case must transmit one-eighth of this total compressive thrust. (a) What is the allowable operating pressure in the case if it ceases to function properly when the maximum shear stress exceeds 63 percent of the uniaxial tensile yield strength? A factor of safety of 1.15 for both the pressure and thrust loads is required. Assume that failure does not occur at the welded seams. (b) In a hydrostatic laboratory test with no thrust, how much pressure would be required in the case to make the maximum shear stress reach 63 percent of the tensile yield?

5-71 A thin-walled cylindrical pressure vessel has an inside radius of 10 in. and a wall thickness of $\frac{1}{4}$ in. Initially it contains a gas at atmospheric temperature and pressure (60°F and 14.7 psia). The temperature of the gas is then increased to a temperature t. Express the approximate maximum tensile stress in the wall as a function of the temperature t. Assume a constant-volume heating process.

5-72 Two flanged half cylinders 24 in. long and 15 in. in diameter are joined by bolting the flanges together using six $\frac{1}{2}$-in.-diameter bolts on each side. Determine the increase in axial tensile stress in the bolts if the internal pressure in the assembly is increased 100 psi.

Problem 5-72

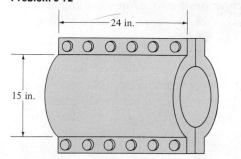

5-73 How many $\frac{1}{2}$-in.-diameter bolts would be required for the assembly of Prob. 5-72 if an increase in internal pressure of 200 psi is not to cause an increase in tensile stress in the bolts of more than 28,000 psi?

5-74 A conduit is to be formed of wooden slats held together by steel bands 1 in. wide and $\frac{1}{16}$ in. thick. The conduit has a diameter of 20 in. and carries water at a pressure of 50 psi. Find the required spacing of the bands if the tensile stress in them may not exceed 30,000 psi.

Problem 5-74

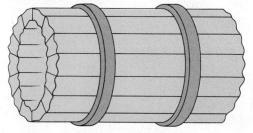

5-11 / Deformation of members in pure tension or compression

The preceding portions of this chapter have dealt with the stresses caused by simple centroidal loads. An equally important effect of any loading is the deformation it causes. The remaining portion of the chapter will deal with deformation in axially loaded members.

In Chapter 2 we defined axial strain for an infinitesimal line element as deformation per unit length. Thus,

$$\varepsilon = \frac{de}{dL} \tag{5-8}$$

where dL is an infinitesimal length in the direction of the strain ε, and de is the elongation or contraction of that length. Consider the axially loaded member of Fig. 5-29. From Eq. 5-8 the axial deformation of the infinitesimal element becomes

$$de = \varepsilon \, dL \tag{5-9}$$

Suppose that we desire to determine the deformation along the $\bar{y}$ axis of a single fiber having a length L and a cross-sectional area $dx\,dz$. This finite deformation can be obtained by summing the infinitesimal longitudinal deformation throughout the length L of the fiber. From Eq. 5-9 the summation becomes

$$e_y = \int de_y = \int_{over\,L} \varepsilon_y \, dy = \int_{y_1}^{y_2} \varepsilon_y \, dy \tag{5-10}$$

Figure 5-29

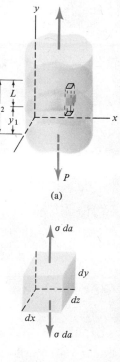

(a)

(b)

Equation 5-10 represents the deformation in the y direction for a single fiber. In order to obtain the deformation of a particular fiber, we must know how its strain ε_y varies over the length L; that is, we must express the strain ε_y as a function of y. Once this strain function has been obtained, the integration process of Eq. 5-10 can be completed to obtain the desired deformation. The same basic procedure of summation of infinitesimal deformations can also be employed in determining the deformation of fibers in the x and z directions.

Since our primary objective is to relate *load* to deformation, the strain is usually expressed in terms of the applied load. This can be done if the load-stress and stress-strain relationships are known. The only specific load-stress relationship we have derived so far is that for the case of pure tension or pure compression, namely, Eq. 5-2. For the special case in which the member has a *constant* cross section and the material is *linear* elastic,

$$\varepsilon = \frac{\sigma}{E} = \frac{P}{A}\frac{L}{E}$$

and Eq. 5-10 becomes

$$e = \frac{PL}{EA} \tag{5-11}$$

where L is the length of the loaded member, A is its cross-sectional area, and E is the elastic modulus of the material.

Although this equation was derived for a special case, it is such a common case that it deserves your special attention. You will use it in the majority of the deformation problems you encounter.

A time-honored way of remembering the equation for such deformation is to rearrange it thus:

$$e = \frac{PL}{EA} \tag{5-12}$$

and observe that the equation for the deformation of an axially loaded member of constant cross section composed of a linear elastic material is given by the word "PLEA."

EXAMPLE 5-15

The control cable for the rudder of an airplane is made of 3-mm-diameter stranded-steel wire and has a total length of 10 m. Determine the elongation of the wire for a pull of 1600 N. The modulus of elasticity of the wire may be assumed to be 200×10^9 Pa.

Solution:

This is a simple centroidal load problem. The cable is uniform in cross section and is subject to the same pull throughout its length. Thus,

$$e = \frac{PL}{EA} = \frac{1600 \times 10}{200 \times 10^9 \times \pi/4 \times 9 \times 10^{-6}} = 0.0113 \text{ m} = 11.3 \text{ mm}$$

Check stress

$$\sigma = \frac{P}{A} = \frac{1600}{\dfrac{\pi}{4}(0.003)^2} = 226 \text{ MPa}$$

The stress is probably alright for elastic behavior of steel.

EXAMPLE 5-16

In Fig. 5-30(a) are given the dimensions of a homogeneous right circular cone resting on its base. The specific weight of the material is w N/m³. Assuming that the material acts elastically and that the height h is much greater than the diameter b, determine the longitudinal deformation of the axis of the cone due to its own weight.

Solution:

Figure 5-30(b) shows a free-body diagram for the part of the cone above a section at a distance y from the apex. Since the cone is homogeneous, the loading is centroidal. Let V_y denote the volume of

Figure 5-30

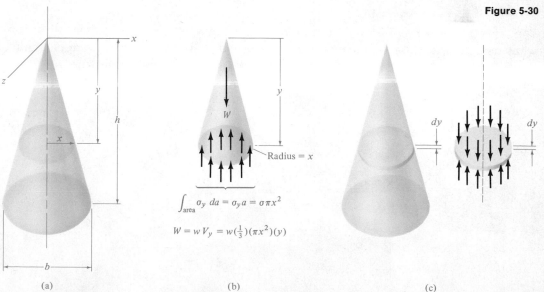

$$\int_{\text{area}} \sigma_y \, da = \sigma_y a = \sigma \pi x^2$$

$$W = w V_y = w(\tfrac{1}{3})(\pi x^2)(y)$$

(a) (b) (c)

the upper section of the cone, and σ_y denote the stress at y. Here, σ_y is assumed to be constant over the entire area of the section. Therefore from Fig. 5-30(b), we have

$$\sum \overset{\downarrow\ +}{F_y} = 0$$

$$wV_y - \int_{area} \sigma_y \, da = 0$$

$$wV_y - \sigma_y a = 0$$

Hence, the load-stress relationship is

$$\sigma_y = \frac{wV_y}{a} = \frac{w\left(\frac{1}{3}\right)\pi x^2 y}{\pi x^2} = \frac{wy}{3}$$

Since the material acts elastically and the loading is uniaxial, the load-strain relationship is

$$\varepsilon_y = \frac{\sigma_y}{E} = \frac{wy}{3E}$$

Then from Eq. 5-10 the load-deformation relationship becomes

$$e_y = \int_0^h \varepsilon_y \, dy = \int_0^h \left(\frac{wy}{3E}\right) dy = \frac{w}{3E}\int_0^h y \, dy = \frac{wh^2}{6E}$$

where the deformation is in units of length.

EXAMPLE 5-17

The dimensions in Fig. 5-31 are those before the loads were applied. In which direction and through what distance in inches will the plate on the right end of the pipe move upon application of the loads? Neglect localized stress and strain concentrations and the weight of the structure. The compressive stress-strain curves for the steel and concrete are shown in Fig. 5-32(a).

Figure 5-31

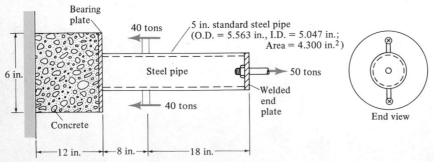

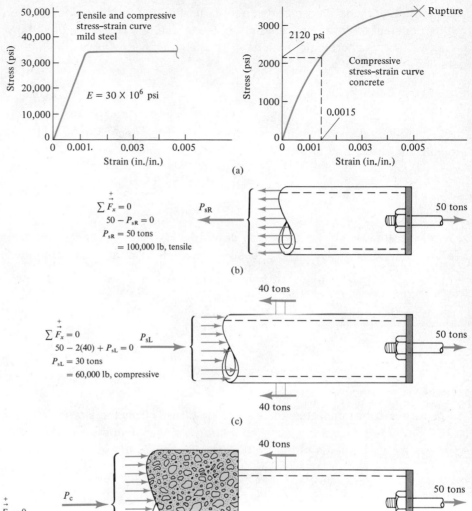

Figure 5-32

Solution:

Replacing y with x in Eq. 5-10 we have

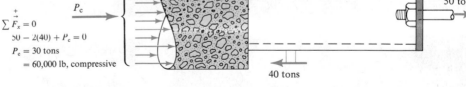

$$e_x = \int_{\substack{\text{over} \\ \text{length}}} \varepsilon_x \, dx$$

Since the strain in a member depends on the load, dimensions, and material, the strain ε_x will not be constant over the entire length of the structure. Therefore, it will be necessary to determine the strain in the concrete, in the left 8-in. portion of the steel pipe, and in the right 18-in. portion of the pipe separately. Thus,

$$e_{total} = \int_{concrete} \varepsilon_c \, dx_c + \int_{\substack{left \\ steel}} \varepsilon_{sL} \, dx_{sL} + \int_{\substack{right \\ steel}} \varepsilon_{sR} \, dx_{sR} \tag{a}$$

Each individual section being considered meets the necessary conditions of loading and geometry for uniform stress and strain. So equation (a) becomes

$$e_{total} = \varepsilon_c \int_0^{12} dx_c + \varepsilon_{sL} \int_0^8 dx_{sL} + \varepsilon_{sR} \int_0^{18} dx_{sR}$$

$$= \varepsilon_c(12) + \varepsilon_{sL}(8) + \varepsilon_{sR}(18) \tag{b}$$

By static analyses of the free-body diagrams in Fig. 5-32(b), (c), and (d), the loads P_c, P_{sL}, and P_{sR} are determined as shown. The stresses in the individual sections are now found as follows:

$$\sigma_c = \frac{P_c}{A_c} = \frac{60,000}{(\pi/4)(6^2)} = 2,120 \text{ psi (compression)}$$

$$\sigma_{sL} = \frac{P_{sL}}{A_{sL}} = \frac{60,000}{4,300} = 13,950 \text{ psi (compression)}$$

$$\sigma_{sR} = \frac{P_{sR}}{A_{sR}} = \frac{100,000}{4,300} = 23,250 \text{ psi (tension)}$$

The strain in the concrete can be obtained directly from the stress-strain curve. Since the stresses in the steel are below the proportional limit, the strains in the steel can be obtained by dividing the stresses by the modulus of elasticity, 30×10^6 psi. The strains are as follows:

$$\varepsilon_c = 0.0015 \text{ in./in. (compression)}$$

$$\varepsilon_{sL} = 0.000465 \text{ in./in. (compression)}$$

$$\varepsilon_{sR} = 0.000775 \text{ in./in. (tension)}$$

Substituting these values in Eq. (b), we obtain

$$e_{total} = -(0.0015)(12) - (0.000465)(8) + (0.000775)(18)$$

$$= -0.0180 - 0.00372 + 0.01392$$

$$= -0.00780 \text{ in.}$$

The end plate moves to the left, since the net deformation is compressive.

PROBLEMS

5-75 A 1-in.-diameter stranded-steel cable is used in a mine hoist. Determine the total elongation in a length of 800 ft when the cable is subjected to a total tensile force of $2\frac{1}{2}$ tons. The modulus of elasticity may be assumed to be 30 million psi.

5-76 A hollow tube 10 m long is subjected to an axial tensile load of 400,000 N. Its outer diameter is 100 mm; its inner diameter is 80 mm and it is composed of a material with a modulus of elasticity of 200×10^3 MPa. Find the elongation caused by the load.

5-77 A cable with a cross-sectional area of 0.2 in.^{2} weighs 3.0 lb/ft. A length of 100 ft is suspended by one end. How much will it deform under its own weight? $E = 30 \times 10^6$ psi.

5-78 How many feet of unstretched cable in Prob. 5-77 would be necessary to reach from the top to the bottom of a 1200-ft shaft?

5-79 A $\frac{1}{2}$-in.-diameter steel rod stretches 0.0020 in. in a length of 8 in. Determine the load.

5-80 A load of 2900 lb applied to a $\frac{1}{2}$-in.-diameter aluminum test specimen caused an elongation of 0.0030 in. in a 2-in. gage length. Determine the modulus of elasticity of the material.

5-81 A $\frac{3}{4}$-in.-diameter brass rod was subjected to a total axial load of 18,000 lb. Determine the elongation in an 8-in. gage length and the change in diameter of the rod.

5-82 A tensile member of a truss is a simple eye-bar, 2.5 m long, with one pin joint at each end. By mistake the bar is made 0.5 mm too short. The truss is left outside overnight, and the eye-bar is heated uniformly indoors. In the morning, the truss is at $+5°F$ and the eye-bar is at $+85°$. The eye-bar is tried again and is now too long. For how many minutes must the engineer wait before the bar will fit if the outside air remains at $+5°F$ and the bar cools down uniformly at a rate of 1°F every 5 min? The coefficient of thermal expansion is 8×10^{-6} m/m/°F.

5-83 The aluminum and steel rods in the illustration are fastened together. Each has an area of 4 in.^{2}. Also, $E_s = 30 \times 10^6$ psi and $E_a = 10 \times 10^6$ psi. The

tensile stress in the steel is 10,000 psi after the loads are applied. Find the total elongation of the composite rod.

Problem 5-83

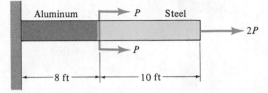

5-84 For the material of the centroidally loaded member in the illustration, $E = 10 \times 10^6$ and $\mu = \frac{2}{5}$. (a) Determine the total change in each dimension of the member. (b) What is the percent change in volume? Assume elastic behavior.

Problem 5-84

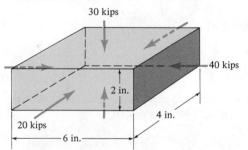

5-85 A homogeneous isotropic bar is made of a material whose approximate stress-strain law is $\sigma = k\varepsilon^{1/2}$, where $k = 10^9$. The original dimensions of the bar are $25 \times 50 \times 500$ mm. (a) If the bar is subjected to a centroidal longitudinal tensile load of 100,000 N, what is the change in length? (b) What is the change in volume, if $\mu = \frac{1}{4}$?

5-86 The member in the illustration is composed of a section of steel and a section of aluminum. If P is 40,000 N, what is the overall elongation due to P and neglecting the weight?

5-87 The member in the illustration is composed of a section of steel and a section of brass. The overall elongation is not to exceed 1.25 mm. The elongation of the brass should be four times that of the steel. Assuming that both materials behave elastically, determine the length of each section.

Problem 5-86   **Problem 5-87**   **Problem 5-88**   **Problem 5-89**

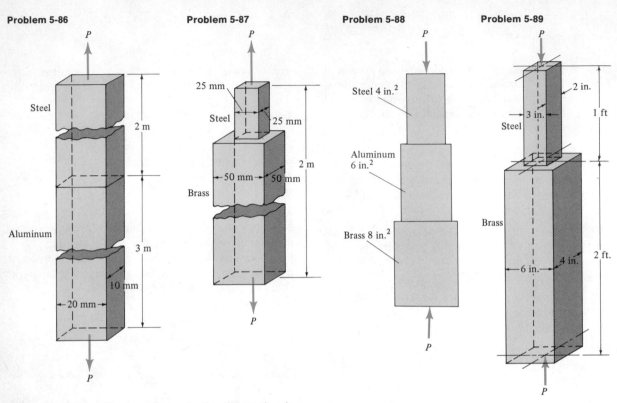

5-88 The composite bar shown in the illustration is made up of a section of steel ($E_s = 30 \times 10^6$ psi), a section of aluminum ($E_a = 10 \times 10^6$ psi), and a section of brass ($E_b = 15 \times 10^6$ psi). The axial deformation of the brass section is half that of the aluminum section, and the axial deformation of the aluminum section is half that of the steel section. The overall length is 4 ft. (a) Determine the length of each section. (b) If the total deformation is 0.020 in., what is the magnitude of P? Assume elastic behavior and neglect stress concentrations.

5-89 When the load P is applied to the composite member in the illustration, the overall contraction of the member is 0.003 in. Determine: (a) the maximum shear stress in the steel; (b) the longitudinal strain in the brass.

5-90 A steel tube 6 ft long having an internal diameter of $\frac{1}{2}$ in. and an external diameter of $\frac{3}{4}$ in. is welded to the end of a steel tube having an internal diameter of $\frac{3}{4}$ in. and an external diameter of 1 in. The length of the second tube is 4 ft. Determine the total elongation if an axial load of 2400 lb is applied to the assembly.

5-91 A 6-mm-diameter steel rod 3 m long is attached to the end of a brass tube with an internal diameter of 6 mm and wall thickness of 1.5 mm. What load will be required to stretch the assembly 0.25 mm if the brass tube is 2 m long?

5-92 Derive an expression for the total elastic deformation of a homogeneous bar of constant cross-sectional area A, length L, and specific weight w due to its own weight when suspended vertically from one end. Compare this result with that obtained if one assumes that the total weight of the bar is concentrated at its center of gravity.

5-93 Determine the overall elongation that the member in the illustration will undergo due to its own weight. Assume elastic behavior and a modulus of elasticity of 10 million psi. The specific weight of the material is $\frac{1}{8}$ lb/in.^{3}.

5-94 The isolator shown in the illustration is composed of an outer steel ring, a disk of hard rubber, and an inner steel core. If the isolator is supported at its outer ring and loaded longitudinally at its inner

Problem 5-93

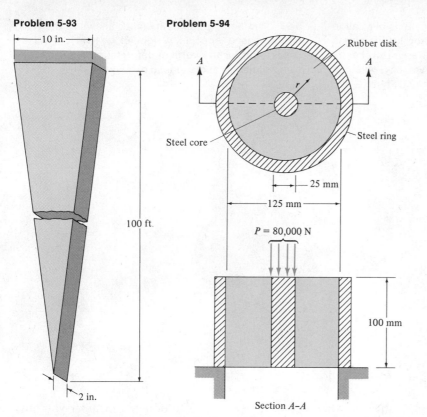

Problem 5-94

Rubber disk

A

A

r

Steel core

Steel ring

25 mm

125 mm

$P = 80,000$ N

100 mm

Section A–A

core, express the shear stress in the rubber as a function of the distance r from the center of the core.

5-95 For the isolator in the illustration for Prob. 5-94 determine the vertical displacement of the core (a) if the rubber has a linear τ-γ relationship with $G = 140$ MPa; (b) if the rubber has a τ-γ relationship given by $\tau = K\gamma^{1/2}$, where $K = 7 \times 10^6$.

5-96 The mounting device shown in the illustration is composed of two parallel steel plates and a block of hard rubber. A horizontal force of 20,000 lb is applied to the upper plate. Determine the horizontal displacement of the upper plate relative to the lower

Problem 5-96

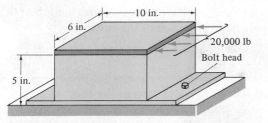

one: (a) if the rubber has a linear τ-γ relationship with a shear modulus of 2×10^4 psi; (b) if the rubber has a τ-γ relationship given by $\tau = K\gamma^{1/2}$, where $K = 10^3$.

5-97 The prismatic rod shown in the illustration, which has a length L, a constant cross-sectional area A, and a uniform density ρ, is rotating at a constant angular velocity ω rad/sec. An element of the rod with volume $A\,dl$ undergoes an inward radial acceleration

Problem 5-97

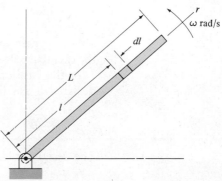

of $I\omega^2$, where I is the distance from the axis of rotation to the element. Find (a) an expression for the stress at any position along the rod and (b) the overall change of length of the rod due to its angular velocity. Assume elastic behavior and a modulus of elasticity of E psi.

5-98 The homogeneous conical member shown in the illustration has a specific weight of w lb/in.^{3}. Find an expression for the vertical deformation caused by its own weight. Assume elastic behavior.

Problem 5-98

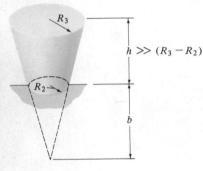

5-99 The homogeneous member shown in the illustration has a specific weight of w N/m^3. Find an expression for the vertical deformation caused by its own weight. Assume elastic behavior.

Problem 5-99

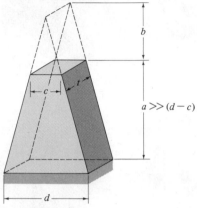

5-12 / Design for several criteria

Up until now when you have been asked to design a member you have been asked to design it for one criterion only, that is, to ensure that it is sufficiently strong. It is often of equal importance to ensure that the member is sufficiently stiff to carry the required load without excessive deformation.

Since the allowable stress in a member is governed by the material of which the member is to be composed and the allowable deformation in a member is determined by the use to which the member is to be put, it would be an unusual coincidence if both criteria were to be satisfied exactly by the same design. Consequently, it is best to make two completely independent designs: one to assure adequate strength (based on allowable stress) and one to assure adequate stiffness (based on allowable deformation). The results of these designs should be compared and the member selected that satisfies both criteria.

In the case of members carrying axial centroidal loads, the strength design can usually be accomplished by the use of the equation:

$$\sigma = \frac{P}{A} \tag{5-2}$$

The design for elastic stiffness will usually come from the equation:

$$e = \frac{PL}{EA} \tag{5-12}$$

Before you use either equation be sure it applies to your particular problem.

If the required load, the preferred material, and the allowable stress and deformation are known it is possible to solve each equation for A. The two values for A are compared and the greater is chosen.

For problems involving checking a member of known dimensions to determine the permissible load, the procedure is reversed. Equation 5-2 is solved for P using the specified stress and the known dimensions. Equation 5-12 is solved for P using the specified deformation and the known dimensions. The two values of P are compared and the lesser is chosen.

In this connection it is appropriate to comment on the notion of allowable stress and allowable deformation. These quantities are usually given to the engineer by means of codes or specifications and have a safety factor already built in. If you are working with allowables, you usually do not apply an additional safety factor to the loads as this will result in overdesign.

You should also be warned that you have merely seen the tip of the iceberg. In actual engineering practice, you will find that you must frequently design for many more criteria than two. Additional criteria that you may encounter are different kinds of stress, that is, shear and tension, or shear and bearing. Less theoretical criteria include such things as size, appearance, ease of fabrication, and economy. A good designer must juggle many such considerations and contrive to satisfy them all.

EXAMPLE 5-18

A tension member 8 ft long must carry an axial load of 40,000 lb as part of a large roof truss. It will be built of structural steel. The building code specifies an allowable tensile stress of 20,000 psi and an allowable deformation of 0.005 in. per foot of length. What is the minimum area that the member may have?

Solution:

We note that there are two independent criteria, stress and deformation; consequenctly, we shall make two separate designs.

For strength we substitute the allowable stress into Eq. 5-2.

$$\sigma = \frac{P}{A}$$

$$20,000 = \frac{40,000}{A}$$

$$A = 2 \text{ in.}^2$$

For deformation we note that the allowable deformation for an 8-ft member is 0.04 in. and substitute it in Eq. 5-12. For structural steel $E = 30 \times 10^6$ psi.

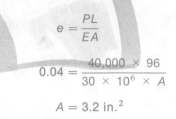

$$e = \frac{PL}{EA}$$

$$0.04 = \frac{40,000 \times 96}{30 \times 10^6 \times A}$$

$$A = 3.2 \text{ in.}^2$$

In order to satisfy both criteria, an area of 3.2 in.² must be provided since the area of 2 in.² would result in a deformation greater than the allowable amount.

PROBLEMS

5-100   An aluminum tension member 4 m long must carry an axial load of 40,000 N. The specifications are: allowable stress, 60 MPa; allowable deformation, 1 mm/m of length; $E = 7 \times 10^{10}$ Pa. Find the minimum area required.

5-101   A steel tension member must carry a 35-kip load. The allowable stress is 20,000 psi, the modulus of elasticity is 27×10^6 psi, and the allowable deformation is 0.008 in./ft. Find the minimum required area. Is it possible to determine the length?

5-102   A cable has a cross-sectional area of 600 mm². The allowable stress is 100 MPa, the allowable deformation is 2 mm/m, and $E = 60,000$ MPa. Determine the allowable load and the actual stress and deformation.

5-103   Determine the maximum allowable load on the $1\frac{5}{8}$ by $1\frac{5}{8}$-in. timber member if the shear stress parallel to the grain is not to exceed 100 psi and if the maximum

Problem 5-103

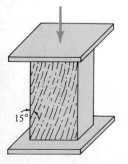

normal stress parallel to the grain is not to exceed 1000 psi.

5-104   The steel member in the illustration is centroidally loaded with a load of 500,000 N. If the allowable stresses are $\sigma_t = 10 \times 10^7$ N/m² and $\tau = 4 \times 10^7$ N/m², what is the minimum value the dimension c may have?

Problem 5-104

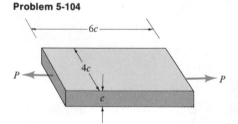

5-105   The member in the illustration is centroidally loaded and has the following specifications: the allowable stresses are $\sigma_t = 10,000$ psi, $\tau = 4,000$ psi; the elongation may not exceed 0.02 in. Assuming elastic behavior and a modulus of elasticity of 30×10^6 psi, determine the maximum allowable value of P.

5-106   The structure in the illustration is made of a material that has an ultimate tensile stress of 60,000 psi, an ultimate shear stress of 25,000 psi, a tensile yield strength (0.2 percent) of 35,000 psi, and a modulus of elasticity of 30×10^6 psi. If the overall elongation of member AB is not to exceed 0.12 in. and a factor of safety of 3 is required, determine the minimum required cross-sectional area of member AB.

Problem 5-105   **Problem 5-106**

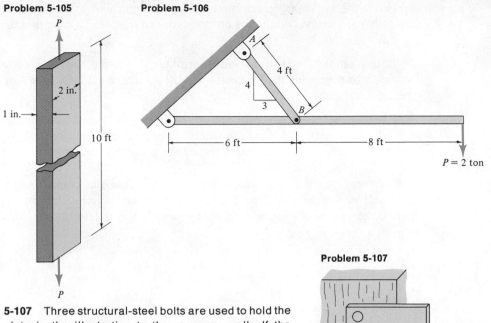

5-107 Three structural-steel bolts are used to hold the plate in the illustration to the masonry wall. If the allowable stress in shear is 12,000 psi and the allowable stress in bearing is 18,000 psi, determine the appropriate diameter for the bolts. The plate is $\frac{3}{8}$ in. thick.

Problem 5-107

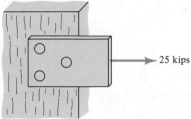

5-13 / Statically indeterminate axially loaded members

In all of the previous problems in this chapter, it was possible to determine the loads and stresses in the members by utilizing equations of equilibrium; that is, the members were statically determinate. There are, however, numerous structures for which the equilibrium equations *alone* are not sufficient to determine the loads. Such structures are *statically indeterminate*. In order to solve for the loads and stresses in members of such structures, it becomes necessary to supplement the equilibrium equations with additional relationships based on any conditions of restraint that may exist. These conditions of restraint usually govern the geometry of the deformation of the member.

Throughout the remainder of this text we shall encounter various types of statically indeterminate members and structures such as beams, shafts, trusses, and some axially loaded members. Each statically indeterminate problem has its own peculiarities as to its method of solution, but there are some general ideas that are common to the solution of most such types of problems.

1 Write the appropriate equations of equilibrium and examine them carefully to ascertain whether or not the problem is statically determinate or indeterminate.

2 If the problem is statically indeterminate, examine the kinematic restraints to determine the necessary conditions that must be satisfied by the deformation(s) of the member(s).

3 Express the required deformations in terms of the loads. When enough of these additional relationships have been obtained, they can be adjoined to the equilibrium equations and the problem can then be solved.

Not every problem will follow this set procedure but usually these steps will lead to the solution. The following examples give some indication of the variety of situations that may be statically indeterminate.

EXAMPLE 5-19

A very stiff bar of negligible weight is suspended horizontally by two vertical rods, as in Fig. 5-33(a). One of the rods is of steel and is $\frac{1}{2}$ in. in diameter and 4 ft long; the other is of brass and is $\frac{7}{8}$ in. in diameter and 8 ft long. If a vertical load of 6000 lb is applied to the bar, where must it be placed in order that the bar will remain horizontal? Assume that the rods behave elastically and that the bending of the bar is insignificant. Also, take E_{st} as 30×10^6 psi and E_b as 14×10^6 psi.

Solution:

Two independent equations of static equilibrium may be written for the free-body diagram of Fig. 5-33(b). Two possible equations are

Figure 5-33

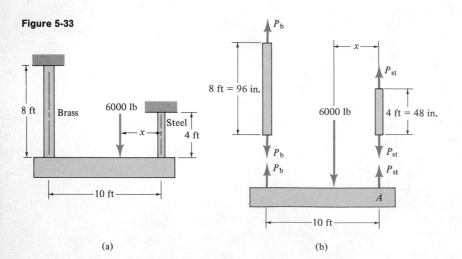

(a)   (b)

$$\overset{\uparrow +}{\sum} F_y = 0 \qquad\qquad \overset{\curvearrowright +}{\sum} M_A = 0$$

$$P_{st} + P_b - 6000 = 0 \qquad P_b(10) - 6000x = 0$$

Since no more *independent* equations of equilibrium can be written and there are three unknown quantities, the structure is statically indeterminate. One additional independent equation is needed. The problem requires that the bar remain horizontal. Therefore, the rods must undergo equal elongations, and we have

$$e_b = e_{st}$$

$$\int_0^{96} \varepsilon_b \, dL_b = \int_0^{48} \varepsilon_{st} \, dL_{st}$$

If the strains are uniform in each rod and the rods act elastically, the last equation becomes

$$\varepsilon_b(96) = \varepsilon_{st}(48)$$

$$\frac{\sigma_b}{E_b}(96) = \frac{\sigma_{st}}{E_{st}}(48)$$

$$\frac{P_b}{A_b}\frac{1}{E_b}(96) - \frac{P_{st}}{A_{st}}\frac{1}{E_{st}}(48) \tag{a}$$

Equation (a) is an *independent* equation relating the load in the brass to the load in the steel. When this equation is solved simultaneously with the equilibrium equations, the results are as follows:

$$P_b = 2510 \text{ lb} \qquad P_{st} = 3490 \text{ lb} \qquad x = 4.175 \text{ ft}$$

EXAMPLE 5-20

The aluminum rod in Fig. 5-34(a) is firmly welded to the ceiling and to the 150-lb block. The rod has no load at the temperature at which it is installed, and the entire weight of the block is carried by the floor. (a) If the temperature of the rod is decreased 30°F at a uniform rate, what force will be exerted on the floor by the block? (b) How much must the temperature decrease to cause the rod to pick up the 150-lb block so that it will be 0.10 in. above the floor? Neglect the weight of the rod.

Solution:

(a) Figure 5-34(b) shows the one available equilibrium equation when the temperature is decreased. Since this equation has two unknowns, P_f and P_r, an additional relationship is necessary. A deformation relationship can best be visualized and derived by assuming that the rod is *detached* at the top and that free thermal contraction occurs, as indicated in Fig. 5-34(c). The shortening e_t in the rod would then be as follows (see the Appendix for the coefficient of thermal expansion):

$$e_t = \alpha(L)(\Delta t) = 13 \times 10^{-6}(20)(12)(-30) = -0.0936 \text{ in.}$$

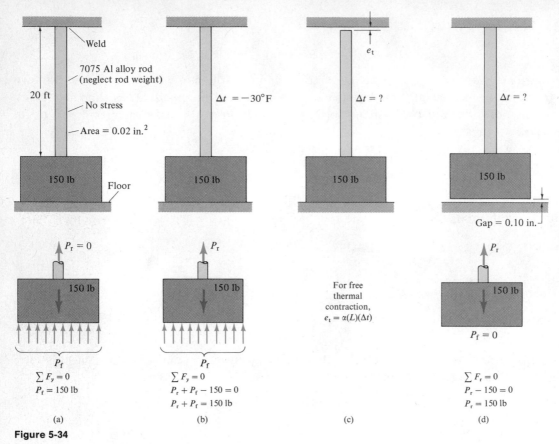

Figure 5-34

We can now determine the load P_r in the rod that would be necessary to pull the top of the rod up 0.0936 in. to where it is actually fastened. Thus,

$$e_t = \varepsilon L = \left(\frac{\sigma}{E}\right)L = \frac{P_r L}{AE} = \frac{P_r(240)}{(0.02)10^7} = 0.0936 \text{ in.}$$

$$P_r = 77.8 \text{ lb}$$

The magnitude of P_r is the amount by which the original load on the floor is reduced. Substituting this load back in the equilibrium equation in Fig. 5-34(b), we get

$$P_f = 150 - 77.8 = 72.2 \text{ lb}$$

(b) If the rod undergoes sufficient thermal contraction to raise the weight, the rod will then support the entire 150 lb, as shown in Fig. 5-34(d). Hence, the rod must first contract an amount e_w to remove the weight from the floor, and then contract an additional 0.10 in. The total contraction becomes

$$e_t = -e_w + (-0.10)$$

$$e = \quad \alpha(L)(\Delta t) = -\frac{P_r L}{AE} + (-0.10)$$

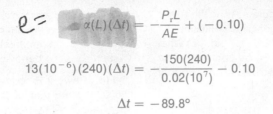

$$13(10^{-6})(240)(\Delta t) = -\frac{150(240)}{0.02(10^7)} - 0.10$$

$$\Delta t = -89.8°$$

It is very important that you be able to physically visualize all the deformation relationships. You may find it helpful to draw "before and after" pictures.

PROBLEMS

5-108 Three pieces of the same size steel cable are used to support a 40,000 N vertical load as shown. Assuming the cables act elastically, determine the tension in each cable.

Problem 5-108

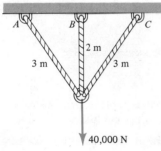

5-109 A relatively rigid beam is pivoted at *a* and is supported by two similar cables attached at *b* and *c*. Assuming the cables are elastic and the beam is rigid, determine the tension in each cable due to the application of the 10,000 lb load.

Problem 5-109

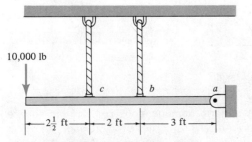

5-110 The square prismatic bar with fixed ends, shown in the illustration, carries an axial load *P* of

20,000 N. (a) What are the reactions at the ends? (b) How far will plane *m-n* be moved downward by the load? Assume that $E = 210 \times 10^9$ Pa.

Problem 5-110

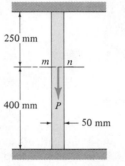

5-111 The bars *A* and *C* in the illustration are made of steel ($E = 30 \times 10^6$), and bar *B* is made of brass ($E = 15 \times 10^6$). Each bar has a cross-sectional area of 4 in.^{2}. What load *P* will cause the stress in the brass to be equal to the stress in the steel?

Problem 5-111

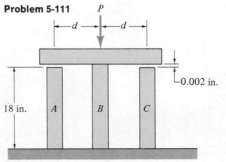

5-112 The load *P* placed on the short reinforced concrete pier in the illustration causes the stress in the

steel to reach 24,000 psi. (a) Calculate P and the stress in the concrete at this load if the concrete conforms to a linear stress-strain law with $E = 2 \times 10^6$ psi. (b) What is the value of P and the stress in the concrete if the concrete has the compressive stress-strain curve of Fig. 5-32(a)?

Problem 5-112

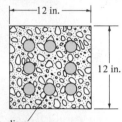

$\frac{7}{8}$ in. diam. steel bars

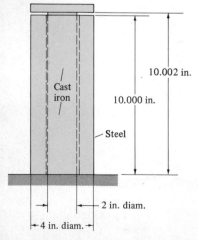

5-113   The structure shown in the illustration is composed of a circular shell of steel and a circular core of cast iron. Initially, the shell is 10 in. long and the core is 10.002 in. long. Also, $E_s = 30 \times 10^6$, $E_{c.i.} = 15 \times 10^6$, and $\mu_s = \mu_{c.i.} = \frac{1}{4}$. If a compressive load of 50,000 lb is

Problem 5-113

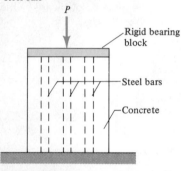

applied to the structure by means of a flat plate, determine (a) the maximum compressive stress in the steel and (b) the maximum shearing stress in the cast iron. Assume elastic behavior.

5-114   Rework Prob. 5-113 for a compressive load of 5000 lb rather than 50,000 lb.

5-115   Refer to Prob. 5-110. If the member is made of mild steel with a yield-point stress of 210 MPa and $P = 920$ kN, determine the reaction at each end.

5-116   As indicated in the illustration, two pieces of 12 mm-diameter high-strength steel cable attached to rigid end plates can compress the mild steel ($\sigma_{yp} = 21 \times 10^7$ N/m^2) bar by means of turnbuckles. The original length of the bar is 1.5 m and its area is 1300 mm^2. If the turnbuckles are tightened 2.0 mm, what will be the contraction of the bar? What will be the stress in the cables?

Problem 5-116

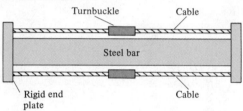

5-117   Same as Prob. 5-116 except the turnbuckles are tightened 12 mm. Assume the cables act elastically.

5-118   The two members in the illustration are fastened together and to the walls. If the members are stressfree before they are loaded, what will be the stress in each after the two 50-kip loads are applied? Neglect stress concentrations at the ends, and assume elastic behavior.

Problem 5-118

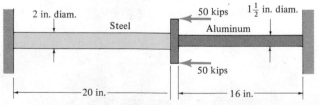

5-119   A uniform homogeneous rod, 25 mm square and 5 m long, is firmly anchored at each end in a vertical position while unstressed. An axial load of 250,000 N is then applied downward at a position 3 m from the bottom of the rod. Determine the reaction at each end, if the material of the rod has a $\sigma - E$ curve approximated by that shown in the figure for Prob. 10-49. You may need a trial-and-error solution.

5-120 A 7075 T6 aluminum alloy bar having a cross-sectional area of 3 in.² is placed between two rigid immovable supports while 29.98 in. long and at a temperature of −20°F. The supports are 30.00 in. apart. The bar is then heated uniformly. (a) At what temperature will the gap first close? (b) Determine the stress in the bar when the temperature reaches 90°F.

The coefficient of thermal expansion is 12×10^{-6} in./in./°F.

5-121 What will the pointer in the illustration read on the scale after the temperature of both the steel bar and the brass bar has been increased by 100°F? The coefficients of thermal expansion are $\alpha_s = 6.5 \times 10^{-6}$ and $\alpha_b = 10.5 \times 10^{-6}$. The top of pointer is attached to the plate allowing the pointer to move either right or left.

Problem 5-121

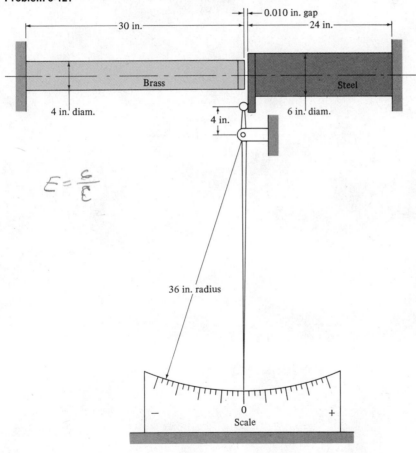

Chapter 6
Torsion

Upon completion of this chapter you will be able to:

1 Determine the torsional load on a circular shaft when given the strain distribution over a cross section of the loaded shaft and a stress-strain relation for the material of which it is composed.
2 Find the stresses or angle of twist resulting from a given torque acting upon a circular shaft.
3 Find the torque causing a given stress or twist of a circular shaft.
4 Design a circular shaft for given criteria of allowable stress and angle of twist.
5 Calculate stresses caused by combinations of axial loads and torques on circular shafts.
6 Determine loads and stresses in statically indeterminate bodies loaded in torsion.
7 Calculate stresses in thin-walled torsion members, both circular and noncircular.

6-2 / Introduction

In Chapter 5 we considered members subjected to axial loads. First, the basic load-stress relationship $P = \int_{\text{area}} \sigma \, da$ was established by using the method of sections. We found that before this relationship could be evaluated, it was necessary to determine the stress distribution over the area in question by investigating the strain distribution and relating the stress to the strain. Then the procedure for deriving load-deformation relationships for axially loaded members was illustrated. This chapter will present a similar treatment of members subjected to torsion by loads that tend to twist the members about their longitudinal centroidal axes.

Most of this chapter deals with members in the form of concentric circular cylinders, solid and hollow, subjected to torques about their longitudinal geometric axes. Although this may seem like a somewhat special case, a quick reflection would show that many torque-carrying engineering members are cylindrical in shape. Examples are drive shafts, bolts, and screwdrivers. A brief discussion of noncylindrical members is also presented. The last section on hollow, thin-walled torsion members introduces the concept of shear flow, which will be utilized in subsequent chapters.

6-3 / Kinematic response of circular cylindrical members

Before we attempt to derive any load-stress or load-deformation relationships for torsion members, we shall investigate the kinematic response of a circular cylindrical member subjected to a pure torque about its longitudinal axis. The most logical way to make this investigation would be to go into the laboratory and twist a cylindrical member. However, since it is impractical for everyone to do this, we shall attempt to describe and simulate the experimental conditions so that you may "observe" the results. When we refer to the surface of a cylinder or to a cylindrical surface in subsequent explanations, we will mean the long curved longitudinal surface, and not a transverse plane surface.

Let us first circumscribe two circumferential lines in parallel planes and a longitudinal line on the surface of a cylindrical bar made of a homogeneous material. The planes are at right angles to the longitudinal axis of the cylinder and are some distance L apart. A grid pattern of small rectangles whose sides are parallel and perpendicular to the longitudinal axis is then scribed on the surface between these circumferential lines. Such a pattern of lines is shown in Fig. 6-1(a), where it is assumed that the dimensions of the rectangles are very small in comparison with the overall dimensions of the bar. This assumption will minimize the effect of the

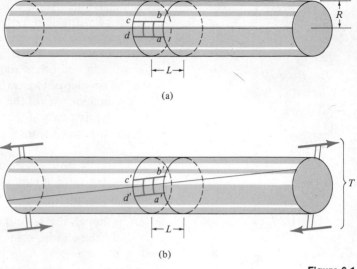

(a)

(b)

Figure 6-1

curved surface on our observations. Let us now apply a torque T to the bar somewhat beyond the scribed lines (to eliminate localized effects due to load within the length L), and observe its effects on the grid pattern. The outlines of the rectangles on the twisted bar are shown in Fig. 6-1(b), where the amount of twist is greatly exaggerated.

The following important observations can be noted from this experiment:

1 The distance L between the outside circumferential lines does not change significantly as a result of the application of the torque. However, the rectangles become parallelograms whose sides have the same length as those of the original rectangles.

2 The circumferential lines do *not* become zigzag; that is, they remain in parallel planes.

3 The original straight parallel longitudinal lines, such as *ad* and *bc*, remain parallel to each other but do *not* remain parallel to the longitudinal axis of the member. These lines become helices (a helix is the path of a point that moves longitudinally and circumferentially along the surface of a cylinder at a uniform rate).

Although hindsight is easier than foresight, these results might have been anticipated from the geometry and loading conditions. For example, since the member was not subjected to any axial loading, we might expect its length to remain unchanged. Also, if we draw a free-body diagram of any portion of the member L units long, we see that it is loaded in exactly the same manner as any other portion of the member. Hence, we would expect the kinematic response of the member to be uniform with respect to the length.

If the experimental procedure were repeated on cylindrical bars of different radii and other homogeneous materials, it would be found that while the *amount* of twist would perhaps vary from test to test, the geometric characteristics of the deformations would be the same in all cases.

From the first two of the geometric observations, it is logical to draw the following conclusions:

1   An element on the surface of a cylindrical member subjected to pure torque is in a state of *pure shear*. There are no significant axial deformations of an element oriented parallel to the longitudinal axis.

2   Even though the second observation is only a *surface* observation of bars of different radii, it may be concluded that the bar does not warp; that is, *plane cross sections* perpendicular to the longitudinal axis of the bar *remain plane* even after the bar is twisted.

Let us now investigate more thoroughly the third observation. In order to obtain some geometric relationships, we redraw the grid pattern on a "flattened out" surface, such as that shown in Fig. 6-2. The circumferential lines $a'b'$ and $d'c'$ are parallel and straight, and the "flattened" helices, such as $d'a'$ and $c'b'$, become parallel straight lines inclined at an angle ϕ with the longitudinal axis of the bar. The shear deformation occurring over the length L is denoted by e_s, and the shear deformation occurring over the length ΔL is Δe_s. By similar triangles, we have

$$\tan \phi = \frac{e_s}{L} = \frac{\Delta e_s}{\Delta L}$$

If we now limit ϕ to very small values, we can say that

$$\tan \phi = \frac{\Delta e_s}{\Delta L} \approx \sin \phi \approx \phi$$

Then, if we think of making the rectangles smaller and smaller, we obtain by definition

Figure 6-2

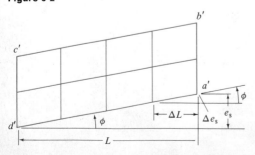

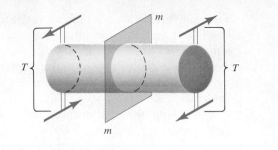

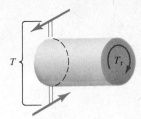

(a)   (b)

Figure 6-3

$$\lim_{\Delta L \to 0} \frac{\Delta e_s}{\Delta L} = \frac{de_s}{dL} = \gamma$$

From this result we arrive at a third conclusion, which follows:

3   In a cylindrical member subjected to pure torque the helix angle ϕ can be interpreted as the *maximum* shear strain at the surface of the member. This value is the same at all surface points within the length L.

These three conclusions will be utilized quite extensively in the succeeding sections. However, it is imperative that you realize that if the geometry of the member or the manner of loading were significantly altered, any one or all of the previously discussed observations and conclusions may no longer be true. Once again, we see that the characteristics of the strain in a member are governed primarily by its geometry and the manner in which the load is applied.

6-4 / Basic load-stress relationship for cylindrical torsion member

The preceding deformation analysis has led us to certain conclusions concerning the response of a circular cylindrical member subjected to a pure torque about its longitudinal geometric axis. Now, with the aid of these ideas, we shall write the basic load-stress relation for a cylindrical torsion member.

The cylindrical member in Fig. 6-3(a) is in static equilibrium. Let us pass a cutting plane mm perpendicular to the longitudinal axis of the member, and then separate the two portions. When the entire member is in equilibrium, each portion must also be in equilibrium. Figure 6-3(b) shows a free-body diagram of the left-hand portion, where T is the resultant external torque on this portion and T_r is the resultant *internal* resisting torque that is exerted *on* this portion *by* the right-hand portion over the area of the member cut by plane mm.

How is this resultant internal resisting torque produced? We can answer the question by recalling the first two conclusions in section 6-3. From these we know that an element oriented parallel to the longitudinal axis is in *pure shear* and also that any plane section perpendicular to the longitudinal axis remains plane. We combine these two ideas and use the cylindrical coordinates x, ρ, and θ to show the forces on a three-dimensional infinitesimal element, as in Fig. 6-4. This element is in a state of pure shear, and we can show that $\tau_{x\theta}$ is equal to $\tau_{\theta x}$ in the following manner. Taking moments of the forces on the element about the *o-o* line, we obtain

$$\sum M = 0$$

$$(\tau_{\theta x}\, da)\rho\, d\theta - (\tau_{x\theta}\, da)\, dx = 0$$

$$(\tau_{\theta x}\, dx\, d\rho)\rho\, d\theta - (\tau_{x\theta}\rho\, d\theta\, d\rho)\, dx = 0$$

$$\tau_{\theta x} = \tau_{x\theta}$$

In the succeeding analyses, we shall refer to this shear stress simply as τ. You should realize that this shear stress exists simultaneously on

Figure 6-4

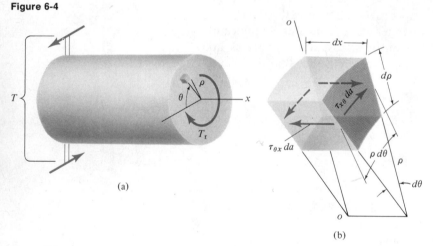

(a)

(b)

Figure 6-5

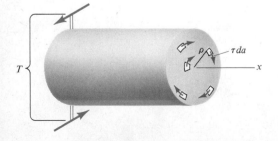

areas in longitudinal and transverse planes; that is, it acts parallel and perpendicular to the longitudinal axis of the cylinder.

Now we can say that T_r is the resultant torque produced by an infinite number of infinitesimal *shear* forces acting on the plane area perpendicular to the axis of the cylinder. Such shear forces are shown in Fig. 6-5. When moments are taken about the x axis, we get

$$\sum M_x = 0$$

$$T - \int_{\text{area}} (\tau \, da)\rho = 0 \qquad (6\text{-}1)$$

$$T = \int_{\text{area}} \tau \rho \, da$$

This is the basic load-shear stress relationship for any cylindrical torsion member. The integral is the total internal resisting torque acting on the cut cross-sectional area.

Of course, as was the case in Chapter 5 for axially loaded members, before this basic load-stress relationship can be of practical value, we must be able to perform the necessary integration. So it will be necessary to determine the *shear-stress distribution* over the cross-sectional area in question. In the next section we shall first determine the *strain distribution*, and shall then relate the stress to the strain to get the desired stress distribution.

EXAMPLE 6-1

What torsional load will cause a stress of 70,000 psi at the outer surface of an aluminum alloy shaft with a diameter of $\frac{1}{2}$ in. if the stress distribution is assumed to be as shown in Fig. 6-6(a)?

Solution:

The torsional load, or torque, is

$$T = \int_{\text{area}} \tau \rho \, da$$

The given stress distribution must be expressed in mathematical form in terms of ρ. Since the stress-distribution "curve" has a slope change at $\rho = 0.0625$ in., we must derive two expressions for τ. The conditions are shown in Fig. 6-6(b), and we must write an expression for each interval, as follows:

For $0 \leq \rho_1 \leq 0.0625$, by similar triangles,

$$\frac{\tau_{\rho_1}}{\rho_1} = \frac{40,000}{0.0625}$$

$$\tau_{\rho_1} = 640,000\rho_1$$

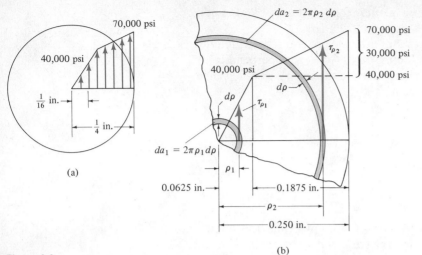

Figure 6-6

For $0.0625 \le \rho_2 \le 0.250$, by similar triangles,

$$\frac{\tau_{\rho_2} - 40,000}{\rho_2 - 0.0625} = \frac{30,000}{0.1875}$$

$$\tau_{\rho_2} = 160,000(\rho_2 - 0.0625) + 40,000$$

Therefore,

$$T = \int_0^{0.0625} (640,000\rho_1)(\rho_1)(2\pi\rho_1\,d\rho_1)$$

$$+ \int_{0.0625}^{0.250} [160,000(\rho_2 - 0.0625) + 40,000]\rho_2(2\pi\rho_2\,d\rho_2)$$

$$= \left[640,000(2\pi)\,\frac{\rho_1^4}{4}\bigg|_0^{0.0625} \right] + \left[160,000(2\pi)\left(\frac{\rho_2^4}{4} - 0.0625\,\frac{\rho_2^3}{3}\right) \right]$$

$$+ 40,000(2\pi)\,\frac{\rho_2^3}{3}\bigg|_{0.0625}^{0.250}$$

$$= 15.3 + 655.7 + 1288.5 = 1959.5 \text{ in.-lb}$$

PROBLEMS

6-1 to 6-7 For each of the shafts, the cross sections of which are shown in the illustrations, the shear stress distribution has been determined to be as indicated.

Find an expression in each case for the torque in terms of the radius r, $\tau_{\max}$, and other constants.

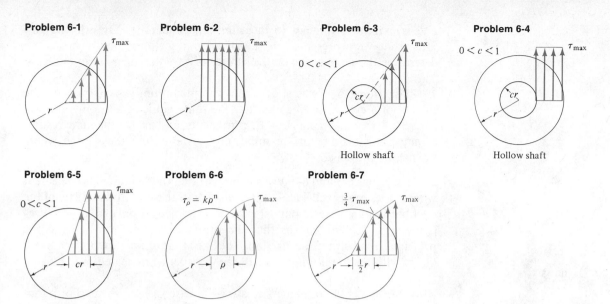

Problem 6-1

Problem 6-2

Problem 6-3

$0 < c < 1$

Hollow shaft

Problem 6-4

$0 < c < 1$

Hollow shaft

Problem 6-5

$0 < c < 1$

Problem 6-6

$\tau_\rho = k\rho^n$

Problem 6-7

$\frac{3}{4}\tau_{max}$

6-8 to 6-10 If solid circular shafts of radius r were made from materials whose shear stress-strain curves are shown in the illustrations, find the pure torque in each case that will cause a maximum shear strain γ_r in the shaft. Assume in each case that the shear strain varies linearly from zero at the center to a maximum γ_r at the outer surface, according to the relation $\gamma_\rho = k\rho$. Express answers in terms of the values τ_r, r, and n; and also in terms of γ_r, r, G, and n.

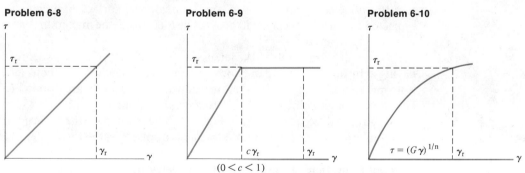

Problem 6-8

Problem 6-9

$(0 < c < 1)$

Problem 6-10

$\tau = (G\gamma)^{1/n}$

6-5 / Shear-strain distribution in cylindrical torsion member

Having established the basic load-stress relationship for a cylindrical torsion member, we shall now determine how the shear strain varies throughout the member. Once again we shall simulate a laboratory experiment in order to observe and interpret the results therefrom.

As was pointed out in Chapter 2, shear strain is an angular distortion

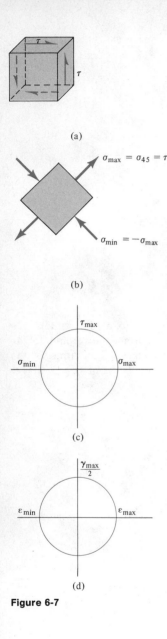

(a)

$\sigma_{max} = \sigma_{45} = \tau$

$\sigma_{min} = -\sigma_{max}$

(b)

τ_{max}

σ_{min} σ_{max}

(c)

$\dfrac{\gamma_{max}}{2}$

ε_{min} ε_{max}

(d)

Figure 6-7

and is extremely difficult to measure experimentally. By utilizing two ideas that were presented in the preceding sections, however, we can determine the shear strain on the surface of a torsion member in this way:

1 In section 6-3 it was shown that the maximum shear strain on the surface of a cylindrical member has the same value at all points on that surface. So we may make our measurements, and expect the same results, at *any* point on the surface sufficiently distant from the section where the torque is applied.

2 In sections 6-3 and 6-4 it was demonstrated that a rectangular element whose sides are parallel and perpendicular to the longitudinal axis of the member is in a state of pure shear and is subjected only to shear forces (see Fig. 6-4). This means that the principal normal stresses and the principal *axial* strains occur along the axes that are oriented at an angle of 45° with the longitudinal axis of the member as indicated in Fig. 6-7.*

Also, Mohr's circle of plane stress for the case of *pure* shear shows that the maximum shear stress, the maximum tensile stress, and the maximum compressive stress all have the same magnitude. Also, the values of the axial *strains* along the principal axes will be the coordinates of the extremities of Mohr's circle of plane *strain*, and the *diameter* of this circle will represent the magnitude of the *maximum* shear strain. Thus, if we measure the *principal axial* strains at any point on the surface of a cylinder subjected to pure torque, we will in effect be measuring the maximum *shear* strain on that surface.

For our experimental work, let us use a hollow cylindrical member having arbitrary inside and outside radii and made of a homogeneous and isotropic material. Two electrical resistance strain gages[†] are bonded to both the outer and inner surfaces of the member along lines inclined 45° to the longitudinal axis, as shown in Fig. 6-8. The cylinder is then placed in a torsion machine whereby a pure torque is applied. If the resulting axial strains measured by the gages are converted to shear strains of the cylinder, we would find that for any particular value of torque the ratio of shear strain to radius is a constant; that is,

$$\frac{\gamma_o}{r_o} = \frac{\gamma_i}{r_i} = \text{const.}$$

This relationship would be true, not just for elastic strains, but for any strain ordinarily encountered in such an engineering member.

* The third principal axis on a three-dimensional element is directed perpendicular to the free surface of the cylindrical torsion member, and the value of the third principal stress is, of course, zero.

† See Chapter 15 for description and theory of electrical resistance strain gages.

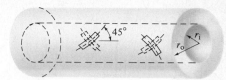

Figure 6-8

If the experimental procedure were repeated on several cylinders of different homogeneous and isotropic materials and with various inside and outside radii, similar results would be obtained. From these experiments it can be concluded that the shear *strain* in a cylindrical torsion member is directly proportional to the radius. In mathematical form this becomes

$$\gamma \propto \rho$$

$$\gamma_\rho = k\rho \tag{6-2}$$

where γ_ρ is the shear strain at any arbitrary distance ρ from the geometric center of the member, and k is the ratio of γ_o to r_o for any particular applied torque.

At first glance, the experimental indication of a linear shear strain distribution might appear to be a rather fortuitous and unexpected result. However, with adequate hindsight, one can make rather convincing arguments based on geometrical symmetry as to why this result should occur. Such an argument will be demonstrated in the next chapter on bending, but our "experiment" will suffice for this discussion of torsion.

Now that we have established both the basic load-shear stress relationship, given by Eq. 6-1, and an expression for the shear-strain distribution, given by Eq. 6-2, we can derive specific relationships between shear stress and torque for cylindrical torsion members of different sizes and materials. In studying the following cases, your attention should be concentrated on the *derivation procedure* as well as the results obtained.

Elastic behavior   The most common condition encountered in engineering is elastic behavior, for which it is assumed that a linear stress-strain relationship exists. In the case of shear stress and shear strain, this linear relationship is of the form

$$\tau = \gamma G$$

where G has been defined as the shear modulus of elasticity. To the engineer this relationship means that the shear stress-strain curve is a straight line with a slope G.

Since we have already shown that the shear strain varies linearly with the radius, it follows that for the special case of a linear stress-strain relationship, the shear stress must also vary linearly with the radius. The distributions for shear strain and *elastic* shear stress are shown in Fig.6-9(a) and (b), respectively.

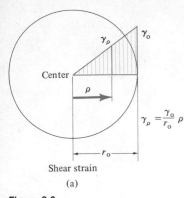

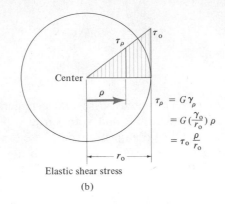

Shear strain (a)   Elastic shear stress (b)

Figure 6-9

Applying the relationships in Fig. 6-9 to the basic load-stress relationship given by Eq. 6-1, we obtain

$$T = \int_{\text{area}} \tau_\rho \rho \, da$$

$$= \int_{\text{area}} \left(\frac{\tau_o}{r_o} \rho \right) \rho \, da$$

$$= \frac{\tau_o}{r_o} \int_{\text{area}} \rho^2 \, da$$

$$T = \frac{\tau_o}{r_o} J \quad \text{or} \quad \tau_o = \frac{T r_o}{J} \tag{6-3}$$

Also,

$$\tau_\rho = \tau_o \frac{\rho}{r_o} = \frac{T\rho}{J} \tag{6-3a}$$

where τ_o is the maximum shear stress at the outer surface of the cylinder; r_o is the outside radius; J is the centroidal polar moment of inertia* of the cross section of the cylinder (either hollow or solid); and τ_ρ is the shear stress at any distance ρ from the center. Equation 6-3 is popularly known as the *elastic torsion formula*. Although this is a very useful relationship, it is often used *incorrectly* in cases for which it is not valid. Be careful that you do not make this very common but serious mistake. It would be well for you to review the assumptions and limitations made in the derivation.

Ideally plastic behavior   The elastic torsion formula Eq. 6-3 is applicable for all materials as long as the shear stress–shear strain relationship is linear. In the case of a brittle material this linearity usually exists almost up to rupture of the material. However, for a ductile material the

* For geometrical properties of various areas refer to Table A-2 of the Appendix.

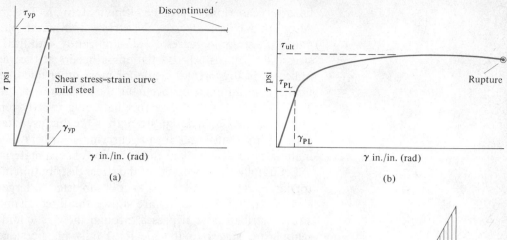

(a)

(b)

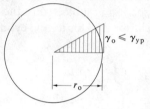

Strain

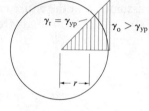

Strain

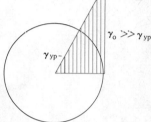

Strain

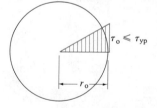

Stress
elastic

(c)

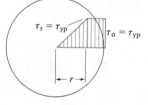

Stress
plastic–elastic

(d)

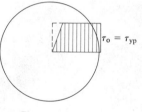

Stress
idealized fully
plastic

(e)

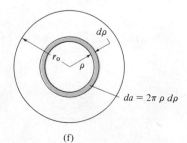

(f)

Figure 6-10

linearity ceases long before the ultimate strength of the material is reached. Also, for ductile materials, the stress-strain curve often becomes quite flat as the strain increases quite significantly with little or no increase in stress. For mild steel, this flattening occurs at the yield-point stress as indicated in Fig. 6-10(a), while for most other ductile materials the flattening is more likely to occur at the ultimate strength as indicated in Fig. 6-10(b). In either case this flattening usually represents an upper bound for the usable design strength of the ductile material. Accordingly, since many shafts and torque-carrying members are made of ductile materials, we shall look at the behavior of such materials.

Let us follow the progress of the stress distribution in a *solid* cylindrical torsion member of mild steel as it is subjected to larger and larger shear strains. In Fig. 6-10(c)–(e) are shown the three distinctive stages of the stress distribution as it passes through the elastic and elastic-plastic, or semiplastic, stages to the idealized fully plastic stage, for which it is assumed that the entire cross section is subjected to the yield-point shear stress. The basic load-stress relationship for a *solid* cylindrical member in the idealized *fully plastic* condition becomes

$$T_{fp} = \int_{area} \tau \rho \, da$$

Since $\tau = \tau_{yp}$ at all points and the expression for da is as shown in Fig. 6-10(f),

$$T_{fp} = \tau_{yp} \int_0^{r_o} \rho(2\pi\rho \, d\rho)$$

$$T_{fp} = \tau_{yp} \frac{2\pi}{3} r_o^3$$

This relationship is often given in the equivalent form

$$T_{fp} = \frac{4}{3} \frac{\tau_{yp}}{r_o} J \tag{6-4}$$

A comparison of Eqs. 6-3 and 6-4 shows that for a *solid* cylindrical member of mild steel, the torque required for the fully plastic condition is four-thirds of the maximum torque permitted for elastic behavior; that is, the torque at which τ_o first reaches τ_{yp} is only three-fourths of the torque required for the entire cross section to reach τ_{yp}.

EXAMPLE 6-2

Derive an expression for the maximum torque that can be carried by a hollow circular shaft of outer radius R and inner radius r and made of an elastic material with an elastic limit shear stress of τ_{EL}.

Solution:

Since the shaft is to behave elastically the stress distribution will be that shown in Fig. 6-10(c). For maximum elastic torque the stress at the outer edge will be τ_{EL}. Placing this information in Eq. 6-3, we get

$$\tau_{EL} = \frac{TR}{J}$$

$$T = \frac{\tau_{EL}J}{R}$$

Now we must evaluate J. Using the definition of J,

$$J = \int \rho^2 \, dA = \int_r^R \rho^2(2\pi\rho \, d\rho) = \frac{2\pi\rho^4}{4} \Bigg]_r^R$$

$$J = \frac{\pi}{2}R^4 - \frac{\pi}{2}r^4 = \frac{\pi}{2}(R^4 - r^4) = J_{\text{solid}} - J_{\text{hole}}$$

Putting this result in our previous equation for T, we find

$$T = \frac{\tau_{EL}\pi(R^4 - r^4)}{2R}$$

PROBLEMS

6-11   A solid circular cylindrical shaft is 400-mm long and has a 40-mm diameter. What is its value of J (polar second moment) if (a) it behaves elastically, and (b) it behaves ideally fully plastic?

6-12   A hollow cylindrical shaft is 100 in. long and has an inside diameter of 5 in. and an outside diameter of 6 in. What is its value of J (polar second moment) if (a) it behaves elastically, and (b) it behaves ideally fully plastic?

6-13   A solid circular cylindrical shaft has a radius r and length L. If a longitudinal hole of radius cr $(0 < c < 1)$ is drilled so that the shaft becomes hollow, express the percent decrease in J as a function of c. If the elastic shear stress is inversely proportional to J, what is the percent change in the shear stress due to the hole?

6-14   Under certain conditions the polar moment of inertia J of a hollow thin-walled circular cylinder may

be approximated as $J = 2\pi r^4 h$, where h is the ratio of the thickness to the outside radius. (a) If $h = 0.10$, what percent difference is introduced in the polar moment of inertia by using the approximate formula? (b) If the elastic shear stress is inversely proportional to J, what is the percent difference in the elastic stress when the approximate value of J is used instead of the exact value?

6-15   A solid circular cylindrical shaft of length L and radius r is subjected to a pure torque T. If its radius were doubled, for a fixed allowable maximum shear stress, what would be the corresponding percent increase in load-carrying capacity (torque) if (a) the shaft material behaves elastically and (b) the shaft material behaves ideally fully plastic?

6-16   (a) Prove that the elastic strength (torque-carrying capacity) of a solid shaft is reduced by only $\frac{1}{16}$ if an axial hole having a diameter equal to one-half the outside diameter is drilled through the shaft.

(b) What percent of the cost is saved if it is assumed that the cost varies linearly with the weight? Assume elastic action only.

6-17 How much is the elastic strength (torque-carrying capacity) of a solid steel shaft reduced by boring an axial hole through the center, if the area of the hole is two-thirds of the original shaft area?

6-18 Same as Prob. 6-17 except assume fully plastic behavior rather than elastic.

6-19 Find the fully plastic torque in inch-pounds for a hollow shaft with outside and inside diameters of 6 in. and 4 in., respectively. The material of the shaft is ideally elastic-plastic (flat-top shear stress-strain curve) and has a shear ultimate stress of 20,000 psi.

6-20 A solid circular shaft made of mild steel is twisted by pure torque until 60 percent of the cross-sectional area yields. What is the required torque, in terms of the yield-point of the material and the radius of the shaft?

6-21 The inner ring of the torsilastic spring is stationary while the outer ring is twisted by the loads P as shown. What is the shear stress developed at the interface between the inner ring and the rubber core? At the interface between the outer ring and the core?

Problem 6-21

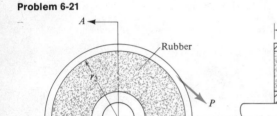

Section AA

6-6 / Shear stresses in circular shafts

While the general load-stress relationship of Eq. 6-1 is applicable to cylindrical torsion members made of any material, it is important to realize that the engineer is seldom called upon to solve such general cases. The vast majority of torsion problems that you will be asked to solve will deal with circular cylindrical shafts composed of materials with a linear elastic behavior when stressed below the proportional limit. For such situations the shearing stress is given by Eq. 6-3, which we rewrite in the form

$$\tau_{\max} = \frac{Tr}{J} \tag{6-5}$$

where T is the applied torque, r is the outside radius of the shaft, and J is the polar moment of inertia of the cross section. In using this "torsion formula" great care must be exercised to ensure that proper units are used for each of the quantities involved.

Another fairly common situation likely to be encountered by the engineer is that in which the shaft is made of a rather ductile material having a reasonably flat stress-strain curve at the ultimate shear stress level τ_{ult}. For such situations, the relationship between the applied torque and the maximum shear stress before rupture is given by a relationship similar to Eq. 6-4, although that particular result is applicable only to solid circular shafts.

The following examples illustrate the use of these ideas.

EXAMPLE 6-3

A cylindrical steel shaft 50 mm in diameter is subjected to a torque that causes a maximum stress of 60 MPa. Find the torque.

Solution:

The body in question is cylindrical and is composed of homogeneous isotropic material stressed below its elastic limit, so Eq. 6-5 is applicable.

The polar moment of inertia for a circle is

$$J = \int_0^r \rho^2 \, dA = \int_0^r \rho^2 (2\pi\rho \, d\rho) = \frac{2\pi\rho^4}{4} \bigg]_0^r = \frac{\pi r^4}{2}$$

The maximum stress occurs at the outer edge of the shaft so that

$$\tau_{max} = \frac{Tr}{J}$$

$$60 \times 10^6 = \frac{T(0.025)}{\pi/2(0.025)^4} = \frac{T}{2.454 \times 10^{-5}}$$

$$T = 1473 \text{ N} \cdot \text{m}$$

While we have shown the integration for J in this example, it is unnecessary to repeat it each time. It is quite appropriate and very convenient to remember that for a solid circular cross section

$$J = \frac{\pi r^4}{2}$$

EXAMPLE 6-4

Suppose that a new type of plastic has the following properties: Each of its stress-strain relationships is linear to rupture, and the rupture stresses are 3000 psi in tension, 4000 psi in compression, and 5000 psi in shear. At what speed (rpm) would a solid shaft of this material rupture if it were $\frac{1}{4}$ in. in diameter and were delivering $\frac{1}{4}$ horsepower? Assume a pure torque loading.

Solution:

The pure torque loading will result in a pure shear state of stress at any point in the shaft. So the maximum shear, tensile, and compressive stresses are equal for any value of the torque (recall Mohr's stress circle for pure shear). Therefore the shaft will rupture in tension when all these stresses reach 3000 psi at the outer surface. Since the stress-strain relationship is linear to rupture, the elastic torsion formula is applicable and the torque required for rupture is

$$T = \frac{\tau J}{r} = \frac{(3000)(\pi/32)(\frac{1}{4})^4}{\frac{1}{8}} = 9.22 \text{ in.-lb} = 0.768 \text{ ft-lb}$$

We know that power equals torque times angular velocity. Thus,

$$\text{power} = \text{ft-lb/min} = (\text{ft-lb})(\text{rad/min})$$

$$\left(\frac{1}{4} \text{ hp}\right)\left(33{,}000 \frac{\text{ft-lb}}{\text{min}} \Big/ \text{hp}\right) = (0.768 \text{ ft-lb})(n \text{ rpm})(2\pi \text{ rad/rev})$$

The speed at rupture is

$$n = 1713 \text{ rpm}$$

A simpler conversion from torque to horsepower may be obtained by combining all the conversion factors so that

$$T \text{ (in.-lb)} = \frac{63{,}000 \text{ hp}}{n(\text{rpm})}$$

EXAMPLE 6-5

A solid $1\frac{1}{2}$-in.-diameter shaft is made of a ductile material whose shear stress-shear strain curve is similar to that of Fig. 6-10(b) with $\tau_{PL} = 12{,}000$ psi and $\tau_{ult} = 30{,}000$ psi. If the shaft is rotating at 500 rpm, what is the maximum horsepower the shaft can transmit if (a) failure is based on initiation of inelastic behavior, and (b) failure is based upon rupture of the shaft? Use a factor of safety of 1.5.

Solution:

(a) For elastic behavior, the elastic torsion formula (Eq. 6-5) is applicable. Thus, for a maximum elastic shear stress of 12,000 psi the failure torque would be

$$T_f = \frac{\tau_{PL}}{r} J = \frac{(12{,}000)(\pi/2)(1.5/2)^4}{(1.5/2)}$$

$$= 7952 \text{ in.-lb}$$

The allowable torque is

$$T_a = \frac{T_f}{FS} = \frac{7952}{1.5} = 5301 \text{ in.-lb}$$

$$\text{power} = \text{torque} \times \text{angular velocity}$$

$$= 5301 \text{ in.-lb} \times 500 \text{ rpm} \times 2\pi \text{ rad/rev}$$

$$= 1.665 \times 10^7 \text{ in.-lb/min}$$

$$1 \text{ horsepower} = 33,000 \text{ ft-lb/min} = 12(33,000) \text{ in.-lb/min}$$

Thus

$$\text{Max hp} = \frac{1.665 \times 10^7 \text{ in.-lb/min}}{(12)(33,000) \text{ in.-lb/min/hp}} = 42 \text{ hp}$$

(b) If failure is based upon rupture of the ductile material, then in all likelihood the fully plastic condition will exist at rupture. Thus, for the *solid* shaft, Eq. 6-4 is applicable if we use the ultimate shear stress of 30,000 in place of τ_{yp}. Then, the failure torque would be

$$T_f = \frac{4}{3} \frac{(30,000)}{(1.5/2)} \frac{\pi}{2} \left(\frac{1.5}{2}\right)^4$$

$$= 26,500 \text{ in.-lb}$$

The allowable torque is

$$T_a = \frac{T_F}{FS} = \frac{26,500}{1.5} = 17,666 \text{ in.-lb}$$

and the horsepower is

$$\text{Max hp} = \frac{(17,660)(500)(2\pi)}{(12)(33,000)} = 140 \text{ hp}$$

PROBLEMS

6-22 A shearing stress of 8000 psi is developed at the outside of a hollow steel shaft having an inside diameter of 4 in. and an outside diameter of 6 in. Determine the total torque transmitted by the cross section.

6-23 A 4-in.-diameter circular steel shaft is subjected to a torque that produces a maximum shearing stress of 10,000 psi. (a) Determine the torque transmitted by the entire cross section. (b) Determine the torque transmitted by a thin ring having a mean diameter of 16 in. and the same area as the original shaft if the maximum shear stress is 10,000 psi.

6-24 A 20-mm-diameter solid steel shaft is subjected to a pure torque of 100 N·m. What maximum shear stress is developed? What will be the maximum tensile stress?

6-25 Same as Prob. 6-24 except the torque is 500 N·m and the shaft is 20-mm diameter and is made of mild steel with $\tau_{yp} = 10^8$ pa.

6-26 A 1-in.-diameter maple dowel is used as a torsion member. What torque can it carry if the allowable stresses are: tension and compression 1600 psi and shear (parallel to the grain) 400 psi? The grain of the dowel is longitudinal.

6-27 A shaft 50 mm in diameter is made of a brittle material with an allowable shear stress of 100 MPa and an allowable tensile stress of 70 MPa. If the stress-strain relationship is linear, what is the maximum horsepower the shaft may safely transmit when it is rotating at 200 rpm?

6-28 A shaft 2 in. in diameter is made of a very ductile material with an ultimate shear stress of 20,000 psi and an ultimate tensile stress of 40,000 psi. What is the maximum horsepower the shaft may safely transmit when rotating at 200 rpm? Failure is probably due to excessive inelastic deformation.

6-29 A hollow shaft with an outside radius of 2 in. and an inside radius of 1 in. is rotating at 500 rpm. Using a factor of safety of 2, determine the maximum horsepower the shaft may safely transmit (a) if the shaft is 24 in. long, and is made of a linear brittle material for which $\sigma_{ult} = 40,000$ psi tension, $\sigma_{ult} = 80,000$ psi compression, and $G = 4 \times 10^6$ psi, and failure is due to rupture; (b) if the shaft is 24 in. long, and is made of a mild steel for which $\tau_{yp} = 15,000$ psi, $\sigma_{yp} = 30,000$ psi, and $G = 12 \times 10^6$ psi, and failure is due to excessive inelastic deformation.

6-30 An electrical resistance axial strain gage is attached to the surface of a $\frac{3}{4}$-in. steel shaft so that the gage axis makes an angle of 60° with the shaft axis. When the shaft is rotating at 1720 rpm, the strain gage reads 1200 μin. tensile (the electrical leads are taken away from the rotating shaft through slip rings).

Note 5-7

Problem 6-31

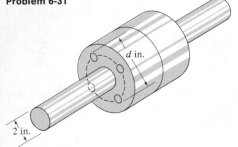

What horsepower is the shaft delivering? Assume only pure torque loading on the shaft and elastic behavior of the steel.

6-31 Two solid steel shafts 2 in. in diameter are to be joined by means of a flanged coupling with four $\frac{1}{4}$-in.-diameter steel bolts equally spaced as shown in the illustration. Determine the diameter d of the circle through the bolts required to make the coupling and the shafts equally strong in resisting torsion. Assume that the shafts are elastic and that the allowable shear stress for steel is 8000 psi.

6-32 Same as Prob. 6-31 except that the shaft and bolts are made of a ductile material whose ultimate shear stress is 34,000 psi.

6-33 Two steel shafts 250 mm in diameter are placed end to end and welded around the outside with a bead 10 mm thick. Find the maximum torque that the resultant shaft will transmit if the shearing stress in the weld is not to exceed 42 MPa, and determine the maximum stress in either shaft at a distance of 0.4 m from the weld.

6-34 Two 100-mm-diameter shafts are connected by a pair of flanges that are held together with 20-mm-diameter structural steel bolts. If the maximum shearing stress in the shaft is 80 MPa, determine the number of bolts required to transmit the corresponding torque. The bolts are set on a bolt circle having a diameter of 150 mm.

6-35 Two shafts of equal length are made of the same material except one is solid and the other is hollow. If the two shafts weigh exactly the same (i.e., same amount of metal in each), which one has the larger J? Which can carry the larger torque?

6-36 Two separate shafts are each carrying a torque of 100 ft-lb. One shaft is solid with a $\frac{3}{4}$-in. diameter, and the other is hollow with an outer diameter (OD) of $1\frac{1}{2}$ in. and an inner diameter (ID) of $1\frac{1}{4}$ in. Which shaft has the greater τ_{max} assuming the same material for each?

6-7 / Angular deformation or angle of twist

The angle of twist in a torsion member is defined as the relative rotation between two transverse plane cross sections. In Fig. 6-11(a) the left-hand end of the member is assumed to be fixed and the right-hand end is caused

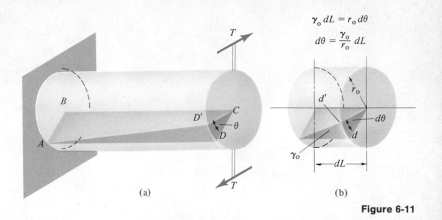

(a) (b)

Figure 6-11

to rotate by some torque T. The surface $ABCD$, which was originally a *longitudinal plane*, becomes the helical surface $ABCD'$. The important feature of this distortion is that the radial line CD remains a straight radial line in the position CD', even after twisting. This result is to be expected, since we have already shown that the shear strain varies linearly with the radius. The angle θ is the angle of twist between the left-hand and right-hand ends of the member.

In order to evaluate θ for any cylindrical torsion member, let us consider a portion of the member having an infinitesimal length dL, as shown in Fig. 6-11(b). For small values of shear strain γ_o, the arc length dd' can be expressed as

$$dd' = \gamma_o\, dL = r_o\, d\theta$$

Hence,

$$d\theta = \frac{\gamma_o}{r_o}\, dL$$

where $d\theta$ is the angle of twist occurring over the length dL. A finite angle of twist occurring over a finite length can be obtained by direct integration. Thus,

$$\theta = \int_{\text{over } L} \frac{\gamma_o}{r_o}\, dL \tag{6-6}$$

This relationship is geometrically valid for all values of γ_o ordinarily encountered in engineering torsion members.

To relate the angle of twist θ to the applied torque T we must first relate the shear strain γ_o to the corresponding shear stress τ_o. Then, by using the appropriate torque-stress relation we can obtain the desired relationship between the applied torque and the angle of twist. In a laboratory experiment in which we were to measure the torque and the angle of twist continuously, the relationship would generally take the form of the curve indicated in Fig. 6-12. The relationship would be linear as long as the material behaved elastically but would become nonlinear

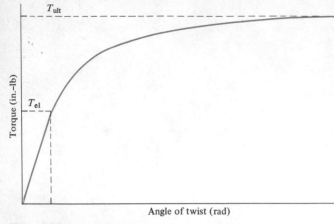

Figure 6-12

as the material began to behave plastically. For a very ductile material the ultimate torque can be calculated by assuming the fully plastic condition to exist at rupture.

When a member subject to torsion has a uniform cross section, is composed of a homogeneous isotropic material, and behaves elastically, we have an important special case. Under these circumstances, the strain at the outer surface is given by

$$\gamma_o = \frac{\tau_o}{G} = \frac{Tr_o}{JG}$$

Putting this expression for strain into Eq. 6-6 gives us

$$\theta = \int_L \frac{Tr_o \, dL}{GJr_o}$$

$$\theta = \frac{TL}{GJ} \tag{6-7}$$

This equation can be used to calculate the angle of twist for elastic shafts, which likely will constitute the majority of members actually encountered in engineering practice.

It is instructive to examine the two simplified equations for elastic deformation that we have obtained thus far for axially loaded and torsion members, respectively:

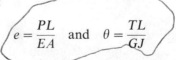

$$e = \frac{PL}{EA} \quad \text{and} \quad \theta = \frac{TL}{GJ}$$

In both cases, L is the length of the member. P and T both represent load, E and G are material properties and are determined by the material of which the member is made, while A and J reflect the dimensions and shape of the cross section. Noticing this strong similarity may help you to remember and use both equations correctly.

EXAMPLE 6-6

A uniform steel shaft 20 ft long and 4 in. in diameter is observed to twist 0.06 rad when transmitting power at 270 rpm. Determine the torque and the power it is transmitting. Assume elastic behavior.

Solution:

Since the shaft is uniform, isotropic, and elastic, we may use the simplified equation (Eq. 6-7) for angle of twist. For steel, $G = 12 \times 10^6$ psi, and J for a 4-in.-diameter circle is 8π in.[4]

$$\theta = \frac{TL}{GJ}$$

$$0.06 = \frac{T \times 240}{(12 \times 10^6)(8\pi)}$$

$$T = 24,000\pi = 75,398 \text{ in.-lb}$$

$$\text{power} = \text{torque} \times \text{angular velocity}$$

$$= 75,398 \times 270 \times 2\pi$$

$$= 128 \times 10^6 \text{ in.-lb/min}$$

$$1 \text{ hp} = 33,000 \text{ ft-lb/min} = 12(33,000) = 396,000 \text{ in.-lb/min}$$

Thus, the power transmitted by the shaft is

$$\text{hp} = \frac{128 \times 10^6}{396 \times 10^3} = 323 \text{ hp}$$

EXAMPLE 6-7

The member in Fig. 6-13 is subjected to the torques shown. Through how many degrees and in what direction will the left-hand end twist with respect to the fixed right-hand end? The stresses have already been checked, and elastic action is assured. Neglect any stress concentrations at the joints.

Solution:

The basic relationship between strain and twist is

$$\theta = \int_{\text{over } L} \frac{\gamma_0}{r_0} \, dL$$

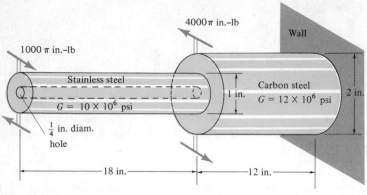

Figure 6-13

In this elastic case, $\gamma_o = \tau_o/G$ and $\tau_o = Tr_o/J$. So $\gamma_o = Tr_o/JG$ and

$$\theta = \int_{\text{over } L} \frac{T \, dL}{GJ}$$

In Fig. 6-13, T, G, and J are not constants or continuous functions of L from one end of the member to the other. In this case, we must find the twist of each part separately. Then

$$\theta = \int_0^{18} \frac{T_{ss} \, dL}{G_{ss} J_{ss}} + \int_0^{12} \frac{T_{cs} \, dL}{G_{cs} J_{cs}}$$

For the stainless-steel part, $T_{ss} = +1000\pi$ in.-lb and is constant for all sections of the 18-in. length. The values of G_{ss} and J_{ss} are also constant throughout that length. For the carbon-steel part, T_{cs} has a constant value of -3000π in.-lb (plus torque is assumed to act clockwise when you look in toward the wall). If you are not sure about T_{cs}, pass a plane through the carbon-steel part at right angles to its longitudinal axis and draw a free-body diagram. The values of G_{cs} and J_{cs} are also constant for the carbon-steel part. Therefore,

$$\theta = \frac{1000\pi}{(10 \times 10^6)(\pi/32)[1^4 - (\frac{1}{4})^4]} \int_0^{18} dL - \frac{3000\pi}{(12 \times 10^6)(\pi/32)(2^4)} \int_0^{12} dL$$

$$\theta = +0.0576 - 0.0060 = +0.0516 \text{ rad}$$

$$= \frac{180}{\pi}(+0.0516) = +2.95°$$

The plus sign means that the net twist of the left-hand end is clockwise when you look in toward the wall.

PROBLEMS

6-37 Determine the horsepower that a 10-in.-diameter turbine shaft 10 ft in length will transmit at 200 rpm if the maximum shearing stress is 15,000 psi. How much will the shaft twist if it is made of steel?

6-38 What must be the length of a steel rod 5 mm in diameter so that it can be twisted through one complete revolution without exceeding a shear stress of 70 MPa?

6-39 A solid circular steel shaft is 2.5 m long and has a radius of 80 mm. If a torque of 4.0×10^4 N·m is applied, what will be the angular twist of the shaft?

6-40 A steel shaft 50 mm in diameter is subjected to a pure torque that twists the shaft through an angle of 0.12 rad in a length of 4 m. Assuming elastic behavior, find the maximum tensile stress in the shaft.

6-41 A tubular steel shaft 2 m long has an outside diameter of 50 mm and an inside diameter of 40 mm. It is to be used as an elastic torsion spring. What is the ratio of applied torque to angle of twist?

6-42 A solid circular shaft having a radius of 50 mm and a length of 250 mm is twisted until its maximum shear strain is 0.001 m/m (rad). Determine the angle of twist and the required torque if the shaft is made of mild steel with $\tau_{yp} = 105$ MPa.

6-43 Solve Prob. 6-42 if the maximum shear strain is 0.02 rad rather than 0.001 rad. Find the torque in Newton-meters.

6-44 A solid circular shaft with a uniform diameter of 2 in. is 12 ft long. At its midpoint 65 hp is delivered to the shaft by means of a belt passing over a pulley. This power is used to drive two machines, one at the left end consuming 25 hp and one at the right end consuming the remaining 40 hp. Determine (a) the maximum shearing stress in the shaft and (b) the relative angle of twist between the two extreme ends of the shaft. The shaft turns at 400 rpm, and the material is steel for which $G = 12 \times 10^6$ psi.

6-45 How much will the right end of the shaft in the illustration twist relative to the fixed left end if $G_s = 12 \times 10^6$ psi and $G_b = 5 \times 10^6$ psi? Assume elastic behavior and neglect stress concentrations.

6-46 How much will the right end of the shaft in the illustration twist relative to the fixed left end if $G_s = 12 \times 10^6$ psi and $G_a = 4 \times 10^6$ psi? Assume elastic behavior and neglect stress concentrations.

6-47 Find the angle of twist of the left end of the member in the illustration with respect to the fixed right end if $G_s = 83,000$ MPa and $G_a = 26,000$ MPa.

Problem 6-45

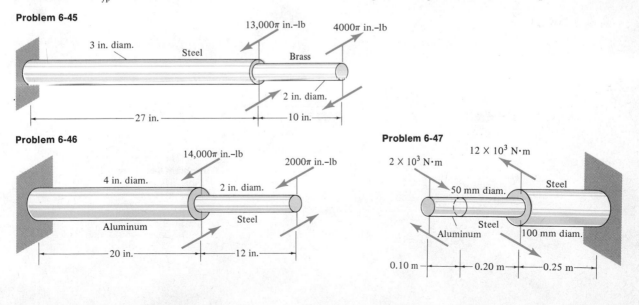

Problem 6-46

Problem 6-47

Assume elastic behavior and neglect stress concentrations.

6-48 Find the angle of twist of the left end of the member in the illustration with respect to the fixed right end if $G_s = 12 \times 10^6$ psi and $G_b = 6 \times 10^6$ psi. Assume elastic behavior and neglect stress concentrations.

6-49 For the shaft shown in the illustration, $G_b = 6 \times 10^6$ psi and $G_a = 4 \times 10^6$ psi. Determine (a) the maximum shear stress in the brass, (b) the maximum shear stress in the aluminum, and (c) the total angle of twist.

6-50 The circular stepped shaft in the illustration is composed of a section of steel ($G_s = 12 \times 10^6$ psi) and a section of brass ($G_b = 6 \times 10^6$ psi). The overall angular deformation from end to end is not to exceed 5°. The angle of twist of the brass section should be one-fourth that of the steel section. Assuming that the elastic limits are not exceeded, determine the maxi-

mum torque that this shaft can carry between its ends. Neglect stress concentrations at the junction.

6-51 The 4-in.-diameter shaft shown in the illustration consists of brass and steel sections rigidly connected. Determine the maximum allowable torque that may be applied as indicated if the angle of twist at the free end is not to exceed 0.05 rad.

6-52 A stepped steel shaft consists of a 4-ft length of 3-in. diameter and a 3-ft length of 2-in.-diameter solid shafting. The maximum allowable angle of twist in the 7-ft length is 0.04 rad. Determine the maximum torque that shaft is permitted to transmit.

6-53 A solid circular shaft is made of a material having a shear stress-strain relationship of $\tau = k\gamma^{1/2}$, where $k = 600,000$. The shaft is 120 in. long and 3 in. in diameter, and is subjected to a torque of 200,000 in.-lb. Assuming that the strain varies linearly from the center of the shaft, determine the angular twist due to the torque.

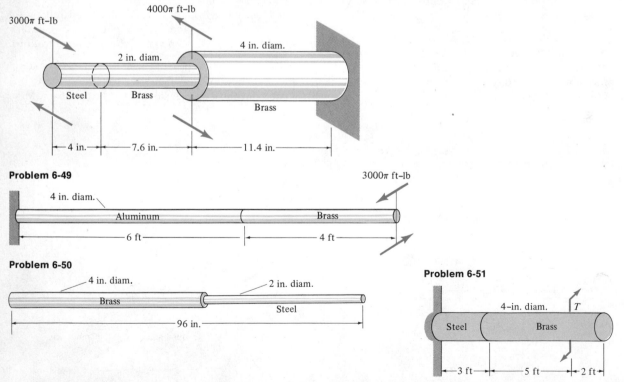

Problem 6-48

3000π ft–lb

4000π ft–lb

2 in. diam.

4 in. diam.

Steel   Brass

Brass

4 in.   7.6 in.   11.4 in.

Problem 6-49

4 in. diam.

3000π ft–lb

Aluminum   Brass

6 ft   4 ft

Problem 6-50

4 in. diam.

2 in. diam.

Brass   Steel

96 in.

Problem 6-51

4–in. diam.   T

Steel   Brass

3 ft   5 ft   2 ft

6-54 A hollow steel shaft that is 3 m long must transmit a torque of 20×10^3 N·m. The total angle of twist in this length must not exceed 2.5°, and the allowable shear stress is 9×10^7 Pa. What is the inside and outside diameter of a shaft that meets these requirements simultaneously?

6-55 The elastic circular shaft in the illustration carries a uniformly distributed torque of T in.-lb/in. along its entire length. (a) What is the maximum shearing stress at a section $L/3$ from the fixed left end? (b) What is the total relative angle of the twist between its ends? What type of real machine part might have a uniformly distributed torque such as this?

Problem 6-55

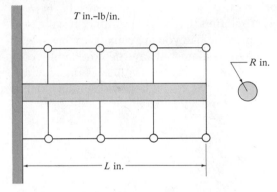

T in.-lb/in.

R in.

L in.

6-56 The circular shaft of constant radius shown in the illustration is acted upon by a torque that varies from zero at the right end to T in.-lb/in. at the wall. Assume a linear shear stress-strain relationship for the material. What is the total angle of twist of the right end of the shaft in terms of T, r, L, and G? What is the maximum shear stress in the shaft?

Problem 6-56

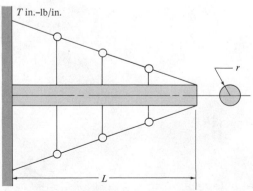

T in.-lb/in.

r

L

6-57 Solve Prob. 6-42 if the shaft is made of the material for which the stress-strain relationship is shown in the illustration.

Problem 6-57

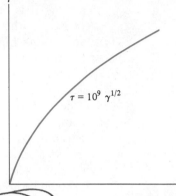

τ

$\tau = 10^9 \, \gamma^{1/2}$

γ

6-58 The composite shaft shown in the illustration is made up of a section of steel, a section of aluminum, and a section of brass. The total length is 6 ft. Also, $G_s = 12 \times 10^6$, $G_a = 4 \times 10^6$, and $G_b = 6 \times 10^6$ psi. The angular deformation of each section should be the same, and the total deformation from end to end is to be 3°. Determine the length of each section and the torque T. Assume elastic behavior and neglect stress concentrations.

Problem 6-58

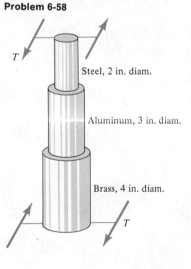

T

Steel, 2 in. diam.

Aluminum, 3 in. diam.

Brass, 4 in. diam.

T

6-8 / Design of circular shafts for multiple criteria

In Chapter 5 you learned that it was often necessary to design an axially loaded member for more than one criterion. The same is true of torsion members. It is quite possible for a member loaded in torsion to twist excessively without being overstressed. Excessive twist is particularly ruinous in machinery where it results in a phenomenon called whipping which can be very destructive.

In order to design a circular shaft that is neither overstressed nor subject to excessive deformation you usually need to make two independent designs, one based on the allowable stress, the other on the allowable angle of twist. Similarly, to find the safe load for a shaft of known dimensions, you must make two separate calculations for T, one based on stress and the other on twist. Comparison of the results of your computations should lead to a solution that satisfies both criteria.

It should be noted in this connection that you are usually unable to vary the length of the member. Length is almost always determined by considerations of function and space rather than mechanics. Consequently, your shaft designs should be restricted to finding the suitable radius unless you are specifically asked to find a length.

EXAMPLE 6-8

Design a steel shaft 10 ft long to transmit 1050 hp at 225 rpm. The allowable shear stress is 12,000 psi and the overall angle of twist may not exceed 3°.

Solution:

First, determine the torque.

$$T = \frac{63,000 \text{ hp}}{\text{rpm}} = \frac{(63,000)(1050)}{225} = 294,000 \text{ in.-lb}$$

Next, design for the allowable stress of 12,000 psi assuming elastic behavior.

$$\tau_{\text{All}} = \frac{Tr}{J} = \frac{Tr}{(\pi/2)(r^4)} = \frac{2T}{\pi r^3}$$

$$r^3 = \frac{2T}{\tau_{\text{All}}\pi} = \frac{2(294,000)}{12,000\pi} = 15.6 \text{ in.}^3$$

$$r = 2.5 \text{ in.}$$

Now design for the allowable twist assuming elastic behavior and noting that $3° = 3\pi/180$ rad.

$$\theta = \frac{TL}{GJ}$$

$$\frac{3\pi}{180} = \frac{294,000(120)}{12 \times 10^6 J}$$

$$J = \frac{294(120)}{12 \times 10^3} \times \frac{180}{3\pi} = 56.15$$

$$J = \frac{\pi r^4}{2} = 56.15$$

$$r^4 = \frac{112.3}{\pi} = 35.74$$

$$r = 2.4$$

Thus, the required radius is 2.5 in. since the smaller value of 2.4 would result in a shear stress greater than the allowable value.

PROBLEMS

6-59 A steel shaft 10 ft long and 2 in. in diameter must not twist more than 0.06 rad and the maximum stress must not exceed 11,000 psi. Determine the maximum power that this shaft will transmit at 270 rpm.

6-60 A brass shaft 12 ft long is to transmit 1050 hp at 225 rpm without having the angle of twist exceed 1.2° or having the shear stress exceed 12,000 psi. What minimum diameter of steel shaft is required?

6-61 Determine the minimum diameter of a steel shaft 12 ft long that will transmit a torque of 12,500 ft-lb with a maximum shear stress of 12,000 psi. and an angle of twist not to exceed 2.5° in the 12-ft length.

6-62 Find the minimum diameter of a brass shaft that will meet the specifications indicated in Prob. 6-61.

6-63 A 2024-T3 aluminum alloy tube 2 m long is to transmit a torque of 160 N·m with a maximum shear stress of 70 MPa and a maximum angle of twist of 10°. If the outside diameter of the tube is 50 mm, determine the minimum wall thickness required.

6-64 A certain shaft is to transmit a torque of 175 N·m with a factor of safety of 2 with respect to failure by yielding and to twist not more than 4.8° in a length of 3 m. Determine the minimum required diameter of the shaft if it is made of (a) structural steel, (b) high-carbon steel SAE 1090, (c) aluminum alloy 6061-T6.

6-65 A hollow circular shaft 1 m long is rotating at 1750 rpm. Its outside diameter is 0.1 m and its inside diameter is 0.05 m. Based on a factor of safety of 3, determine the maximum horsepower the shaft may safely transmit if the shaft is made of a linear brittle material, for which $G = 70 \times 10^9$; $\sigma_{ult} = 420$ MPa, tension; and $\sigma_{ult} = 560$ MPa, compression; and the overall angular deformation may not exceed 2°.

6-66 Solve Prob. 6-65 if the shaft is made of a material whose τ-γ curve is shown in the illustration and the angular deformation is not to exceed 10°. Also the shaft is 18 in. long. The OD is 4 in. and the ID is 2 in.

6-67 The solid circular shaft shown in the illustration is rotating at 630 rpm and is made of a linear homogeneous material that has an ultimate tensile stress of 60,000 psi, an ultimate shear stress of 40,000 psi, and a shear modulus of 12×10^6. If the overall angle of twist is not to exceed 0.12 rad, and a factor of safety of 3 is required, what is the maximum horsepower the shaft may safely transmit?

Problem 6-66

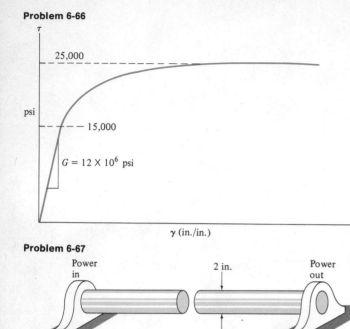

Problem 6-67

6-9 / Summary of cylindrical torsion members

When determining the mode of failure of any load-carrying member, you should keep clearly in mind the mechanical properties of the material involved and the existing state of stress in the member. Perhaps no other simple structural member illustrates the importance of the state of stress better than a cylindrical torsion member.

Consider, for example, a cylindrical torsion member made of a very ductile and homogeneous material, such as mild steel. If the member were twisted until it ruptured, the rupture would result primarily from the shear stresses on a plane perpendicular to the axis of the member. For a simple experiment, take a "Tootsie Roll" and twist it until it breaks in two, and observe the characteristics of the rupture.

As another observation, at one time or another you have probably seen someone break a wooden broom handle or mop handle by twisting it too much. The handle probably "split" along the grain, which usually runs somewhat parallel to the longitudinal axis. This splitting was caused by a combination of the shear stresses parallel to the longitudinal axis and the relative weakness of the wood in shear parallel to its grain.

On the other hand, consider a torsion member made of a brittle material, such as gray cast iron, which is relatively much weaker in tension than in shear or compression. If the member were twisted until it ruptured, the rupture would occur along a helical plane inclined approximately 45° to the longitudinal axis, since the maximum tensile stresses occur on this plane. For a simple test, twist a piece of blackboard chalk, and observe the rupture. Try to apply *pure* torque without bending the piece.

As a final consideration, hollow thin-walled cylindrical members are often used to transmit torque. If there is too much torque and the wall is too thin, the compressive stresses on a 45° plane may cause the wall to buckle and collapse. Try this experiment with a soda straw.

So far in this chapter we have considered circular cylindrical members subjected *only* to pure torques about their longitudinal axes. In engineering structures it is very common to encounter members that are subjected to axial, torsion, and bending loads simultaneously. Even though we have not yet studied bending, our present knowledge permits us to consider members subjected to a combination of axial and torsion loads. The simplest way to analyze such a member is to utilize the principle of superposition; that is, we can superpose the effects due to torsion onto the effects due to axial loading, or vice versa. Under conditions of combined loading, the principle of superposition may be used to determine either stresses or strains. The order of superposition is unimportant, as long as the deformations are small. This will usually be the case if the stresses are less than the elastic limit. We shall consider only the superposition of elastic stresses and deformations in this text.

Finally, since we have analyzed both the stresses and deformations occurring in cylindrical torsion members, we can combine this knowledge with the equations of static equilibrium to consider some statically indeterminate torsion members. The following examples illustrate the application of some of the ideas that have been discussed in this section.

EXAMPLE 6-9

The hollow closed cylinder of Fig. 6-14(a) is 10 in. in diameter and 12 ft long and has a wall thickness of $\frac{1}{8}$ in. It is subjected to an internal pressure of 60 psi, an axial tensile thrust of 8000 lb, and a torque of 1000 ft-lb. Find the maximum principal stress and the absolute maximum shearing stress and the orientation of each. Assume that the action is elastic, and neglect localized stress concentrations at the ends. Take J as $2\pi tr^3$ (approximation for thin-walled circular section).

Solution:

Any element on the *outside* surface would have the same state of stress as element A in Fig. 6-14(a). We may superpose the effects of internal pressure, torque, and axial load to get the complete state of stress. Since the longitudinal stress σ_L due to internal pressure and the stress σ_a due to axial load are in the same direction, we may add these two algebraically. Thus,

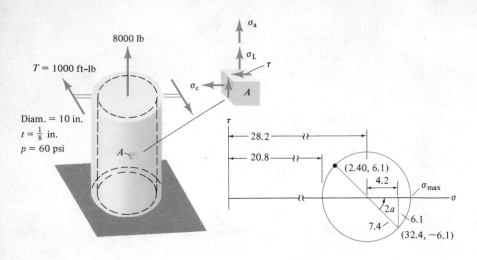

(a)   (b)

Figure 6-14

$$\sigma_L + \sigma_a = \frac{pD}{4t} + \frac{P}{A}$$

$$= \frac{(60)(10)}{(4)(1/8)} + \frac{8000}{\pi(10)(1/8)} = 3240 \text{ psi, tensile}$$

The circumferential stress σ_c is

$$\sigma_c = \frac{pD}{2t} = \frac{(60)(10)}{(2)(1/8)} = 2400 \text{ psi, tensile}$$

The shear stress τ due to torque is

$$\tau = \frac{Tr_o}{J} = \frac{(1000)(12)(5)}{2\pi(1/8)(5^3)} = 610 \text{ psi}$$

From Mohr's stress circle in Fig. 6-14(b), in which one unit represents 100 psi we get

$$\sigma_{max} = 2820 + 740 = 3560 \text{ psi}$$

$$2\alpha = \tan^{-1} \frac{6.1}{4.2} = 55.4°$$

$$\alpha = 27.7°$$

The maximum principal stress acts in a direction making an angle of 27.7° counterclockwise with the longitudinal axis of the cylinder. Since the three principal stresses at any point on the outside surface are +3560 psi, +2080 psi, and 0 psi, the absolute maximum shear stress acts on a plane that bisects the angle between the planes of the +3560 psi and 0 psi stresses. This shear stress is equal to

$$\tau_{max} = \tfrac{1}{2}(\sigma_{max} - \sigma_{min})$$
$$= \tfrac{1}{2}(3560 - 0) = 1780 \text{ psi}$$

Note: On the *inside* surface of the cylinder, the third principal stress may be considered to be −60 psi (due to the internal pressure). The value of σ_{max} would, however, actually be smaller if the inside radius of the cylinder were used in the torsion formula. So, actually, the maximum shear stress on an element on the *inside* surface of this thin-walled vessel would have nearly the same value as was found for the stress on the outside surface.

EXAMPLE 6-10

In the solid steel shaft in Fig. 6-15(a), what is the maximum shear stress? Neglect stress concentrations and assume elastic action.

Solution:

Consider the free-body diagram of Fig. 6-15(b). If moments are taken about the longitudinal axis,

$$\sum M_o = 0$$
$$400\pi - T_L - T_R = 0$$

Since elastic action is assumed, $\tau = Tr/J$. Also, the left-hand and right-hand sections have the same values for r and J. Therefore,

$$400\pi = \frac{\tau_L J}{r} + \frac{\tau_R J}{r} = (\tau_L + \tau_R)\frac{J}{r} \tag{a}$$

This is the only statics equation available, and it has two unknowns. So we must relate the unknowns with a deformation equation. Geometric compatibility requires that the left-hand portion of the shaft twist the same amount as the right-hand portion; that is, θ_L must be equal to θ_R. For elastic action and where T, G, and r_o are constant for each side, Eq. 6-7 gives

$$\frac{\tau_L L_L}{G_L r_L} = \frac{\tau_R L_R}{G_R r_R}$$
$$\tag{b}$$
$$\tau_L = \frac{L_R}{L_L}\tau_R$$

Figure 6-15

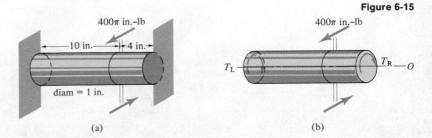

(a)   (b)

Combining equations (a) and (b), we obtain

$$\frac{400\pi r}{J} = \frac{L_R}{L_L}\tau_R + \tau_R$$

The maximum shear stress is

$$\tau_R = \frac{400\pi(0.5)}{\pi(1^4/32)(1 + 4/10)} = 4570 \text{ psi}$$

Also

$$\tau_L = \frac{4}{10}(4570) = 1828 \text{ psi}$$

PROBLEMS

6-68 A solid steel shaft 2 in. in diameter and 25 in. long transmits power by means of a pure torque at a rate of 396,000 in.-lb/sec when turning at 3150 rpm. If $G = 12 \times 10^6$ psi, determine (a) the magnitude and direction of the maximum tensile stress and (b) the total angle of twist. Assume elastic behavior.

6-69 For the hollow member in the illustration, find the magnitudes and directions of the principal stresses and the maximum shear stress. Assume elastic behavior, and neglect localized stresses at the ends.

Problem 6-69

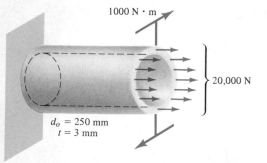

1000 N · m

20,000 N

$d_o = 250$ mm
$t = 3$ mm

Notes 5-11 → 5-15 (Answer)

6-70 (a) Determine the maximum principal stress in the composite member in the illustration. Where does this stress occur, and what is its orientation? (b) Determine the maximum shear stress in the member. Where does this shear stress occur, and what is its orientation? Neglect stress concentrations at the joint and at the fixed end and assume elastic behavior.

6-71 A stud bolt 25 mm in diameter is to be tightened with a 70 N·m couple. If the axial tensile force in the unthreaded shank of the bolt at the time of the development of the maximum torque is 30,000 N, determine the maximum tensile and shear stresses in the bolt. Neglect the stresses in the threaded portion of the bolt.

6-72 The engine of a light airplane is delivering 90 hp directly to the propeller, which is turning at 2360 rpm during takeoff. The forward thrust developed by the propeller is 500 N. How big must the solid propeller shaft be if the allowable stress in tension is 90 MPa and that in shear is 50 MPa?

Problem 6-70

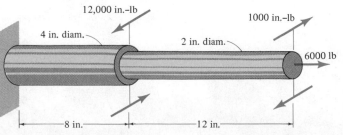

12,000 in.-lb

1000 in.-lb

4 in. diam.

2 in. diam.

6000 lb

8 in.

12 in.

6-73 What would be the reading of a single axial strain gage, in microinches per inch, if it were attached to the surface of a solid circular 2024-T3 aluminum alloy member that is 3 in. in diameter and subjected to an axial pull of 2 tons and a torque of 3 ft-tons. The gage axis makes an angle of 30° with the longitudinal axis of the member. The loads are in such directions that each alone would cause a tensile strain in the gage. $E = 10.6 \times 10^6$ psi and $\mu = \frac{1}{3}$.

6-74 Derive an expression for the maximum shear stress in the steel portion of the member shown in the illustration. Assume that there is elastic behavior and that the member is stressfree before being loaded. Neglect stress concentrations.

Problem 6-74

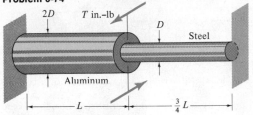

6-75 The tip of the weightless pointer in the illustration moves 25 mm on the scale as a result of the application of the torque. What is the maximum shearing stress in the steel if $G_s = 84 \times 10^9$ Pa and $G_a = 28 \times 10^9$ Pa?

6-76 Between two rigid parallel vertical plates that are 40 in. apart, two rods are to be welded with their axes on the same line. One rod is $1\frac{1}{2}$ in. in diameter and 30 in. long, and the other rod is 1 in. in diameter and 10 in. long. First the two rods are welded at the plates so that the other two ends come together 30 in. from one plate or 10 in. from the other plate. Before these two free ends are welded together, the free end of the 30-in. rod is twisted 4°; and the weld is then made

Problem 6-77

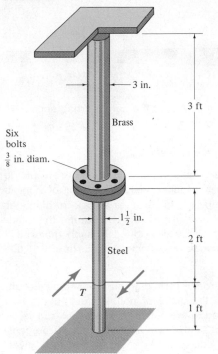

while a torque holds this twist in the larger rod. When the weld is completed, the torque is removed and the composite member is allowed to come to equilibrium. What is the resulting maximum shear stress in each rod if $G = 5.9 \times 10^6$ psi for the material of both rods? Assume elastic action and neglect localized stress concentrations.

6-77 The brass and steel torsion members are fixed at the top and bottom and bolted at the flange with six bolts. The bolts are each $2\frac{1}{2}$ in. from the center of the shafts. What maximum torque T can be applied if the average shearing stress in the bolts must not exceed 14,000 psi?

Problem 6-75

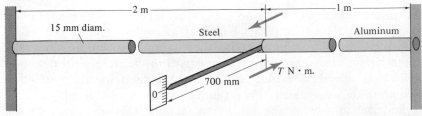

Problem 6-78

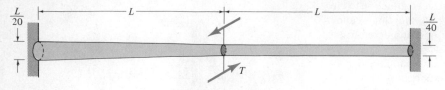

6-78 What is the twist at the center cross section of the shaft in the illustration? Assume elastic action, and express your answer in terms of T, L, and G; G is the same for the entire shaft.

6-79 A hollow circular aluminum shaft with an outside diameter of 4 in. and an inside diameter of 2 in. is continuously attached to the outside of a 2-in.-diameter steel shaft for its entire length. Determine the maximum shearing stress developed in each material if the composite shaft is subjected to a torque of 10,000 ft-lb.

6-80 A brass rod $1\frac{1}{4}$ in. in diameter is used as a shaft. Determine the thickness of hollow steel tubing that must be bonded to the outside of the brass rod (a) in order to increase the torsional strength 50 percent, (b) in order to decrease the angle of twist for a given torque 50 percent. Assume that the maximum allowable shearing stress in the brass is 8000 psi. and in the steel is 10,000 psi and that $G_s = 2G_B$.

6-81 In order to increase the torsional strength of a steel rod 2 in. in diameter, a $\frac{1}{16}$ in. thick brass tube is rigidly attached as indicated in the figure. Determine the increase in the torque required to produce a maximum stress of 8000 psi. in the brass in the center section. Assume $G_s = 2G_B$.

Problem 6-81

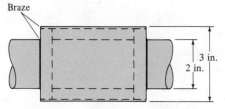

Braze

3 in.
2 in.

6-82 Determine the maximum shearing stress developed in the composite shaft shown in the figure if the unit is subjected to a torque of 11,000 N·m. Assume $G_s = 2G_B = 3G_A$.

Problem 6-82

75 mm OD 2024 T-3 aluminum alloy tube

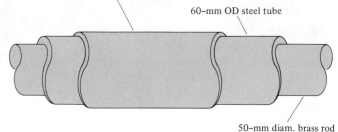

60-mm OD steel tube

50-mm diam. brass rod

6-10 / Noncylindrical torsion members

In general, the analysis of a noncylindrical torsion member is far more complicated than that of a circular cylindrical member. The major difficulty lies in determining the shear-strain distribution, since the discussion that we have presented in regard to the strain distribution in cylindrical members is not applicable to torsion members of other geometric cross sections. For example, we have concluded that in a cylindrical member plane transverse sections remain plane and the shear strain varies linearly from the geometric center. A simple experiment shows that these conclusions are not true for a torsion member having a rectangular cross section.

In Fig. 6-16(a) is shown a rectangular bar on which has been drawn a grid pattern of small rectangles. When the bar is twisted by a pure torque, the grid pattern assumes the form indicated in Fig. 6-16(b). Several observations can be made from a close examination of the pattern.

1 The *transverse* lines do not remain straight when the torque is applied. Therefore, we can expect that transverse planes will not remain plane; that is, the transverse planes warp.

2 The maximum distortion, and therefore the maximum shear strain, occurs at the middle of each *longer* side. Also, the largest distortion on the shorter side occurs at its middle.

3 No distortion, and therefore no shear strain, occurs at the corners of the member. Of course, this result could be anticipated from the analysis of a free-body diagram of a corner element. Such an element has two free surfaces that are perpendicular to each other. For it to be in equilibrium, no shear force can exist on any other face of the element.

Several ingenious methods have been devised to determine the shear-strain distribution in noncircular torsion members. Perhaps the foremost of these is the *membrane analogy*.* However, the mathematics required to pursue this discussion to any meaningful conclusions is beyond the level of this book. The solutions of many problems for solid noncircular torsion members can be found in more advanced books.†

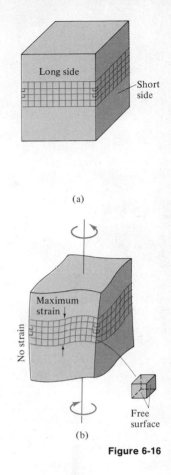

Figure 6-16

6-11 / Hollow thin-walled torsion members

In the case of some noncircular torsion members, a careful study of the geometric and equilibrium requirements can lead to an *approximate* relationship between torque and shear stress. One such example is a hollow thin-walled member whose wall thickness is very small compared with its other dimensions; the wall thickness is not necessarily constant along the entire circumference. A portion of such a member of arbitrary shape and variable thickness is shown in Fig. 6-17(a). A free-body diagram of an element of the wall is shown in Fig. 6-17(b). For this element we

* See *Advanced Mechanics of Materials*, 2d ed., by F. B. Seely ahd J. O. Smith (New York: Wiley, 1952).

† See *Theory of Elasticity*, 2d ed., by S. Timoshenko and J. N. Goodier (New York: McGraw-Hill, 1951).

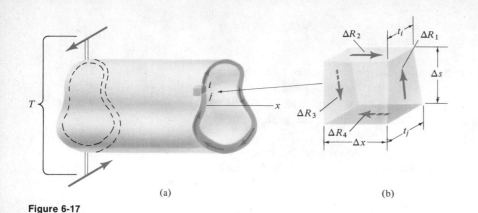

(a) (b)

Figure 6-17

know that no forces exist on the free inside and outside surfaces, and we make the simplifying *assumption* that no significant normal forces are present on the other four faces. This will lead to no serious error, provided that the member is not twisted too severely.

In the analysis of thin sections subjected to torsion or bending, it is often convenient to introduce a quantity called *shear flow*, denoted by q. The shear flow at any point in the thin section is defined as the longitudinal shear force across the thickness of the section per unit of length parallel to the longitudinal axis of the member. For example, at i in Fig. 6-17, the average intensity of the longitudinal shear force per unit longitudinal length would be $\Delta R_2/\Delta x$, and the shear flow across this thickness of section would be

$$q_i = \lim_{\Delta x \to 0} \frac{\Delta R_2}{\Delta x} = \frac{dR_2}{dx}$$

Similarly, the shear flow across the thickness at j is

$$q_j = \lim_{\Delta x \to 0} \frac{\Delta R_4}{\Delta x} = \frac{dR_4}{dx}$$

Within the limits of our original assumption, equilibrium in the x direction requires that $\Delta R_2 = \Delta R_4$. So it follows that $q_i = q_j$. Also, since points i and j were chosen arbitrarily, we can conclude that *the shear flow has the same value at any point around the circumference of the wall of a thin-walled torsion member*. Note that this statement does *not imply* that the *shear stress* is constant around the circumference of the cross section.

We shall now express the shear flow for a thin-walled torsion member in terms of the applied torque. In Fig. 6-18(a) and (b) are shown the location of an infinitesimal element of the wall and the forces acting on it. Here the forces parallel to the x axis are expressed in terms of the shear flow. If moments are taken about the a axis and counterclockwise moments are considered positive, we get

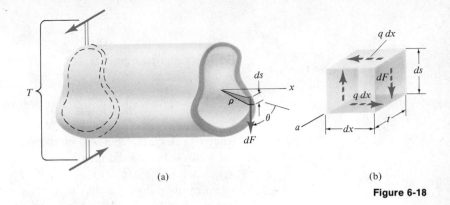

(a)   (b)

Figure 6-18

$$\sum M_a = 0$$

$$(dF)(dx) - (q\,dx)\,ds = 0$$

$$dF = q\,ds \qquad (6\text{-}8)$$

Now, if we take moments about the longitudinal x axis of the member in Fig. 6-18(a) and substitute the value of dF from Eq. 6-8, we obtain a relationship between the shear flow and the torque. Thus,

$$\sum M_x = 0$$

$$T - \int \rho(dF \sin\theta) = 0 \qquad (6\text{-}9)$$

$$T = \int_{\substack{\text{around} \\ \text{circumference}}} \rho q\,ds\,\sin\theta$$

where θ is the angle between the tangential force dF and the radius ρ.

The indicated integration operation requires some advanced theories of calculus.* We can, however, interpret the integral in a practical manner as follows: Since q is constant,

---

* In vector calculus the expression inside the integral would be written as the vector product $\rho \times (d\mathbf{s})$; and by definition its magnitude would be equal to the area of a parallelogram with sides of length ρ and ds. This area would be equal to twice the area of the shaded triangle in Fig. 6-18(a). In rectangular coordinates, with $\rho = \mathbf{i}0 + \mathbf{j}y + \mathbf{k}z$ and $d\mathbf{s} = \mathbf{i}0 + \mathbf{j}\,dy + \mathbf{k}\,dz$,

$$\int_{\substack{\text{around} \\ \text{circumference}}} \rho \times d\mathbf{s} = \mathbf{i}\oint_{\substack{\text{around} \\ \text{circumference}}} z\,dy - y\,dz$$

From Green's lemma,

$$\oint_{\substack{\text{around} \\ \text{circumference}}} z\,dy - y\,dz = \iint_{\substack{\text{over} \\ \text{enclosed area}}} 2\,dy\,dz = 2A$$

where A is the area enclosed by the wall of the member.

$$T = q \int_{\substack{\text{around} \\ \text{circumference}}} \rho(ds \sin \theta) \tag{6-10}$$

But $\rho(ds \sin \theta)$ represents twice the area of the shaded triangle in Fig. 6-18(a). Hence,

$$\int_{\substack{\text{around} \\ \text{circumference}}} \rho(ds \sin \theta) = 2A \tag{6-11}$$

where A is the area enclosed by the wall of the member, that is, the area of the hollow part of the member. It may be argued that A should be the area enclosed by the "mean" wall. However, if the "mean" area were significantly different from the internal area, the member would probably not be classified as *thin-walled*, and our entire discussion would not be applicable. Also, by using the inside area the result will be on the conservative side. By combining Eqs. 6-10 and 6-11, we have

$$q = \frac{T}{2A} \tag{6-12}$$

It is significant to recall that the derivation of Eq. 6-12 was based solely on equilibrium requirements. No reference was made to the mechanical properties of the material involved or to the *actual* shear stresses and strains existing in the wall of the member.

A so-called approximate or average shear stress at a point in the wall can be evaluated by dividing the shear flow by the wall thickness at that particular point. Thus, the average shear stress at i in Fig. 6-17 is

$$\tau_{i \, av} = \frac{T}{2At_i} \tag{6-13}$$

Although this is only an approximate relationship between torque and shear stress, it is sufficiently accurate to make it applicable to many design problems. For your own benefit you should compare the results found by using Eq. 6-3 with those found by using Eq. 6-13 for a thin-walled cylindrical tube. Note that for this case Eq. 6-13 can be written as follows:

$$\tau_{av} = \frac{T}{2At} \frac{r}{r} = \frac{Tr}{2tAr}$$

where $2tAr$ is an approximate value of J which was used in Example 6-9.

PROBLEMS

6-83 A thin-walled circular brass tube with a wall thickness of 1.2 mm is to be used to transmit 50π N·m of torque. Assuming that the tube does not buckle and that the shear stress is not to exceed 42 MPa, determine the required diameter of the tube.

6-84 A thin-walled elliptical tube is used to transmit 90 N·m of torque. The cross section of the tube is shown in the illustration. What is the shear stress in the wall of the tube?

Problem 6-84

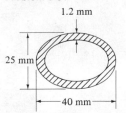

6-85 An extruded aluminum tube has the cross section shown in the illustration. A piece of this tubing is to be used to transmit a torque. If the shear stress may not exceed 10,000 psi, what is the maximum permissible torque based on a factor of safety of 3?

Problem 6-85

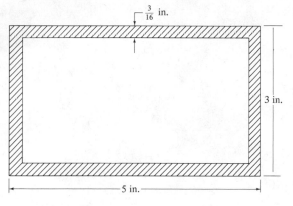

6-86 A thin-walled tube has a square cross section and a wall thickness of 0.04 in. Approximately what torque can a 1 × 1 in. (outside) tube transmit if the shear stress in the wall is not to exceed 4500 psi?

6-87 A steel tube with closed ends has an outside diameter of 60 mm and a wall thickness of 1.2 mm. If it is subjected to an internal gage pressure of 2.8 MPa and a torque of 50π N·m, determine the magnitude and direction of the maximum principal stress.

6-88 A piece of circular thin-walled tubing of 1 in. OD and 0.06 in. thickness carries gas under a pressure of 300 psig. Due to faulty installation, the tube is twisted so that it is subjected to a constant torque of 120π in.-lb. Determine the maximum shear and tensile stresses in the tubing.

Chapter 7
Beam stresses

7-1 / Objectives

Upon completion of this chapter you will be able to:

1 Draw shear and moment diagrams for statically determinate beams carrying combinations of concentrated and uniformly distributed loads.
2 Calculate normal stress due to bending.
3 Design a member to carry a bending load.
4 Solve problems dealing with eccentric loads.
5 Calculate longitudinal and transverse shear stress.
6 Calculate stresses due to combined loadings of axial forces, torques, and bending moments.

7-2 / Introduction

Probably the most common type of structural member is the beam. A beam may be defined as a member whose length is relatively large in comparison with its thickness and depth, and which is loaded with transverse loads that produce significant bending effects as opposed to twisting or axial effects (refer to Fig. 1–5). In actual structures beams can be found in an infinite variety of sizes, shapes, and orientations as, for example the curved member *abc* and the straight member *bde* in Fig. 7-1. For the present discussion, we shall consider straight beams oriented so that their length is horizontal, leaving the discussion of curved beams to Chapter 11.

Beams are generally classified according to their geometry and the manner in which they are supported. Geometrical classification includes such features as the shape of the cross section, whether the beam is straight or curved, whether the beam is tapered or has a constant cross section, and other features that will be considered later. Beams may be readily classified according to the manner in which they are supported. Some types that occur in ordinary practice are shown in Fig. 7-2, the names of some of these being fairly obvious from direct observation. Note that the beams in (d), (e), and (f) are statically indeterminate.

A beam can be further classified according to the type of load or loads it is carrying. For instance, a cantilever beam may be carrying a uniformly distributed load. In that case the beam might be classified as a uniformly loaded cantilever beam. Further extension of this method of classification is possible, and if carried to extremes, the names become rather complex and lengthy. In this text, however, we shall merely classify beams according to their geometry and support.

As the first step in determining the load-carrying capacity of a beam,

Figure 7-1

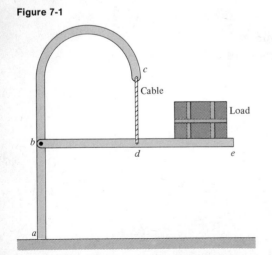

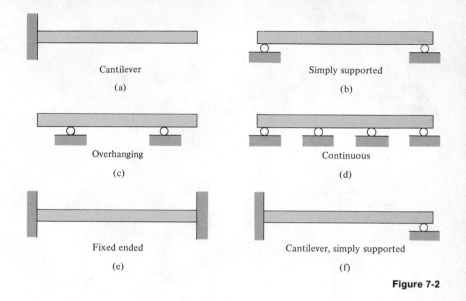

Cantilever

(a)

Simply supported

(b)

Overhanging

(c)

Continuous

(d)

Fixed ended

(e)

Cantilever, simply supported

(f)

Figure 7-2

we must analyze the internal shear and bending moment reactions throughout the length of the beam. Then we shall proceed to develop relationships between the applied loads and the resulting stresses and deformation of the beam. This chapter will be concerned primarily with the load-stress relationships, while the following chapters will deal with the load-deflection relationships and the solution of statically indeterminate beams.

At this point it would be worthwhile for you to briefly review sec. 1–3 dealing with internal reactions, since those ideas are the backbone of the succeeding discussion.

7-3 / Basic relationships between shear force and bending moment for beams

Sign conventions for beams Consider now a simply supported beam, such as that shown in Fig. 7-3(a). This beam carries some arbitrary continuously* distributed load expressed as w pounds per unit length, where w is a function of x.

A free-body diagram of a portion of the beam extending x_1 units from the left-hand end is shown in Fig. 7-3(b), while a similar diagram for a portion having a length of $x_1 + \Delta x$ units is shown in (c). In Fig. 7-3(d)

---

* This assumption of continuity is made here for convenience and will subsequently be dropped to allow discontinuities such as concentrated loads and loading jumps.

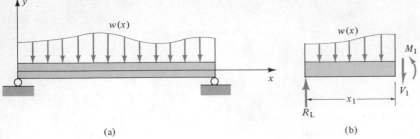

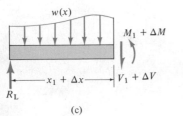

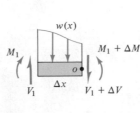

Figure 7-3

is a free-body diagram of the Δx portion of the beam. Note that the moments and shears in (d) are consistent with those in (b) and (c).

In American engineering practice it is customary to assume that a *positive* bending moment is one that causes a beam to bend *concave upward*. Another way of saying the same thing is as follows: A positive bending moment produces tension in the lower fibers of the beam and compression in the upper fibers. A negative moment produces the opposite effects. Hence, *all* the bending moments in Fig. 7-3(b)–(d) are considered positive.

The directions of the shear forces in Fig. 7-3(b)–(d) are also customarily assumed to be positive; that is, a positive shear is one that acts downward on the right-hand side of the section or upward on the left-hand side.

In general the internal shear force V and the internal bending moment M depend upon the value of x so that

$$V = V(x) \quad \text{with} \quad V_1 = V(x_1)$$
$$M = M(x) \quad \text{with} \quad M_1 = M(x_1)$$

(7-1)

The shear-force function $V(x)$ and the bending-moment function $M(x)$ depend upon the external support reactions and the loading function $w(x)$, and these relationships can be found by applying the equilibrium requirements to a section (or several sections) of the beam such as that in Fig. 7-3(b). We shall now see that these three functions are related to each other in a very specific manner.

Basic equations for beams For equilibrium in the y direction in Fig. 7-3(d),

$$\overset{\downarrow+}{\Sigma} F_y = 0$$

$$(V_1 + \Delta V) + w(x')\Delta x - V_1 = 0$$

where x' is some value of x between x_1 and $x_1 + \Delta x$. Hence

$$\frac{\Delta V}{\Delta x} = -w(x')$$

Taking the limit as Δx approaches zero, we have

$$\frac{dV}{dx} = -w(x) \tag{7-2}$$

Thus, at any point in a beam *where the loading is continuous*, the rate of change of the shear force is equal to the negative loading $w(x)$; the negative sign indicates that the shear decreases when the loading acts downward, as assumed here.

Taking moments about point o in Fig. 7-3(d), we have

$$\Sigma M_o = 0$$

$$(M_1 + \Delta M) + w(x')\Delta x\, \alpha\, \Delta x - M_1 - V_1 \Delta x = 0$$

$$V_1 = \frac{\Delta M}{\Delta x} + w(x')\alpha\, \Delta x$$

where α is some number between 0 and 1. Now, taking the limit as Δx approaches zero and omitting the subscript, we obtain

$$V = \frac{dM}{dx} \tag{7-3}$$

Thus, at any point along a beam *where the loading is continuous*, the shear is equal to the derivative of the moment at the corresponding point.

Writing Eq. 7-3 in another form and integrating between two points 1 and 2 at distances x_1 and x_2 from the left-hand end of the beam, we have

$$V\, dx = dM$$

$$\int_{x_1}^{x_2} V\, dx = \int_{x_1}^{x_2} dM = M_2 - M_1 = \Delta M_{1,2} \tag{7-3a}$$

Thus, the integral $\int_{x_1}^{x_2} V\, dx$ represents the *change* in moment between points 1 and 2 (if there are no jump discontinuities in the moment due to concentrated bending couples).

Finally, from Eq. 7-2 we see that wherever the loading is continuous, V will be differentiable and thus

$$\frac{d^2M}{dx^2} = \frac{dV}{dx} = -w(x) \tag{7-4}$$

The writing of shear and moment equations for specific loadings will be covered in the next chapter, where they will be seen to be extremely important in finding deflections.

7-4 / Conventional shear-force and bending-moment diagrams for beams

Shear and moment diagrams   A *shear diagram* is a graphical plot of how the internal shear force V varies throughout the length of the beam, that is, a plot of the shear function $V(x)$. Similarly, a *moment diagram* is a graphical plot of how the internal bending moment M varies throughout the length of the beam, that is, a plot of the bending-moment function $M(x)$. Such diagrams give a graphical picture of how the loading affects the beam and indicate those critical points where the bending and/or shear effects are likely to cause failure of the beam.

Equations 7-2 to 7-4 are quite useful in drawing shear and moment diagrams since Eq. 7-3 relates the slope of the moment diagram to the value of the shear, and Eq. 7-3a relates an area on the shear diagram to the *change* in the value of the moment. The example problems will illustrate the use of these relationships.

While it is possible to obtain shear and moment diagrams by writing and plotting the shear and moment equations as will be done in Chapter 8, it is also possible to obtain the diagrams by means of free-body analysis.

1   Determine all the external support reactions for the beam.

2   Set up axes for the shear and moment diagrams preferably directly beneath a scale drawing of the beam. Draw vertical lines on the diagram at all points where concentrated loads or external reactions are applied or where the loading changes in any abrupt manner. These will locate the discontinuities or jumps in the diagrams.

3   By using FBDs determine the value of the shear and the moment at two or more places between each pair of discontinuities. Values should be obtained at locations "very close" to the points of discontinuity.

4   Plot the points resulting from the evaluation and draw the curve by segments.

EXAMPLE 7-1

Plot the shear diagram for the beam shown in Fig. 7-4(a).

Solution:

The first step is to find the reactions by taking moments about the right end in Fig. 7-4(b).

$$\overset{+\curvearrowright}{\sum} M = 0$$

$$16R_1 - (400 \times 8)(12) - (6000)(8) = 0$$

$$R_1 = 5400 \text{ lb}$$

Taking $\sum F_y = 0$, we find that

$$5400 + R_2 - 3200 - 6000 = 0$$

$$R_2 = 3800 \text{ lb}$$

The second step is shown in Fig. 7-5. Note that the discontinuities will occur at the left end because of the external reaction R_1, at the midpoint because of both the 6000-lb load and the abrupt discontinuance of the distributed loading, and at the right end because of the reaction R_2.

Referring to step 3 in section 7-4, we see that we shall need free-body diagrams cut near each reaction and on each side of the 6000-lb load. In addition we shall draw a free-body diagram cutting the beam 4 ft from the left end. These diagrams are shown in Fig. 7-6.

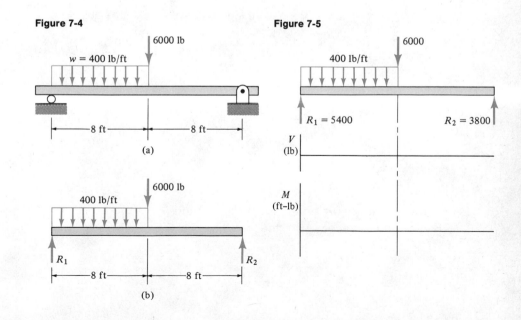

Figure 7-4

6000 lb

$w = 400$ lb/ft

8 ft — 8 ft

(a)

6000 lb

400 lb/ft

R_1 — R_2

8 ft — 8 ft

(b)

Figure 7-5

6000

400 lb/ft

$R_1 = 5400$ $R_2 = 3800$

V
(lb)

M
(ft–lb)

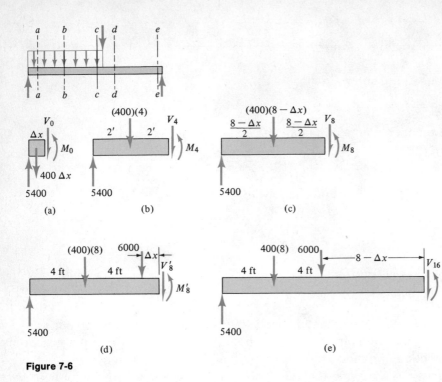

Figure 7-6

Evaluating V and M from Fig. 7-6(a) by taking moments at the right end, we find

$$\overset{+\uparrow}{\sum} F_y = 0 \quad \text{gives } 5400 - 400\,\Delta x - V_0 = 0$$

$$\overset{+\curvearrowright}{\sum} M_R = 0 \quad \text{gives } 5400\,\Delta x - \frac{400\,\Delta x^2}{2} - M_0 = 0$$

To get "very close" to the left reaction, we let $\Delta x \rightarrow 0$ and we find that

$$V_0 = 5400$$

$$M_0 = 0$$

These are the values we shall plot at the left end ($x = 0$).

Repeating this process with the free body in Fig. 7-6(b), we find

$$\overset{+\uparrow}{\sum} F_y = 0$$

$$5400 - 400(4) - V_4 = 0$$

$$V_4 = 3800 \text{ lb}$$

$$\overset{\curvearrowright(+\downarrow)}{\sum} M_R = 0$$

$$5400(4) - 400(4)(2) - M_4 = 0$$

$$M_4 = 18,400 \text{ ft-lb}$$

From Fig. 7-6(c), which represents a section cut just slightly to the left of the midpoint of the beam, we find

$$\overset{+\uparrow}{\sum} F_y = 0$$

$$5400 - 400(8 - \Delta x) - V_8 = 0$$

Letting $\Delta x \rightarrow 0$

$$V_8 = 5400 - 3200 = 2200 \text{ lb}$$

Also

$$\overset{\curvearrowright(+\downarrow)}{\sum} M_R = 0$$

$$5400(8 - \Delta x) - 400(8 - \Delta x)\left(\frac{8 - \Delta x}{2}\right) - M_8 = 0$$

Letting $\Delta x \rightarrow 0$

$$M_8 = 5400(8) - \frac{400(8^2)}{2} = 30,400 \text{ ft-lb}$$

From Fig. 7-6(d), which represents a section cut just slightly to the right of the discontinuity at the midpoint, we find

$$\overset{+\uparrow}{\sum} F_y = 0$$

$$5400 - 400(8 + \Delta x) - 600 - V_8' = 0$$

Letting $\Delta x \rightarrow 0$

$$V_8' = 5400 - 3200 - 6000 = -3800 \text{ lb}$$

Also

$$\overset{\curvearrowright(+\downarrow)}{\sum} M_R = 0$$

$$5400(8 + \Delta x) - 3200(4 + \Delta x) - 6000 \Delta x + M_8' = 0$$

Letting $\Delta x \rightarrow 0$

$$M_8' = 5,400(8) - 3,200(4) = 30,400 \text{ ft-lb}$$

Note that the shear at the midpoint of the beam has been found to have two values: $V_8 = 2200$ lb and $V_8' = -3800$ lb. This will result in a discontinuity, or jump, in the shear diagram. On the other hand, the value for the moment was found to be the same for both Fig. 7-6(c) and Fig. 7-6(d). Thus the moment diagram will not jump at the midpoint of the beam but, as we shall see, will change shape.

Last, from Fig. 7-6(e) we find

$$\overset{+\uparrow}{\sum} F = 0$$

$$5400 - 3200 - 6000 - V_{16} = 0$$

$$V_{16} = -3800 \text{ lb}$$

$$\overset{\curvearrowright+}{\sum} M_R = 0$$

$$5400(16 - \Delta x) - 3200(12 - \Delta x) - 6000(8 - \Delta x) - M_{16} = 0$$

Letting $\Delta x \to 0$

$$M_{16} = 5400(16) - 3200(12) - 6000(8) = 0$$

The last moment computation was not really necessary since we know from equilibrium considerations that the moment at that point must be zero. However, the computation does serve as a check on our work.

Summarizing our findings to this point and considering x to be the distance from the left end, we make the following table:

x	V_x (kips)	M_x (ft-kips)
0	5.4	0
4	3.8	18.4
$8 - \Delta x$	2.2	30.4
$8 + \Delta x$	-3.8	30.4
16	-3.8	0

To construct the shear and moment diagrams we first plot the information from our table. The shear values are plotted in Fig. 7-7(b) and the moment values in Fig. 7-7(d). We have lined them up with the load diagram 7-7(a).

To complete the construction of the shear and moment diagrams, we must join these plotted points with some type of curve or line. But what is the shape of the connecting curves or lines? This is where we can make use of the relations given by Eqs. 7-2 to 7-4. Equation 7-2 says that if the loading is constant (uniform), the shear diagram will have a constant slope; that is, the shear "curve" will be a straight line. In this problem the left half of the beam has a constant (uniform) loading of 0.4 kips/ft, while the right half has a constant loading of 0 kips/ft.

Thus, we join the points on the left half of the shear diagram with a straight line having a negative slope of 0.4 kips/ft. Similarly, we join the points on the right half with a straight line with a zero slope, that is, a horizontal line. The completed shear diagram is shown in Fig. 7-7(c).

As for the moment diagram, Eq. 7-3 says that the curve on the moment diagram will have a slope

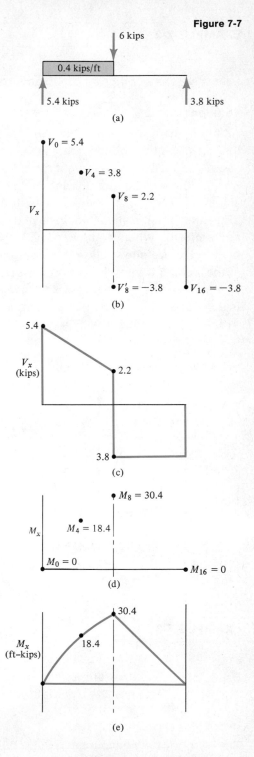

Figure 7-7

equal to the value of the shear. Therefore, for the left half of the beam, the moment curve will initially have a positive slope since $V_0 = 5.4$, but its *slope* will gradually *decrease* since the values of the shear are decreasing as we move from left to right. Thus, this portion of the moment diagram is indeed a "curve" as shown in Fig. 7-7 (e). For the right half of the beam, we already know that the value of the shear is a constant -3.8 kips. Thus the moment curve for this portion will have a constant negative slope, that is, a straight line as shown in Fig. 7-7 (e).

7-5 / Shear diagrams by inspection

The preceding example demonstrates procedures for calculating values of shear and moment for a specific location and producing shear and moment diagrams from that information. This method for obtaining shear and moment diagrams, while useful and easily understood, is a bit tedious.

Most people who do much work with shear and moment diagrams draw shear diagrams by "inspection" and obtain moment diagrams by a graphical integration of the shear diagrams.

The following examples demonstrate the inspection method for shear diagrams. One simply moves from left to right along the beam plotting upward external forces as positive, downward external forces as negative.

EXAMPLE 7-2

Draw the shear diagram for Example 7-1 without recourse to equations or free-body diagrams.

Solution:

All the loads are shown in Fig. 7-8(a). To obtain the shear diagram, we move across the beam from left to right, plotting the forces as we come to them, taking upward forces as positive, downward as negative. Our train of thought runs thus: The first force we encounter is the reaction of 5400 lb, so we plot the +5400 lb at the left end of the diagram. We then have a uniformly distributed load of 400 lb/ft. That means that for every foot we go along the beam, the shear must decrease by 400 lb. Thus after 1 ft, the shear would have decreased by 400 lb and be at 5000. After 2 ft, it would be at 4600 and so forth. When we have gone 8 ft, the shear will have decreased by 8 × 400 or 3200 lb and be at 2200 lb. We show this with a straight sloping line beginning at 5400 and running down to 2200 lb at the point 8 ft from the left end. At this point there is a 6000-lb concentrated load so the diagram "jumps" vertically downward 6000 lb to a value of −3800 lb. In the last 8 ft there are no loads, so there is a constant shear over that portion of the beam resulting in a horizontal line on the shear diagram. At the end a vertical reaction of 3800 lb would cause an upward "jump," thus closing the diagram.

Figure 7-8

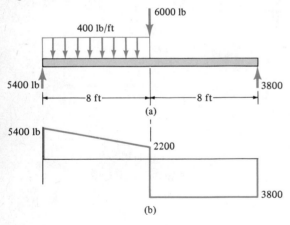

(a)

(b)

EXAMPLE 7-3

Draw the shear diagram for the beam in Fig. 7-9(a) without recourse to equations or free-body diagrams. (Note that the external reactions have already been determined.)

Solution:

The first force we encounter is the 1000-lb reaction, so we plot + 1000 at the left end of the diagram.

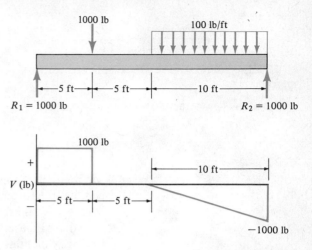

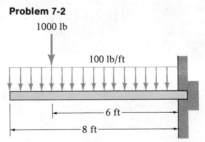

Figure 7-9

Nothing happens for 5 ft, so we have a constant shear over that distance. At 5 ft there is a downward force of 1000 lb. That "jumps" the shear down to the zero axis where it remains until the 10-ft point, where a uniform load of 100 lb/ft begins. That means that for every foot we move along the axis, the shear will decrease by 100 lb. At 1 ft beyond the midpoint of the beam the shear curve will be 100 lb below the axis; at 2 ft beyond it will be 200 lb below, and so on for 10 ft. At the end of 10 ft, the curve is 1000 lb below, and at that point the 1000-lb upward force causes a jump that closes the diagram.

PROBLEMS

7-1 to 7-6 Draw the shear diagrams for the beams shown. Give the numerical values for the shear at all discontinuity points.

Problem 7-1

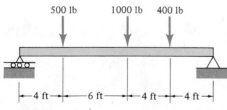

Problem 7-3

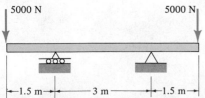

Problem 7-2

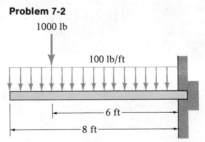

Problem 7-4

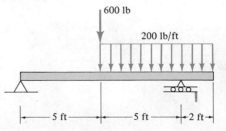

Problem 7-5

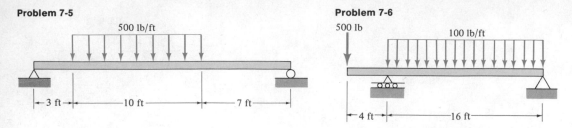

500 lb/ft

|←3 ft→|←———10 ft———→|←——7 ft——→|

Problem 7-6

500 lb

100 lb/ft

|←4 ft→|←————16 ft————→|

7-6 / Moment diagrams by graphic integration

Moment diagrams may be obtained from shear diagrams by using a combination of simple geometry and basic calculus.

A review and reexamination of some of the concepts in section 7-3 are in order. Figure 7-10(a) shows a loaded beam and Fig. 7-10(b) is the free-body diagram of a slice cut from it. (Note that these are the same as Figs. 7-3(a) and (d), respectively.)

Taking moments about point o in Fig. 7-10(b) and letting $\Delta x \to 0$ gives

$$V = \frac{dM}{dx} \qquad [7\text{-}3]$$

Writing Eq. 7-3 in another form and integrating between two points 1 and 2 at distances x_1 and x_2 from the left-hand end of the beam, we have

$$V\,dx = dM$$

$$\int_{x_1}^{x_2} V\,dx = \int_{x_1}^{x_2} dM = M_2 - M_1 = \Delta M_{1,2} \qquad [7\text{-}3a]$$

Thus, the integral $\int_{x_1}^{x_2} V\,dx$ represents the *change* in moment between points 1 and 2 (if there are no jump discontinuities in the moment due to concentrated bending couples).

Let us look at the implication of this statement in connection with shear and moment diagrams. Figure 7-11 shows a simple beam and Fig. 7-11(b) is its shear diagram.

Observe that on the shear diagram the quantity $V\,dx$ is the area on the shear diagram over the distance dx. (This is still true when V is not constant if V is taken as the average shear in the dx interval.)

Figure 7-10

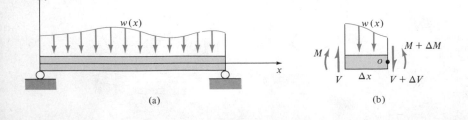

(a)

(b)

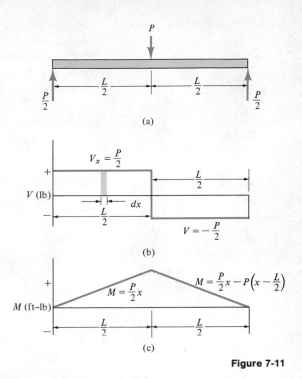

Figure 7-11

This allows us to state Eq. 7-3a in the following words: The area on the shear diagram over a given interval is equal to the *change* in moment over that interval, and the change is the same algebraically (signwise) as the shear area, that is, positive for area above the axis, negative for area below the axis.

Now let us look at the implications for an indefinite integration of Eq. 7-3.

$$\int V\,dx = \int dM = M \qquad (7\text{-}5)$$

This says that the integral of the shear equation is the moment equation. Thus the moment equation is of one order higher than the shear equation. If the shear equation is of zero order (a horizontal line), the moment equation will be of the first order (a sloping line). If the shear equation is of the first order (a sloping line), the moment equation will be of the second order (parabolic).

Finally, consider again Eq. 7-3.

$$V = \frac{dM}{dx}$$

As we have already observed, this tells us that the value of the shear at any point is the slope of the moment curve at that point. A maximum (or minimum) value of the moment can thus be expected to occur at any point(s) at which the *slope* of the moment curve is zero (horizontal), that is, where $V = 0$ and the shear diagram crosses the axis.

The following examples will demonstrate the use of these concepts in obtaining moment diagrams by the graphical integration of shear diagrams.

EXAMPLE 7-4

Draw the moment diagram for the beam in Example 7-3.

Solution:

Figure 7-12 shows the beam and its shear diagram as obtained in Example 7-3. The area under the shear diagram is calculated as indicated in Fig. 7-12(b).

From equilibrium, note that the moment at the left end must be 0. Next, the shear diagram plus the area statement tells us that the change in moment over the next 5 ft is $+5000$. Therefore, the moment at the end of 5 ft is $0 + 5000$, or $+5000$ ft-lb. The area under the shear curve over the next 5 ft is 0. Therefore, the moment at the end of 10 ft is $5000 + 0 = 5000$ ft-lb. Over the last 10 ft there is an area under the shear curve of -5000 so the moment at the end of that 10 ft is $5000 - 5000 = 0$. Over the first 5 ft the shear is a constant so the moment diagram has a constant slope; that is, a straight line connects our first two points. In the next interval the shear is 0 so the moment curve between the second and third points has a zero slope. In other words, the moment is constant. Over the last interval the shear diagram is a straight line with negative slope. Since such a line is the derivative of a parabola with negative curvature, we know our moment curve must be just that (a parabola with negative curvature). Connecting the points in this fashion results in the diagram in Fig. 7-13.

If values at some intermediate points are desired they may be obtained by calculating the areas

Figure 7-12

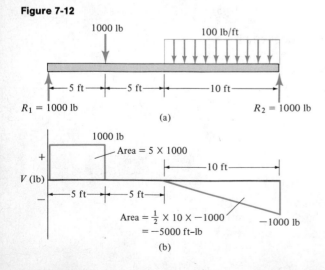

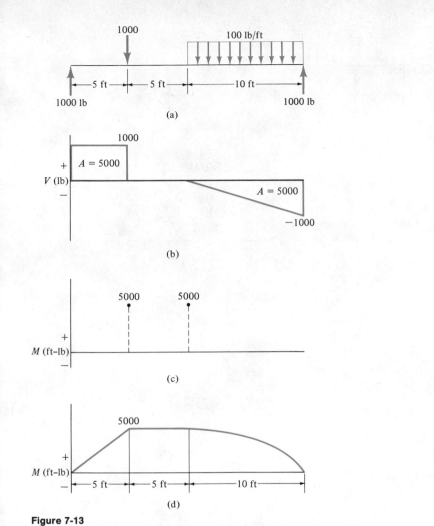

Figure 7-13

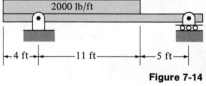

Figure 7-14

under the shear diagram for the appropriate locations. For example, at $x = 15$ ft, the moment would be

$$M_{15} = M_{10} + \text{area under shear diagram between } x = 10 \text{ and } x = 15$$
$$= 5000 - \tfrac{1}{2}(5)\,(500)$$
$$= 5000 - 1250 = 3750 \text{ ft-lb}$$

EXAMPLE 7-5

For the beam in Fig. 7-14, sketch the complete shear and moment diagrams, labeling the values at points of discontinuity. Also, determine the value of the maximum moment and locate the section at which it occurs.

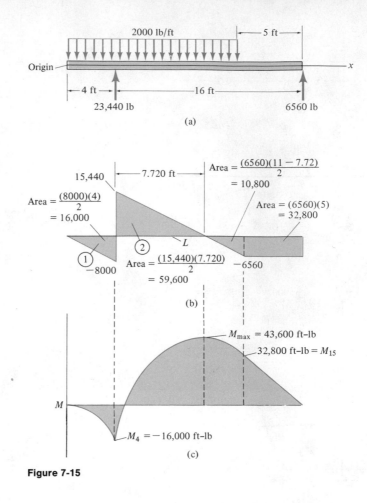

Figure 7-15

Solution:

Figure 7-15(a) shows the loading and reactions on the beam and Fig. 7-15(b) shows the shear diagram obtained by inspection. You should satisfy yourself that this diagram is correct. The calculations for the areas on the shear diagram are also shown in Fig. 7-15(b).

The moment at the left end is known to be zero from equilibrium considerations. Between the left end and the left support, the area under the shear curve is a negative 16,000, thus the change in moment is − 16,000 ft-lb. The moment at the left support (M_4) is

$$M_4 = 0 - 16,000 = -16,000$$

The shear diagram for this region is a straight line with a negative slope so the moment diagram is a parabola with a negative curvature.

The maximum moment will occur where the shear diagram crosses the axis (where the shear is zero). To locate this point, we observe that the line has a slope of 2000 lb/ft. Thus by ratio and proportion

$$\frac{15,440}{L} = \frac{2000}{1}$$

$$L = \frac{15,440}{2000} = 7.72 \text{ ft}$$

The point of maximum moment is 7.72 ft to the right of the reaction.

The change in the moment between the reaction and the point at which maximum moment occurs is given by the area under the shear curve between the two points or 59,600 ft-lb. Thus, the maximum moment is M_4, the moment at the left reaction, plus this change in the moment.

$$M_{max} = -16,000 + 59,600 = 43,600 \text{ ft-lb}$$

The line connecting the moment at the reaction and the maximum moment is a parabola with a zero slope at the point of maximum moment.

The moment (M_{15}) at the point where the uniform loading ceases is obtained by using the area under the shear diagram between the axis crossing and the point in question, noting that this is a negative area.

$$M_{15} = 43,600 - 10,800 = 32,800 \text{ ft-lb}$$

This point lies on the same parabola as the preceding two points since there was a continuous straight line in the shear diagram for the entire region.

The moment (M_{20}) at the right end is obtained by using the last area on the shear diagram.

$$M_{20} = M_{15} + \text{area}$$
$$= 32,800 - 32,800 = 0$$

The last segment of the shear diagram is a horizontal line, that is, a negative constant. The last segment of the moment diagram is therefore a straight line with a negative slope.

Since the diagram has closed to the known value (zero) of the moment at the right end, all calculations would appear to be correct.

EXAMPLE 7-6

Draw the shear and moment diagrams for the cantilever beam shown in Fig. 7-16(a).

Solution:

First we note that this is a cantilever beam, so the external reactions at the wall will be an upward force (R_w) and a clockwise couple (M_w). These reactions are evaluated by means of the equilibrium equations as follows:

$$\overset{+\uparrow}{\sum} F = 0$$

$$-300(8) - 2000 + R_w = 0$$

$$R_w = 4400 \text{ lb}$$

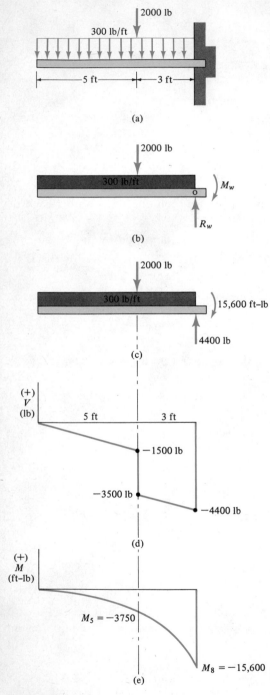

(a)

(b)

(c)

(d)

(e)

Figure 7-16

$$\overset{\curvearrowleft{+}}{\sum} M_o = 0$$

$$+300(8)(4) + 2000(3) - M_w = 0$$

$$M_w = +15,600 \text{ ft-lb}$$

These results are shown in Fig. 7-16(c).

Next, we lay out the axes for our shear and moment diagrams noting that we may expect discontinuities at the concentrated load and at the wall.

The shear diagram begins at zero since the force at the free end is zero. It goes downward at the rate of 300 lb/ft for 5 ft. Just to the left of the 2000-lb load the value of the shear is -1500 lb. The addition of the 2000-lb load pushes it further into the negative direction to a value of -3500 lb. During the next 3 ft, the shear continues to drop at 300 lb/ft to a value of -4400 lb. The upward reaction at the wall closes the shear to zero. Since the beam carries a uniform load, we connect values we have plotted with straight lines as shown in Fig. 7-16(d).

The moment diagram begins at zero at the left end. The change in moment between the left end and the 2000-lb load is the area of the triangular portion of the shear diagram and is negative, thus

$$M_5 = 0 - \tfrac{1}{2}(5)(1500) = -3750$$

The change in moment over the next 3 ft is the area of the negative trapezoid, thus

$$M_8 = -3,750 - \tfrac{1}{2}(3500 + 4400)(3)$$

$$= -15,600$$

Thus the moment diagram closes to the value of the previously computed couple reaction. The plotted points are connected by segments of two parabolas with negative slopes increasing in magnitude as we move from left to right. This results in parabolas that are concave downward as shown in Fig. 7-16(d). (Note that the slope of the parabola at the left end is zero.)

Strictly speaking it was unnecessary to calculate the reactions for this problem, but, as you can see, it provides a nice check on the work to do so.

PROBLEMS

7-7 A beam 15 ft long is loaded so that its shear diagram is as shown. Draw its load diagram and its moment diagram. Determine the simple reactions at $x = 5$ and $x = 15$ and the maximum moment.

Problem 7-7

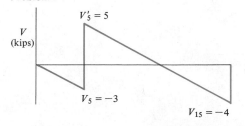

7-8 A cantilever beam 3 m long is loaded so that its moment diagram is as shown. Draw the shear diagram and the load diagram. Determine the reactions.

7-9 to 7-14 Draw the moment diagrams for the beams in Probs. 7-1 to 7-6. Give values at all maxima and all discontinuities.

Problem 7-8

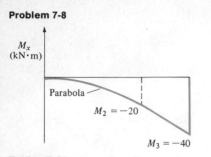

Parabola
$M_2 = -20$
$M_3 = -40$

7-15 to 7-28 Draw the shear and moment diagrams for the beams in Probs. 7-15 to 7-28. Give values at all maxima and all discontinuities.

Problem 7-15

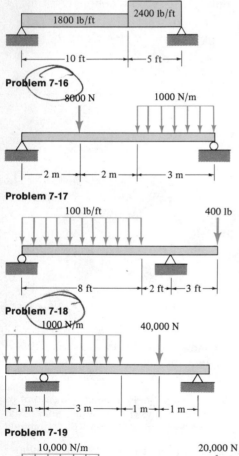

2400 lb/ft
1800 lb/ft
├─10 ft─┤─5 ft─┤

Problem 7-16

8000 N
1000 N/m
├─2 m─┼─2 m─┼─3 m─┤

Problem 7-17

100 lb/ft
400 lb
├──8 ft──┤─2 ft─┼─3 ft─┤

Problem 7-18

1000 N/m
40,000 N
├1 m─┼──3 m──┼1 m┼1 m┤

Problem 7-19

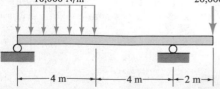

10,000 N/m
20,000 N
├──4 m──┼──4 m──┼2 m┤

Problem 7-20

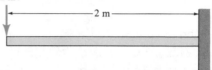

3 kips
1 kip/ft
3 kips
├─4 ft─┼──8 ft──┼─4 ft─┤

Problem 7-21

12 kN
├───────2 m───────┤

Problem 7-22

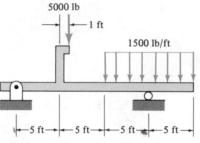

5000 lb
1 ft
1500 lb/ft
├─5 ft─┼─5 ft─┼─5 ft─┼─5 ft─┤

Problem 7-23

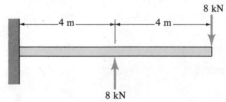

8 kN
├────4 m────┼────4 m────┤
8 kN

Problem 7-24

2 kips
5 ft
3 ft
2 kips

Problem 7-25

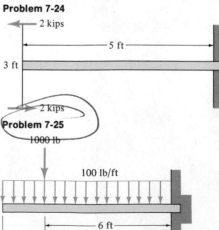

1000 lb
100 lb/ft
├────6 ft────┤
├──────8 ft──────┤

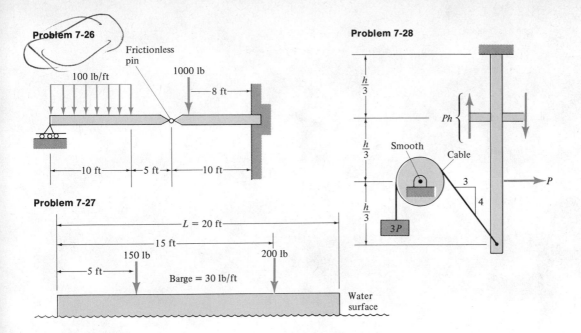

Problem 7-26

Frictionless pin

100 lb/ft

1000 lb

8 ft

10 ft

5 ft

10 ft

Problem 7-27

$L = 20$ ft

15 ft

150 lb

200 lb

5 ft

Barge = 30 lb/ft

Water surface

Problem 7-28

$\dfrac{h}{3}$

$\dfrac{h}{3}$

$\dfrac{h}{3}$

Smooth

Cable

Ph

3

4

P

$3P$

7-7 / Pure bending of beams with symmetrical sections

Initial restrictions   Generally, when a beam having an arbitrary cross section is subjected to transverse loads, the beam will bend. In addition, twisting and buckling may occur, and a problem that includes the combined effects of bending, twisting, and buckling can become a complicated one. In order that we may investigate the bending effects alone, we shall place certain restrictions on the geometry of the beam and on the manner of loading.

First we shall assume that the beam is straight, has a constant cross section, and is made of a homogeneous material, and that its cross section *has a longitudinal plane of symmetry*. Second, we shall assume that the *resultant* of the *applied loads lies in this plane of symmetry*. These conditions will eliminate the possibility that the beam will twist. Later we shall see that these strict requirements can be lessened somewhat, but for the present it is convenient to assume that the above conditions exist. Also, we shall assume for the present that the geometry of the overall member is such that bending and *not* buckling is the primary mode of failure. In Fig. 7-17 are shown several possible shapes of beams, each of which has a longitudinal plane of symmetry.

We know that the internal reactions on any cross section may consist of a resultant normal force, a resultant shear force, and a resultant couple. In order that we may examine the bending effects alone, we shall restrict the loading to one for which the resultant normal and shear forces are zero on any section perpendicular to the longitudinal axis of the member.

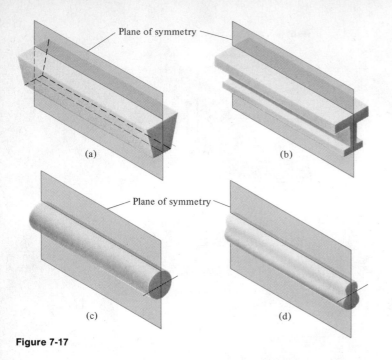

Plane of symmetry

(a)

(b)

Plane of symmetry

(c)

(d)

Figure 7-17

The zero shear force implies that the bending moment is the same at every cross section of the beam; that is, $dM/dx = 0$. We may visualize this beam, or some portion of the beam, as being loaded only by pure couples at its ends, remembering that these couples are assumed to be applied in the plane of symmetry. When a member is so loaded, it is said to be in *pure bending* and the plane of symmetry is called the *plane of bending*. The problem as now stated has sufficient symmetry to permit reasonable arguments about the deformation of the beam.

Longitudinal deformation   Figure 7-18(a) shows one portion of a homogeneous straight beam with a longitudinal plane of symmetry. For convenience a rectangular cross section was selected. At *AB*, *CD*, and *EF* are represented three equally spaced plane sections perpendicular to the longitudinal axis of the beam. Figure 7-18(b) shows the shape of the longitudinal plane of symmetry of the beam after the application of the couples at its ends, with the deformation greatly exaggerated. The portion of the beam between the plane sections at *AB* and *CD* was originally geometrically identical in length and cross section with the portion between the sections at *CD* and *EF*. Also, each of these portions is loaded in exactly the same manner, that is, by pure couples. Therefore, in order that each of these portions may be geometrically compatible with its adjacent portions in the deformed state, we may expect that these portions would undergo similar compatible deformations. The same

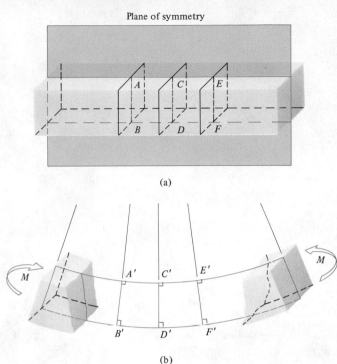

Figure 7-18

argument can be extended to the entire length of the beam (except of course to a part in the immediate vicinity of the externally applied loads). We can now draw several important conclusions in regard to the deformation in Fig. 7-18(b), which is characteristic of the deformation of any beam having a symmetrical cross section and subjected to pure bending.

1 Plane sections originally perpendicular to the longitudinal axis of the beam *remain plane* and perpendicular to the longitudinal axis after bending; that is, in Fig. 7-18(b), the cross sections at $A'B'$, $C'D'$, and $E'F'$ do not become curved or warped.

2 In the deformed beam, the planes of these cross sections have a common intersection; that is, any line originally parallel to the longitudinal axis of the beam becomes an arc of a circle.

Note that the analysis so far has dealt only with the deformation in any plane parallel to *the vertical plane of symmetry*. No remarks have yet been made in regard to the deformations perpendicular to the plane of symmetry. They will be considered presently.

Let us now examine more closely the deformation of an element of the beam having a length Δx. A small portion of the unloaded beam is shown

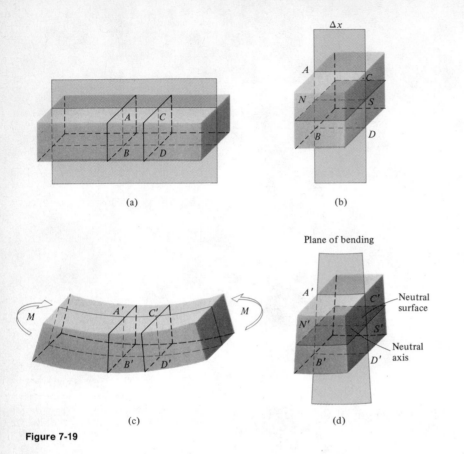

Figure 7-19

in Fig. 7-19(a), and the unloaded element is shown in (b). Also, a part of the deformed beam and the deformed element are shown in Fig. 7-19(c) and (d), where the deformation is greatly exaggerated.

Under the influence of the couples M, the upper fibers of the beam are shortened while the lower fibers are elongated. However, there is one surface (containing the arc $N'S'$) in which the fibers undergo *no* elongation or contraction. This longitudinal surface is called the *neutral surface*. For a beam with a symmetrical cross section and under pure bending, the neutral surface is perpendicular to the longitudinal plane of bending. Also, the intersection of the neutral surface with the longitudinal plane of bending of a beam, arc $N'S'$ in Fig. 7-19(d), is called the *neutral axis* of the beam.

Although we have now defined the neutral surface and the neutral axis for a beam, we have not as yet attempted to determine their locations in the beam. This will be done shortly, but for the present we shall assume that the neutral surface can be located.

In Fig. 7-20(a), u represents the distance between the neutral axis NS

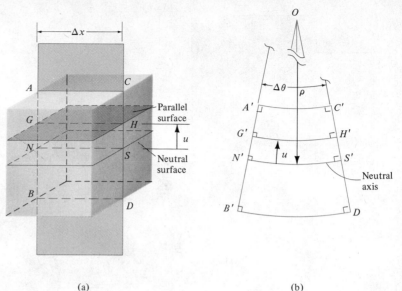

(a)   (b)

Figure 7-20

and some other parallel line GH in the plane of symmetry. We now make the assumption that the distance between these lines in the deformed beam, Fig. 7-20(b), will not be significantly different from the original distance in the unloaded beam. Thus, if the radius of curvature of $N'S'$ is ρ, that of $G'H'$ will be $\rho - u$. From the definition of neutral surface and the geometry of Fig. 7-20(b), we have

$$\Delta x = N'S' = \rho(\Delta\theta)$$

Also, the deformation of a fiber whose original position was GH becomes

$$G'H' - GH = (\rho - u)(\Delta\theta) - \Delta x = (\rho - u)(\Delta\theta) - \rho(\Delta\theta) = -u(\Delta\theta)$$

By definition, the axial strain of the fiber GH becomes

$$\varepsilon = \frac{-u(\Delta\theta)}{\Delta x} = \frac{-u(\Delta\theta)}{\rho(\Delta\theta)} = -\frac{u}{\rho} \tag{7-6}$$

Since for pure bending the radius of curvature is constant for the entire length of the beam, we see that *the longitudinal axial strain is directly proportional to the distance u from the neutral surface.* The negative sign indicates compressive strain for a positive value of u (fibers above the neutral surface), and tensile strain for a negative value of u (fibers below the neutral surface). It is assumed, of course, that the radius of curvature is positive (the beam is concave upward).

Transverse deformation Before concluding this section, we shall consider briefly the deformations perpendicular to the longitudinal plane

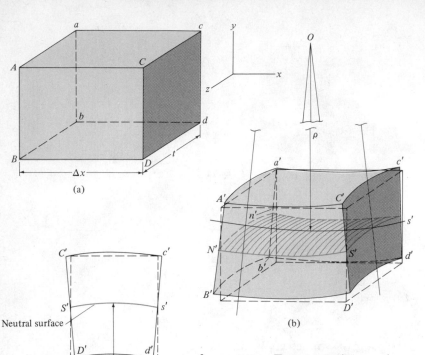

(a)

(b)

Neutral surface

(c)

Figure 7-21

of symmetry. For convenience a beam with a rectangular cross section will again be used for the discussion. See the sections *ABba* and *CDdc* in Fig. 7-21(a).

We have already seen that the longitudinal axial strain is directly proportional to the distance from the neutral surface, which is represented in Fig. 7-21(b) by the shaded surface $N'S's'n'$. We know that an elongation in a longitudinal direction will result in a contraction in the transverse direction because of the Poisson effect. Thus the fibers above the neutral surface will be elongated in a direction perpendicular to the plane of symmetry, and those below will be shortened. If the longitudinal strain is denoted by ε_x, the strain ε_z in the z direction beomes

$$\varepsilon_z = -\mu\varepsilon_x = -\mu\left(-\frac{u}{\rho}\right) = \mu\left(\frac{u}{\rho}\right) = \frac{u}{\rho/\mu}$$

Hence, the original rectangular cross section *CDdc* deforms as shown in Fig. 7-21(c), and $S's'$ has a radius of curvature of ρ/μ.

The neutral surface actually undergoes a double curvature, as indicated in Fig. 7-21(b); strictly speaking, therefore, our previous analysis is valid only for the plane of symmetry. In applying the theory of this section in succeeding sections, however, we shall neglect the curvature due to Poisson's effect, and we shall assume that the longitudinal axial strain throughout the length and thickness of the beam is proportional to the distance from the neutral surface. The existence of linear strain distribution can be easily verified by a simple laboratory experiment in which several strain gages are used on a beam.

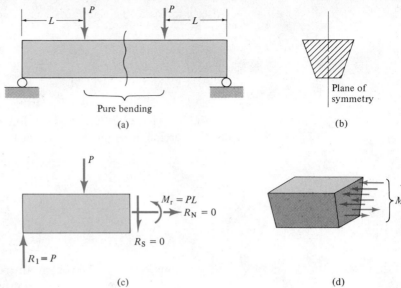

Pure bending

(a)

Plane of symmetry

(b)

$M_r = PL$

$R_N = 0$

$R_S = 0$

$R_1 = P$

(c)

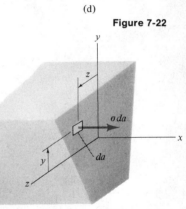

M_r

(d)

Figure 7-22

7-8 / Basic load-stress relationship for bending

The entire analysis in the preceding section is based on a member subjected to pure bending. This condition exists when the only resultant internal reaction is a pure couple. For example, in Fig. 7-22(a), the portion of the beam between the two equal loads P is in pure bending, because the *resultant* normal and shear forces on any cross section in this portion will be zero. The resultant internal reactions on a cross section are shown in Fig. 7-22(c). There is a resultant bending moment, and this moment must be the net effect of the normal tensile and compressive forces exerted by the individual fibers of the beam. Thus, as shown in Fig. 7-22(d), normal stresses do exist on each cross section, but from the previous deformation analysis and the fact that there is no resultant shear force, we can assume that there are no shear stresses on the cross section. Normal stresses due to bending are commonly called *flexure stresses*.

Now, although we do not yet know the location of the neutral axis, we shall arbitrarily establish a reference coordinate system in which the x axis is taken as the neutral axis. As shown in Fig. 7-23, the xy plane is then the plane of symmetry of the beam, and the neutral surface lies in the xz plane. Since the resultant internal moment M_r is actually a moment about the z axis, we have

Figure 7-23

$$M_r = \sum M_z \quad \overset{\curvearrowleft +}{}$$

$$M_r = -\int_{\substack{\text{cross} \\ \text{section}}} y(\sigma\, da) \qquad (7\text{-}7)$$

Also, if forces to the right are considered positive,

$$R_N = \overset{+}{\underset{\rightarrow}{\sum}} F_x = 0$$

$$R_N = \int_{\substack{\text{cross} \\ \text{section}}} \sigma \, da = 0 \tag{7-8}$$

Finally, we know that the resultant moment about the y axis is zero. If counterclockwise moments about this axis are considered positive, we get

$$\sum M_y = 0$$

$$\sum M_y = \int_{\substack{\text{cross} \\ \text{section}}} z(\sigma \, da) = 0 \tag{7-9}$$

Equations 7-7 and 7-9 represent the requirements that must be satisfied by the stress distribution over each cross section of the beam. Actually, because of the symmetry of this problem with respect to the xy plane, Eq. 7-9 will automatically be satisfied whenever a beam with a symmetrical cross section is loaded in the plane of symmetry.

Having expressed the strain distribution by Eq. 7-6 and the load-stress requirements by Eqs. 7-7 and 7-8 we are now ready to determine specific load-stress relationships. Of course, any such relationships will be dependent on the stress-strain relationship for the material involved. The following important special cases are intended to illustrate the procedure for deriving specific load-stress relationships.

Case I: Elastic bending of beams with symmetrical sections Elastic behavior of flexure members plays an important role in design procedure. For instance, a loaded beam that has not been stressed beyond the elastic limit of the material will exhibit no permanent "sag" when the loads are removed. Also, for most materials, elastic deformations are very small, and the deflection of an elastic beam will therefore be small. In many design problems small elastic deformations can be tolerated but large inelastic deformations would produce failure. For these reasons we are quite justified in studying elastic behavior of beams in considerable detail.

We have already shown in section 7-7 that the longitudinal axial *strain* varies linearly from the neutral surface. Such a distribution can be illustrated graphically by means of a diagram such as that in Fig. 7-24, where the upper fibers at section cc are in compression and the lower fibers are in tension. The strain at any point at a distance y from the neutral surface can be determined by the following simple ratio:

$$\frac{\varepsilon}{y} = \frac{\varepsilon_{\max}(\text{tension})}{b} = \frac{\varepsilon_{\max}(\text{compression})}{a} \tag{7-10}$$

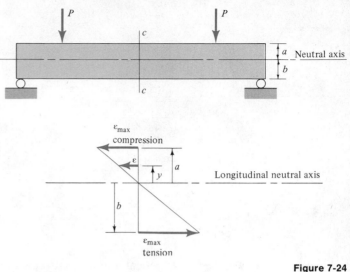

Figure 7-24

So far, no restriction on the behavior of the material has been imposed. Therefore, this strain distribution implies neither elastic nor inelastic behavior, and it is usually assumed that this distribution exists at least until the beam has been deformed beyond use.

For a uniaxial state of stress, elastic response of a homogeneous isotropic material usually manifests itself in a linear stress-strain relationship having the form

$$\sigma = E\varepsilon \qquad (7\text{-}11)$$

Since the axial strain in a beam varies linearly from the neutral surface, elastic behavior will be chacterized by a *linear stress distribution* because the stress is simply the strain multiplied by the modulus of elasticity. However, it must be remembered that in a bent beam some of the fibers are in tension and others are in compression. Fortunately, the modulus of elasticity for most structural materials is the same in tension as in compression. Figure 7-25(a) shows the stress distribution for this case, while Fig. 7-25(b) shows the stress distribution for the case where the moduli differ. We shall now continue our development for the former case.

Substituting the relationship of Fig. 7-25(a) in Eq. 7-8, we obtain

$$\int_{\substack{\text{cross} \\ \text{section}}} \sigma \, da = 0$$

$$\int ky \, da = k \int_{\substack{\text{cross} \\ \text{section}}} y \, da = 0 \qquad (7\text{-}12)$$

Since k is not zero, the integral, which represents the first moment of the cross-sectional area about the neutral surface, must be zero. This require-

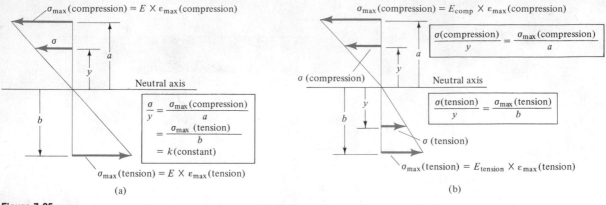

Figure 7-25

ment implies that the neutral surface of the beam must pass through the horizontal centroidal axis of the cross section. Thus, *for elastic action the strain and stress vary linearly from the transverse centroid of the cross section.*

Substitution in Eq. 7-7 yields the following result:

$$M_r = -\int_{\substack{\text{cross} \\ \text{section}}} y(\sigma \, da) = -\int_{\substack{\text{cross} \\ \text{section}}} y(ky \, da)$$

$$M_r = -k \int_{\substack{\text{cross} \\ \text{section}}} y^2 \, da \tag{7-13}$$

Since we have just seen that y is measured from the centroidal axis, the integral in Eq. 7-13 represents the second moment of the cross-sectional area about its horizontal centroidal axis. This second moment is commonly called the rectangular moment of inertia although no masses are involved, only cross-sectional areas. These second moments of area should not be confused with mass moments of inertia encountered in dynamics.

If the second moment of area is denoted by I_c, the equation becomes

$$M_r = -kI_c \tag{7-14}$$

The minus sign merely indicates that a positive bending moment M_r produces compression in the upper fibers and tension in the lower fibers. In particular, from Fig. 7-25(a) we have

For upper fibers: For lower fibers:

$$\frac{\sigma_{\max}(\text{compression})}{a} = k \qquad \frac{\sigma_{\max}(\text{tension})}{b} = k$$

$$\sigma_{\max}(\text{compression}) = \frac{M_r a}{I_c} \qquad \sigma_{\max}(\text{tension}) = \frac{M_r b}{I_c} \tag{7-15}$$

In general,

$$\frac{\sigma}{y} = k$$

$$\sigma = -\frac{M_r y}{I_c} \qquad (7\text{-}15a)$$

Equations 7-15 and 7-15a are various forms of the famous and widely used (and misused) *elastic flexure formula*. It would be well for you to go back over the derivation and note the limitations that were imposed in the procedure.

Case 2: Ideally plastic behavior It was pointed out at the beginning of the preceding derivation that the strain distribution of Fig. 7-24 is independent of the behavior of the material. Since many structural members are made of ductile materials, particularly mild steel, we shall devote some time to considering ideally plastic behavior. The essential characteristic of a very ductile material is that the uniaxial stress/strain curve becomes flat-topped as the strain becomes relatively large. In the case of mild steel this flattening occurs at the yield-point stress as indicated in Fig. 7-26(a), whereas for most other ductile materials it occurs at the ultimate stress as indicated in Fig. 7-26(b).

Let us follow the progress of the stress distribution as a beam of mild steel is subjected to larger and larger strains. In Fig. 7-26(c), (d), and (e) are shown three distinctive stages of the stress distribution as the material passes through the elastic, the elastic-plastic, and the fully plastic stages. For the *idealized* fully plastic condition, all the fibers above the neutral surface (wherever it may be) are subjected to a *uniform* compressive stress σ_{ypc}, and the lower fibers are subjected to a *uniform* tensile stress σ_{ypt}. If compressive stress is denoted by a minus sign, Eq. 7-8 becomes

$$\int_{\substack{\text{cross}\\\text{section}}} \sigma \, da = 0$$

$$\int_{\substack{\text{upper}\\\text{area}}} (-\sigma_{ypc}) \, da + \int_{\substack{\text{lower}\\\text{area}}} \sigma_{ypt} \, da = 0$$

If the yield-point stress is the same for tension and compression, we obtain

$$\sigma_{yp}(\text{upper area} - \text{lower area}) = 0$$

Hence, the area above the neutral surface must be equal to the area below the neutral surface. In other words, for fully plastic action the neutral surface splits the cross section in half, areawise. The position of the neutral surface depends on the geometry of the cross section; the neutral surface *may or may not* pass through the centroidal axis.

The moment-stress relationship can be obtained from Fig. 7-26(f). The resultant forces F_c and F_t act at the centroids of the corresponding distributed forces. Thus,

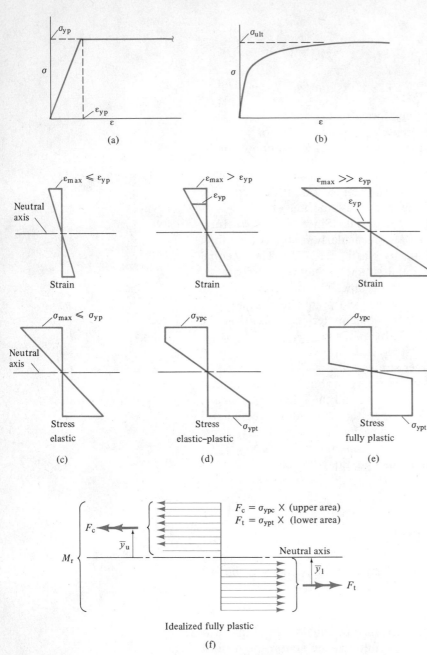

Figure 7-26

$$M_r = [\sigma_{ypc} \times (\text{upper area}) \times (\bar{y}_u)] + [\sigma_{ypt} \times (\text{lower area}) \times (\bar{y}_l)] \quad (7\text{-}16)$$

where $\bar{y}_u$ and $\bar{y}_l$ are the upper and lower centroidal distances, respectively, and each is measured from the fully plastic neutral surface.

EXAMPLE 7-7

A beam is fabricated of three 6 × 1-in. ductile steel plates welded together to form a symmetrical I-shaped section, as shown in Fig. 7-27(a). If the material has an elastic limit of 25,000 psi and an ultimate stress of 40,000 psi, determine (a) the maximum elastic bending moment and (b) the fully plastic ultimate bending moment, assuming an idealized stress distribution.

Solution:

(a) From Fig. 7-27(a), the second moment of area (or the moment of inertia) about the centroidal axis is

$$I_c = \frac{(1)(6^3)}{12} + 2\left[\frac{(6)(1^3)}{12} + (6)(1)(3.5^2)\right]$$

$$= 18 + 148 = 166 \text{ in.}^4$$

For elastic behavior the flexure formula is applicable. Since the maximum stress (tensile or compressive) occurs at the outside fibers, the use of Eq. 7-15 gives

$$M_{\text{elastic}} = \frac{\sigma_{\max}I_c}{y_{\max}} = \frac{(25,000)(166)}{4} = 1,037,500 \text{ in.-lb}$$

(b) For the fully plastic case, the idealized stress distribution is shown in Fig. 7-27(b). This distribution can be thought of as four uniformly distributed forces, each of which is equal to the ultimate stress multiplied by the corresponding cross-sectional area. Hence, assuming that the lower fibers are in tension, we obtain

$$M_{\text{ult}} = F_1(3.5) + F_2(1.5) + F_3(1.5) + F_4(3.5)$$
$$= 2[F_1(3.5) + F_2(1.5)] = 2(40,000)[6(3.5) + 3(1.5)]$$
$$= 2,040,000 \text{ in.-lb}$$

Figure 7-27

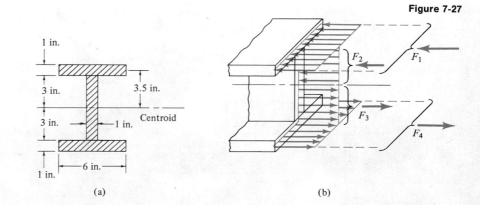

(a)   (b)

Note: In this particular example, both the neutral surface for elastic action and that for plastic action pass through the geometric center. In general, this condition will *not* be true. Also note that in this case the ultimate bending moment for fully plastic behavior is approximately twice the maximum elastic bending moment even though the ultimate stress is not twice the elastic limit stress. Thus if failure is based on rupture rather than the *initiation* of inelastic behavior, the allowable bending moment for this beam can be twice as large.

PROBLEMS

7-29 to 7-32 For each of the beam cross sections shown, locate the neutral surface for (a) elastic bending, and (b) fully plastic bending.

Problem 7-29   **Problem 7-30**

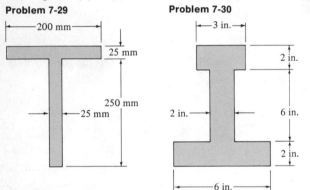

Problem 7-31

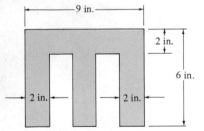

Problem 7-32

7-33 Which of the two cross sections shown in the illustration would be better for use as an elastic beam? Prove your answer.

Problem 7-33

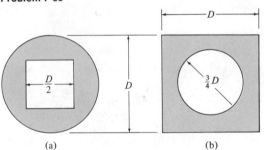

7-34 The elastic beam in the illustration has an I of 164 in.^{4} about its centroidal axis. Under the load P at point A, 2 in. from the top of the beam, there is a longitudinal strain of -0.0005 in./in.; and at point B, 2 in. from the bottom of the beam, there is a strain of $+0.00075$ in./in. The beam is made of a material with modulus $E = 10 \times 10^6$. What is the amount of the load P? Neglect the localized effect of the load.

Problem 7-34

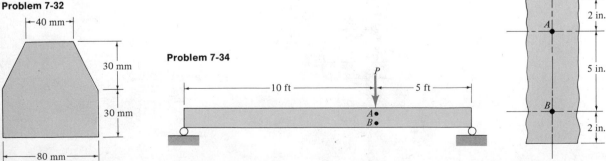

7-35 and 7-36 The illustration shows the cross section of a mild-steel beam subjected to pure bending about the axis *xx*. Determine the ratio M_{fp}/M_{el}.

Problem 7-35

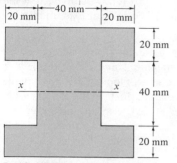

Problem 7-36

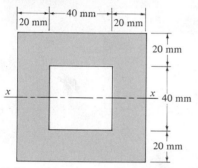

7-37 What is the ratio of the ideally fully plastic moment M_{fp} to the maximum elastic bending moment bending M_{el} for a beam of mild steel having a rectangular cross section?

7-38 What is the ratio of the ideally fully plastic bending moment to the maximum elastic bending moment for a beam of mild steel with a circular cross section subjected to pure bending?

7-39 What is the ratio of the ideally fully plastic bending moment to the maximum elastic bending moment for a beam of mild steel subjected to pure bending and having a cross section in the shape of an upright isosceles triangle?

7-40 What is the ratio of the ideally fully plastic bending moment to the maximum elastic bending moment for a beam of mild steel subjected to pure bending and having an elliptical cross section *c* in. deep (major axis) and *b* in. wide (minor axis)?

7-41 Determine the fully plastic ultimate bending moment for a beam having the cross section shown

in the illustration, if the ultimate stress for the ductile material is 60,000 psi.

Problem 7-41

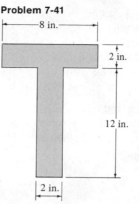

7-42 The illustration shows the cross section of a beam made of a very ductile material that has an ultimate stress of 100 MPa. What is the fully plastic ultimate bending moment for this section?

Problem 7-42

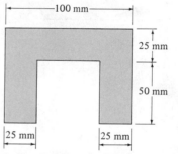

7-43 A simply supported beam of negligible weight is made of a material that has a yield-point stress and has the cross section shown in the illustration. The beam is 12 ft long and is loaded with a single concentrated load at the middle of the span. When this load reaches 20,000 lb, the beam yields and

Problem 7-43

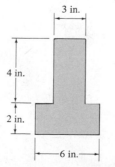

ideally fully plastic behavior occurs. What is the yield-point stress of the material?

7-44 The steel beam in part (a) of the illustration has the cross section shown in part (b). Determine the maximum allowable value of P for (a) failure due to the initiation of inelastic behavior, and (b) failure due to excessive deformation resulting from the formation of a fully plastic hinge with $\sigma_{yp} = 200$ MPa.

7-45 If the beam in part (a) of the illustration is made of a material that has the stress-strain curve shown in part (b), what load W will cause complete failure of the beam? Describe the failure.

7-46 The beam shown in part (a) of the illustration is deformed by couples M into the shape shown in part (b) so that there is a strain of 0.0015 in./in. at point A. The beam is made of a material whose stress-strain curve is shown in part (c). Assuming that the strains vary linearly from the neutral axis, determine the value of M.

7-47 A beam having a length L and a rectangular cross section 2c in. in height and b in. in width is subjected to a pure bending moment M. The tensile and compressive stress-strain law for the material is approximated by $|\sigma| = \lambda |\varepsilon|^{\frac{1}{3}}$. In an attempt to calculate the maximum flexure stress, a designer uses

Problem 7-44

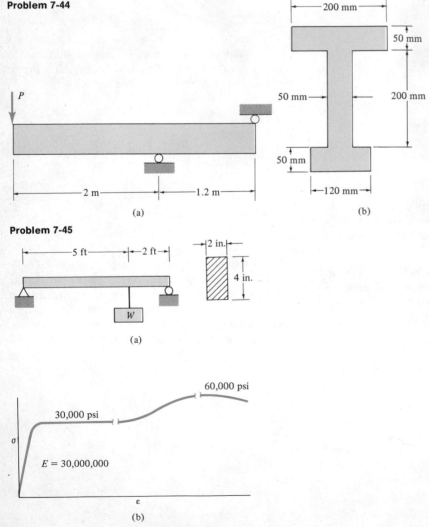

(a)

(b)

Problem 7-45

(a)

(b)

Problem 7-46

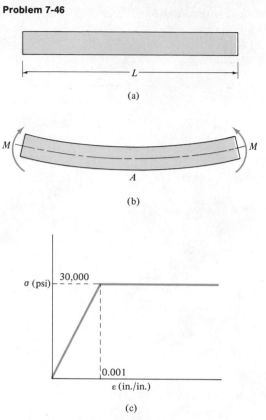

(a)

6 in.

2 in.

(b)

(c)

the relationship $\sigma = k(Mc/I)$. What should be the value of k for this application?

7-48 A rectangular beam 8 in. wide and 12 in. deep has a nonlinear stress distribution given by $\sigma =$ $ky^{\frac{1}{3}}$ for both the upper and lower fibers of the beam, the upper fibers being in compression and the lower in tension. What is the magnitude of the pure bending moment producing this stress if $k = 4000$?

7-9 / The use of the elastic flexure equation

The Elastic Flexure formula, Eq. 7-15a, or simply the flexure equation, is usually written in the form

$$\sigma = \frac{My}{I} \qquad\qquad [7\text{-}15a]$$

and is the most widely used relationship for analyzing stresses in beams. Many of the problems in section 7-8 were designed to make you sensitive to situations to which the flexure equation does not apply. However, you should be aware that it is applicable in the vast majority of beam problems and you should become adept at using it properly. The following examples and problems will deal with situations where it does apply and will help you develop fluency in its use.

EXAMPLE 7-8

A beam carries a maximum moment of 6300 N·m. It is composed of a circular pipe with an outer diameter of 75 mm and an inner diameter of 50 mm. Compute the maximum stress developed. Assume the beam is elastic.

Solution:

The maximum stress will be developed at the outer edge of the pipe, so:

$$\sigma_{max} = \frac{M_{max} r_o}{I} = \frac{M_{max} r_o}{(\pi/4)(r_o^4 - r_i^4)}$$

$$= \frac{6300 \times 0.0375}{(\pi/4)[(0.0375)^4 - (0.025)^4]} = \frac{236.250}{1.246 \times 10^{-6}} = 189.6 \text{ MPa}$$

EXAMPLE 7-9

The beam shown in Fig. 7-28(a) has the cross section shown in Fig. 7-28(b). Find the maximum tensile stress and the maximum compressive stress. Assume elastic action.

Solution:

The first step is to draw the shear and moment diagrams for the beam. These are shown in Fig. 7-28(c) and (d). The moment value is slightly adjusted to allow for errors due to round-off in the reactions.

The neutral axis of this beam coincides with the horizontal centroidal axis of the cross section, so our next step is to locate the centroid and compute the moment of inertia about the centroidal axis.

To locate the centroid we recall:

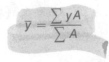

$$\bar{y} = \frac{\sum yA}{\sum A}$$

Measuring y from the bottom

$$\bar{y} = \frac{(6 \times 1)(8.5) + (8 \times \frac{3}{4})(4)}{(6 \times 1) + (8 \times \frac{3}{4})} = \frac{(6)(8.5) + (6)(4)}{6 + 6}$$

$$= \frac{75}{12} = 6.25$$

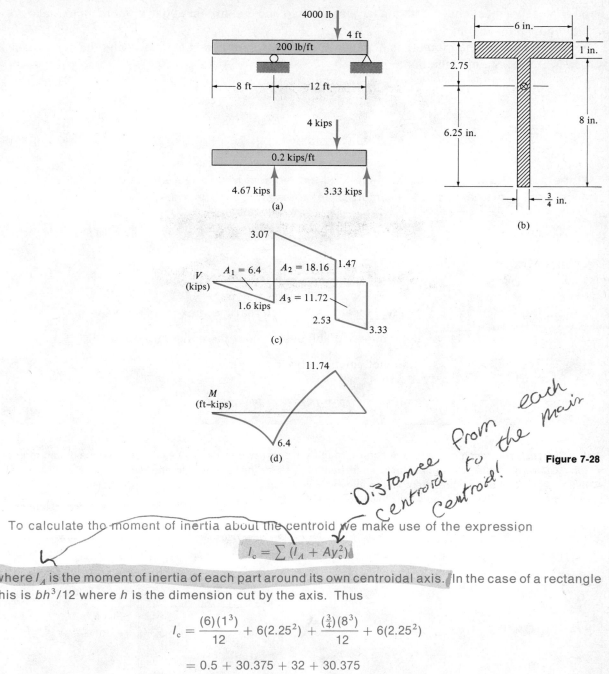

Figure 7-28

To calculate the moment of inertia about the centroid we make use of the expression

$$I_c = \sum (I_A + Ay_c^2)$$

where I_A is the moment of inertia of each part around its own centroidal axis. In the case of a rectangle this is $bh^3/12$ where h is the dimension cut by the axis. Thus

$$I_c = \frac{(6)(1^3)}{12} + 6(2.25^2) + \frac{(\frac{3}{4})(8^3)}{12} + 6(2.25^2)$$

$$= 0.5 + 30.375 + 32 + 30.375$$

$$= 93.25 \text{ in.}^4$$

Now we are ready to compute stresses. Since the cross section is not symmetrical about the neutral axis, we must check both the largest positive and the largest negative moment to make sure of finding both maximum stresses.

At the left support the moment is negative, so the tensile stress is at the top and the compressive stress is at the bottom of the beam.

$$\sigma_t = \frac{M_1 y_t}{I} = \frac{(6.4)(12)(2.75)}{93.25} = 2.26 \text{ ksi}$$

$$\sigma_c = \frac{M_1 y_c}{I} = \frac{(6.4)(12)(6.25)}{93.25} = 5.15 \text{ ksi}$$

At the point of application of the 4-kip load the moment is positive, so the tensile stress occurs at the bottom of the beam and the compressive stress occurs at the top.

$$\sigma_t = \frac{M_2 y_t}{I} = \frac{(11.74)(12)(6.25)}{93.25} = 9.44 \text{ ksi}$$

$$\sigma_c = \frac{M_2 y_c}{I} = \frac{(11.7)(12)(2.75)}{93.25} = 4.15 \text{ ksi}$$

Comparing the stresses at the two locations, we see that the maximum stresses are

$\sigma_t = 9.44$ ksi at the bottom of the beam under the concentrated load

$\sigma_c = 5.15$ ksi at the bottom of the beam over the support

PROBLEMS

7-49 The cast-iron machine part, a section through which is shown in the illustration, acts as a beam resisting a *positive* bending moment. If the allowable tensile stress is 10,000 psi and the allowable compressive stress is 35,000 psi, what maximum moment may be applied to the member? Assume elastic behavior.

7-50 An elastic straight beam has the constant cross section shown in the illustration. If a *negative* bending

Problem 7-49

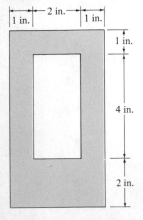

Problem 7-50

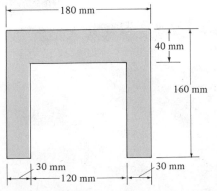

Problem 7-53

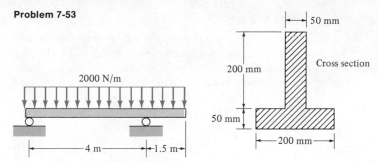

moment of 50,000 N·m acts about its horizontal centroidal axis, what are the maximum tensile and compressive flexure stresses?

7-51 What maximum elastic tensile flexure stress will occur in a simply supported timber beam 14 ft long with a rectangular cross section 6 in. wide by 10 in. deep? It is uniformly loaded with 300 lb/ft. Exactly where does this stress occur?

7-52 Determine the maximum allowable load which a Douglas fir beam $1\frac{5}{8}$ in. by $3\frac{5}{8}$ in. in cross section will carry at the center of an 8-ft span if the long dimension is (a) vertical, (b) horizontal.

7-53 Determine the maximum flexural stress in the beam shown and state where it occurs.

7-54 The beam in Prob. 7-16 has the cross section shown in Prob. 7-53. Find the maximum tensile and compressive flexure stresses.

7-55 The beam in Prob. 7-53(a) is composed of a Douglas fir 4 × 12 in. Find the maximum flexural stress.

7-56 The beam in Example 7-5 is an S 10 × 25.4. Find the maximum flexural stress, using the Appendix to determine properties of areas.

7-57 A steel rod 2 in. in diameter is used as a cantilever beam as shown in the illustration. If the allowable stress is 20,000 psi, what is the maximum allowable value for P?

Problem 7-57

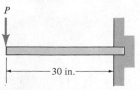

7-58 The beam in Prob. 7-14 is an S 5 × 14.75. What is the maximum flexural stress?

7-59 The beam in Prob. 7-11 is a C 380 × 50.4 placed with the 380-mm dimension horizontal and the flat side at the bottom. Determine the maximum tensile and compressive flexure stress in the beam (note: $S_y = I_y/x_{max}$).

7-60 The cross section in Prob. 7-50 is for the beam in Prob. 7-17. Find the maximum compressive flexure stress in the beam.

7-61 How long can a cantilevered beam of square cross section 4 × 4 in. be if it must support a 1000-lb load at its end and the bending stress cannot exceed 24,000 psi?

7-62 The elastic beam shown in the illustration is a rolled-steel W 10 × 100. Including the weight of the beam itself (100 lb/ft), determine the magnitude and location of the maximum flexure stress.

Problem 7-62

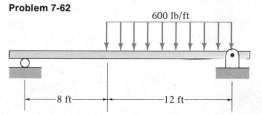

7-63 The elastic beam shown in the illustration is a rolled-steel W 12 × 120. What is the maximum value

Problem 7-63

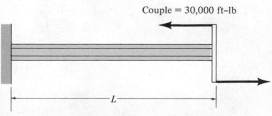

L can have, if the flexure stress is not to exceed 5000 psi anywhere in the beam? Do not neglect the weight of the member (120 lb/ft).

7-64 For the elastic beam shown in the illustration, determine the magnitude and location of (a) the maximum tensile flexure stress, (b) the maximum compressive flexure stress.

pressive flexure stress.

7-65 For the elastic beam AB in the illustration, determine the magnitude and location of (a) the maximum tensile flexure stress, (b) the maximum compressive flexure stress.

Problem 7-64

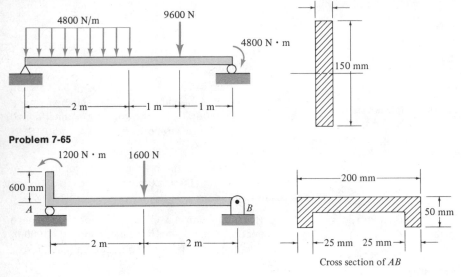

Problem 7-65

Cross section of AB

7-10 / The design of beams

The selection of an appropriate cross section to carry a given bending load is an extremely important problem. It is somewhat facilitated by the use of the notion of section modulus.

The section modulus of an area with respect to an axis is the moment of inertia of the area about that axis divided by the perpendicular distance from the axis to the most distant point on the cross section. It is commonly denoted by S, so that

$$S = \frac{I}{y_{max}}\qquad(7\text{-}17)$$

If we rearrange Eq. 7-15 making use of the section modulus concept, we see that for a design based upon elastic behavior:

$$\sigma_{max} = \frac{My_{max}}{I}$$

$$\sigma_{max} = \frac{M}{S}\qquad(7\text{-}18)$$

$$S = \frac{M}{\sigma_{max}} \qquad (7\text{-}18a)$$

Thus, the required section modulus for a beam may be found by dividing the maximum moment by the allowable stress. This has no particular advantage for a geometric section such as a circle or a rectangle but it is very convenient for the fabricated sections that are commonly used for beams. Appendices in this book will supply you with section modulus information for a selection of rolled-steel sections.

Although the beam must be designed so that none of the allowable stresses are exceeded, the initial design of a homogeneous beam is usually based on allowable tensile stress. It may be modified to take into account the effect of shear and compressive stress if such a modification is seen to be necessary.

The actual design of a beam is quite simple, although occasionally tedious. The process is as follows:

1 From the loadings that the beam must carry, draw the shear and moment diagrams.

2 From the diagram determine the maximum moment.

3 Divide the maximum moment by the allowable tensile stress to find the required section modulus.

4 Select a cross section that provides the section modulus. Check the shearing stress (discussed in section 7-13), and for a brittle material, make sure the section is not overstressed in compression.

You may have been wondering what to do about the weight of the beam itself. In the case of a beam of metal or wood, the weight of the beam is usually neglected since it is small compared to the superimposed loads. In the case of concrete, an initial guess is made at the size and a weight of 150 lb/ft^3 is included in the calculations. Needless to say, such an estimate must be checked and possibly modified when the actual size is determined.

EXAMPLE 7-10

Select a wide-flange (W) beam or an American Standard (S) beam to carry the loads shown in Example 7-9. The maximum allowable stress is 18 ksi. Use the tables in the Appendix.

Solution:

Since both shapes are symmetric about their centroids, we need only be concerned about the numerical maximum moment; thus the section modulus required is

$$S = \frac{M}{\sigma} = \frac{11.74 \times 12}{18} = 7.81 \text{ in.}^3$$

Referring to the tables and looking at the S_x columns (this presumes the beam will be placed as shown in the tables with web vertical), we find that the following beams satisfy our requirements.

S 6 × 17.25	$S_x = 8.77$
W 5 × 16	$S_x = 8.51$
W 6 × 16	$S_x = 10.2$

Since the second number is the weight per foot, the W beams are lighter, therefore cheaper. The W 6 × 16 is probably the best choice since it provides more strength at the same cost.

PROBLEMS

7-66 Find the section modulus of a circle about its diameter.

7-67 Find the section modulus of a rectangle about an axis bisecting its long sides.

7-68 to 7-70 Find the section moduli of the cross sections of Probs. 7-29 to 7-31 for elastic bending.

7-71 to 7-82 Using an allowable tensile stress of 10,000 psi or 70 MPa, select appropriate W and S-shape rolled-steel sections to use for the beams in Probs. 7-7 to 7-18. Use the appendices in this book for information on section moduli.

7-11 / Combined axial and bending loads

When a member is subject to simultaneous axial and bending loads, the normal stresses due to the two loadings are calculated separately and added algebraically so that the total normal stress due to the combination is

Figure 7-29

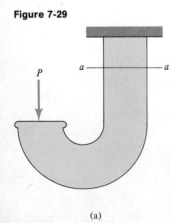

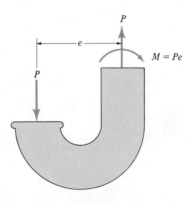

(a)   (b)

$$\sigma = \frac{P}{A} \mp \frac{My}{I} \qquad (7\text{-}19)$$

This sort of combination loading frequently results from a push or pull applied at a point other than the centroid of the cross section. Figure 7-29(a) shows a member subjected to such a loading. Figure 7-29(b) shows the internal loading on the cross section at *a-a*. It consists of load *P* placed at the centroid and a bending couple of magnitude *Pe*. We can now proceed with the application of Eq. 7-19.

As was first mentioned in Chapter 6, this idea of superposing the effects of combined loadings is valid for elastic behavior and we shall restrict our applications to such situations.

EXAMPLE 7-11

The hook shown in Fig. 7-30(a) has the cross section shown in Fig. 7-30(b). Find the maximum tensile and compressive stresses occurring normal to section *a-a*.

Solution:

The first step is to draw a free-body diagram showing the reactions on section *a-a*. These consist of a force acting through the centroid of the cross section and a couple. We recall that the centroid of a triangle lies at a point one-third of the distance from the base to the apex. In this case this is 1 in. from the inner edge of the cross section. Placing the 2-kip force reaction at the centroid results in an eccentricity of 4 in., giving us a moment reaction of 8 in.-kips as shown in Fig. 7-30(c).

The maximum stresses will occur at the extreme distances from the neutral axis which, in this case, passes through the centroid of the cross section.

Since the moment will cause tension at the inner edge, the stress at the inner edge will be

Figure 7-30

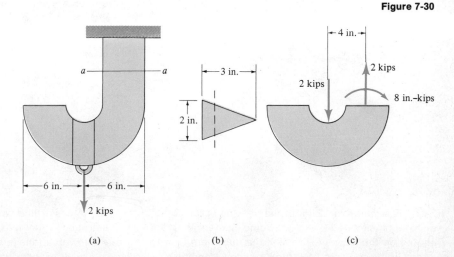

(a)   (b)   (c)

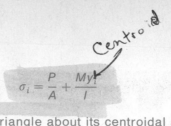

Centroid

$$\sigma_i = \frac{P}{A} + \frac{My}{I}$$

Recalling the moment of inertia of a triangle about its centroidal axis to be $bh^3/36$, we may substitute as follows:

$$\sigma_i = \frac{2}{(3)(2)/2} + \frac{(8)(1)}{(2)(3)^3/36} = \frac{2}{3} + \frac{8}{3/2}$$

$$= \frac{2}{3} + \frac{16}{3} = 6 \text{ ksi tension}$$

The moment causes compression at the outer edge while the axial load causes tension at all locations on the cross section, thus the stress at the outer edge is

$$\sigma_o = \frac{P}{A} - \frac{My_o}{I}$$

Using the values for moment of inertia and area from the earlier computation gives

$$\sigma_o = \frac{2}{3} - \frac{(8)(2)}{3/2} = \frac{2}{3} - \frac{32}{3} = \frac{-30}{3}$$

$$\sigma_o = 10 \text{ ksi compression}$$

PROBLEMS

7-83 Assuming elastic behavior, determine the maximum normal stress in the short block in the illustration. Express your answer in terms of P and d, and indicate where this stress occurs.

Problem 7-83

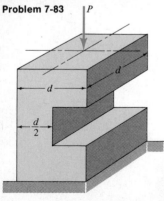

7-84 A circular rod 2 in. in diameter is subjected to an axial tensile load P and a bending moment M. What should be the ratio of P to M if the maximum tensile stress in the rod is to be three times the magnitude of

the maximum compressive stress? Assume elastic behavior.

7-85 Determine the maximum normal stress in the member shown in the illustration. Assume elastic behavior, and neglect stress concentrations at the wall.

Problem 7-85

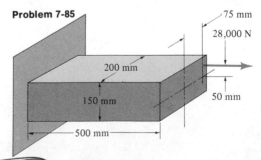

7-86 A cast frame has the cross section shown in the illustration. What maximum force P may be applied if the allowable tensile stress at section AA is 5000 psi and the allowable compressive stress is 15,000 psi?

Problem 7-86

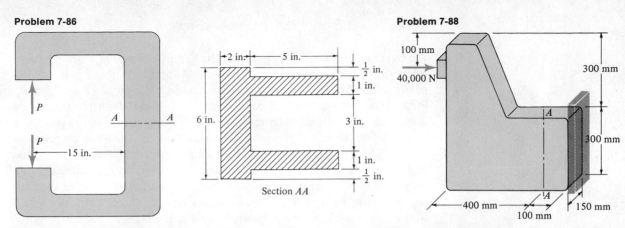

Problem 7-88

Section *AA*

7-87 A concrete retaining wall is 12 ft high, 4 ft thick, and 30 ft long. The material that it restrains stands to a height of 5 ft on one side of the wall and may be assumed to develop a pressure against the wall that varies linearly from zero at the 5-ft elevation to an intensity of 900 lb/ft² at the base. Concrete weighs 150 lb/ft³. Determine the stress distribution at the base of the wall. Give your answer in the form of a sketch.

7-88 Find the maximum compressive stress developed on plane *AA* by the 40,000-N load acting on the bumper shown.

7-89 A cast frame is to be built with a triangular cross

section at *A-A* as shown in the illustration. If the load *P* is 10,000 lb and the tensile stress may not exceed 5000 psi determine the dimensions of the cross section. You may need a trial and error solution.

7-90 Determine the maximum stress in the base of the concrete wall shown. The water develops a pressure varying linearly from zero at the surface to a maximum at the base. Concrete weighs about 23.5 kN/m³ and water about 9.8 kN/m.³

7-91 Find the normal stresses on a horizontal plane at points *A*, *B*, *C*, and *D* of the member shown. The weight of the member is negligible.

Problem 7-89

Problem 7-90

Problem 7-91

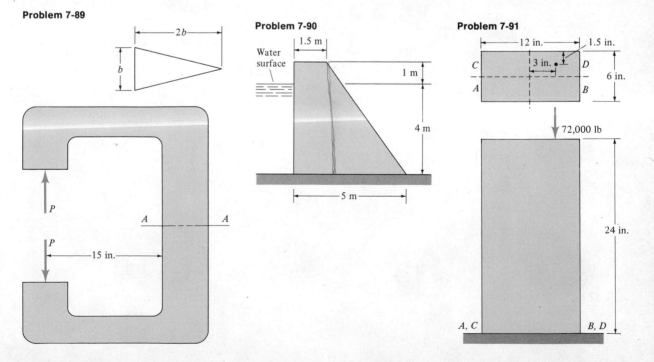

7-12 / Combined shear and bending

Up to this point we have considered the flexure stresses in a beam caused by the internal bending moment reaction M and have ignored the effects of the internal shear force V. Strictly speaking, the presence of the shear force and the resulting shear stresses and shear deformation would invalidate some of our statements in section 7-7 in regard to the geometry of the deformation and the resulting axial strain distribution represented by Eq. 7-6. Plane sections would no longer remain plane after bending, and the geometry of the actual deformation would become considerably more involved. Fortunately, for a beam whose length is large in comparison with the dimensions of the cross section, the deformation effect of the shear force is relatively small; and it is *assumed* that the longitudinal axial strains are still distributed in the same manner as for pure bending. When this assumption is made, the load-stress relationships developed in section 7-7 are considered valid. Experience and experi-

Figure 7-31

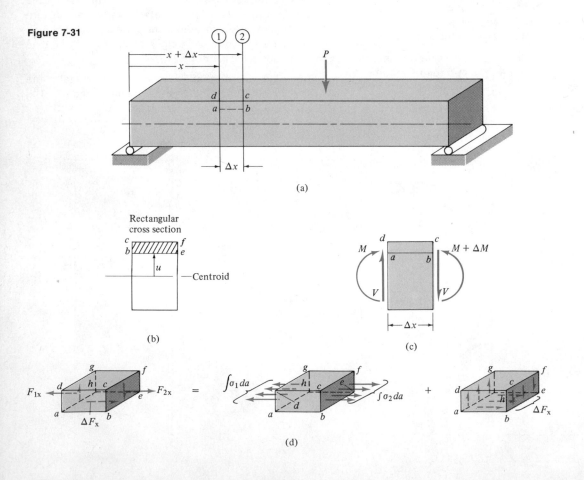

(a)

(b)

(c)

(d)

mental work indicate that this assumption is sufficiently accurate for most practical purposes.

The following question immediately arises: When are the shearing effects so large that they cannot be ignored as a design consideration? It is difficult to answer this question. Probably the best way to begin is to try to approximate the shear stresses on the cross section of the beam. This can be done by examining the necessary equilibrium requirements.

Figure 7-31(a) shows a simply supported beam with a single transverse load P. Once again, for convenience, a rectangular cross section has been chosen, as shown in Fig. 7-31(b). Figure 7-31(c) shows the free-body diagram of that portion of the beam between sections ① and ② in Fig. 7-31(a). The vertical shear forces on these sections produce a change in bending moment denoted by ΔM in (c). Figure 7-31(d) shows a free-body diagram of a slice of the beam extending between sections ① and ② and having its lower surface *abeh* at a distance u above the neutral surface of the beam.

Since there is a change in bending moment between sections ① and ②, it is reasonable to expect that the normal stresses σ_1 and σ_2 on the areas *ahgd* and *befc* may be different. Therefore, the resultant horizontal force F_{1x} on the area *ahgd* would be different in magnitude from the resultant force F_{2x} on the area *befc*. Equilibrium would require the existence of a horizontal shear force ΔF_x on the surface *abeh*. This horizontal shear force can be expressed in terms of the normal stresses as follows:

$$\overset{+}{\overset{\rightarrow}{\sum}} F_x = 0$$

$$F_{2x} + \Delta F_x - F_{1x} = 0$$

$$\Delta F_x = F_{1x} - F_{2x} = \int_{ahgd} \sigma_1 \, da - \int_{befc} \sigma_2 \, da$$

$$\Delta F_x = \int_{befc} (\sigma_1 - \sigma_2) \, da \tag{7-20}$$

The existence of such longitudinal shear forces can easily be illustrated by means of a simple experiment. Suppose that several planks or slabs are stacked one on top of the other without fastening them together, as shown in Fig. 7-32(a). When a bending load is applied, the stack will deform as indicated in Fig. 7-32(b). Since the slabs were free to slide on one another, the ends do not remain even but become staggered. Each of the slabs behaves as an independent beam, and the total resistance to bending of n slabs is approximately n times the resistance of one slab alone. If the experiment is repeated after the slabs have been fastened together so as to prevent their sliding on one another, the entire assembly will behave as a single beam having a thickness equal to n times the thickness of one slab. In the case of elastic action, the bending resistance of the assembly will be approximately n^3 times the bending resistance of one slab. From this simple experiment you can see the importance of a beam being able to resist longitudinal shear forces so that this "slipping" will not occur.

In order to evaluate the longitudinal shear force as given by Eq. 7-20, it is usually *assumed* that the normal stresses σ_1 and σ_2 on their respective areas are given by the *elastic* flexure formula (Eq. 7-15). For this assumption, expressions for σ_1 and σ_2 are

$$\sigma_1 = -\frac{My}{I} \qquad \sigma_2 = -\frac{(M + \Delta M)y}{I} \tag{7-21}$$

where y is measured from the neutral surface, and I is the second moment of the *entire* cross-sectional area about its horizontal centroidal axis.

Substituting these expressions for σ_1 and σ_2 in Eq. 7-20, we obtain

$$\Delta F_x = -\int_{befc} \left[\frac{My}{I} - \frac{(M + \Delta M)y}{I} \right] da$$

$$= \frac{\Delta M}{I} \int_{befc} y \, da \tag{7-22}$$

The integral, which will be denoted by Q, represents the first moment of area *befc* about the centroidal axis of the entire cross section. This area *befc* is merely that *part* of the total cross section which lies above the surface on which the longitudinal shear force acts.

We shall now reintroduce the concept of shear flow, which was first discussed in connection with thin-walled torsion members. Shear flow, denoted by q, is defined as the longitudinal shear force across the thickness of the cross section per unit length of the beam. According to this definition, the shear flow across the thickness of the beam at any cross section is

$$q = \lim_{\Delta x \to 0} \frac{\Delta F_x}{\Delta x} \tag{7-23}$$

If the value of ΔF_x given by Eq. 7-22 is substituted, we have

$$q = \lim_{\Delta x \to 0} \frac{\Delta M}{\Delta x} \frac{Q}{I} = \frac{dM}{dx} \frac{Q}{I}$$

$$q = \frac{VQ}{I} \tag{7-24}$$

where V is the *resultant vertical* shear force at the cross section under consideration.

The expression in Eq. 7-24 represents the shear force per unit length of the beam across the thickness of the beam at the cross section under consideration. For example, suppose that it is desired to determine the shear flow across the geometric middle of the portion of a beam shown in Fig. 7-33(a). The procedure would be as follows.

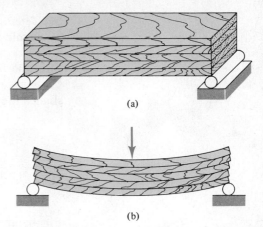

(a)

(b)

Figure 7-32

1 Determine the vertical shear force V from the free-body diagram in Fig. 7-33(b). Thus, $V = R_1 - P_1$.

2 Locate the centroid of the entire cross section. In this case the distance from the top to the centroid is $h/3$.

3 The distance from the top to the geometric middle of the depth is $h/2$.

4 The moment of inertia, or the second moment, of the *entire* triangular cross section about the *centroid* is $bh^3/36$.

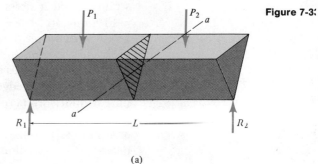

Figure 7-3:

(a)

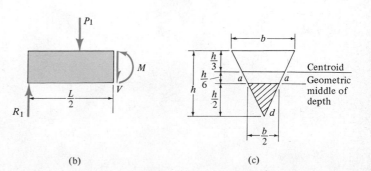

(b)

(c)

5 Determine Q. The area *above* the specified longitudinal surface *aa* at the middle of the depth of the beam is the *unshaded* area in Fig. 7-33(c). However, since the first moment of the entire area about the centroid is zero, the first moment of the unshaded area about the centroid is the same (except for sign) as that of the shaded area below line *aa*. For the shaded area,

$$Q = \frac{1}{2}\left(\frac{b}{2}\right)\left(\frac{h}{2}\right) \times \left(\frac{h}{6} + \frac{1}{3}\frac{h}{2}\right)$$

$$= \left(\frac{1}{8}bh\right)\left(\frac{1}{3}h\right) = \frac{1}{24}bh^2$$

Finally, the shear flow across line *aa* is

$$q = \frac{VQ}{I} = \frac{(R_1 - P_1)(\frac{1}{24}bh^2)}{bh^3/36} = \frac{3}{2}\frac{(R_1 - P_1)}{h} \text{ lb/in.}$$

7-13 / Average longitudinal and transverse shear stresses

You should note that the development of the shear-flow equation (Eq. 7-24) was based solely *on equilibrium requirements and the elastic flexure formula.* No consideration was given to the manner in which the shear strain and shear stress are distributed throughout the beam.

From the free-body diagram in Fig. 7-34, most of which is the same as Fig. 7-31(d), we see that shear forces exist on both the longitudinal surface *abeh* and the transverse surfaces *ahgd* and *befc*. These shear forces will produce shear stresses on their respective surfaces. Also, any infinitesimal element having an edge in common with line *be* has equal shear stresses on two mutually perpendicular surfaces; that is, $\tau_l = \tau_t$. Thus, we see that simultaneous *and equal* longitudinal and transverse shear stresses exist in a beam acted on by transverse forces. However, it is difficult to determine the distribution of these stresses throughout the length, depth, and thickness (width) of the beam.

Figure 7-34

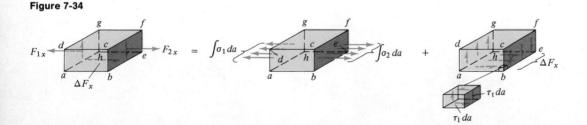

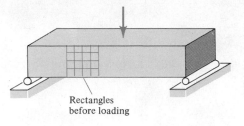

Rectangles
before loading

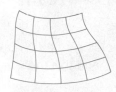

After loading
(distortion exaggerated)

Figure 7-35

It is relatively easy to show experimentally by means of a grid pattern on the side of a *rectangular* beam, as in Fig. 7-35, that the maximum distortion—and, therefore, the maximum shear strain—occurs at the neutral axis of bending, while no distortion occurs at the free edges of the beam. However, this observation serves only as a *qualitative* analysis of the distribution of the shear strain. We still do not know the exact variation of the shear stresses through the beam, but it is usually assumed that the shear stresses are constant across the thickness of the beam. For example, in Fig. 7-34 we assume that the shear stresses at b have the same value as those at e and at all points on the line between b and e. We are *not* saying that the stresses at b are equal to those at c, since our previous experiment has indicated that the shear stresses are not constant throughout the vertical depth of the beam.

We have already argued that the longitudinal shear stress τ_l at any point is equal to the transverse shear stress τ_t at the point. If the shear force per unit length of the beam at any particular point is given by the shear flow (Eq. 7-24), and the thickness of the beam is denoted by t, then the average longitudinal or transverse shear stress across the thickness of the beam is

$$\tau_{l(\text{av})} = \tau_{t(\text{av})} = \frac{q}{t} = \frac{VQ}{It} \qquad (7\text{-}25)$$

This expression for the shear stress is consistent with the previously noted observation of Fig. 7-35 in that for a *rectangular* cross section the shear stress becomes maximum at the centroidal axis (neutral surface) and is zero at the outer edges, since Q behaves in this manner (see Example 7-12, which follows).

Although the shear flow found by Eq. 7-24 *always* is maximum at the centroid and zero at the edges, it is important to note that the shear stress found by Eq. 7-25 may or may not be maximum at the centroid. Where the thickness of the member varies throughout the depth, the position of the maximum transverse shear stress depends on the shape of the section. For instance, it is left as an exercise for the student to show that the maximum shear stress for a triangular cross section, like that in Fig. 7-33, occurs at the geometric middle of the depth, and *not* at the centroid.

EXAMPLE 7-12

Determine an expression for the shear-stress distribution on a rectangular cross section of height h and thickness t.

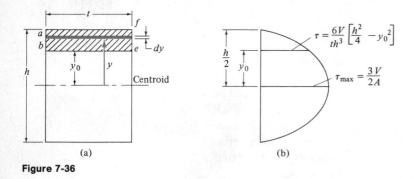

       (a)   (b)

Figure 7-36

Solution:

Refer to Fig. 7-36. If it is assumed that the shear stresses are uniform across the thickness, the shear stress at a surface at a distance y_0 from the centroidal axis is, by Eq. 7-25,

$$\tau = \frac{VQ}{It} = \frac{V}{It} \int_{\substack{\text{area} \\ abef}} y\,da$$

$$= \frac{V}{(th^3/12)t} \int_{y_0}^{h/2} y(t\,dy) = \left(\frac{12V}{t^2h^3}\right)t\left.\frac{y^2}{2}\right|_{y_0}^{h/2}$$

$$\tau = \frac{6V}{th^3}\left[\frac{h^2}{4} - y_0^2\right]$$

From this last expression we get the following results: at $y_0 = h/2$, $\tau = 0$; at $y_0 = 0$, or at the centroidal axis, τ has its maximum value, which is $3V/2A$. Also, since τ is a function of y_0^2, the distribution is parabolic, as shown in Fig. 7-36(b).

EXAMPLE 7-13

A 20-ft-long standard steel beam W 8 × 40 is simply supported and carries a uniformly distributed load of 400 lb/ft including the weight of the beam. Determine (a) the largest flexure stress in the beam and (b) the largest transverse shear stress in the beam.

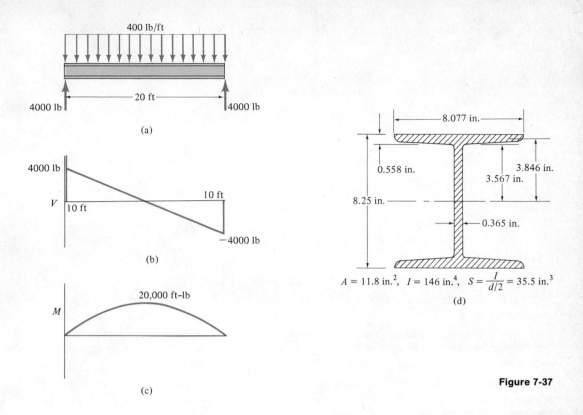

Figure 7-37

Solution:

The shear and moment diagrams are shown in Fig. 7-37(b) and (c), respectively, indicating a maximum shear force of 4000 lb at either end and a maximum bending moment of 20,000 ft-lb at the mid span.

The cross-sectional properties for the beam are found from Table A-3 of the Appendix and are indicated in Fig. 7-37(d).

(a) The largest flexure stress will occur in the outer fibers of the beam at the mid span. Thus

$$\sigma_{max} = \frac{Md/2}{I} = \frac{M}{S}$$

$$= \frac{(20,000)(12)}{35.5} = 6760 \text{ psi}$$

with tension in the lower fibers and compression in the upper fibers.

(b) The transverse shear stress is given by Eq. 7-25 as

$$\tau = \frac{VQ}{It}$$

Now $V_{max} = 4000$ lb and occurs at either end; Q_{max} occurs at the centroid of the cross section where

t also happens to be a minimum. Thus, τ_{max} occurs in the middle fibers of the beam at either end of the beam. Thus

$$Q_{max} = Q_{flange} + Q_{half\ web}$$
$$= (8.077)(0.558)(3.846) + (0.365)(3.567)(3.567/2)$$
$$= 17.334 + 2.322 = 19.656 \text{ in.}^3$$

and

$$\tau_{max} = \frac{VQ}{It} = \frac{(4000)(19.656)}{(146)(0.365)}$$
$$= 1475 \text{ psi}$$

This result is somewhat typical in that ordinarily the largest transverse shear stress in a beam will be considerably smaller than the largest flexure stress.

PROBLEMS

7-92 to 7-95 For each of the cross sections shown in the illustrations for Probs. 7-29 to 7-32, respectively, determine Q_{max}.

7-96 A cast-iron beam has the T-shaped section shown in the illustration. If this beam transmits a vertical shear of 102 kips, find the average transverse shear stress at each of the levels indicated. Sketch a plot showing how the shear stress varies throughout the height of the beam.

Problem 7-96

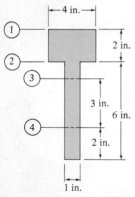

7-97 An elastic homogeneous straight beam that carries a constant shear force V has the constant triangular cross section shown in the illustration. Express the average transverse shear stress as a function of the distance y from the vertex of the triangle.

Also, show that the maximum shear stress occurs at $y = h/2$ and is given by $\tau_{max} = 3V/2A$, where A is the cross-sectional area.

Problem 7-97

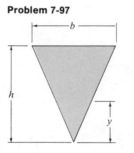

7-98 In a beam whose cross section is shown in the illustration, where would the maximum horizontal shear stress occur in the cross section? Prove your answer.

7-99 Find the ratio of the maximum shearing stress in an elliptical beam to the average shearing stress in the same beam.

7-100 An elastic homogeneous straight beam carries a constant shear force V and has a constant circular cross section. Express the average transverse shear stress as a function of the distance y_0 from the centroid of the cross section. Also, show that the maximum shear stress occurs at $y_0 = 0$ and that its value is $\tau_{max} = 4V/3A$, where A is the cross-sectional area.

Problem 7-98   **Problem 7-100**

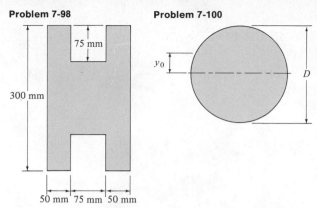

7-101 A beam is made of six planks nailed together to form the cross section shown in the illustration. Each nail is capable of resisting 120 lb of shear force. If the section is carrying a shear force of 3000 lb, determine the required longitudinal spacing of the nails connecting A with B or C.

Problem 7-101

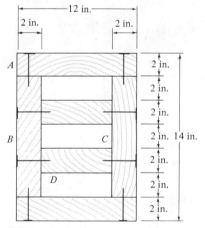

7-102 A simply supported beam of 6-m span carries a uniformly distributed load of 10,000 N/m. The cross section is 50 × 200 mm. What is the maximum horizontal shear stress that is developed and exactly where does it occur?

7-103 A beam 15 ft long has a support 3 ft from each end. A uniformly distributed load of 960 lb/ft acts on the overhangs only. The cross section of the beam is a hollow symmetrical rectangle, 6 in. wide by 8 in. deep outside and 4 in. wide by 4 in. deep inside. Assuming that the material is linearly elastic, (a) find the maximum tensile stress and its location. (b) Find the maximum horizontal shear stress and its location.

7-104 A steel beam 30 ft long has simple supports at the right-hand end and 10 ft from the left-hand end. It has a uniform load of 400 lb/ft over the right 15 ft of the beam and a load of 600 lb/ft on the overhang. In addition, a 4000-lb concentrated load acts midway between the supports. The standard cross section is an S 10 × 25.4. (a) Assuming elastic behavior, determine the maximum tensile and compressive flexure stresses in the beam, and show where they occur. (b) Find the largest horizontal shear stress in the beam.

7-105 A simply supported, uniformly loaded timber beam 16 ft long has a nominal rectangular cross section 4 in. wide and 12 in. deep. If the maximum allowable flexure stress is 1200 psi and the maximum allowable horizontal shear stress is 100 psi, what is the maximum uniformly distributed load w lb/ft that the beam can carry? Refer to Appendix A-11.

7-106 A rectangular timber beam is to be used to support a uniform load of 1500 N/m over a 3-m span. Determine an acceptable nominal size of timber if the flexure stress is not to exceed 10 MPa and the horizontal shear stress is not to exceed 0.5 MPa. See Appendix.

7-107 For the beam of Prob. 7-53, determine the value and location of the largest transverse shear stress.

7-108 For the beam of Prob. 7-65, determine the value and location of the largest transverse shear stress.

7-109 A rectangular timber 4 in. wide by 10 in. deep is to be used as a cantilever beam to carry a uniform load of 100 lb/ft which includes the beam weight. If the allowable flexure stress is 1200 psi and the allowable horizontal shear stress is 80 psi, how long can the beam be?

7-110 An elastic cantilever beam has the loading and dimensions shown in the illustration. Sketch the shear and bending-moment diagrams, and find (a) the location and value of the maximum tensile flexure stress; (b) the location and value of the maximum compressive flexure stress; (c) the position and value of the maximum transverse shear stress.

7-111 For the elastic beam shown in the illustration, determine the magnitude and location of (a) the maximum tensile flexure stress, (b) the maximum compressive flexure stress, and (c) the maximum transverse shear stress.

Problem 7-110

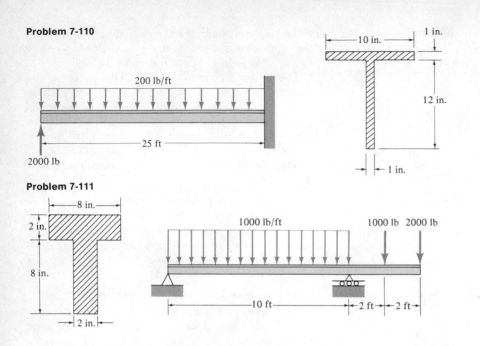

200 lb/ft

2000 lb

25 ft

10 in.

1 in.

12 in.

1 in.

Problem 7-111

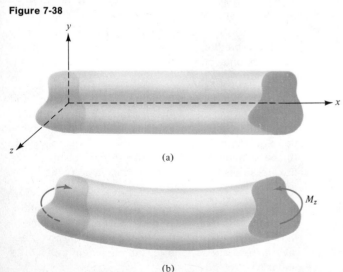

8 in.

2 in.

8 in.

2 in.

1000 lb/ft

1000 lb   2000 lb

10 ft

2 ft

2 ft

7-14 / Elastic bending of unsymmetrical sections

So far we have dealt only with beams having a longitudinal plane of symmetry and whose bending loads are applied in this plane of symmetry. We shall now see that for elastic behavior these strict requirements of symmetry can be somewhat relaxed.

Consider the beam of arbitrary cross section shown in Fig. 7-38(a), where the x axis is the longitudinal centroidal axis of the beam and the

Figure 7-38

y

x

z

(a)

M_z

(b)

y and z axes are the *principal* centroidal axes of the cross section. For this choice of axes, we have

$$\int_{\substack{\text{cross} \\ \text{section}}} y \, da = 0 \qquad \int_{\substack{\text{cross} \\ \text{section}}} z \, da = 0 \qquad \int_{\substack{\text{cross} \\ \text{section}}} yz \, da = 0 \qquad (7\text{-}26)$$

Let us assume that a *pure couple* about *either principal* axis is applied to the beam, as in Fig. 7-38(b), where for convenience a couple about the z axis was chosen. Under this loading condition, the arguments in section 7-8 are still reasonable and plane sections originally perpendicular to the x axis will remain plane after loading. Hence, the longitudinal axial strains will vary linearly from some neutral surface.

Now it will be assumed that the beam is *elastic*, homogeneous, and isotropic. The axial stress is then given by the expression in Fig. 7-25(a), namely,

$$\sigma = ky \qquad (7\text{-}27)$$

Substituting this expression in Eqs. 7-7 to 7-9, we have

$$\sum M_z = M_z$$

$$M_z = -\int_{\text{area}} y(\sigma \, da) = -\int_{\text{area}} y(ky \, da)$$

$$= -k \int_{\text{area}} y^2 \, da = -kI_z$$

$$k = -\frac{M_z}{I_z}$$

$$\sigma = -\frac{M_z y}{I_z} \qquad (7\text{-}28)$$

Also,

$$\sum F_x = R_N = 0$$

$$= \int_{\text{area}} \sigma \, da = \int_{\text{area}} (ky) \, da = k \int_{\text{area}} y \, da = 0$$

by Eq. 7-26. Finally,

$$\sum M_y = 0$$

$$= \int_{\text{area}} z(\sigma \, da) = \int_{\text{area}} z(ky \, da) = k \int_{\text{area}} yz \, da = 0$$

by Eq. 7-26.

Thus, if the bending couple is applied about the z axis, we see that the

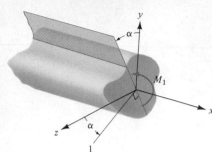

Figure 7-39

elastic axial stress will be given by Eq. 7-28 and that the neutral surface will lie in the xz plane. In a similar manner, if a couple is applied about the y axis, the stress will be

$$\sigma = -\frac{M_y z}{I_y} \qquad (7\text{-}29)$$

By using Eqs. 7-28 and 7-29, we are now able to handle the more general problem in which the couple producing the bending moment acts about some axis that is not parallel to either principal axis. Such a condition is illustrated in Fig. 7-39, where once again the y and z axes are *principal centroidal* axes and the couple acts about the 1 axis. From elementary mechanics we know that M_1 can be resolved into components about the y and z axes. These components are

$$M_z = M_1 \cos \alpha \qquad M_y = M_1 \sin \alpha$$

The stresses produced by these component couples can be calculated by applying Eqs. 7-28 and 7-29. The resultant stress at any point on the cross section is then obtained by superposition of the bending stresses produced by each of the component couples.

You should note that throughout the discussion of elastic bending of beams with unsymmetrical sections, it was assumed that the bending moment was produced by a *pure couple*. In general, if there is a resultant shear force on any cross section resulting from transverse loading of the beam, the bending will be accompanied by twisting of the beam. The problem then becomes one of combined bending and torsion of an unsymmetrical section. Such problems are discussed in Chapter 11.

7-15 / Summary of beam stresses

The preceding sections have dealt with methods of evaluating the flexure stresses and the average *elastic* transverse and horizontal shear stresses in beams subjected to various loading conditions. Also, it has been shown that at any particular point in a beam the stress on an element may be composed of longitudinal normal stresses (flexure stresses) and transverse shear stresses. Under this condition, the state of stress is biaxial, and the maximum normal and shear stresses will act in directions that are neither longitudinal nor transverse. Hence, it is often necessary to employ Mohr's circle in order to completely determine the principal stresses at a particular point.

Recall that the flexure stresses are maximum at the outermost fibers and are zero at the neutral surface. On the other hand, the transverse shear stresses are zero at the outside surface and have a maximum value somewhere in the interior. Also, both the flexure and shear stresses vary with the loading on the beam. Hence, the problem of determining the point or points at which the absolute maximum principal normal and shear stresses occur can be a very difficult one that requires careful investigation. In determining maximum stresses in beams, do not be satisfied with your results until you have exhausted all likely combinations of flexure and shear stresses that could give the maximum principal stresses. Often, the construction of shear and moment diagrams and a comparison of the orders of magnitude of the flexure stresses and transverse shear stresses will greatly simplify this problem.

Since members subjected to axial and torsion loads were investigated in previous chapters, we can now handle problems in which axial, torsion, and bending loads are combined. As before, we shall consider only superposition of *elastic* stresses.

EXAMPLE 7-14

A short cantilever beam is shown in Fig. 7-40(a). Determine the principal normal and shear stresses at point A just below the flange, and show them on a sketch. Assume elastic action and neglect any stress concentration at the wall.

Figure 7-40

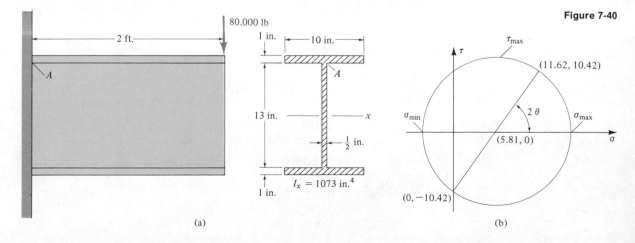

(a)   (b)

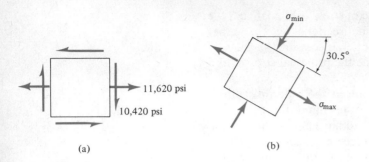

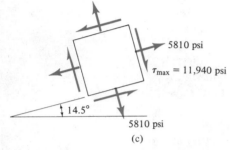

Figure 7-41

Solution:

The bending moment at the left-hand end of the beam is

$$M = (80,000)(2)(12) \text{ in.-lb}$$

and the vertical shear force is $V = 80,000$ lb. Hence, the flexure stress at point A is

$$\sigma = \frac{My}{I} = \frac{(80,000)(2)(12)(6.5)}{1073} = 11,620 \text{ psi (tension)}$$

and the transverse shear stress at A is

$$\tau = \frac{VQ}{It} = \frac{(80,000)[(10)(1)(7)]}{(1073)(1/2)} = 10,420 \text{ psi}$$

The stresses acting on an element at A oriented parallel and perpendicular to the longitudinal axis of the beam are shown in Fig. 7-41(a). Also, Mohr's circle for this state of stress is shown in Fig. 7-40(b). From the geometry of the circle, we have the following results:

$$\sigma_{max} = 17,750 \text{ psi (tension)} \qquad \sigma_{min} = 6130 \text{ psi (compression)}$$

$$\tau_{max} = 11,940 \text{ psi} \qquad \theta = 30.5°$$

Thus, the desired orientations are as shown in Fig. 7-41(b) and (c).

Note: In such a short member as this with its thin-web cross section, the validity of the flexure formula is questionable. Notice, for example, that the shear and normal stresses are of the same order of magnitude.

EXAMPLE 7-15

The dimensions of a T-shaped beam and the loads on it are shown in Fig. 7-42(a) and (b). The horizontal 5000-lb tensile force acts through the centroid of the cross-sectional area. Determine the maximum principal normal stress and the maximum shear stress in the beam, assuming elastic behavior.

Solution:

Sketches of the shear and moment diagrams are shown in Fig. 7-42(c) and (d), respectively. Hence, both the maximum *flexure* stress and the maximum *transverse* shear stress occur at the extreme right-hand end of the beam. Also, the axial stress produced by the tensile load will be uniform throughout the entire beam.

The distance from the bottom of the beam to the centroid of the cross section is found to be

$$\bar{Y} = \frac{(12)(3) + (12)(7)}{12 + 12} = 5 \text{ in.}$$

Also, the moment of inertia or second moment of the area about the centroid is

$$I = \left[\frac{(6)(2^3)}{12} + (12)(2^2) \right] + \left[\frac{(2)(6^3)}{12} + (12)(2^2) \right]$$

$$= 52 + 84 = 136 \text{ in.}^4$$

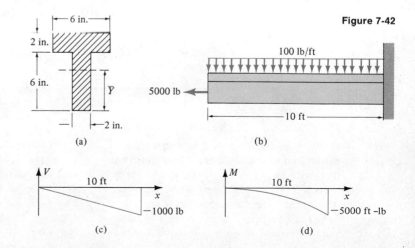

Figure 7-42

(a)

(b)

(c)

(d)

The longitudinal normal stress on the uppermost fibers is

$$\sigma_{up} = \frac{P}{A} + \frac{My_{up}}{I} = \frac{5000}{24} + \frac{(5000)(12)(3)}{136}$$

$$= 208 + 1323 = 1531 \text{ psi (tension)} \quad \leftarrow\Box\rightarrow$$

while the normal stress on the lowest fibers is

$$\sigma_{low} = \frac{P}{A} - \frac{My_{low}}{I} = \frac{5000}{24} - \frac{(5000)(12)(5)}{136}$$

$$= 208 - 2205 = -1997 = 1997 \text{ psi (compression)} \quad \rightarrow\Box\leftarrow$$

The maximum longitudinal or transverse shear stress occurs at the centroid. This stress is

$$\tau = \frac{VQ}{It} = \frac{(1000)[(2)(5)(2.5)]}{(136)(2)}$$

$$= 91.9 \text{ psi} \quad \Box$$

At any point between either outer surface and the centroid, there will be both a normal stress and a shear stress. However, a comparison of their orders of magnitude shows that the normal stresses at the outside far exceed the transverse shear stresses at the centroid. Hence, we can conclude that the critical fibers will be those at the outside surface, where the maximum principal normal stresses will be

$$\sigma_{max} = 1531 \text{ psi (tension)}$$

$$\sigma_{max} = 1997 \text{ psi (compression)}$$

and, recalling Mohr's circle for uniaxial stress, the maximum shear stress will occur at the lower fiber and will be

$$\tau_{max} = \frac{\sigma_{max} - \sigma_{min}}{2} = \frac{1997 - 0}{2} = 999 \text{ psi}$$

This shear stress will act 45° to the longitudinal axis of the beam.

PROBLEMS

7-112 A simply supported beam 8 ft long carries a uniformly distributed load of 1000 lb/ft. The beam has a rectangular cross section 2 in. wide and 6 in. deep. Assuming elastic behavior, determine the maximum normal and shear stresses developed at a point 2 in. below the top surface and 1 ft from the right-hand support. Sketch the planes on which these stresses acts.

7-113 A cylindrical pressure tank has an inside diameter of 20 in. and an outside diameter of 21 in. and an overall length of 200 in. The tank is simply supported at each end, as shown in the illustration. The tank is filled with a gas under a pressure of 100 psig. The total weight of the tank and gas combined is 10,000 lb, and it can be considered uniformly distributed over the entire length of the tank. Deter-

Problem 7-113

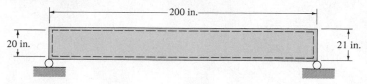

mine the maximum normal stress in the wall of the tank. Suppose the pressure is increased to 400 psig?

7-114   A propeller shaft 2 in. in diameter is subjected to a combination of a twisting torque of 16,000 in.-lb, a bending moment of 6000 in.-lb, and an axial compressive thrust of 15,000 lb. What are the maximum tensile, compressive, and shear stresses in the shaft?

7-115   What should be the thickness t of the cross section in the illustration if the tensile stress at the wall cannot exceed 30 MPa? Use a safety factor of 2.

Problem 7-115

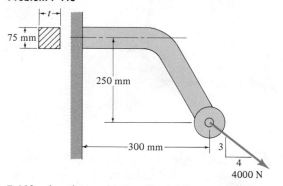

7-116   As shown in the illustration, a circular can-

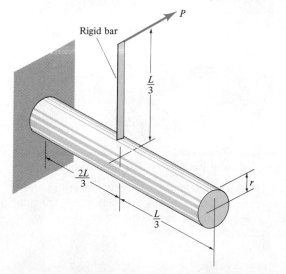

tilever beam of radius r is subjected to a load P that acts perpendicular to the longitudinal axis of the beam. If the material is linearly elastic and the maximum allowable shear stress is τ_{max}, what is the maximum allowable load P in terms of τ_{max}, L, and r?

7-117   The beam shown in the illustration has the dimensions and loading indicated. Sketch the shear and bending-moment diagrams for the horizontal part in terms of P. Determine P if (a) the beam is made of a material for which the allowable elastic stresses are 30,000 psi in tension and 40,000 psi in compression, and (b) failure is due to shear when the maximum elastic horizontal shear stress is 5000 psi. Neglect stress concentration at the wall.

Problem 7-117

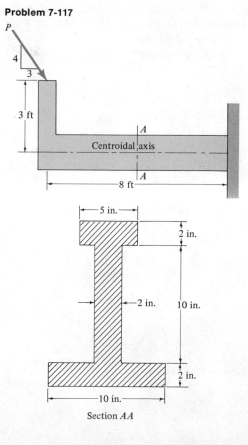

Section AA

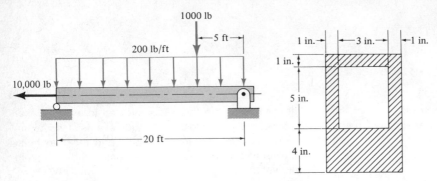

7-118 A loaded beam and its original dimensions are shown in the illustration. The 10,000-lb axial load acts through the centroid of the cross section. Determine the value and location of (a) the maximum normal stress, (b) the maximum longitudinal shear stress, and (c) the maximum shear stress.

7-119 A rigid bar 100 mm long is welded to a solid homogeneous shaft having a length of 200 mm and a diameter of 50 mm. A load of 20 kN is applied as shown in the illustration. Determine the principal stresses and the maximum shear stress at point *A* on top at the extreme left-hand end of the shaft. Assume elastic behavior, and neglect stress concentrations at the wall.

Problem 7-119

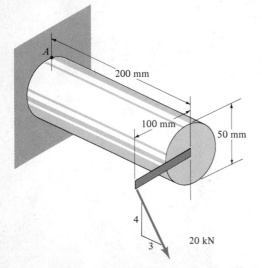

7-120 The member shown in the illustration has a solid circular cross section of radius *r*. It is subjected

to a load of 7000 lb applied as shown. Find the smallest permissible radius if the maximum normal stress in the horizontal portion may not exceed 25,000 psi. Neglect stress concentrations.

Problem 7-120

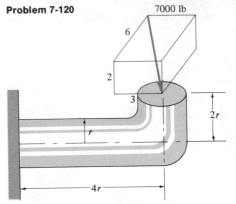

7-121 Find the principal stresses and the maximum shearing stress at point *A* on the surface of the shaft shown.

Problem 7-121

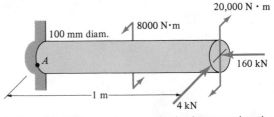

7-122 Find the maximum principal stress developed on the surface of the line shaft shown at section *a-a*.

7-123 Find the maximum load to which the member shown may be subjected if the maximum principal stress (neglecting stress concentration) is not to exceed 20,000 psi.

Problem 7-122

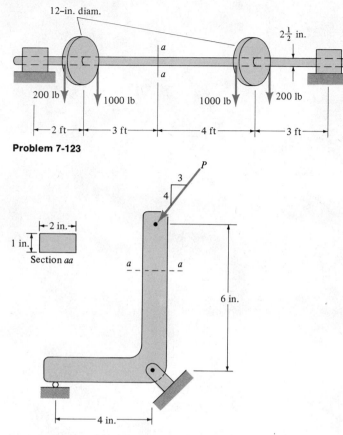

12–in. diam.

$2\frac{1}{2}$ in.

a

a

200 lb

1000 lb

1000 lb

200 lb

2 ft

3 ft

4 ft

3 ft

Problem 7-123

P

3

4

2 in.

1 in.

Section aa

a a

6 in.

4 in.

Chapter 8
Beam deflections I

8-1 / Objectives

Upon completion of this chapter you will be able to:

1 Write shear and moment equations for loaded beams.
2 Calculate the deflections of statically determinate beams by direct
   integration and by superposition.
3 Calculate reactions on statically indeterminate beams by direct
   integration and by superposition.

8-2 / Introduction

In Chapter 7 we were concerned primarily with determining the flexural and transverse shear stresses in straight homogeneous beams of uniform cross section. In this chapter we shall be primarily interested in the deformation of beams and subsequently in the solution of statically indeterminate problems.

Before proceeding further in this chapter, it would be well for you to review section 7-4 dealing with shear and moment diagrams and section 7-7 dealing with the kinematics of beam deformation. Those ideas form the basis for much of what we shall do in this chapter.

8-3 / Shear and moment equations

In the preceding chapter the basic relations for shear and moment equations, Eqs. 7-2, 7-3, and 7-4, were developed but little was done with the equations themselves. Instead by inspection we developed graphical techniques for drawing shear and moment diagrams. While these methods are quick and effective, they are usually easy to use only for loadings that produce straight-line figures in the shear diagram since few people are particularly adept in the calculation of areas under parabolas.

Shear and moment equations, however, are not so limited and may be used to plot shear and moment diagrams for more complex loadings. In addition they are extremely useful in calculating the deflection of a beam.

Shear and moment equations are mathematical expressions for the internal reactions at various locations in a beam in terms of the externally applied loads and supports. Both the shear equations and the moment equations may be obtained from the same free-body diagrams. The steps are as follows:

1   Determine all unknown reactions and decide on a fixed reference point (usually the left end).

2   Draw a FBD of that portion of the beam to the left of the desired section. Dimension it in terms of x, the distance from the fixed reference point. Show the unknowns V_x and M_x acting on the cut section as positive.

3   To obtain the shear equation, take $\sum F_y = 0$ and solve for V_x.

4   To obtain the moment equation, write $\sum M = 0$, taking moments about the right end. An easy way to do this is to multiply each term in the shear equation by its distance from the cut section.

Figure 8-1

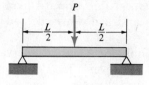

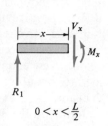

$$0 < x < \frac{L}{2}$$

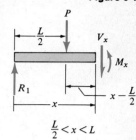

$$\frac{L}{2} < x < L$$

(a)

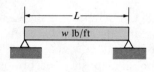

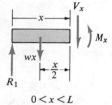

$$0 < x < L$$

(b)

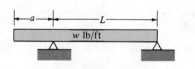

$$0 \le x < a$$

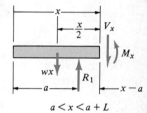

$$a < x < a + L$$

(c)

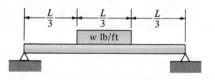

$$0 < x \le \frac{L}{3}$$

$$\frac{L}{3} \le x \le \frac{2L}{3}$$

(d)

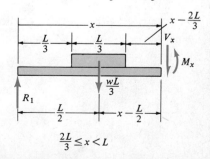

$$\frac{2L}{3} \le x < L$$

Always work from left to right in order to preserve normal sign conventions. In both shear and moment equations, an upward load will result in a positive term in the equations and a downward load will result in a negative term.

Each shear or moment equation is valid only for a limited interval of the beam. New equations must be written whenever a load or reaction is added to, or subtracted from, the free body as you move along the axis of the beam.

The limits on the equations are most conveniently stated as distances from the reference point or origin. All equations for a given beam should be referred to the same origin. *No equation is complete without specifying the interval for which it is applicable.*

Refer to the various beams shown in Fig. 8-1. Each beam is shown with the free bodies necessary to write its shear and moment equations. In all cases, the reference origin is taken at the left end.

In all cases the shear and moment are shown as positive according to the conventions given in the preceding chapter. The following examples will deal more explicitly with writing the shear and moment equations.

EXAMPLE 8-1

The cantilevered beam in Fig. 8-2(a) carries a uniform loading of 300 lb/ft. Write the shear and moment equations for the beam and use them to obtain the shear and moment diagrams.

Solution:

Although it is not always necessary, we begin by finding the external reactions, in this case, the force R and the couple C indicated in Fig. 8-2(b)

$$\sum F = 0$$

$$R - (300 \text{ lb/ft})(12 \text{ ft}) = 0 \qquad R = 3600 \text{ lb}$$

$$\sum M = 0$$

$$C - (300 \text{ lb/ft})(12 \text{ ft})(6 \text{ ft}) = 0$$

$$C = 21{,}600 \text{ ft-lb}$$

A free-body diagram of a portion of the beam x ft long is shown in Fig. 8-2(c) with *internal* reactions V and M, both assumed positive according to our previously adopted sign convention. Equilibrium requires

$$\sum F = 0$$

$$V + (300 \text{ lb/ft})(x \text{ ft}) = 0 \tag{a}$$

$$V = -300x$$

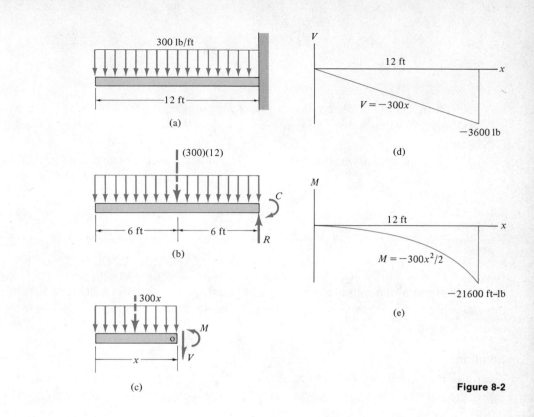

Figure 8-2

$$\sum M_o = 0$$

$$M + (300 \ \text{lb/ft})(x \ \text{ft})(x/2 \ \text{ft}) = 0$$ (b)

$$M = -300x^2/2$$

Equations (a) and (b) are the shear equation and moment equation, respectively. Note that

$$V(0) = 0 \qquad V(12) = -3600 \ \text{lb}$$

$$M(0) = 0 \qquad M(12) = -21,600 \ \text{ft-lb}$$

which agree with the values for the external wall reactions obtained previously. The shear equation is linear in x with

$$\frac{dV}{dx} = \frac{d}{dx}(-300x) = -300$$

$$= -w$$

as required by Eq. 7-2. Thus the shear diagram plots as a straight line with a *negative* slope as shown in Fig. 8-2(d).

The moment equation is quadratic in x and plots as a parabola. Observe that

$$\frac{dM}{dx} = \frac{d}{dx}\left(\frac{-300x^2}{2}\right) = -300x$$

$$= V$$

as required by Eq. 7-3. Hence, the moment diagram starts at $M = 0$ and curves downward with an increasingly *negative* slope, reaching a value of $M = -21,600$ ft-lb at $x = 12$ ft as indicated in Fig. 8-2(e).

Thus, the largest (although algebraically smallest) value of the shear is 3600 lb and that for the moment is 21,600 ft-lb.

EXAMPLE 8-2

For the beam in Fig. 8-3(a), write shear and moment equations for each region. Then sketch the complete shear and moment diagrams, labeling the values at points of discontinuity. Also, determine the value of the maximum moment and locate the section at which it occurs. Note that this is the same beam as in Example 7-5.

Solution:

From the free-body diagram of the entire beam in Fig. 8-4(a), we first find the reactions R_1 and R_2. Using the left-hand end as the origin of the x axis, we obtain the following shear equations by referring to the free-body diagrams in Fig. 8-3(b)–(d):

For $0 \leq x < 4$ $V = -2000x$

For $4 < x \leq 15$ $V = -2000x + 23,440$

For $15 \leq x < 20$ $V = (-2000)(15) + 23,440 = -6560$

Notice that the first two equations represent straight lines each with a negative slope of 2000. The last equation is a straight line with a zero slope. Observe how these three lines are plotted in Fig. 8-4(b) to form the shear diagram. The vertical line in the shear diagram is a jump discontinuity caused by the concentrated reaction.

Again using the same origin and free-body diagrams, we get the following moment equations:

For $0 \leq x < 4$ $M = -2000x(x/2) = -2000x^2/2$

For $4 < x \leq 15$ $M = -2000x^2/2 + 23,440(x - 4)$

For $15 \leq x < 20$ $M = -2000(15)[x - (15/2)] + 23,440(x - 4)$

Notice that the first two equations are those of parabolas and the third is that of a straight line with a negative slope. Verify the fact that $dM/dx = V$ by differentiating each moment equation and comparing the result with the appropriate shear equation.

To obtain the proper shape of each curve in the moment diagram, we can utilize Eq. 7-3. In the

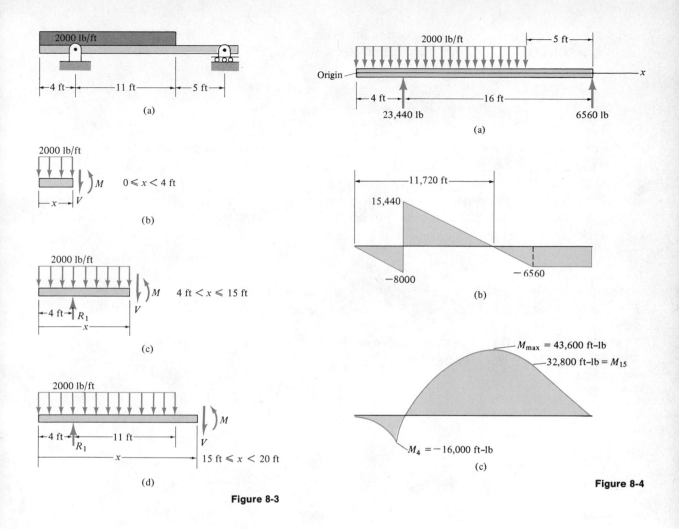

Figure 8-3

Figure 8-4

region where $0 \leq x < 4$, the shear is negative and its value decreases as x increases. Consequently, the slope of the moment diagram should be negative and *decreasing*, and the parabola, therefore, opens concave downward. A similar argument shows that the parabola in the region where $4 < x \leq 15$ is also concave downward, and that the straight line in the region where $15 \leq x \leq 20$ has a constant negative slope.

The location of the maximum moment will be either at a discontinuity or at a point of zero slope on the moment diagram. In the latter case, since $dM/dx = V$, *the maximum moment occurs where the shear is zero*. Therefore, equating the second shear equation to zero and solving for x will give the distance to the maximum moment from the left-hand end. This distance is 11.720 ft. This location can be found just as easily from the geometry of the shear diagram in Fig. 8-4(b) as was done in Example 7-5. You should compare that example with this one.

The value of the shear or moment at any location in the beam can be found by substituting the distance from the origin into the appropriate equation. For instance, the maximum moment can be found by substituting 11.720 ft for x in the second moment equation. The result is

$$M_{max} = -\frac{2000(11.720)^2}{2} + 23,440(11.720 - 4)$$

$$= 43,600 \text{ ft-lb}$$

In working problems and drawing free-body diagrams, it is usually convenient to assume that the shears and moments on the free-body diagrams are positive, without worrying about whether they are actually positive or negative. If you initially assume a quantity to be positive, then a positive answer means that it is positive and a negative answer means that it is negative. Thus, the interpretation of the sign is automatic. If, however, you initially assume that a shear or moment is negative, then you must be very careful to interpret your answer correctly.

PROBLEMS

8-1 to 8-36 For each of the beams shown, derive the shear and moment equations. You may wish to check your work by sketching the shear and moment diagrams.

Problem 8-1

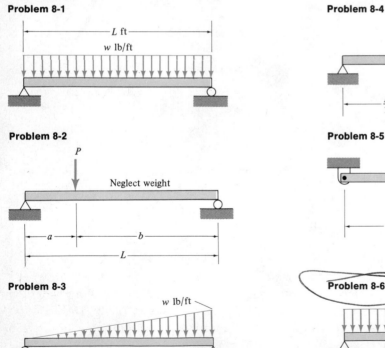

Problem 8-2

Problem 8-3

Problem 8-4

Problem 8-5

Problem 8-6

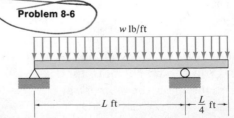

Problem 8-7

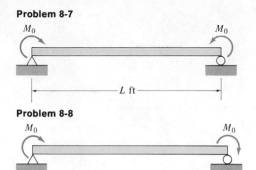

Problem 8-8

M_0 M_0

L ft

Problem 8-9

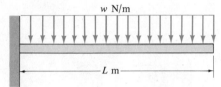

w N/m

L m

Problem 8-10

notes 6-1 w N/m

L m

Problem 8-11

P

L ft

Problem 8-12

P

$\frac{L}{2}$ ft $\frac{L}{2}$ ft

P

Problem 8-13

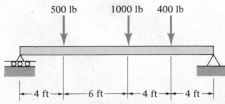

500 lb   1000 lb   400 lb

4 ft — 6 ft — 4 ft — 4 ft

Problem 8-14

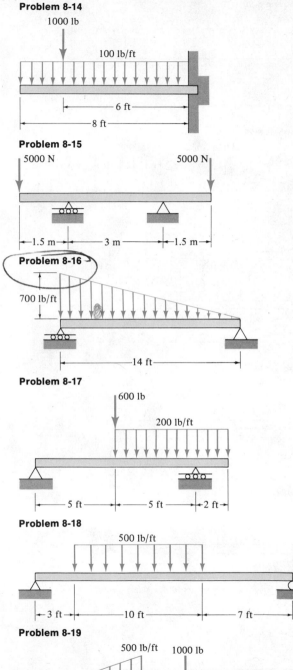

1000 lb

100 lb/ft

6 ft

8 ft

Problem 8-15

5000 N   5000 N

1.5 m — 3 m — 1.5 m

Problem 8-16

700 lb/ft

14 ft

Problem 8-17

600 lb

200 lb/ft

5 ft — 5 ft — 2 ft

Problem 8-18

500 lb/ft

3 ft — 10 ft — 7 ft

Problem 8-19

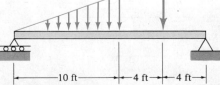

500 lb/ft   1000 lb

10 ft — 4 ft — 4 ft

Problem 8-20

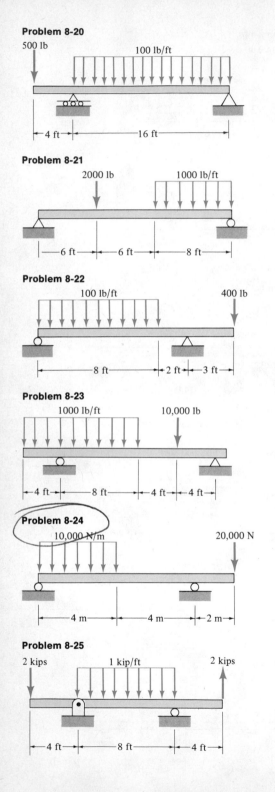

500 lb

100 lb/ft

├ 4 ft ┤├───16 ft───┤

Problem 8-21

2000 lb

1000 lb/ft

├─6 ft─┼─6 ft─┼───8 ft───┤

Problem 8-22

100 lb/ft

400 lb

├────8 ft────┼2 ft┼3 ft┤

Problem 8-23

1000 lb/ft

10,000 lb

├4 ft┼──8 ft──┼4 ft┼4 ft┤

Problem 8-24

10,000 N/m

20,000 N

├──4 m──┼──4 m──┼2 m┤

Problem 8-25

2 kips

1 kip/ft

2 kips

├4 ft┼───8 ft───┼4 ft┤

Problem 8-26

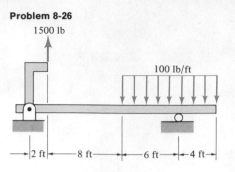

1500 lb

100 lb/ft

├2 ft┼──8 ft──┼─6 ft─┼4 ft┤

Problem 8-27

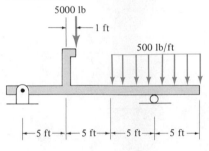

5000 lb

├ 1 ft

500 lb/ft

├─5 ft─┼─5 ft─┼─5 ft─┼─5 ft─┤

Problem 8-28

w lb/ft

w lb/ft

$M = \dfrac{wL^2}{2}$

├──── $\dfrac{3L}{4}$ ft ────┼ $\dfrac{L}{4}$ ft ┼ $\dfrac{L}{2}$ ft ─┤

Problem 8-29

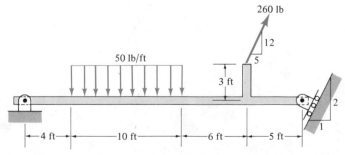

260 lb

12

5

50 lb/ft

3 ft

2

1

├─4 ft─┼────10 ft────┼──6 ft──┼──5 ft──┤

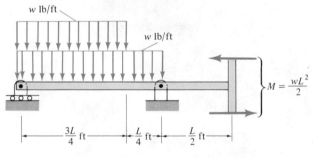

Problem 8-30

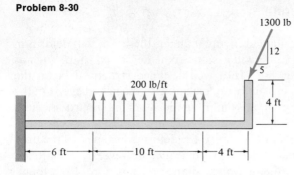

Problem 8-34

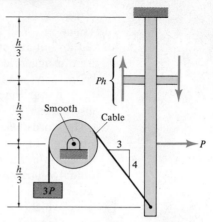

Problem 8-31

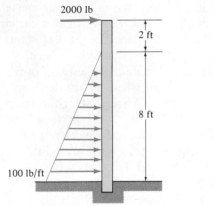

Problem 8-35

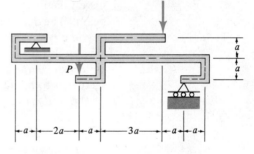

Problem 8-32

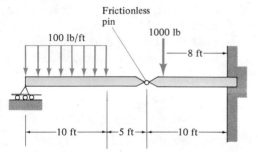

Problem 8-36

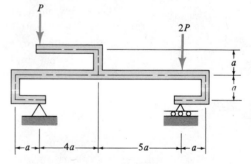

Problem 8-33

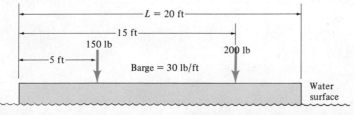

8-4 / General load-deflection relationship

Whenever a real beam is loaded in any manner, the beam will deform in that an initially straight beam will assume some deformed shape such as those exaggerated in Fig. 8-5. In a well-designed structural beam the deformation of the beam is usually undetectable to the naked eye. On the other hand, the bending of a swimming pool diving board is quite observable.

The word *deflection* generally refers to the deformed shape of a member subjected to bending loads. More specifically, however, we shall use deflection in reference to the deformed shape and position of the longitudinal *neutral axis* of a beam. In the deformed condition the neutral axis, which is initially a straight longitudinal line, assumes some particular shape which is called the *deflection curve*. The deviation of this curve from its initial position at any point is called the deflection at that point.

Consider the cantilevered beam of Fig. 8-6(a). For convenience, we have taken the origin of the *x-y* coordinate system at the left end of the beam. The deflected beam is shown in Fig. 8-6(b), where *y* is the deflection at some point *p* of the beam specified either by the horizontal coordinate *x* or by the distance *s* measured along the deflection curve. Obviously the deflection *y* varies with the position *x* (or *s*), and our objective in this section is to relate the deflection *y* to the variable *x* for any given loading on the beam; that is, we wish to determine the load-deflection relationship for a beam.

Figure 8-6(c) shows a small portion *ds* of the beam where θ is the slope and ρ the radius of curvature of the deflection curve at the particular location under consideration.

From elementary calculus, the relationships in Fig. 8-6(d) are valid for any smooth curve, such as the curve formed by the deflected neutral axis (deflection curve). If it is assumed that the slope at any arbitrary point is very small compared with unity, then the relationships in Fig. 8-6(f) are valid approximations. Thus,

$$\theta = \frac{dy}{dx} \qquad \frac{1}{\rho} = \frac{d^2y}{dx^2} = \frac{d}{dx}\left(\frac{dy}{dx}\right) = \frac{d\theta}{dx} \tag{8-1}$$

This assumption implies that the deflection of any point is very small in comparison with the length of the beam, and that the horizontal projected length of the deflection curve is the same as its undeformed length. Such an assumption is quite reasonable when it is realized that relatively few

Figure 8-5

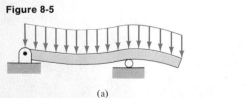

(a)

(b)

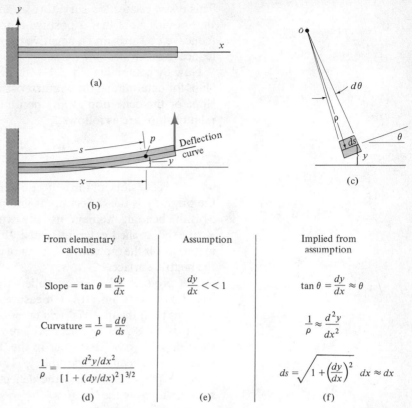

Figure 8-6

engineering members can tolerate large deflections without their failing. There are, of course, exceptions. Even in most of these cases, however, large deflections would usually result in inelastic behavior of the material, and failure would probably be based on some criterion other than deflection.

Since we have already restricted the analysis to small deflections, it is reasonable to assume further that the material behaves elastically; that is, the flexure stress on any fiber is given by the relationship

$$\sigma = -\frac{Mu}{I}$$

where u is the distance from the neutral surface to the fiber. (Recall the sign convention used in the derivation of Eq. 7-15a.) When this assumption is made and Hooke's law is used, Eq. 7-6 yields the result:

$$\varepsilon = -\frac{u}{\rho} = \frac{\sigma}{E} = -\frac{Mu}{EI}$$

$$\frac{1}{\rho} = \frac{M}{EI} \tag{8-2}$$

This result relates the curvature at any point to the bending moment at that point. Note that a positive moment produces a positive curvature (concave upward), and a negative moment produces a negative curvature (concave downward).

Now by combining Eq. 8-1 with Eq. 8-2, we obtain the basic relationships for determining the characteristics of the deflection curve, such as its slope or the deflection at any point, as well as its general shape. These relationships are as follows:

$$\frac{d^2y}{dx^2} = \frac{d\theta}{dx} = \frac{1}{\rho} = \frac{M}{EI} \tag{8-3}$$

where θ is the slope of the deflection curve at any point at distance x from the origin; y is the deflection at any point at distance x from the origin; M is the bending moment, usually expressed as a function of the distance x and the applied external loads; E is the modulus of elasticity of the material; I is the moment of inertia of the entire cross-sectional area about the neutral surface.

The product EI is often called the *flexural rigidity* or the *bending modulus* of the beam. It is a measure of the stiffness of the beam, since it involves both the material and cross section of the beam.

Equation 8-3 is a linear differential equation in the independent variable x. It is of first order in the dependent variable θ and of second order in the dependent variable y. The solution of this equation will give both the proper slope and the deflection of the beam, *provided that the solution satisfies the conditions of restraint imposed by the supports.* In succeeding sections, various techniques for solving Eq. 8-3 for θ and y will be presented.

8-5 / Deflection by direct integration

Since Eq. 8-3 does not contain a first-order term dy/dx, successive ordinary integrations with the appropriate constants of integration will give the slope θ and the deflection y as functions of x. The step-by-step procedure will be outlined and then illustrated by several examples.

1 Set up a reference coordinate system consistent with Eq. 8-3. The origin of a right-hand system at the left-hand end of the beam is one such coordinate system.

2 Derive from equilibrium requirements an expression (or expressions) for the moment as a function of x. In cases where there are discontinuities in the loading, several expressions may be necessary (see Example 8-4, which follows).

3 Determine the bending modulus EI. If the cross section varies with the length, the moment of inertia must be expressed as a function of x. If there are abrupt discontinuities in the cross section or in the material, then several expressions may be necessary.

4 Integrate the equation once for the slope θ, and twice for the deflection y, being careful to *include the constant of integration in each case*. These constants are then evaluated from the conditions imposed on the deflection curve by the supports. A freehand sketch of the deflection curve is often helpful in recognizing the conditions of restraint.

For example, for the beam in Fig. 8-5(a) the deflection must be zero at both the left and right supports, whereas for the beam in Fig. 8-5(b) the deflection must be zero at the left roller support, while both the deflection *and the slope* must be zero at the fixed-wall support. Recognizing the appropriate conditions on the shape of the deflection curve is an important step in determining the correct deflection curve equation for any particular beam.

As the following examples will show, it is sometimes desirable to work the problems using symbols such as P for load and w for load per unit length, rather than numerical values, although this will depend upon the nature of the individual problem. Also, you must be careful to use a consistent set of units for E, I, x, and so on.

EXAMPLE 8-3

Assuming elastic action for the beam shown in Fig. 8-7(a), sketch the deflection curve. Then determine the slope at point B and the deflection at point A. Neglect the weight of the beam itself.

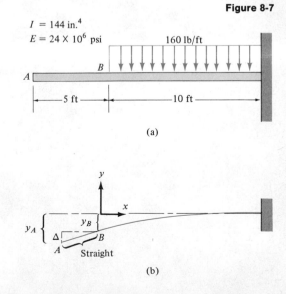

Figure 8-7

$I = 144$ in.^{4}
$E = 24 \times 10^6$ psi
160 lb/ft

(a)

(b)

Solution:

A free-body diagram of any portion of the beam between points A and B will show that there is no bending moment in that portion. Hence, that portion will *not* bend. It will remain straight, and its slope will be equal to the slope at B. The other part of the beam between B and the wall does have an internal bending moment, and this part will deflect into some shape. Therefore the deflection curve for the entire beam will be as shown in Fig. 8-7(b).

For convenience, we shall choose an x-y coordinate system with its origin at the undeformed position of B. Consider a free-body diagram of a portion of the beam between the origin and a point at a distance x from the origin. If counterclockwise moments are positive, and x is expressed in inches,

$$\overset{+}{\sum} M = 0$$

$$\left(\frac{160 \text{ lb/ft}}{12 \text{ in./ft}}\right)(x \text{ in.})\left(\frac{x}{2}\text{ in.}\right) + M = 0$$

$$M = -\frac{160}{24}x^2 \text{ in.-lb}$$

From Eq. 8-3 we have

$$\frac{d^2y}{dx^2} = \frac{d\theta}{dx} = \frac{M}{EI} = -\frac{160}{24}x^2\frac{1}{EI} \tag{a}$$

The restraining wall requires that both the slope and deflection be zero at the wall, where $x = 120$ in. Thus,

$$y_{x=120} = 0 \qquad \theta_{x=120} = 0 \tag{b}$$

Integrating equation (a), we obtain

$$\frac{dy}{dx} = 0 \qquad EI\theta = -\frac{160}{24}\frac{x^3}{3} + C_1 \qquad \Rightarrow \quad EI\frac{dy}{dx} = \frac{-160x^3}{(24)(3)} + C_1 \tag{c}$$

$$EIy = -\frac{160}{24}\frac{x^4}{12} + C_1x + C_2 \tag{d}$$

Using the boundary conditions (b) above, we find that

$$C_1 = \frac{160}{24}\frac{(120)^3}{3} \qquad C_2 = -\frac{160}{24}\frac{(120)^4}{4}$$

Hence, the equations for the slope and the deflection are as follows:

$$EI\theta = \frac{160}{24}\left[-\frac{x^3}{3} + \frac{(120)^3}{3}\right] \tag{e}$$

$$EIy = \frac{160}{24}\left[-\frac{x^4}{12} + \frac{(120)^3}{3}x - \frac{(120)^4}{4}\right] \tag{f}$$

Now, from equation (e), the slope at point B is given by $x = 0$:

$$\theta_B = \left[\frac{160}{(24)(24)(10^6)(144)}\right]\left[-0 + \frac{(120)^3}{3}\right]$$

$$= \frac{160}{(12^2)(10^3)} = 1.11 \times 10^{-3} \text{ rad}$$

By Fig. 8-7(b), the total deflection at A will be

$$y_A = y_B + \Delta$$

From equation (f), for $x = 0$,

$$y_B = \left[\frac{160}{(24)(24)(10^6)(144)}\right]\left[-0 + 0 - \frac{(120)^4}{4}\right]$$

$$= \frac{160}{(16)(10^2)} = -0.1 \text{ in. (downward)}$$

From Fig. 8-7(b) and the fact that θ_B is very small,

$$\Delta = (5)(12)(\theta_B) = 60 \times \frac{160}{(12^2)(10^3)} = 0.0667 \text{ in. (downward)}$$

Hence,

$$y_A = 0.1 + 0.0667 = 0.1667 \text{ in. (downward)}$$

EXAMPLE 8-4

Find an expression for maximum deflection of the straight, homogeneous, elastic beam in Fig. 8-8(a). Sketch the deflection curve.

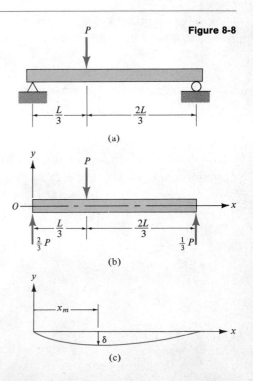

Figure 8-8

(a)

(b)

(c)

Solution:

A free-body diagram of the entire beam is shown in Fig. 8-8(b). Also, the left-hand end of the beam is chosen as the origin of the coordinate system. Since the loading is discontinuous at $x = L/3$, it will be necessary to integrate and solve two sets of equations simultaneously. Thus, for $0 \leq x \leq L/3$, we have

$$M = \frac{2}{3}Px$$

$$EI\theta = \frac{Px^2}{3} + C_1 \tag{a}$$

$$EIy = \frac{Px^3}{9} + C_1x + C_2 \tag{b}$$

Similarly, for $L/3 \leq x \leq L$, we have

$$M = \frac{2}{3}Px - P\left(x - \frac{L}{3}\right) = -\frac{P}{3}(x - L)$$

$$EI\theta = -\frac{P}{3}\frac{(x - L)^2}{2} + C_3 \tag{c}$$

$$EIy = -\frac{P}{3}\frac{(x - L)^3}{6} + C_3x + C_4 \tag{d}$$

Now, to evaluate the four constants of integration, four independent boundary conditions will be needed. Since the deflection of each support must be zero, two boundary conditions are

$$y_{x=0} = 0 \qquad y_{x=L} = 0$$

These conditions yield, from equations (b) and (d), respectively,

$$C_2 = 0 \qquad C_4 = -C_3L \tag{e}$$

The two additional boundary conditions are not quite so obvious as were the first two. Since the beam itself is continuous, equation (b) and equation (d) must give the same value for y at their only common point, at which $x = L/3$. This condition yields

$$C_1 = \frac{PL^2}{3^3} - 2C_3 \tag{f}$$

Also, since the deflection curve is smooth, equations (a) and (c) must give the same slope value at $x = L/3$. Thus,

$$C_1 = -\frac{PL^2}{3^2} + C_3 \tag{g}$$

Equations (e), (f), and (g) yield

$$C_1 = -\frac{5}{3^4} PL^2 \qquad C_2 = 0 \qquad C_3 = \frac{4}{3^4} PL^2 \qquad C_4 = -\frac{4}{3^4} PL^3$$

Finally, the deflection will be a maximum when θ is zero. Hence, from equation (c),

$$0 = -\frac{P}{3} \frac{(x - L)^2}{2} + \frac{4PL^2}{3^4}$$

$$x_{max} = L\left(1 \pm \frac{2}{3}\sqrt{\frac{2}{3}}\right)$$

Only the second root is applicable, and substitution of this root in equation (d) yields

$$\delta = y_{max} = -\frac{PL^3}{EI}\frac{2^4}{3^6}\sqrt{\frac{2}{3}}$$

The deflection curve is shown in Fig. 8-8(c) with the maximum deflection occurring slightly to the right of the load P.

As these examples show, determining the deflection of a beam by means of direct integration is usually a long and tedious process. Fortunately for the practitioner few deflection problems now are done by this manual process. Many commercial programs are available for computers that will literally produce the desired deflection in the twinkling of an eye. Needless to say, these are rapidly becoming the most commonly used methods of finding deflections. However, it is important for you to work through some problems so as to develop a feel for the relations between the mathematical equations and the geometry of the deflection curves.

PROBLEMS

8-37 to 8-44 For each of the beams shown, obtain the equation for the elastic deflection curve. Also, determine the value and location of the maximum deflection. Sketch the deflection curve.

Problem 8-37

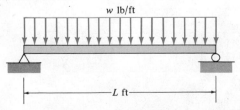

Problem 8-38

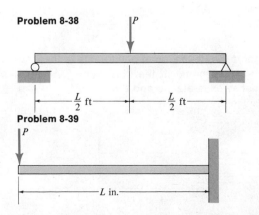

Problem 8-39

Problem 8-40

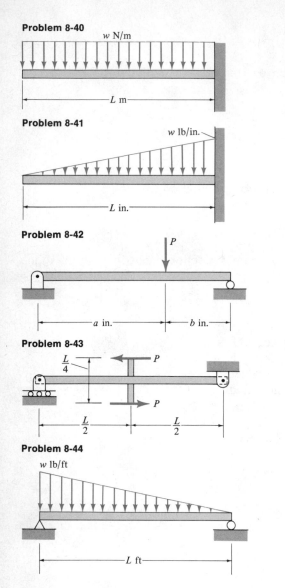

Problem 8-41

Problem 8-42

Problem 8-43

Problem 8-44

8-45   What is the deflection of point A due to the load of 6000 lb on the beam in the illustration? Assume that the behavior is elastic and that $E = 24 \times 10^6$ psi and $I = 240$ in.^{4}.

Problem 8-45

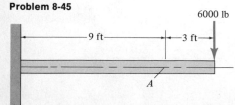

8-46   Determine the deflection of point A in the beam shown in the illustration. Assume that the action is elastic and that $E = 210 \times 10^9$ Pa and $I = 32 \times 10^6$ mm^4.

Problem 8-46

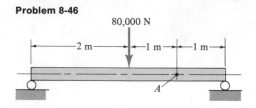

8-47   A hickory diving board 3 by 12 in. in cross section and 12 ft long is built with one end fixed. If a person standing at the middle of the board causes a deflection of 1 in. at the free end, what is the weight of the person? Assume that $E = 2 \times 10^6$.

8-48   In the construction of private dwellings contractors frequently support flooring by arranging the joists in one of the following three ways: (a) 2×8 in. spaced 16 in. on centers, (b) 2×10 in. spaced 16 in. on centers, (c) 2×10 in. spaced 18 in. on centers. If deformation is the major criterion in deciding which arrangement should be used, determine the stiffness ratio of (b) to that of (a), (c) to (a), and (b) to (c). Which is better, (a) or (c)?

8-49 to 8-54   The beam shown in the illustration has a constant cross section and is made of a homogeneous, linearly elastic material. Derive the bending-moment, slope, and deflection equations for the beam. Sketch the bending-moment, slope, and deflection diagrams for the beam.

Problem 8-49

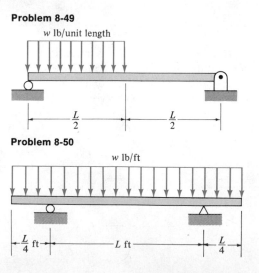

Problem 8-50

Problem 8-51

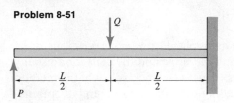

Problem 8-52

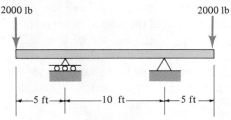

Problem 8-53

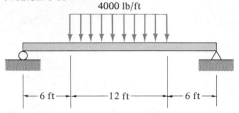

Problem 8-54

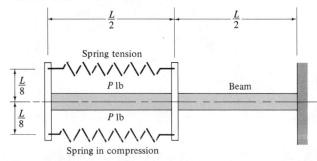

8-55 For the cantilever beam in the illustration, I is 140 in.^{4} and E is 30×10^6 psi. If the load P is 5000 lb, what is the vertical deflection of point A at the extreme right-hand end of the beam? Assume elastic behavior.

Problem 8-55

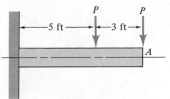

8-56 A circular cantilever beam carries a concentrated load as shown in the illustration. The radius of the beam varies linearly with the length from 50 mm at the right-hand end to 100 mm at the left-hand end. What is the deflection of the right-hand end if $E = 210 \times 10^9$ Pa?

Problem 8-56

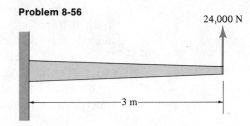

8-57 For the beam shown in the illustration, determine the horizontal and vertical displacement of the point A due to the load P. Assume elastic behavior and small deformations.

Problem 8-57

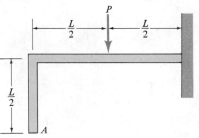

8-58 The beam shown in the illustration is to be used as a weighing device with the load at W and the indicator at P. We wish to calibrate the device so that the weight can be read directly on the indicator scale. Derive an expression between the load W and the movement (in inches) of the pointer assuming E and I to be known. Make a sketch of the shape of the deflected beam.

Problem 8-58

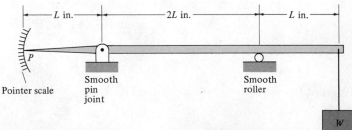

8-6 / Deflection by superposition

When a beam is subjected to several loads placed at various positions along the beam, the problem of determining the slope and the deflection usually becomes quite involved and tedious. This is true regardless of the method used. However, many complex loading conditions are merely combinations of relatively simple loading conditions. If it is assumed that the beam behaves elastically for the combined loading, as well as for the individual loads, the resulting final deflection of the loaded beam is simply the sum of the deflections caused by each of the individual loads. This sum may be an algebraic one or it might be a vector sum, the type depending on whether or not the individual deflections lie in the same plane. The superposition method is illustrated in the following example.

EXAMPLE 8-5

For the beam in Fig. 8-9(a) determine the flexure stress at point A and the deflection of the left-hand end.

Solution:

The stress at point A is a combination of a compressive flexure stress due to the concentrated load and a tensile flexure stress due to the distributed load. Hence,

$$\sigma_A = \frac{M_z(2)}{I_z} - \frac{M_y(3)}{I_y}$$

$$= \frac{[(5)(6)(80^2)/2](2)}{(6)(4^3)/12} - \frac{(600)(80)(3)}{(4)(6^3)/12}$$

$$= 6000 - 2000 = 4000 \text{ psi (tension)}$$

The deflection of a cantilever beam due to a uniformly distributed load w is (by Prob. 8-40)

$$y = \frac{1}{EI_z} \left(-\frac{wx^4}{24} + \frac{wL^3}{6}x - \frac{wL^4}{8} \right)$$

where x is the distance from the free end in inches and w is in pounds per inch. Also, the deflection due to a concentrated load P at its free end is (by Prob. 8-39)

$$z = \frac{1}{EI_y} \left(\frac{Px^3}{6} - \frac{PL^2}{2}x + \frac{PL^3}{3} \right)$$

Thus, at the free end of the beam in Fig. 8-9(b), we have

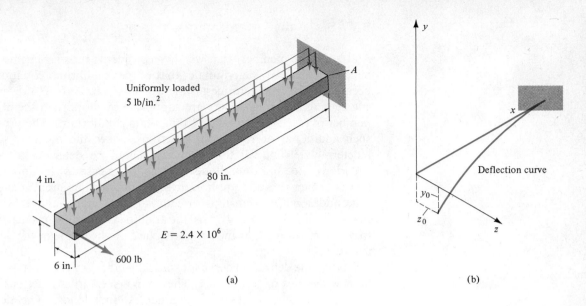

Uniformly loaded
5 lb/in.^{2}

4 in.

80 in.

$E = 2.4 \times 10^6$

6 in.

600 lb

(a)

Deflection curve

y_0

z_0

(b)

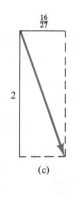

$\frac{16}{27}$

2

(c)

Figure 8-9

$$y_0 = \frac{(30)(80^4)}{[(6)(4^3)/12](8)(2.4)(10^6)} = -2 \text{ in.}$$

$$z_0 = \frac{(600)(80^3)}{[(4)(6^3)/12](3)(2.4)(10^6)} = \frac{16}{27} \text{ in.}$$

The deflection at the free end is the vector sum shown in Fig. 8-9(c).

$$y_0 \longmapsto z_0 = 2.1 \text{ in.}$$

Note: The expression for y_0 and z_0 above could have been obtained directly from Table A-12 of the Appendix.

8-7 / Statically indeterminate beams

All the flexure members that we have considered thus far in this chapter have been statically determinate; that is, the equilibrium equations have been sufficient to determine all the external reactions. However, many engineering flexure members are supported by more than the minimum number of reactions necessary to maintain equilibrium. The presence of such additional or redundant reactions makes the member statically indeterminate, and all the reactions cannot be determined from the equations of equilibrium alone. Just as in the cases of torsion members and axial members, the equilibrium equations must be supplemented with some additional relationships based on the characteristics of the deformation of the member. For beams, the additional relationships usually arise from the restraints that the supports place on the shape of the deflection curve.

Finding the deflection curve for statically indeterminate beams requires no new theories or techniques. The unknown external reactions may be treated simply as ordinary individual external loads. The deflection caused by these external loads can be found by any convenient methods such as direct integration, superposition, or by use of Table A-12. Since the presence of the external reactions places geometrical restrictions on the deflection curve, there will always be a sufficient number of boundary conditions to determine the unknown reactions completely. The following examples illustrate the general procedure.

EXAMPLE 8-6

Determine all the external reactions for the elastic beam shown in Fig. 8-10(a). What is the maximum deflection of the beam?

Solution:
From the free-body diagram of Fig. 8-10(b), we have

$$\overset{+\uparrow}{\sum} F_y = 0$$

$$R_1 + R_2 - wL = 0 \tag{a}$$

$$\overset{\curvearrowleft{+}}{\sum} M = 0$$

$$-R_1 L - M_2 + w\frac{L^2}{2} + M_1 = 0 \tag{b}$$

The couples M_1 and M_2 are present because the beam must have a horizontal tangent at each wall.

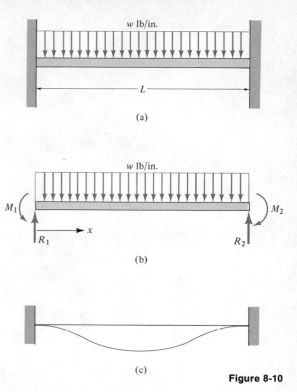

Figure 8-10

Thus, the boundary conditions are:

$$y = 0 \quad \text{at} \quad x = 0 \quad \text{and} \quad x = L$$

$$\theta = 0 \quad \text{at} \quad x = 0 \quad \text{and} \quad x = L$$

The deflection equation can be obtained by direct integration of the moment equation for the beam:

$$M = R_1 x - M_1 - \frac{wx^2}{2}$$

$$EI\theta = R_1 \frac{x^2}{2} - M_1 x - \frac{wx^3}{6} + C_1 \tag{c}$$

$$EIy = R_1 \frac{x^3}{6} - M_1 \frac{x^2}{2} - \frac{wx^4}{24} + C_1 x + C_2 \tag{d}$$

Substitution of the boundary conditions in equations (c) and (d) gives

$$C_1 = 0 \qquad C_2 = 0$$

$$R_1 \frac{L^2}{2} - M_1 L - \frac{wL^3}{6} = 0$$

$$R_1 \frac{L^3}{6} - M_1 \frac{L^2}{2} - \frac{wL^4}{24} = 0$$

From these relationships,

$$R_1 = \frac{wL}{2} \qquad M_1 = \frac{1}{12} wL^2$$

From equations (a) and (b)

$$R_2 = R_1 \qquad M_2 = M_1$$

Of course, from the symmetry of the problem, this last result should have been anticipated. The deflection equation becomes

$$Ely = \frac{wLx^3}{12} - \frac{wL^2 x^2}{24} - \frac{wx^4}{24}$$

The maximum deflection occurs at $x = L/2$ and has the value

$$y = \frac{1}{EI} \left[\frac{wL^4}{96} - \frac{wL^4}{96} - \frac{wL^4}{(16)(24)} \right]$$

$$= -\frac{wL^4}{384EI} \text{ in.}$$

EXAMPLE 8-7

Determine all the external reactions on the elastic beam shown in Fig. 8-11(a). Sketch the deflection curve.

Solution:

The free-body diagram of Fig. 8-11(b) shows that there are three unknowns and only two independent equilibrium equations. However, if R_2 is treated as an ordinary external load, this problem can be conveniently handled by the superposition method.

The deflection equation for the simply supported, uniformly loaded beam shown in Fig. 8-11(c) is

$$Ely_1 = \frac{wLx^3}{12} - \frac{wx^4}{24} - \frac{wL^3 x}{24}$$

Also, from Example 8-4 we know that for a simply supported beam with a downward load P at a distance $L/3$ from the left-hand support, the deflection equation for the portion for which $0 \leq x \leq L/3$ is

$$Ely_2 = \frac{Px^3}{9} - \frac{5}{81} PL^2 x$$

In our present case, $P = -R_2$. Hence, for $0 \leq x \leq L/3$ the deflection of our original beam is

$$Ely = Ely_1 + Ely_2$$

$$= \frac{wLx^3}{12} - \frac{wx^4}{24} - \frac{wL^3 x}{24} - R_2 \frac{x^3}{9} + \frac{5}{81} R_2 L^2 x \qquad \text{(a)}$$

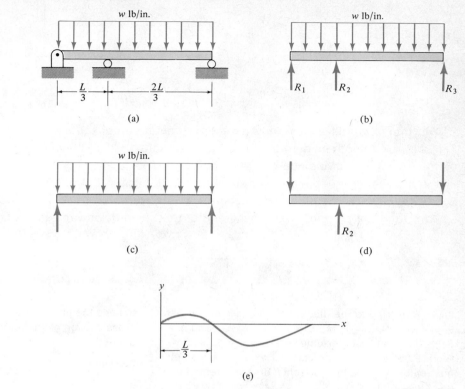

Figure 8-11

For the original beam, the deflection at $x = L/3$ must be zero. Thus,

$$\frac{wL^4}{(12)(27)} - \frac{wL^4}{(24)(81)} - \frac{wL^4}{(24)(3)} - \frac{R_2 L^3}{(9)(27)} + \frac{5}{81}\frac{R_2 L^3}{3} = 0$$

$$-\frac{22}{(24)(81)}wL^4 = -\frac{R_2 L^3 (4)}{(3)(81)}$$

$$R_2 = \frac{11}{16}wL$$

With this value for R_2 we can now determine R_1 and R_3 from the equilibrium requirements of Fig. 8-11(b). Thus,

$$R_1 = \frac{wL}{24} \qquad R_2 = \frac{11wL}{16} \qquad R_3 = \frac{13wL}{48}$$

Substitution of the value of R_2 in equation (a) gives, for $0 \le x \le L/3$,

$$Ely = \frac{wLx^3}{144} - \frac{wx^4}{24} + \frac{wL^3 x}{(9)(144)} \tag{b}$$

Also, by differentiation

$$EI\theta = \frac{wLx^2}{48} - \frac{wx^3}{6} + \frac{wL^3}{(9)(144)} \tag{c}$$

At $x = L/3$ and $x = 0$,

$$EI\theta_{x=L/3} = -\frac{wL^3}{(81)(4)} \qquad EI\theta_{x=0} = \frac{wL^3}{(9)(144)}$$

With these values we can now sketch a qualitative deflection curve (exaggerated) as shown in Fig. 8-11(e). A moment diagram or curvature diagram is usually a great aid in visualizing the shape of the deflection curve since a positive curvature indicates concave upward, etc.

It is interesting to observe that although the deflection of the portion of the beam for which $0 \le x \le L/3$ is upward, the left-hand reaction R_1 is also upward. Thus, this reaction is pushing the beam upward, and not pulling the left-hand end of the beam downward as might be anticipated.

PROBLEMS

8-59 Referring to the illustration, and assuming that $P = 10,000$ N, $L = 2$ m, $E = 100 \times 10^9$ Pa, and $I = 500 \times 10^3$ mm^4, determine the magnitude of M_0 required to keep the longitudinal axis of the pointer horizontal. What is the horizontal displacement of the tip of the pointer?

Problem 8-59

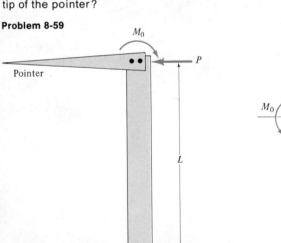

8-60 For the elastic beam shown in the illustration, $L = 14$ ft. The pointer AB is rigidly attached to the beam directly above the right-hand support. With no load on the beam, the tip A of the pointer is at the neutral surface. Determine the moment M_0 necessary to keep the tip of the pointer at the neutral axis of the cross section of the beam when a load P of 5000 lb is applied.

Problem 8-60

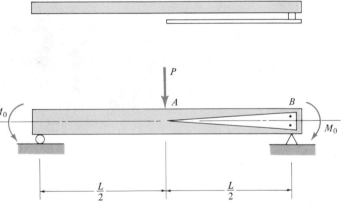

Top view

8-61 The free end of the prismatic, linearly elastic cantilever beam shown in the illustration is supported at its left-hand end by a linear elastic spring with a spring constant of k lb/in. If the spring carries no load when the beam is horizontal, how far will the spring be compressed when the beam is loaded with a uniformly distributed load of w lb/in.?

8-62 An elastic cantilever beam is 10 ft long and made of a rolled-steel equal-leg angle $6 \times 6 \times 1$ (see

Problem 8-61

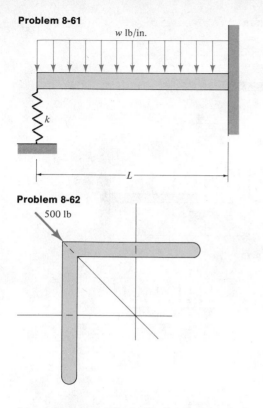

Problem 8-62

500 lb

Problem 8-63

10,000 lb

30°

8 in.   12 in.

6 in.

10 in.

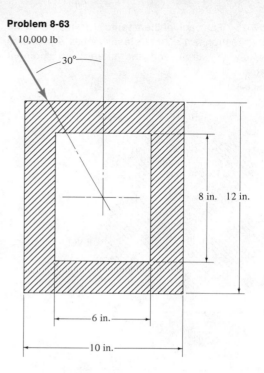

Appendix table). The beam is loaded at its free end as indicated in the illustration. Find the deflection of the free end, if $E = 30 \times 10^6$ psi.

8-63 A simply supported beam 20 ft long and having the cross section shown in the illustration carries a concentrated load of 10,000 lb at its mid span. The load acts in the direction indicated. What is the maximum deflection if $E = 30 \times 10^6$ psi?

8-64 An elastic beam with a rectangular cross section carries a uniformly distributed vertical load of 750 N/m. The beam is 6 m long and is simply supported at its ends at an angle with the vertical axis as shown in the illustration. If $E = 30 \times 10^9$ Pa, determine the magnitude and direction of its deflection at the mid span.

8-65 An elastic beam with a length of L in. and a square cross section carries a uniformly distributed vertical load of w lb/in. If the beam is simply supported at its ends at an angle θ with the horizontal, determine the magnitude and direction of the deflection at the mid span if (a) $\theta = 30°$; (b) $\theta = 45°$; (c) θ is an arbitrary angle such that $0 \leq \theta \leq 90°$.

Problem 8-64

Vertical axis

10 cm

15 cm

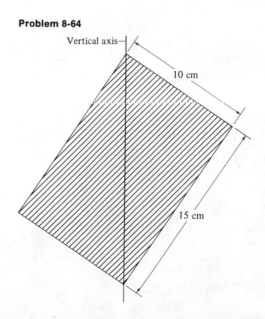

8-66 to 8-75 For each of the beams shown, find all the support reactions. Assume elastic behavior. Sketch the deflection curve and the shear and moment diagrams.

Problem 8-65

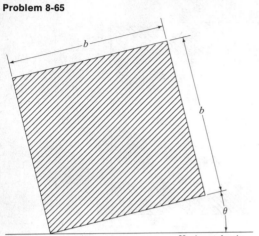

Problem 8-66

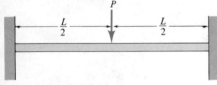

Problem 8-67

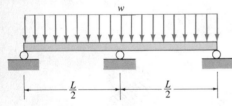

Problem 8-68

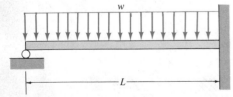

Problem 8-69

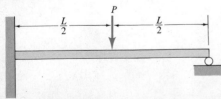

Problem 8-70

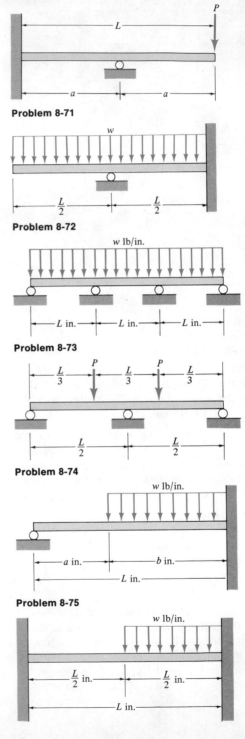

Problem 8-71

Problem 8-72

Problem 8-73

Problem 8-74

Problem 8-75

Chapter 9
Beam deflections II

9-1 / Objectives

Upon completion of this chapter you will be able to:

1 Obtain shear and moment equations for a beam by the use of singularity functions.
2 Use singularity functions to obtain deflections.
3 Obtain deflections by using the moment-area method.

9-2 / Introduction

Chapters 7 and 8 have dealt with both graphical and mathematical methods of relating the loading on a beam to the corresponding internal shear and moment reactions. Also, Chapter 8 established the basic relationship between the *elastic* deflection curve and the bending moment, namely Eq. 8-3, from which we "solved" for the deflection by the tedious process of successive integration.

In this chapter we shall develop somewhat more sophisticated mathematical and graphical methods for solving elastic deflection problems. For those engineers who work frequently with deflection problems, these methods can expedite the solution process considerably.

9-3 / Singularity functions

As Example 8-4 illustrated, when the loading on a beam has a discontinuity such as a concentrated point load, it becomes necessary to handle the problem in two or more parts. This mathematical difficulty can be eliminated by use of the so-called singularity functions.

Recall relations (Eqs. 7-2 to 7-4) between the loading $w(x)$, shear V, and moment M,

$$w(x) = -\frac{dV}{dx} \qquad [7\text{-}2]$$

$$V = \frac{dM}{dx} \qquad [7\text{-}3]$$

$$w(x) = -\frac{d^2M}{dx^2} \qquad [7\text{-}4]$$

which are valid so long as the loading function $w(x)$ is continuous. Briefly, we seek to define some loading functions that can physically represent such things as point loads, jump loading conditions, and point couples but still satisfy Eqs. 7-2 to 7-4. To this end we introduce the Dirac delta function, step function, and doublet function, respectively.

Dirac delta function   The basic idea is to replace the concentrated load P by a high-intensity distributed load P/ε distributed over a very small length ε, so that even in the limit as $\varepsilon \to 0$ the product of P/ε with the length ε remains the finite value P, as illustrated in Fig. 9-1(a) and (b). For convenience we let P equal unity and define the Dirac delta function as

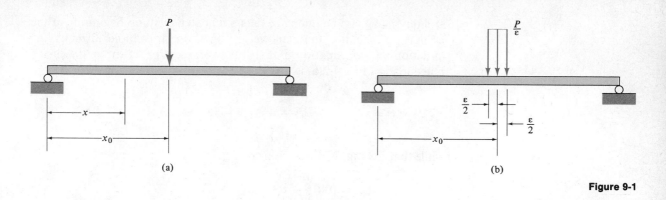

(a)   (b)

Figure 9-1

$$\delta\langle x - x_0\rangle = \lim_{\varepsilon \to 0} \begin{cases} 0 & \text{when} \quad x < \left(x_0 - \dfrac{\varepsilon}{2}\right) \\[2mm] \dfrac{1}{\varepsilon} & \text{when} \quad \left(x_0 - \dfrac{\varepsilon}{2}\right) < x < \left(x + \dfrac{\varepsilon}{2}\right) \\[2mm] 0 & \text{when} \quad x > \left(x_0 + \dfrac{\varepsilon}{2}\right) \end{cases} \qquad (9\text{-}1)$$

Then using this delta function we see that the distribution

$$w(x) = P\delta\langle x - x_0\rangle$$

integrated over the length of the beam l gives

$$\int_0^l P\delta\langle x - x_0\rangle \, dx = \lim_{\varepsilon \to 0} \int_{x_0-(\varepsilon/2)}^{x_0+(\varepsilon/2)} \frac{P}{\varepsilon} \, dx = P$$

Hence the distribution

$$w(x) = P\delta\langle x - x_0\rangle$$

can be used to represent a concentrated load P applied at point x_0.

We note that the delta function is not a function in the usual mathematical sense of continuity and differentiability. However, we justify our introduction of it simply because its formal use leads to results that we can interpret physically, such as the concentrated load discussed above.

Step function We now introduce the second singularity function, the unit step function, defined by

$$u\langle x - x_0\rangle = \int_0^x \delta\langle t - x_0\rangle \, dt \qquad (9\text{-}2)$$

which, from the definition of $\delta\langle t - x_0\rangle$ has the property

$$u\langle x - x_0\rangle = \begin{cases} 0 & x < x_0 \\ 1 & x > x_0 \end{cases} \qquad (9\text{-}3)$$

and physically can be interpreted as a unit step or jump beginning at the point x_0. With this function we are now able to handle formally the initiation and termination of distributed loads. For example, the distributed load of Fig. 9-2(a) is given by

$$w(x) = w_0 u\langle x - x_0\rangle - w_0 u\langle x - x_1\rangle = \begin{cases} 0 & x < x_0 \\ w_0 & x_0 < x < x_1 \\ 0 & x > x_1 \end{cases}$$

while that for Fig. 9-2(b) is given by

$$w(x) = 100\frac{x}{5} - 100\frac{x}{5} u\langle x - 5\rangle = \begin{cases} 20x & x < 5 \\ 0 & x > 5 \end{cases}$$

We note for later use that the definite integral of the product of a continuous function $f(x)$ with the unit step function $u(x - x_0)$ is

$$\int_0^x f(t)u\langle t - x_0\rangle\, dt = \int_{x_0^+}^x f(t)u\langle t - x_0\rangle\, dt$$

$$= u\langle x - x_0\rangle \int_{x_0}^x f(t)\, dt \qquad (9\text{-}4)$$

where x_0^+ means $x_0 + \varepsilon$ as $\varepsilon \to 0$, since $u\langle t - x_0\rangle$ is not defined for $t = x_0$. In particular, if $f(x) = (x - x_0)^n$ we have the useful results

$$\int_0^x (t - x_0)^n u\langle t - x_0\rangle\, dt = u\langle x - x_0\rangle \int_{x_0}^x (t - x_0)^n\, dt$$

$$= u\langle x - x_0\rangle \frac{(x - x_0)^{n+1}}{n + 1} \qquad (9\text{-}5)$$

for $n > -1$.

Figure 9-2

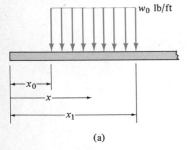

(a)

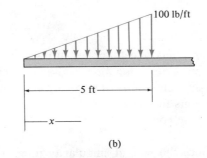

(b)

Doublet function   The delta function enables us to represent a concentrated load as a distribution $P\delta\langle x - x_0\rangle$, and the unit step function enables us to introduce or terminate distributed loads. We now introduce the third singularity function, the doublet function, in order to express concentrated couples in terms of a load distribution function. To this end we visualize the concentrated couple C as a pair of very large concentrated forces C/ε separated by a very small distance ε, as illustrated in

Fig. 9-3, so that the product of C/ε with ε equals the finite number C. For convenience we let C equal unity. Now, if we think of the loads $1/\varepsilon$ as being distributed over the lengths ε, we define the doublet function by

$$\psi\langle x - x_0 \rangle = \lim_{\varepsilon \to 0} \begin{cases} 0 & \text{when} \quad x < (x_0 - \varepsilon) \\[2mm] \dfrac{1}{\varepsilon^2} & \text{when} \quad (x_0 - \varepsilon) < x < x_0 \\[2mm] -\dfrac{1}{\varepsilon^2} & \text{when} \quad x_0 < x < (x_0 + \varepsilon) \\[2mm] 0 & \text{when} \quad (x_0 + \varepsilon) < x \end{cases} \qquad (9\text{-}6)$$

Hence, the distribution corresponding to a concentrated counterclockwise couple C applied at $x = x_0$ is given by

$$w(x) = C\psi\langle x - x_0 \rangle$$

Figure 9-3

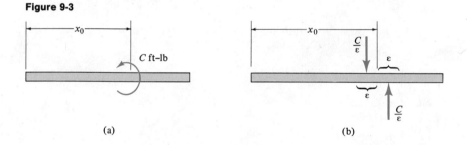

(a) (b)

Integration of the doublet function yields

$$\int_0^x \psi\langle t - x_0 \rangle \, dt = \lim_{\varepsilon \to 0} \begin{cases} 0 & \text{when} \quad x < (x_0 - \varepsilon) \\[2mm] \dfrac{1}{\varepsilon} & \text{when} \quad x = x_0 \\[2mm] 0 & \text{when} \quad x > x_0 + \varepsilon \end{cases}$$

which we recognize as the delta function. Hence

$$\int_0^x C\psi\langle t - x_0 \rangle \, dt = C\delta\langle x - x_0 \rangle \qquad (9\text{-}7)$$

Thus with the doublet function we can represent a loading distribution corresponding to the effect of a concentrated couple; with a delta function we can represent a distribution corresponding to a concentrated load; and with the step function we can introduce or terminate ordinary distributed loads. These representations are illustrated in Fig. 9-4(a)–(c), respectively. The following example illustrates the application of these functions for determining shear and moment equations.

Figure 9-4

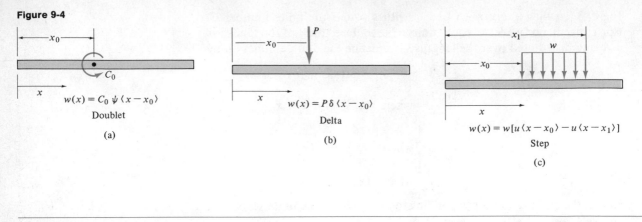

$$w(x) = C_0 \, \psi \, \langle x - x_0 \rangle$$

Doublet

(a)

$$w(x) = P \delta \langle x - x_0 \rangle$$

Delta

(b)

$$w(x) = w[u \langle x - x_0 \rangle - u \langle x - x_1 \rangle]$$

Step

(c)

EXAMPLE 9-1

Using singularity functions, obtain the shear and moment equations for the beam of Example 8-2.

Solution:

Refer to Fig. 8-4(a). Recalling that $w(x)$ is positive when directed downward, we write the loading functions for each of the loadings.

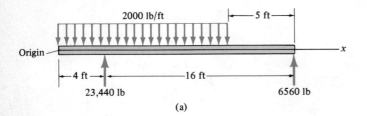

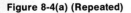

(a)

Figure 8-4(a) (Repeated)

For the uniformly distributed load of 2000 lb/ft which terminates at $x = 15$

$$w(x) = 2000 - 2000u\langle x - 15 \rangle \tag{a}$$

For the left pin reaction located at $x = 4$

$$w(x) = -23{,}440\delta\langle x - 4 \rangle \tag{b}$$

For the right pin reaction located at $x = 20$

$$w(x) = -6560\delta\langle x - 20 \rangle \tag{c}$$

Hence for the whole beam

$$w(x) = 2000 - 2000u\langle x - 15 \rangle - 23{,}440\delta\langle x - 4 \rangle - 6560\delta\langle x - 20 \rangle \tag{d}$$

The boundary conditions are $\quad V = 0 \quad$ at $\quad x = 0 \qquad x = 20^+$

$$M = 0 \quad \text{at} \quad x = 0 \qquad x = 20^+$$

(The 20^+ means that we consider the values just to the right of the pin reaction.)

Now, Eq. 7-2 can be written as

$$dV = -w\,dx$$

whose definite integral is

$$\int_0^{V(x)} dV = -\int_0^x w(t)\,dt$$

$$V(x) = -\int_0^x [2000 - 2000u\langle t - 15\rangle - 23{,}440\delta\langle t - 4\rangle - 6560\delta\langle t - 20\rangle]\,dt$$

$$= -2000x + 2000(x - 15)u\langle x - 15\rangle + 23{,}440u\langle x - 4\rangle + 6560u\langle x - 20\rangle \qquad \text{(e)}$$

where we have utilized Eqs. 9-2 and 9-5 for $n = 0$. We see that $\quad V(0) = 0 \qquad V(20^+) = 0$

as required. Now, Eq. 7-3 can be rewritten as $\quad dM = V\,dx$

whose definite integral is $\qquad \displaystyle\int_0^{M(x)} dM = \int_0^x V(t)\,dt$

Thus, using equation (e),

$$M(x) = \int_0^x [-2000t + 2000(t - 15)u\langle t - 15\rangle + 23{,}440u\langle t - 4\rangle$$

$$+\, 6560u\langle t - 20\rangle]\,dt$$

$$= -2000\frac{x^2}{2} + 2000\frac{(x - 15)^2}{2}u\langle x - 15\rangle + 23{,}440(x - 4)u\langle x - 4\rangle$$

$$+\, 6560(x - 20)u\langle x - 20\rangle \qquad \text{(f)}$$

when we have again used Eq. 9-5. We see that

$$M(0) = 0 \qquad M(20^+) = 0$$

as required. As a check, let us evaluate M at $x = 11.720$ as in Example 8-2.

$$M(11.72) = -2000\frac{(11.720)^2}{2} + 0 + 23{,}440(11.720 - 4)(1) + 0 = 43{,}600 \text{ ft-lb}$$

9-4 / Deflection by use of singularity functions

The singularity functions can be utilized in determining the slope and deflection of beams. One way to do this is to start with the loading function for the entire beam; then, by utilizing the relations

$$w(x) = -\frac{dV}{dx} \qquad\qquad [7\text{-}2]$$

$$V = \frac{dM}{dx} \qquad\qquad [7\text{-}3]$$

$$\frac{M}{EI} = \frac{d\theta}{dx} \qquad\qquad [8\text{-}3]$$

$$\theta = \frac{dy}{dx} \qquad\qquad [8\text{-}1]$$

one can successively integrate each of these equations, and by determining each of the constants of integration from appropriate boundary conditions, eventually obtain a single deflection equation for the entire beam. In practice, however, it is often easier to draw a free-body diagram, write the moment equation for the entire beam in terms of singularity functions, and then integrate once to get the slope θ and once again for the deflection y. Usually this procedure will involve only the integration of the unit step function which was performed in Eqs. 9-4 and 9-5. The following examples serve to illustrate these ideas.

EXAMPLE 9-2

Rework Example 8-4 using singularity functions.

Solution:

Refer to Fig. 8-8. The moment equation for the entire beam can be written in terms of singularity functions as follows:

$$M = \frac{2}{3}Px - P\left(x - \frac{L}{3}\right)u\left\langle x - \frac{L}{3}\right\rangle + \frac{1}{3}P(x - L)u\langle x - L\rangle \qquad (a)$$

where $u\langle x - L/3\rangle$ is the unit step function beginning at $x = L/3$, and so on. Then, from Eq. 8-3 we have

$$EI\frac{d^2y}{dx^2} = EI\frac{d\theta}{dx} = M \qquad (b)$$

with boundary conditions

$$y_{x=0} = 0 \qquad y_{x=L} = 0 \qquad (c)$$

Integrating (a) we have (recall Eqs. 9-4 and 9-5)

$$EI\theta = \frac{Px^2}{3} - \frac{P(x - L/3)^2}{2}u\left\langle x - \frac{L}{3}\right\rangle$$

$$+ \frac{1}{6}P(x - L)^2 u\langle x - L\rangle + C_1 \qquad (d)$$

$$Ely = \frac{Px^3}{9} - \frac{P}{6}\left(x - \frac{L}{3}\right)^3 u\left\langle x - \frac{L}{3}\right\rangle$$

$$+ \frac{1}{18}P(x - L)^3 u\langle x - L\rangle + C_1 x + C_2 \tag{e}$$

The boundary conditions yield

$$y_{x=0} = 0 = 0 - 0 + 0 + 0 + C_2$$

$$C_2 = 0$$

$$y_{x=L} = 0 = \frac{PL^3}{9} - \frac{P}{6}\left(\frac{2}{3}L\right)^3 + 0 + C_1 L + 0$$

$$C_1 = -\frac{5}{3^4}PL^2 \tag{f}$$

Let us check our result by evaluating the deflection at $x = L/3$.

$$Ely_{x=L/3} = \frac{P(L/3)^3}{9} - 0 + 0 - \frac{5}{3^4}PL^2\frac{L}{3}$$

$$= -\frac{4}{3^5}PL^3$$

which is the same value as obtained from equation (b) of Example 8-4.

EXAMPLE 9-3

Determine the maximum deflection of the beam shown in Fig. 9-5(a). $E = 10 \times 10^6$ psi and $I = 14.4$ in.^{4}.

Solution:

For convenience, we shall work the problem using the dimensions in feet. The wall reactions are shown in Fig. 9-5(b). The tricky part of this problem is to express the moment of the distributed load in terms of the step function. Referring to Fig. 9-5(c) we see that the moment (with respect to the right end) of the partial distributed load of the figure is

$$\frac{1}{2}\left(\frac{x-4}{6}100\right)(x-4)\left(\frac{x-4}{3}\right)u\langle x-4\rangle = \frac{100}{36}(x-4)^3 u\langle x-4\rangle$$

With this, the moment equation for the entire beam is

$$M = -2400 + 300x - \frac{100}{36}(x-4)^3 u\langle x-4\rangle \tag{a}$$

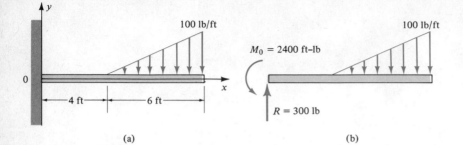

(a) (b)

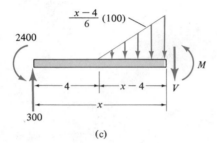

(c)

Figure 9-5

Thus, integrating we obtain

$$EI\theta = -2400x + 150x^2 - \frac{25}{36}(x-4)^4 u\langle x-4 \rangle + C_1 \tag{b}$$

$$EIy = -1200x^2 + 50x^3 - \frac{5}{36}(x-4)^5 u\langle x-4 \rangle + C_1 x + C_2 \tag{c}$$

The wall restrains the beam so that

$$\theta = 0 \qquad y = 0 \quad \text{at} \quad x = 0$$

Hence $C_1 = C_2 = 0$. Therefore the maximum deflection is

$$y_{\text{max}} = y_{x=10} = \frac{1}{EI}\left[-1200(10)^2 + 50(10)^3 - \frac{5}{36}(6)^5\right]$$

$$= -\frac{12^4}{10^7 12^2 14.4}\, 7.1 \times 10^4 = -0.071 \text{ ft}$$

Alternative solution:

The moment equation (a) can be obtained by writing the loading function $w(x)$ for the entire beam and then integrating twice to obtain the appropriate moment equation. Utilizing the definitions of the doublet, delta, and step functions we have from Fig. 9-5(b)

$$w(x) = 2400\psi\langle x - 0\rangle - 300\delta\langle x - 0\rangle + \frac{100}{6}(x-4)u\langle x-4\rangle$$

Integrating, Eqs. 7-2 and 7-3 yield

$$V(x) = -2400\delta\langle x-0\rangle + 300u\langle x-0\rangle - \frac{100}{6}\frac{(x-4)^2}{2}u\langle x-4\rangle + C_1$$

$$M(x) = -2400u\langle x-0\rangle + 300xu\langle x-0\rangle - \frac{100}{6}\frac{(x-4)^3}{6}u\langle x-4\rangle$$

$$+ C_1 x + C_2$$

where we have utilized Eqs. 9-2, 9-5, and 9-7. Boundary conditions require

$$V(10) = 0 \qquad M(10) = 0$$

so that $C_1 = C_2 = 0$. The moment equation is now identical to (a) and the remainder of the solution is the same as before.

PROBLEMS

9-1 to 9-5 Rework Probs. 8-49 to 8-53 using singularity functions.

9-6 and 9-7 Rework Probs. 8-42 and 8-43 using singularity functions.

9-8 and 9-9 The beam shown in the illustration is made of an S 10 × 35 standard steel beam. Determine the deflection at the middle of the beam caused by the applied loads. Neglect the weight of the beam itself.

9-10 and 9-11 The beam shown in the illustration is made of a W 8 × 40 steel beam. Determine the maximum deflection caused by the applied loads. Neglect the weight of the beam itself.

Problem 9-8

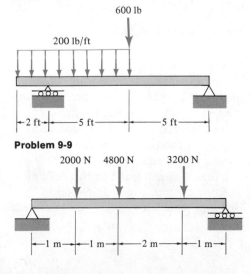

Problem 9-10

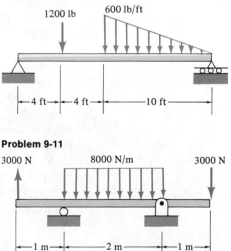

Problem 9-9

Problem 9-11

9-5 / Moment-area method for deflection

In many cases the slope and the deflection of a beam can be determined with relative ease by a semigraphical integration technique, usually called the moment-area method. The method is based on a geometrical interpretation of definite integrals.

Recall Eq. 8-3, which is

$$\frac{d\theta}{dx} = \frac{1}{\rho} = \frac{M}{EI} = \text{curvature}$$

Rewrite this equation in the form

$$d\theta = \frac{1}{\rho} dl = \frac{M}{EI} dl \tag{9-8}$$

where $d\theta$ is the infinitesimal change in the slope of the deflection curve occurring over the infinitesimal length dl. Notice that $(M/EI)\, dl$ can be interpreted as an infinitesimal area on the M/EI (curvature) diagram. Such a curvature diagram can be obtained from the moment diagram simply by dividing the moment by the bending modulus EI. If the beam is homogeneous and has a constant cross section, the moment and curvature diagrams will have the same general shape. For example, see Fig. 9-6(a) and (b).

The finite *change* in slope between any two distinct points on the beam can be obtained by integrating Eq. 9-8 between the corresponding limits. Thus,

$$\Delta\theta_{AB} = \int_A^B d\theta = \int_A^B \frac{M}{EI} dl \tag{9-9}$$

From the previous statements, we see that the definite integral on the right may be interpreted as the area under the curvature curve between points A and B. Positive areas, or those above the x axis, indicate increases in slope; while negative areas, or those below the x axis, indicate decreases in slope. Notice, however, that the integral, and hence the corresponding area, represents the *change* in slope and *not* the slope itself.

The deflection of the beam can be obtained indirectly by considering the *tangential deviation*. The tangential deviation t_{AB} is defined as the distance, measured perpendicular to the undeformed neutral axis, between the point A on the deflection curve and a tangent line to the deflection curve drawn through the point B. For example, see Fig. 9-6(c). From this figure, in which the deflections and slopes are greatly exaggerated, we see that

$$dt_{AB} = l\, d\theta$$

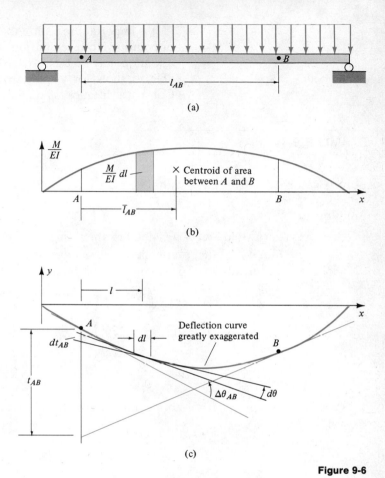

Figure 9-6

Thus,
$$t_{AB} = \int_A^B l\, d\theta = \int_A^B l\frac{M}{EI}\, dl \qquad (9\text{-}10)$$

$$t_{AB} = \bar{l}_{AB}a$$

where $\bar{l}_{AB}$ is the horizontal distance from point A to the centroid of the area a between points A and B on the curvature diagram. Thus, the moment of the area on the curvature diagram is always taken *about the point for which the tangent deviation is being determined*. It is from Eq. 9-10 that the method gets its name moment-area, since the equation represents the first moment of an area.

Since $\bar{l}_{AB}$ is necessarily a positive number, t_{AB} will be positive when the corresponding curvature area a is positive, and negative when the area is negative. A positive value for t_{AB} indicates that point A lies *above* the extended tangent line through point B; and A lies *below* this tangent line if t_{AB} is negative.

It is important to realize that t_{AB} is not the deflection of either point A or point B. However, the following examples illustrate how the tangential deviation may be used to determine the deflection. Also, you should be able to show for youself that t_{AB} will generally *not* be equal to t_{BA}. Construct distance t_{BA} on the curve in Fig. 9-6(c).

EXAMPLE 9-4

Solve Example 8-3 by means of the moment-area method.

Solution:

The beam is redrawn in Fig. 9-7(a) and the shear, moment, curvature, slope, and deflection diagrams are shown in Fig. 9-7(b)–(f), respectively. The beam has zero slope at the right-hand end, and thus the slope at point B will be equal to the *change* in slope between points B and C. Hence,

$$\theta_C - \theta_B = \Delta\theta_{BC} = \text{area of curvature diagram (parabolic spandrel)}$$

$$0 - \theta_B = -\frac{1}{3}(120)\left(\frac{1}{36,000}\right) = -\frac{1}{900} \text{ rad}$$

$$\theta_B = \frac{1}{900} \text{ rad}$$

The tangent at point C is horizontal and thus, in this case, t_{BC} will represent the deflection of point B. Hence,

$$y_B = t_{BC} = \bar{l}_{BC} \times \text{Area } BC = \frac{3}{4}(120)\left(-\frac{1}{900}\right) = -\frac{1}{10} \text{ in. } \downarrow$$

Then, as in Example 8-3, the deflection of the left-hand end is

$$y_A = t_{BC} - 60\,\theta_B$$

$$= -\frac{1}{10} - (60)\left(\frac{1}{900}\right)$$

$$= -\frac{1}{10} - 0.0667 = -0.1667 \text{ in. } \downarrow$$

Aternatively, y_A could have been determined directly by noting that

$$y_A = t_{AC} = \bar{l}_{AC} \times \text{Area } AC$$

$$= \left(60 + \frac{3}{4}\,120\right)\left(-\frac{1}{900}\right)$$

$$= -\frac{150}{900} = -0.1667 \text{ in.}$$

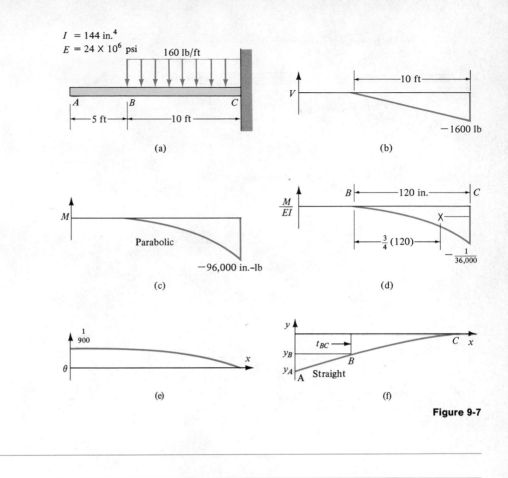

$I = 144$ in.^{4}
$E = 24 \times 10^6$ psi

Figure 9-7

EXAMPLE 9-5

A solid, circular, reduced-section rod shown in Fig. 9-8(a) is simply supported and loaded at its mid span with a load of 600 lb. Assuming that the material is elastic and homogeneous with $E = 10 \times 10^6$, and neglecting stress concentrations at the fillet, find the maximum deflection of the rod.

Solution:

The moment and curvature diagrams are shown in Fig. 9-8(b) and (c), respectively. From the geometry and loading of the rod, it is logical to expect that the maximum deflection will occur at some point in the reduced portion. If we consider some arbitrary point C in this portion, from Fig. 9-8(d) we see that

$$y_C + t_{CA} = \frac{x}{72} t_{BA}$$

$$y_C = \frac{x}{72} t_{BA} - t_{CA} \tag{a}$$

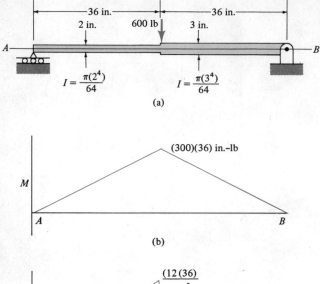

$$I = \frac{\pi(2^4)}{64} \qquad I = \frac{\pi(3^4)}{64}$$

(a)

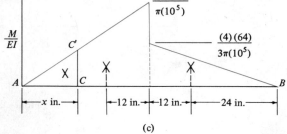

(b)

(c)

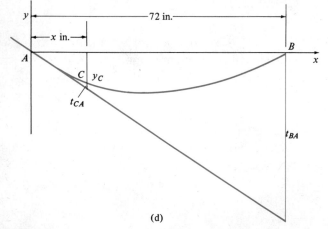

(d)

Figure 9-8

Now, from Fig. 9-8(c),

$$t_{CA} = \text{moment of area } ACC' \text{ about } C$$

$$= \frac{x}{3}\frac{1}{2} \times \frac{x}{36}\frac{(12)(36)}{\pi(10^5)}$$

$$= \frac{2x^3}{\pi(10^5)} \tag{b}$$

Similarly,

$$t_{BA} = (24)\left(\frac{36}{2}\right)\frac{(4)(64)}{3\pi(10^5)} + (48)\left(\frac{36}{2}\right)\frac{(12)(36)}{\pi(10^5)}$$

$$= \frac{(2^{12})(3^2) + (2^9)(3^6)}{\pi(10^5)} \tag{c}$$

Combining equations (a), (b), and (c), we have

$$y_C = \frac{1}{\pi(10^5)}\left[(2^9 + 2^6 \times 3^4)x - 2x^3\right] \tag{d}$$

We know that the maximum deflection occurs where $dy/dx = 0$. Hence,

$$\frac{dy}{dx} = \frac{1}{\pi(10^5)}\left[(2^9 + 2^6 \times 3^4) - 6x^2\right] = 0$$

from which

$$x = (2^3)\sqrt{\frac{89}{6}} = 31 \text{ in.}$$

Substitution of this value in equation (d) gives the desired result. Thus,

$$y_{max} = \frac{1}{\pi(10^5)}\left[(2^9 + 2^6 \times 3^3)31 - 2(31^3)\right]$$

$$= 0.372 \text{ in.}$$

These examples show that the moment-area method is quite advantageous as long as the curvature-diagram geometry is relatively simple. Also it is a convenient way of determining the deflection of a particular point rather than the deflection equation of the entire beam.

PROBLEMS

9-12 to 9-15 Using the moment-area method, determine the maximum deflection for the beam in Probs. 8-37 to 8-40.

9-16 Using the moment-area method, determine the maximum deflection for the beam in Prob. 8-43.

9-17 to 9-19 Rework Probs. 8-45 to 8-47 by the moment-area method.

9-20 Rework Prob. 8-55 by the moment-area method.

9-21 The circular cantilever beam in the illustration carries a concentrated load at its left-hand end. The beam has a varying diameter of 50, 75, and 100 mm. Find the deflection of the left end if $E = 200 \times 10^9$ Pa.

Problem 9-21

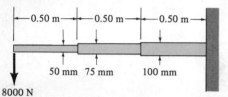

50 mm 75 mm   100 mm

8000 N

9-22 The reduced rectangular-section beam shown in the illustration has a width of 2 in. Neglecting any stress concentrations and assuming that $E = 30 \times 10^6$ psi, find the maximum flexure stress and the maximum deflection.

Problem 9-22

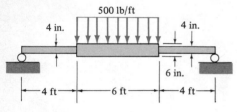

9-23 A reduced circular-section beam has the dimensions and loading shown in the illustration. Neglecting any stress concentrations, find the maximum flexure stress and the deflection at the midpoint. Does this latter value represent the maximum deflection? Assume that there is elastic behavior and that $E = 10 \times 10^6$ psi.

Problem 9-23

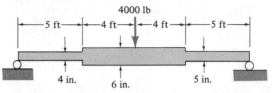

9-24 and 9-25 Using the moment-area method, determine the value and location of the maximum deflection for the beams in Probs. 8-50 and 8-52.

Chapter 10
Buckling

10-1 / Objectives

Upon completion of this chapter you will be able to:

1 Describe buckling.
2 Determine slenderness ratio.
3 Evaluate the effect of end conditions.
4 Use Euler's equation to determine the elastic buckling load for a slender column.
5 Use an Engesser equation to determine the inelastic buckling load for an intermediate column with a known stress-strain diagram.
6 Explain and demonstrate the manner by which empirical column formulas are obtained.
7 Use empirical column formulas to obtain the allowable load on a given axially loaded column.
8 Use empirical column formulas to determine whether or not a given column can safely carry a given eccentric load.

10-2 / Introduction

Buckling is a mode of failure generally resulting from structural instability due to compressive action on the structural member or element involved. Some common engineering structures that are likely to buckle if overloaded are metal building columns, compression members in bridge or roof trusses, the hull of a submarine, the metal skin on aircraft fuselages or wings with excessive torsional and/or compressive loading, any thin-walled torque tube, the thin web of an I beam with excessive shear load, and a thin flange of an I beam subjected to excessive compressive bending effects. Notice from these few examples that buckling is the result of compressive action. Overall torsion or shear, as we learned in previous chapters, may cause a localized compressive action that could lead to buckling.

The distinctive feature of buckling is the catastrophic and often spectacular nature of the failure. The collapse of a column supporting stands in a stadium or the roof of a building usually draws large headlines and cries of engineering negligence. On a less grandiose scale, you can witness and get a better understanding of buckling by trying a few of the tests shown in Fig. 10-1. In Fig. 10-1(a)–(d) are examples of temporary or elastic buckling; while (e)–(h) are examples of permanent or plastic buckling.

Although plates, shells, tubes, and various kinds of structural members have a tendency to buckle under load, this initial study will consider only straight members that are axially loaded and have constant cross section. Other types are usually studied in more advanced courses. The constant cross section will be of such proportions that its principal second moments of area are of the same order of magnitude. Such compression members that are slender enough to make buckling effects important are called *columns*.

In stating that this chapter deals with straight members that are axially loaded and have constant cross section, we do not mean to imply that it is possible in reality to have a column that is dimensionally perfect, of a perfectly homogeneous material, and exactly axially loaded. Actually, the fact that every column has some degree of imperfection in loading, dimensions, and/or material is the *cause* of the initiation of buckling if it is overloaded. If an ideal slender member did exist, and if it were ideally loaded, it would not buckle. Instead, it would fail (yield, fracture, or "flatten out") as the result of excessive compressive stress. It is sometimes convenient to discuss and analyze an ideal column and to apply a controlled "imperfection" in the analysis, as is done in the next section.

We point out that buckling need not occur in every real compression member if the load is increased until something damaging happens. A member will buckle or collapse only if it is slender enough. If the

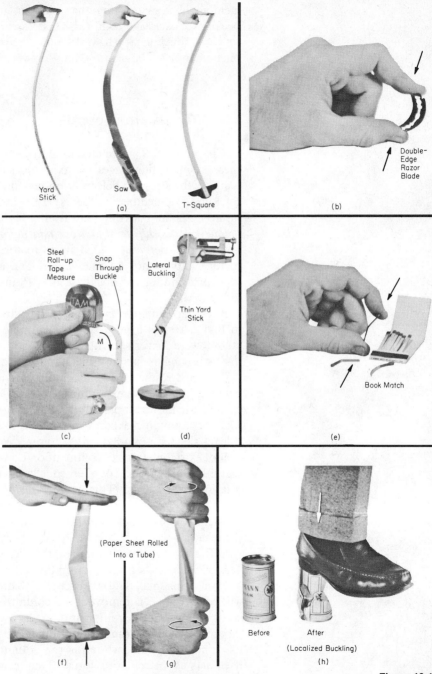

Yard Stick

Saw

T-Square

(a)

Double-Edge Razor Blade

(b)

Steel Roll-up Tape Measure

Snap Through Buckle

M

(c)

Lateral Buckling

Thin Yard Stick

(d)

Book Match

(e)

(Paper Sheet Rolled Into a Tube)

(f)

(g)

Before

After

(Localized Buckling)

(h)

Figure 10-1

member is stocky, rather than slender, it will crush (yield or fracture) because of excessive compressive stress. How slender is slender enough to cause buckling? How do other parameters influence this unique phenomenon? The next sections answer these questions.

10-3 / The nature of buckling

In Chapters 5 to 9, we related load to stress and load to deformation. For these nonbuckling cases of axial, torsional, bending, and combined loading, the stress or deformation was the significant quantity in failure. Buckling of a member is uniquely different in that the quantity significant in failure is the buckling load itself. *The failure (buckling) load bears no unique relationship to the stress and deformation at failure!* Our usual approach of deriving a load-stress relationship and a load-deformation relationship cannot be used here. The approach here will be to find an expression for the buckling load P_f in terms of whatever parameters influence it.

Buckling is unique from our other structural-element considerations in that it results from a state of *unstable equilibrium*. All our previous studies were concerned with elements in stable equilibrium. The following discussion will help you understand the concept of instability.

To visualize the three classes of behavior and the concept of critical buckling load, let us experiment again. Follow Fig. 10-2 carefully. The examples in the upper part with the pipe and ball are simple, but important. The corresponding experiments in the lower part are performed on an "ideal" slender, rounded-end column. We know that an ideal column cannot exist, but we shall comment on how these experiments relate to the real thing. The term "rounded ends" means that there is no resistance to rotation of the ends. This condition may also be called pivoted ends, hinged ends, or pinned ends. It is assumed that this ideal column is slender enough to buckle *elastically*; that is, for any of the loads in Fig. 10-2 the resulting *axial* stress is below the elastic limit.

In Fig. 10-2(a) some axial compressive load P is applied to the column. The column is then given a small (relative to its length) deflection by applying the small lateral force F. If the load P is sufficiently small, when the force F is removed, the column will spring back to its original straight position of equilibrium, just as a ball returns to the bottom of a pipe. In the case of the ball and pipe, gravity tends to restore the ball to its original position, while for the column the elasticity of the column itself acts as the restoring force. This restoring action indicates that the original position is one of *stable* equilibrium in that a small disturbance will tend to disappear.

This same procedure can be repeated for increased values of the load P

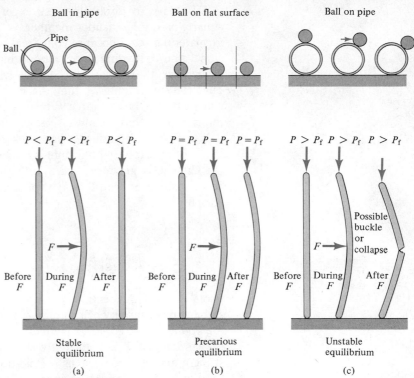

Figure 10-2

until some critical value P_f is reached, as indicated in Fig. 10-2(b). When the column carries this load, and a lateral force F is applied and removed, the column will remain in the slightly deflected position. The elastic restoring force of the column is not sufficient to return the column to its original straight position but is sufficient to prevent excessive deflection of the column. The conditions are like those that exist when a ball is placed on a flat surface. The amount of deflection will depend on the magnitude of the lateral force F. Hence, the column can be in equilibrium in an infinite number of slightly bent positions. By definition, we shall say that for a column carrying an axial load P_f, the straight position is one of *neutral* or precarious equilibrium.

If the column is subjected to an axial load P that exceeds P_f, as indicated in Fig. 10-2(c), and a lateral force F is applied and removed, the column will "snap" to a considerably bent position. That is, the elastic restoring force of the column is not sufficient to prevent a small disturbance from growing into an excessively large deflection. Depending on the value of P, the column either will remain in the bent position or will completely collapse and fracture, just as a ball will roll off a pipe. This type of behavior indicates that for axial loads greater than P_f, the straight position of a column is one of *unstable* equilibrium in that a small disturbance will tend to grow into an excessive deformation.

If you try to duplicate these experiments with a thin yardstick, T-square, or similar slender member, you probably will not need to apply any lateral force F because your "real" column will have material, dimensional, and/or loading imperfections that will serve to initiate the buckling. Remember that a perfect column with perfect loading would never buckle. You might say that the initial deflection of the ideal column caused by the lateral force F in Fig. 10-2 has the same effect as do the imperfections that are always present in real columns.

When "hand" loading a yardstick there is a tendency to reduce the load when unstable equilibrium is reached and noticeable buckling occurs. Buckling tests are more realistic (and dramatic) when performed on a constant-load or dead-load type of machine. Failures are then truly catastrophic.

10-4 / Elastic buckling

From the previous section we see that the central question insofar as the stability or instability of a column is concerned is the relationship between the applied axial compressive load P and the elastic restoring ability of the column. If P is sufficiently small, the column will tend to remain straight and stable; whereas if P is too large, the column will be unable to maintain its straight position and will tend to become unstable and eventually buckle. We seek to determine a critical value for P that will serve to distinguish the safe stable situation from the unsafe unstable situation. There are several widely accepted theories used for determining this critical value, called the theoretical buckling load. In this text we shall use what is usually called the *bifurcation theory*.

Figure 10-3 shows the relation between axial load and unit lateral deflection for an ideal column and a real column for *small elastic* unit deflections Δ/L. For the *ideal* column, when the load P exceeds a certain value P_b, two possible equilibrium positions exist. Either the column can remain straight, in which case $\Delta = 0$; or the column can become bent, in which case $\Delta > 0$. The straight position (② to ④ on the curve) is *unstable*, while the bent position (② to ③ on the curve) is *stable*. Point ②, at which the two branches of equilibrium join, is called a *bifurcation point*.

The shape of the curve for a real column would depend on the degree of the imperfections present. As the imperfections diminish, the curve for a real column approaches the stable branch ①-②-③ of the curve for the ideal column. This is, an *elastic real* column will attain a slightly bent equilibrium position for all axial loads up to and somewhat beyond P_b. However, the theoretical load P_b at the bifurcation point is often considered to be the actual buckling load. Thus, for an *elastic* column, the

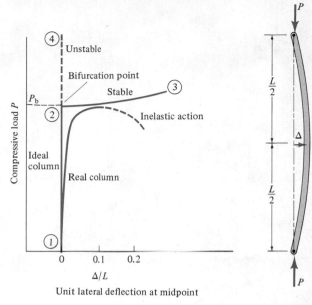

Figure 10-3

failure criterion is based on the assumption that

$$P_f \text{ (buckling)} = P_b \text{ (bifurcation)}$$

Hence, our problem now is to determine the lowest load at which more than one possible equilibrium position can exist. The load P_b is usually called the *Euler elastic buckling load*.

The derivation that follows will be concerned with an ideal column having pivoted ends. The influence of other types of end conditions and restraints will be considered afterward.

Figure 10-4(a) is a sketch of an ideal column with a centroidal axial load P that is just sufficiently large to hold the column in a *slightly* deflected position after a lateral load has been applied and removed. The column is represented diagrammatically in Fig. 10-4(b). The deflection (greatly exaggerated) is assumed to be very small, for two reasons. First, this assumption implies that the load P is very close to the bifurcation load P_b; second, the small deflection permits us to use the elastic deflection equations derived in Chapter 8. In effect, we are assuming that the member is in a state of neutral equilibrium. Figure 10-4(c) is really the same as (b) except that the column is turned on its side so that the chosen x and y axes will look more familiar to you. This rotation will also make it easier for you to see the application of the now familiar general equation of the elastic curve for bending (Eq. 8-3).

From the free-body diagram of Fig. 10-4(d), the moment M_x at any point at a distance x from the end is $-Py$. The negative sign is in agree-

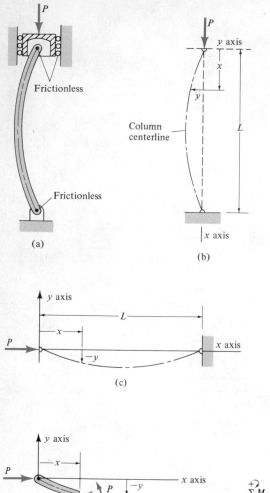

Frictionless

Frictionless

(a)

Column
centerline

x axis

(b)

(c)

Free body diagram

(d)

$$\overset{+\curvearrowleft}{\Sigma M} = 0$$
$$P(-y) - M_x = 0$$
$$M_x = P(-y)$$

Figure 10-4

ment with our usual sign convention for beams since y is negative (downward). Therefore, from Eq. 8-3,

$$EI\,\frac{d^2y}{dx^2} = M_x = -Py$$

or

$$EIy'' + Py = 0 \tag{10-1}$$

This is a linear second-order differential equation with constant *positive* coefficients. Notice that P is the applied axial load and the product EI

is a measure of the elasticity of the column. The problem is now to determine the relation between P, E, and I for which this equation will have admissible solutions satisfying the boundary conditions of the column.

There are several formal methods for obtaining the general solution of such an equation, but careful consideration shows that the solution for y must be of such a form that its second derivative differs from y by only a *negative* constant. The most convenient functions that meet this requirement are the elementary trigonometric functions. Hence, a general solution is

$$y = C_1 \sin kx + C_2 \cos kx \qquad (10\text{-}2)$$

This solution not only must satisfy the differential equation, but must also satisfy the physical boundary conditions. Specifically, y must be zero when x is equal to both zero and L since the ends have no deflection. The first boundary condition gives

$$y_0 = 0 = C_1 \sin k0 + C_2 \cos k0$$

Hence, $C_2 = 0$ and Eq. 10-2 becomes

$$y = C_1 \sin kx \qquad (10\text{-}3)$$

Substituting for y in Eq. 10-1, we get

$$EI(-C_1 k^2 \sin kx) + PC_1 \sin kx = 0$$

from which we obtain

$$C_1 \sin kx(-EIk^2 + P) = 0 \qquad (10\text{-}4)$$

This equation will be satisfied if $C_1 = 0$ or if $k = 0$. In either case, $y \equiv 0$. Hence, $y \equiv 0$ is a possible solution of Eq. 10-1. This solution represents the *unstable* straight position of the column.

Equation 10-4 will also be satisfied if $(-EIk^2 + P) = 0$. Then

$$k^2 = \frac{P}{EI} \qquad k = \perp \sqrt{\frac{P}{EI}}$$

and from Eq. 10-3 a possible second solution of Eq. 10-1 is

$$y = C_1 \sin \sqrt{\frac{P}{EI}} x \qquad [10\text{-}3a]$$

where C_1 cannot be zero. Now, in order that this solution will satisfy our second boundary condition, for which $y = 0$ and $x = L$, we must have

$$y_L = 0 = C_1 \sin \sqrt{\frac{P}{EI}} L$$

But since $C_1 \neq 0$, then

$$\sqrt{\frac{P}{EI}}\,L = n\pi \quad (n = 1, 2, \ldots) \tag{10-5}$$

Hence, the *smallest* load P for which more than one possible equilibrium position can exist is the one for which $n = 1$. The load corresponding to $n = 1$ is, by definition, the Euler elastic buckling load. Thus

$$P_b = P_f = \frac{(\pi 1)^2 EI}{L^2} = \frac{\pi^2 EI}{L^2} \tag{10-6}$$

The loads corresponding to the higher values of n also are bifurcation loads, indicating that there are an infinite number of bifurcation points. By definition, *the Euler load is the load corresponding to the lowest bifurcation point.* Equation 10-6 is commonly called the *Euler buckling equation.*[*]

It is significant here that we cannot evaluate C_1. We only make use of the fact that it is not zero. Not being able to determine C_1 means, physically, that the deflection of the column in its state of neutral equilibrium cannot be determined uniquely from our equation of equilibrium (Eq. 10-1) and the associated boundary conditions. The reason for this is that certain geometric approximations were made in deriving Eq. 8-3 (refer to section 8-4).

10-5 / Discussion of elastic buckling

Equation 10-6 is often expressed in either of the following more convenient forms:

$$P_f = \frac{\pi^2 E(Ar^2)}{L^2} = \frac{\pi^2 EA}{(L/r)^2}$$

or

$$\frac{P_f}{A} = \frac{\pi^2 E}{(L/r)^2} \tag{10-7}$$

where A is the cross-sectional area of the column, and r is the radius of gyration about the bending or buckling axis.

It is very important to notice that the derivation in section 10-4 was based on the general *elastic* deflection equation $EIy'' = M_x$. Therefore, Eq. 10-7 is applicable only when E is a constant, that is, when the buckling load P_f causes an axial stress below the *proportional limit* at the initiation of buckling. From Eq. 10-7 we see that the buckling load P_f (or the unit buckling load P_f/A), is dependent on the stiffness E of the material, and

[*] Leonard Euler, a famous Swiss mathematician, first derived this equation in a paper published in 1759.

the dimensions L and r. The ratio L/r is defined as the *slenderness ratio* of the column.

You must develop a physical feel for the slenderness ratio. As you gain experience with engineering columns, you will acquire this feel. You will then realize that a slenderness ratio greater than 200 is for members that are so slender that they are of little use as load-carrying engineering members in compression. Members with slenderness ratios less than about 30 are so stocky that they have little or no tendency to buckle and are not considered columns. Therefore, most engineering compression members, if they are considered columns, are of intermediate slenderness. The limiting values of about 30 to about 200 are for real materials whose stiffness E is between approximately 1 million and 30 million psi, or 7,000 and 210,000 MPa, as is the case for most common engineering materials.

Figure 10-5 is an example of Eq. 10-7 plotted for a common engineering material, structural steel. When the slenderness ratio is over 100, the equation is valid because the unit buckling load is below the proportional limit of the material and E is a constant. For values of L/r greater than 200, the curve is dashed indicating that it is unlikely that an engineering member would be this slender. The fact that the curve is dashed for values of L/r below 100, or for unit loads above the proportional limit, is more important. A column for which $L/r = 80$ would buckle under an axial load *less* than the "Euler load" (load predicted from the Euler equation). It would buckle *before* the Euler load was reached because the stiffness of the material would be reduced when the unit buckling load exceeded the proportional limit and the material began to yield.

Obviously, the range for the slenderness ratio from 30 to 100 is very important from a practical engineering standpoint. A rational approach to the determination of the buckling loads for columns with slenderness ratios in this range, which buckle inelastically, is presented later in this

Figure 10-5

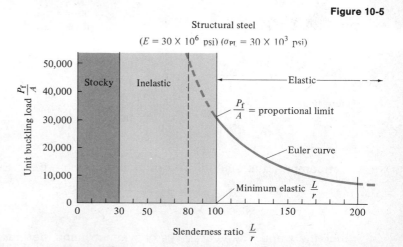

Structural steel
$(E = 30 \times 10^6 \text{ psi}) (\sigma_{PL} = 30 \times 10^3 \text{ psi})$

chapter. There are also many empirical approaches to the design of columns for this range of slenderness ratios. These are discussed in section 10-9.

Before Eq. 10-6 or 10-7 is used, one question still to be answered is: What value of the moment of inertia I or the radius of gyration r should be used in evaluating the buckling load? The value of I in Eq. 10-6 is the one carried directly through the derivation from Eq. 10-1, and therefore, from Eq. 8-3, this equation in turn is based on the elastic flexure formula. The value of I is therefore the second moment of the cross-sectional area with respect to the axis about which the beam bends, or in the case of a column, the axis about which the column buckles. From the Euler equation (Eq. 10-6 or Eq. 10-7), we see that the buckling load P_f will be least when I is least. Thus, assuming no resistance or equal resistance to bending in all directions at the ends, a column will bend about its axis of least moment of inertia. Since $r = \sqrt{I/A}$, this axis will also be the axis of least radius of gyration. Of course, if the column is restrained in such a manner that buckling occurs about a particular axis, then the value of I with respect to that axis must be used.

EXAMPLE 10-1

A structural-steel strut with rounded ends has the rectangular cross section shown in Fig. 10-6, and is 10 in. long. At what compressive axial load will the strut fail? The modulus of the steel is 30 million psi.

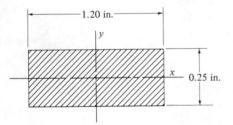

Figure 10-6

Solution:

First compute the slenderness ratio to determine if the Euler formula is applicable.

$$I_x = \frac{(1.20)(0.25)^3}{12} \qquad I_y = \frac{(0.25)(1.20)^3}{12}$$

Since the x and y axes are axes of symmetry and therefore principal axes of the cross section, I_x is the least I of the section. Since $r = \sqrt{I/A} = 0.0722$ in., $L/r = 10/0.0722 = 138$. From Fig. 10-5, for steel, this value of L/r is well into the Euler range. This slender compression member will therefore fail by elastic buckling.

By Eq. 10-7,

$$\frac{P_f}{A} = \frac{\pi^2 E}{(L/r)^2} = \frac{\pi^2(30{,}000{,}000)}{138^2} = 15{,}550 \text{ psi}$$

Hence,

$$P_f = (15{,}500)(1.20)(0.25) = 4665 \text{ lb}$$

Thus, $P_f/A = 15{,}550$, which is well below the proportional limit. The member is therefore behaving elastically.

EXAMPLE 10-2

If an $\angle$ 76 × 51 × 9.5 structural-steel angle, for which the cross section is shown in Fig. 10-7, must support 64,000 N in axial compression, what is the maximum length that it can have? Assume hinged ends.

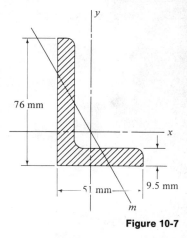

Figure 10-7

Solution:

From a steel handbook, the least radius of gyration for the angle cross section is $r_m = 10.92$ mm, and the area is 1116 mm^2. The m axis in this case is a principal axis and lies in the plane of the cross section. If we assume that the Euler equation is applicable, we obtain

$$\frac{P_f}{A} = \frac{\pi^2 E}{(L/r)^2}$$

$$\frac{64{,}000}{1116 \times 10^{-6}} = \frac{\pi^2(209 \times 10^9)}{(L/0.01092)^2} = 57.35 \times 10^6$$

$$L = 2.071 \text{ m}$$

Check the slenderness ratio to see if our assumption is really okay. Since $L/r = 2.071/0.01092 = 190$, it is well into the Euler range. Also, $P_f/A = 57.35$ MPa, which is well below the proportional limit for steel.

10-6 / End conditions

The Euler equation (Eq. 10-7) was arbitrarily derived for the case of rounded ends. This is sometimes referred to as the fundamental case. An engineering compression member may have end conditions that approximate those in the fundamental case of rounded ends, but more often the ends would have a different degree of fixity. The effect of the end fixity on the buckling load can be determined if the end conditions approximate any of those shown in Fig. 10-8(a)–(e) or if an intelligent guess can be made for k in Fig. 10-8(f).

The length L assumed in the derivation of the Euler equation is the distance covered by one loop of the sine curve of Eq. 10-3a when $n = 1$ in Eq. 10-5. Physically, the length used in the Euler equation is the distance between points of zero moment (inflection points). Only in the fundamental case in Fig. 10-8(a) will the total length L of the member be equal to the distance L_e between inflection points, sometimes called the effective length. Substituting the distance L_e expressed as a function of L in the Euler equation (Eq. 10-6) for each of the cases in Fig. 10-8, we get the following results:

For $L_e = L$, as in the fundamental case or as in Fig. 10-8(a),

$$P_f = \frac{\pi^2 EI}{L_e^2} = \frac{\pi^2 EI}{L^2}$$

For $L_e = L/2$, as in Fig. 10-8(b) or (c),

$$P_f = \frac{\pi^2 EI}{(L/2)^2} = \frac{4\pi^2 EI}{L^2}$$

For $L_e = 2L$, as in Fig. 10-8(d),

$$P_f = \frac{\pi^2 EI}{(2L)^2} = \frac{1}{4}\frac{\pi^2 EI}{L^2}$$

For $L_e = 0.7L$, as in Fig. 10-8(e),

$$P_f = \frac{\pi^2 EI}{(0.7L)^2} = \frac{2\pi^2 EI}{L^2}$$

For $L_e = kL$, as in Fig. 10-8(f),

$$P_f = \frac{\pi^2 EI}{(kL)^2} = \frac{1}{k^2}\frac{\pi^2 EI}{L^2}$$

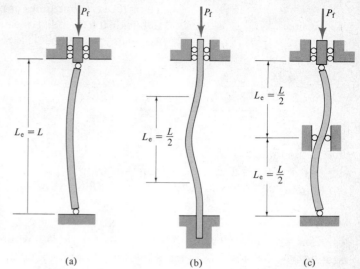

(a)   (b)   (c)

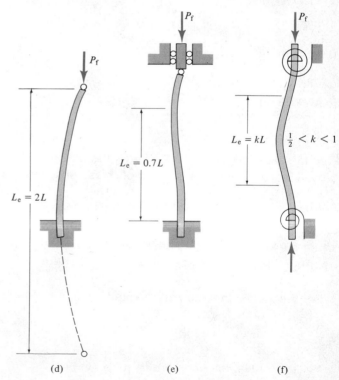

(d)   (e)   (f)

Figure 10-8

Thus, it takes four times as much load to buckle a fixed-ended column as to buckle a free-ended one of the same overall length; and only one-fourth as much load to buckle a column with one fixed end and one free end.

EXAMPLE 10-3

What is the least thickness a rectangular wood plank 4 in. wide can have, if it is used for a 20-ft column with one end fixed and one end pivoted, as in Fig. 10-8(e), and must support an axial load of 1000 lb? Use a factor of safety of 5. The modulus of the wood is 1.5 million psi.

Solution:

For these end conditions $L_e = 0.7L$. Also, $I = bt^3/12$, where b is the 4-in. breadth and t is the desired least lateral dimension. Since P_f is the allowable load times the safety factor,

$$P_f = \frac{\pi^2 EI}{(0.7L)^2} = 5000 = \frac{\pi^2(1.5)(10^6)(4)t^3/12}{[(0.7)(12)(20)]^2}$$

Hence, $t = 3.06$ in.

EXAMPLE 10-4

Find the buckling load for the rectangular steel connecting rod in Fig. 10-9. Assume that the material behaves elastically for all values of stress encountered.

Solution:

If it is assumed that the bearings are frictionless, there is no resistance to buckling about the x axis. When buckling about the z axis is considered, it may be assumed that there is no side play in the bearings, and that the ends are fixed. Therefore, the load required to buckle the rod about each axis will be determined.

For the x axis, the ends are pivoted, as in Fig. 10-8(a), and $L_e = L$. Also, I is a maximum. Hence,

$$P_x = \frac{\pi^2 EI_x}{L^2} = \frac{\pi^2(30,000,000)(3/8)(1^3)/12}{16^2} = 36,800 \text{ lb}$$

For the z axis, the ends are fixed, as in Fig. 10-8(b), and $L_e = L/2$. Also, I is a minimum. Hence,

$$P_z = \frac{\pi^2 EI_z}{(L/2)^2} = \frac{\pi^2(30,000,000)(1)(3/8)^3/12}{(16/2)^2} = 20,600 \text{ lb}$$

The rod will buckle first about the z axis at a load of 20,600 lb.

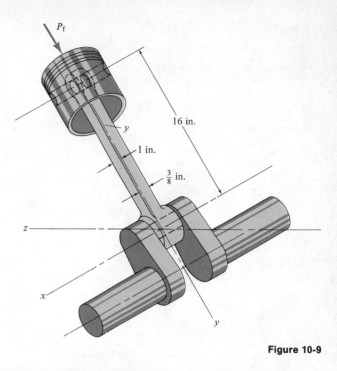

P_f

16 in.

y

1 in.

$\frac{3}{8}$ in.

z

x

y

Figure 10-9

PROBLEMS

10-1 to 10-6 A 10-ft-long column with completely hinged ends has the cross section shown. Determine the slenderness ratio for the column.

Problem 10-1

4 X 8 timber
(nominal size)

Problem 10-2

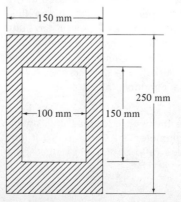

Notes

$\leftarrow$ 6-5

6 in.

Problem 10-3

5 mm

75 mm

Problem 10-4

150 mm

250 mm

100 mm

150 mm

Problem 10-5

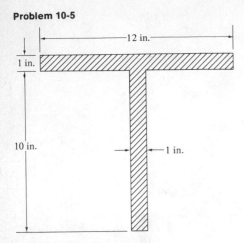

Problem 10-6

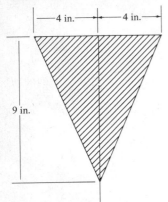

Problem 10-9

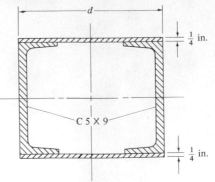

10-7 How long is a round, hinged-ended column 3 in. in diameter if the slenderness ratio is 100?

10-8 What is the slenderness ratio of a column whose effective length is 20 ft? The column is a W 8 × 40 structural-steel section.

10-9 Two plates, each $\frac{1}{4}$ in. thick, are welded to two standard channels in the positions shown in the illustration so that the elements act as a unit. If the built-up member is used as a column, approximate the distance d in order that the column will have an equal tendency to buckle about any axis of the cross section. Neglect the I for the channels about their vertical centroidal axes.

10-10 What must be the wall thickness of a hollow tube whose outside diameter is 2.5 in. if it is 7 ft long and has a slenderness ratio of 120 when used as a hinged-ended column?

10-11 Show that a square cross section is a better column section than a circular one having the same area.

10-12 Find the theoretical buckling load of a wood yardstick $\frac{1}{4}$ in. thick and $1\frac{1}{2}$ in. wide. Assume that the wood is linearly elastic and has a modulus of 2 million psi, and that the ends are frictionless.

10-13 Find the minimum required diameter of a solid, round structural-steel strut that has hinged ends, is 120 cm long, and must support a compressive load of 20,000 N. Use a factor of safety of 3.

10-14 A hollow magnesium tube having an outside diameter of $1\frac{1}{2}$ in. and a wall thickness of $\frac{1}{16}$ in. is used for a compression member with hinged ends to support 2400 lb. How long can it be, assuming elastic behavior and a factor of safety of 1.5? Magnesium has a modulus of 6.5×10^6 psi.

10-15 Find the dimensions of a hinged-ended timber column with a square cross section that is 12 ft long and is to carry an axial compressive load of 20,000 lb. The timber behaves linearly and has an elastic modulus of 1.8 million psi. Use a factor of safety of 3.

10-16 Find the maximum allowable load that a rectangular (40 × 20 mm) aluminum alloy hinged-ended column 1 m long can carry if the elastic modulus of the aluminum alloy is 7.2×10^{10} N/m². Assume linear behavior, and use a factor of safety of 2.5.

10-17 What axial compressive load will cause a 10-ft length of standard 2-in. steel pipe to buckle elastically? Actual dimensions of a 2-in. pipe are: OD = 2.375 in.; ID = 2.067 in. Assume frictionless ends.

10-18 to 10-20 What is the minimum slenderness ratio

at which elastic buckling can occur for pivoted columns of the materials whose compression stress-strain curves are shown in the illustration?

Problem 10-18

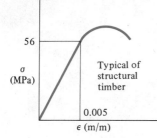

Problem 10-19

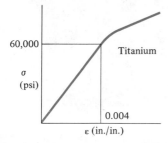

Problem 10-20

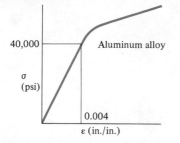

10-21 Same as Prob. 10-1 except the column is 16 ft long with a support of its midpoint.

10-22 Same as Prob. 10-2 except the column is 10 ft long and has one end fixed and the other end hinged.

10-23 Same as Prob. 10-3 except the column is 5 ft long with one fixed end and the other end free.

10-24 Same as Prob. 10-4 except the column is 14 ft long with both ends fixed.

10-25 Compare the Euler buckling loads of the following two very slender round columns: (a) manufactured plastic with modulus of 2×10^9 N/m², cross-sectional area of A, length of l, fixed ends; and (b) clear spruce wood with modulus of 9×10^9 N/m², cross-sectional area of A, length of l, with one free end.

10-26 With a factor of safety of 5, how heavy a platform can a 160-lb flagpole sitter place atop a 40-ft-high, constant-cross-section flagpole made from a standard 4-in. steel pipe ($l = 7.233$ in.⁴ and $r = 1.51$ in.)? One end of the pole is firmly fixed in the ground. The platform and the sitter are assumed to remain as axial loads. Neglect the weight of the pole itself, and assume elastic behavior.

10-27 A 200-lb student is climbing an aluminum alloy flagpole 60 ft tall to place a derby hat on its top. The pole has an inside diameter of 3.5 in. and an outside diameter of 4 in. Can he get to the top before the pole buckles? If not, how high can he go? Neglect the weight of the pole itself, and assume that he is a contortionist and is able to keep his center of gravity along the axis of the pole. The modulus of the aluminum is 10×10^6 million psi.

10-28 An aluminum yardstick (rectangular cross section $\frac{1}{4} \times 1\frac{1}{4}$ in.) having frictionless ball-and-socket joints at its ends is held between two rigid walls. If the yardstick is unstressed at room temperature, find the temperature change required to buckle the stick. It is assumed that the stick will buckle elastically when the buckling load is reached. Coefficient of thermal expansion for aluminum is 12.8×10^{-6} in./in./°F, and the modulus is 10.6 million psi.

10-29 As shown in the illustration, four high-strength steel wires 129 mm² in cross section and 3 m long are used to guy the 2.4 m mast whose cross section is 50×50 mm extruded magnesium. The turnbuckles on the guy wires are tightened so that the tension in each wire is negligible. The turnbuckles have a thread pitch of 1.6 threads per mm; E for magnesium is 45×10^9 Pa. How many additional turns of each turnbuckle (simultaneously) will buckle the mast? Assume elastic behavior.

10-30 Determine the necessary minimum *width* and *thickness* of each of the two 4 ft long members AB in the illustration to prevent elastic buckling. Drum diameter is 2 ft; crank radius is 3 ft. Pull P exerted on the crank is 180 lb and the weight of the rock is negligible in comparison to the tension in the cord. The material is wood with a modulus of 2 million psi.

Problem 10-29

Problem 10-31

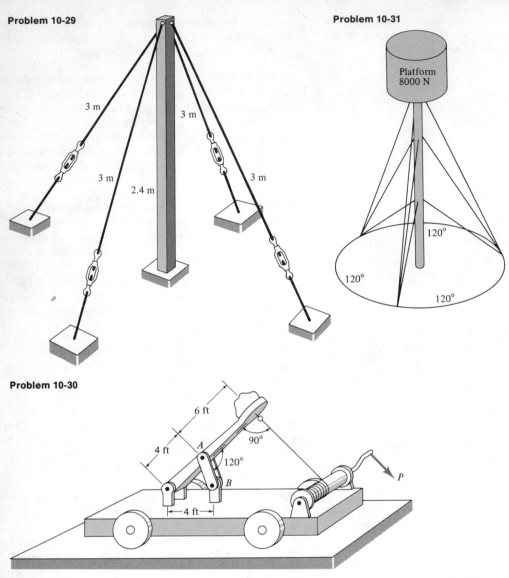

Problem 10-30

10-31   The 25-m steel mast with the 8000-N platform on top is guyed at the third-points as shown in the illustration. If the cross section of the mast is hollow and circular with $R_o/R_i = 1.2$, what must be the diameter in order to prevent elastic buckling? Assume zero load in the guy wires.

10-32   In the structure shown in part (a) of the illustration, the member AB is made of a material that has the compression stress-strain curve shown in part (b). What is the minimum allowable area of a square cross section of member AB, if buckling is the mode of failure? Verify your answer.

10-33   A structural-steel pivoted-ended column 5 m long has two lateral supports at the midpoint, as shown in the illustration. If the cross section is as shown, what is the value of the critical buckling load? The elastic limit of the steel is 270 MPa and $E = 210 \times 10^9$ Pa.

10-34   A vertical bar is embedded at both its ends

Problem 10-32

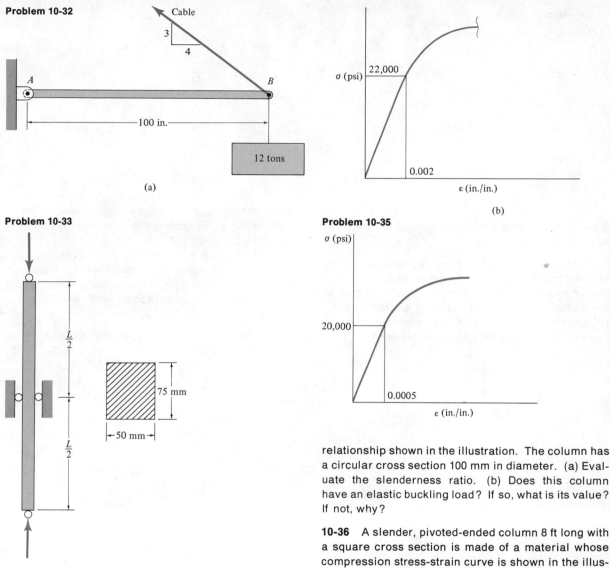

(a)

(b)

Problem 10-33

Problem 10-35

while unstressed at a temperature of 70°F. The bar is 20 ft long and has an effective radius of gyration of 1.2 in. When the bar cools to 40°F, it has a tensile stress of 4000 psi. Find the elastic modulus of the material and the temperature at which the bar will buckle. The coefficient of thermal expansion is 1.3×10^{-5} in./in./°F.

10-35 A pivoted-ended column 3 m long is made of a material that has the compressive stress-strain relationship shown in the illustration. The column has a circular cross section 100 mm in diameter. (a) Evaluate the slenderness ratio. (b) Does this column have an elastic buckling load? If so, what is its value? If not, why?

10-36 A slender, pivoted-ended column 8 ft long with a square cross section is made of a material whose compression stress-strain curve is shown in the illustration. Determine the minimum area this column must have so that it will not fail by elastic buckling.

10-37 An aluminum pole is 13 m long and has a hollow circular cross section with a 150-mm outside diameter and a 140-mm inside diameter. One end of the pole is rigidly fastened in the ground, and the other end is free. What is the maximum height on this pole at which a 10,000-N platform can be mounted without causing the pole to buckle? What is the weight of the heaviest platform that could be mounted on the top?

Problem 10-36

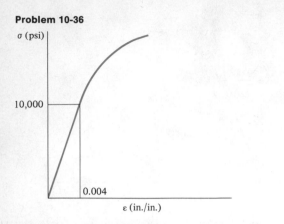

Assume that the center of gravity of the platform is along the axis of the pole.

10-38 Four $4 \times 4 \times \frac{1}{2}$-in. equal-leg structural-steel angles 32 ft long are riveted together to form a single unit, as shown in the illustration. The ends are pivoted and frictionless. What is the buckling load for this unit? What is the buckling load if the rivets are removed and each angle acts as a separate column?

Problem 10-38

10-39 The wing strut shown in the illustration has a standard $2\frac{1}{2}$-in. streamlined section with the following properties: $I_{11} = 0.4063$ in.^{4}; $I_{22} = 0.1018$ in.^{4}; $A = 0.3773$ in.^{2} On a hard landing the resultant load is 1000 lb. What factor of safety does the wing strut have? The strut is pinned about the axis 2-2 and fixed about the axis 1-1. The strut is of alloy steel with a modulus of 30 million psi.

10.40 Same as Prob. 10-35 except that one end is fixed and the other free.

Problem 10-39

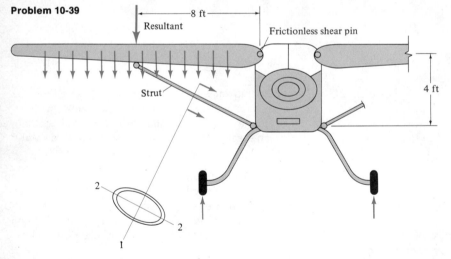

10-7 / Inelastic buckling

Our previous discussions have been concerned with members that buckle at a stress level below the proportional limit. These columns have relatively large slenderness ratios. As mentioned earlier, a great variety of engineering compression members have intermediate slenderness ratios, and many of these do not buckle until *after* the proportional limit has been exceeded. Thus, from an engineering point of view, there is an important class of members between those having very low slenderness ratios (stocky), which fail in direct compression with no buckling effects, and those having high slenderness ratios, which buckle *elastically*. For example, a structural-steel member for which L/r is 80 would buckle before the Euler load could be reached, as shown in Fig. 10-5, because the material stiffness would decrease when the axial stress exceeded the proportional limit.

To determine a rational inelastic buckling load we must take into account the reduced stiffness, or the reduced slope of the stress-strain curve. Consider an ideal column, Fig. 10-10(a), having an intermediate slenderness ratio and pivoted ends, that is to be loaded in small increments up to its critical buckling load P_f, which occurs after the proportional-

Figure 10-10

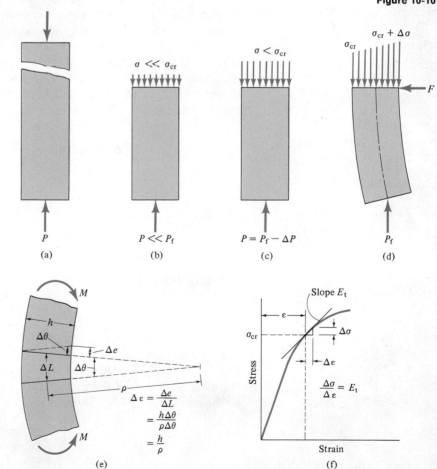

limit stress has been exceeded. In Fig. 10-10(b) and (c) are shown *uniform* stress distributions resulting from the increasing axial load.

Assume that as the last increment of axial load ΔP is applied, a *small* lateral force F is also applied. This simultaneous application of both an axial force and a lateral force is necessary to make sure that the longitudinal compressive strains in the member continue to increase. With this type of loading program, the stress at any point in the column will *not decrease* at any time. On the other hand, since the stresses are in the inelastic range, if the axial load were first increased to cause the critical stress and then the lateral load were applied, the fibers on the convex side would stretch slightly and thus undergo a *decrease* in stress ("unloading") and there would be an accompanying increase in the modulus back to approximately the elastic value. This second loading program leads to a "double modulus" theory. We shall develop the theory based on the first loading assumption of monotonic increasing stress.

With our assumption of no decrease in stress, the distribution will be as shown in Fig. 10-10(d), where the stress $\Delta\sigma$ is that due to the bending effects. The additional strain induced by the bending is relatively very small, and it is assumed that plane cross sections remain plane. Also, since $\Delta\sigma$ is small, it is assumed that this increase in stress is directly proportional to the increase in strain. That is, $\Delta\sigma = (\Delta\varepsilon)E_t$, where E_t is the tangent modulus for $\sigma = \sigma_{cr}$ in Fig. 10-10(f). From Fig. 10-10(e) we see that the deformation due to bending is Δe, and the corresponding strain is

Figure 10-11

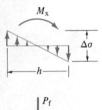

$$\Delta\varepsilon \ (\text{bending}) = \frac{\Delta e}{\Delta L} = \frac{h\,\Delta\theta}{\rho\,\Delta\theta} = \frac{h}{\rho}$$

Hence,

$$\Delta\sigma = (\Delta\varepsilon)E_t = \frac{h}{\rho}\,E_t \tag{10-8}$$

A free-body diagram of the deflected column will be the same as that of Fig. 10-4(d). Thus,

$$M_x = -P_f y \tag{10-9}$$

In order to express the bending moment M_x in terms of the stress, we can think of the net stress distribution as being composed of two separate ones, as shown in Fig. 10-11. There is a *uniform* stress distribution due to the axial load P_f, and there is a *linearly varying* one due to the bending or the eccentricity of P_f. Then, as in the elastic flexure formula,

$$\frac{\Delta\sigma}{2} = \frac{M(h/2)}{I}$$

or

$$\Delta\sigma = \frac{M_x h}{I}$$

$$M_x = \frac{(\Delta\sigma)I}{h} \tag{10-10}$$

Then, from Eqs. 10-8 and 10-9,

$$M_x = \frac{h}{\rho} E_t \frac{I}{h} = \frac{E_t I}{\rho} = -P_f y \tag{10-11}$$

Finally, using the same approximation for $1/\rho$ as in section 8-4, we have

$$\frac{d^2 y}{dx^2} \approx \frac{1}{\rho} = -\frac{P_f y}{E_t I}$$

$$E_t I y'' + P_f y = 0 \tag{10-12}$$

This is the same as Eq. 10-1 for elastic buckling, but here we have the tangent modulus E_t instead of the elastic modulus E (review section 4-4 for a discussion of E_t). Solving Eq. 10-12 just as we did Eq. 10-1, we find that the lowest inelastic buckling (bifurcation) load is

$$P_f = \frac{\pi^2 E_t I}{L^2} \tag{10-13}$$

or

$$\frac{P_f}{A} = \frac{\pi^2 E_t}{(L/r)^2} \tag{10-14}$$

This relationship is commonly called the *tangent modulus equation* or the *Engesser equation*, after the man who first proposed it.

The result based on the double modulus theory is a much more complicated expression, which gives a somewhat higher value for the buckling load than does the tangent modulus formula. Which assumption most closely approximates what actually happens in a column? This question was not fully answered until 1946, when F. R. Shanley* explained that the load based on the double modulus theory was an upper limit, and the tangent modulus theory more nearly predicted the buckling load of an actual column.

10-8 / Discussion of the Engesser equation

Although Eq. 10-14 is very similar to the Euler equation (Eq. 10-7), its application is considerably more difficult. The main problem is to find

* The column paradox, *Journal of Aeronautical Sciences*, 13:678 (Dec. 1946).

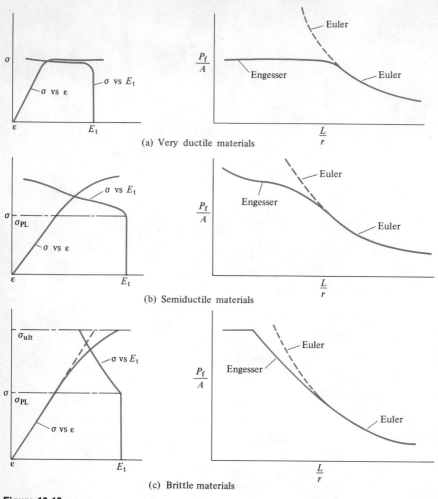

(a) Very ductile materials

(b) Semiductile materials

(c) Brittle materials

Figure 10-12

the proper value for E_t. Since E_t is the slope of the stress-strain curve for the unit axial stress $\sigma_{cr} = P_f/A$, it may be necessary in some problems to solve Eq. 10-14 by trial and error. In such a problem it is convenient to have a plot of σ verses E_t for the material of the particular column. Some sample σ versus E_t curves are shown in Fig. 10-12.

On the other hand, if an analytical expression for the stress-strain curve is known, then E_t can be obtained as a function of the stress by the following relationship:

$$E_t = \frac{d\sigma}{d\varepsilon} \quad \text{(slope)}$$

or
$$\frac{1}{E_t} = \frac{d\varepsilon}{d\sigma} \quad \text{(evaluated at } \sigma_{cr})$$

Then E_t in Eq. 10-14 can be expressed explicitly in terms of σ_{cr}.

In conclusion, if we use the Euler equation to predict the buckling load for unit axial stresses (P_f/A) below the proportional limit and we use the Engesser equation for unit stresses above the proportional limit, we can then predict the buckling load for all intermediate slenderness ratios (between the stocky members and the extremely slender members). Various plots of unit stress versus slenderness ratio are shown in Fig. 10-12. The dashed portion of the Euler curve is included merely to illustrate the significant effect that inelastic behavior can have on the buckling load as compared with elastic behavior.

The tangent modulus equation resulted from the solution of a second-order linear differential equation similar to that used for elastic behavior. Therefore, the discussion of end conditions in section 10-6 is applicable to the Engesser equation, as well as to the Euler equation.

EXAMPLE 10-5

A round tube, with 1-in. OD and $\frac{7}{8}$-in. ID, is 8 in. long and is made of a material whose stress-strain diagram is approximated in Fig. 10-13. If one end is fixed and one end is free, what is the axial compressive buckling load?

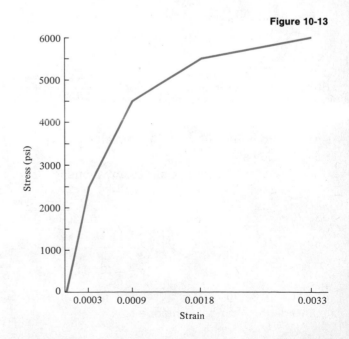

Figure 10-13

Solution:

The area is 0.184 in.^{2} and $I = (\pi/64)(d_o^4 - d_i^4)$. The radius of gyration is

$$r = \sqrt{\frac{I}{A}} = 0.331 \text{ in.}$$

and the effective slenderness ratio is

$$\frac{L_e}{r} = \frac{2L}{r} = \frac{16}{0.331} = 48.4$$

For the given end conditions, the tangent modulus equation becomes

$$\frac{P_f}{A} = \sigma_{cr} = \frac{\pi^2 E_t}{(L_e/r)^2}$$

It is not necessary to worry about whether we should use the Engesser equation or the Euler equation. The Engesser equation will automatically become the elastic Euler equation if the slenderness ratio dictates use of the elastic modulus E. For the computed value of L_e/r,

$$\sigma_{cr} = \frac{\pi^2}{(L_e/r)^2} E_t = 0.00422 E_t \qquad \text{(a)}$$

Since we do not have an analytical expression for E_t as a function of σ_{cr} for the material, we must work from the given experimental stress-strain curve. The curve has four distinct slopes, each of which is the value of E or E_t for some range of values of σ_{cr}. A trial-and-error solution is employed here.

For the initial (elastic) range, the slope is

$$E = \frac{2500}{0.0003} = 8.33 \times 10^6$$

Substituting in equation (a), we get

$$\sigma_{cr} = (0.00422)(8.33 \times 10^6) = 35,200 \text{ psi}$$

As seen from Fig. 10-13, the initial value of E is not valid above 2500 psi and certainly not anywhere near 35,200 psi.

For the second range of stress, the slope is

$$E_{t1} = \frac{4500 - 2500}{0.0009 - 0.0003} = 3.33 \times 10^6$$

and

$$\sigma_{cr} = (0.00422)(3.33 \times 10^6) = 14,050 \text{ psi}$$

As seen from Fig. 10-13, E_{t1} is valid only between 2500 and 4500 psi and not near 14,050 psi.

For the third range, the slope is

$$E_{t2} = \frac{5500 - 4500}{0.0018 - 0.0009} = 1.11 \times 10^6$$

and

$$\sigma_{cr} = (0.00422)(1.11 \times 10^6) = 4680 \text{ psi}$$

As seen from Fig. 10-13, E_{t2} is valid between 4500 and 5500 psi, and this range includes the value 4680 psi. Therefore, the critical stress at buckling is 4680 psi. Since $\sigma_{cr} = P_f/A$, the buckling load is

$$P_f = \sigma_{cr}A = (4680)(0.184) = 862 \text{ lb}$$

PROBLEMS

10-41   Rework Example 10-5 if the tube is pivoted at both ends.

10-42   Rework Example 10-5 if the tube is 12 in. long instead of 8 in.

Problem 10-46

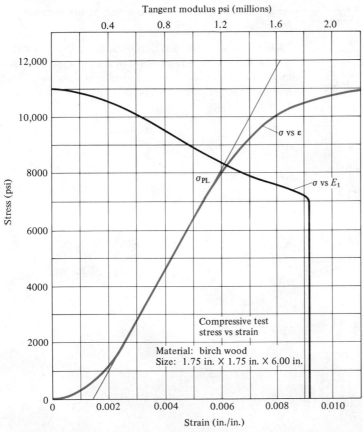

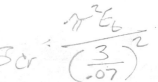

$$\sigma_{cr} = \frac{\pi^2 E_b}{\left(\frac{3}{.07}\right)^2}$$

10-43 A thin compressive member with fixed ends is made of the material as in Example 10-5. The member is 3 ft long and has a solid 1 × 2-in. rectangular cross section. Determine the buckling load.

10-44 A thin compressive member is made of the material in Example 10-5. What is the smallest slenderness ratio this member may have for the member to be considered a column rather than a stocky compressive member?

10-45 A thin compressive member is made of the material in Example 10-5. What is the largest slenderness ratio this member may have if inelastic rather than elastic buckling is to be the mode of failure?

10-46 How long can a 2 × 2 (nominal) wood compression strut be if it is to support an axial load of 11 kips and is made of the material whose stress-strain and tangent modulus curves are shown in the illustration? Assume pivoted ends and a factor of safety of 2.

10-47 How much axial compression load can a wood dowel 2 in. in diameter and 30 in. long support if made of the material in Prob. 10-46? One end is fixed and the other pivoted.

10-48 How much axial compression load can a dowel 2 in. in diameter support if the ends are fixed and the

Problem 10-49

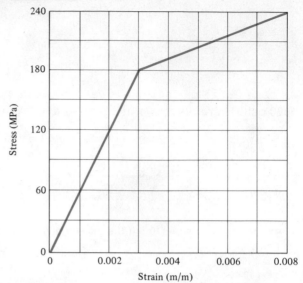

length is 3 ft? The dowel is made of the material in Prob. 10-46.

10-49 The stress-strain curve for an aluminum alloy can be approximated as shown in the illustration. How much axial load can a 1.25-m column made of the material support if the cross section is 50 × 55 mm?

Problem 10-51

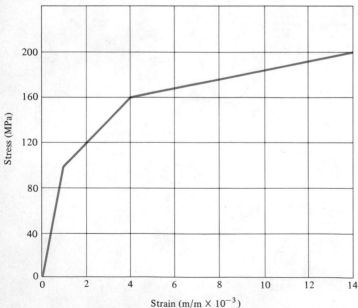

The ends are fixed. *Verify the applicability of any formula used.*

10-50 For the material of Prob. 10-49, for what range of slenderness ratios is inelastic buckling likely to occur?

10-51 A slender member of 75-mm diameter is pinned at both ends and loaded in axial compression. The member is 1 m long and is made of an alloy whose properties are approximated in the illustration. Using a factor of safety of 2, find the maximum allowable load this member can safely carry before buckling.

10-52 A hollow member, 100 in. long and circular in cross section, is rigidly fixed at both ends and loaded in axial compression with a load of 325,000 lb. The member is to be proportioned so that its inside radius will be three-fourths of its outside radius. The compressive properties of the material are shown in the illustration for Prob. 10-57. Determine the minimum permissible dimensions of the member.

10-53 A rounded-end column with a cross section 1 in. square is 15 in. long. If the material of the column has a stress-strain curve that can be approximated by the expression $\varepsilon = \sigma^2/50,000,000$, what is the buckling load?

10-54 What is the buckling load of an $\angle\,89 \times 64 \times 6.4$ angle (see the Appendix for properties), used as a compression strut 1.5 m long with fixed ends, if the stress-strain law of the material is $\sigma = 4 \times 10^8 \varepsilon^{1/3}$?

10-9 / Empirical column formulas

The column theory for inelastic behavior has not been too well understood by engineers until recent years. It has therefore been common engineering practice for many years to use empirical relationships for designing columns with intermediate slenderness ratios. One reason for their popularity is that it is usually simpler to use them than the rigorously derived theoretical formulas.

An empirical relationship is generally based on data obtained by measuring the buckling loads in a series of tests on experimental columns with various slenderness ratios. The unit buckling loads corresponding to various slenderness ratios are computed and plotted resulting in a display of points similar to that in Fig. 10-14.

Figure 10-14

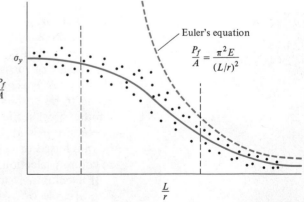

Euler's equation

$$\frac{P_f}{A} = \frac{\pi^2 E}{(L/r)^2}$$

The dashed lines show the approximate boundaries of the three basic modes of failure: stocky members whose mode of failure is primarily crushing and may be predicted by the yield stress of the material, slender columns whose mode of failure is buckling and can be predicted by Euler's equation, and intermediate columns whose failure is the result of a combination of the two actions and is frequently described by an empirical equation.

After such a plot is developed, curve fitting methods are applied and a convenient and mathematically simple expression fitting the results, that is, an empirical formula, is developed.

After the constants in an empirical equation are determined to get the best fit of the resulting curve and the experimental points, a liberal factor of safety is usually applied to make the expression suitable for use as a design formula. The inherent unstable nature of the buckling phenomenon, as discussed earlier, causes a considerable scatter in test data for columns, even those from very carefully controlled laboratory tests. This condition necessitates much larger safety factors in columns than are usually necessary in most other structural elements.

It is important to note three facts about the empirical formulas:

1   Each empirical column formula is good only for the material for which it was derived.

2   Each formula is good only between the limits given with it. Beyond those limits the curve no longer matches the data.

3   Most empirical formulas have a safety factor built in so that they produce an allowable load rather than a critical one. If in doubt, examine the yield stress of the material versus the stress given by the formula.

Figure 10-15 gives the empirical equations, their limits, and the curves they represent for three common engineering materials. Equations for additional materials can be found in most engineering handbooks. Handbooks also frequently provide solutions to the equations for various values of L/r.

Since these column formulas are empirical and since most cross sections used for columns are not simple geometric forms like circles and squares, the design of most columns is actually a sophisticated trial-and-error process. A reasonable cross section is selected and checked. If it is found to be unsuitable the information gained is used in the selection of another cross section for the next trial, and so on. Since this is a reiterative procedure it is ideally suited for computer solution, and programs may be purchased or devised which perform the considerable labor involved in column selection.

If a column is to be of a geometric rather than a composite shape (i.e., a circle or a square rather than a channel or wide-flange section)

Steel*

$$C_c = \sqrt{\frac{2\pi^2 E}{\sigma_y}}$$

σ_y = yield stress

$$\frac{L}{r} < C_c$$

$$\sigma_{All} = \frac{\sigma_y}{F.S.}\left[1 - \frac{(L/r)^2}{2C_c^2}\right]$$

$$F.S. = \frac{5}{3} + \frac{\frac{3}{8}(L/r)}{C_c} - \frac{1}{8}\left(\frac{L/r}{C_c}\right)^3$$

$$C_c < \frac{L}{r} < 200$$

$$\sigma_{All} = \frac{12\pi^2 E}{23(L/r)^2}$$

$\frac{L}{r} > 200$ is not allowed.

Aluminum† (Alloy 6061-T6)

$$\frac{L}{r} \leq 9.5 \qquad \sigma_{All} = 19 \text{ ksi} = 131 \text{ MPa}$$

$$9.5 < \frac{L}{r} < 66 \qquad \sigma_{All} = \left[20.2 - 0.126\left(\frac{L}{r}\right)\right] \text{ ksi}$$

$$= \left[139 - 0.868\left(\frac{L}{r}\right)\right] \text{ MPa}$$

$$66 \leq \frac{L}{r} \qquad \sigma_{All} = \frac{51,000}{(L/r)^2} \text{ ksi}$$

$$= \frac{372 \times 10^3}{(L/r)^2} \text{ MPa}$$

C_1, C_2, C_3, and C_4 are constants determined for the specific alloy and unit system.

Timber‡

$$\frac{L}{r} < 124 \qquad \sigma_{All} = \frac{\pi^2 E}{2.727(L/r)^2}$$

$\frac{L}{r} > 124$ is not allowed.

σ_{All} may not exceed allowable stress parallel to grain for timber used.

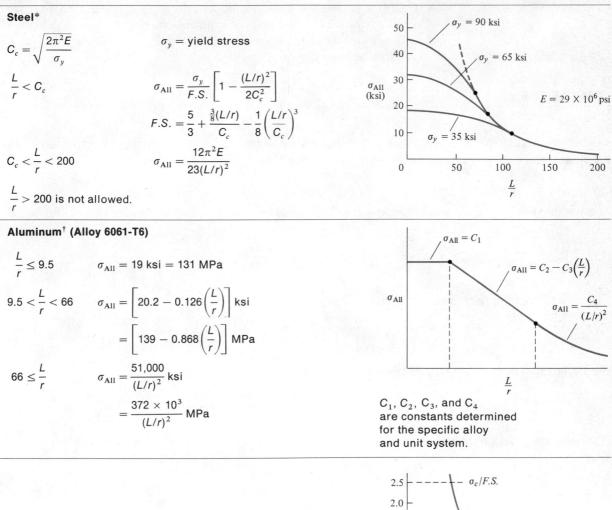

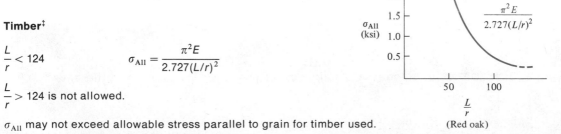

(Red oak)

---

* Specifications from American Institute of Steel Construction.
† *Specifications for Aluminum Structures,* Aluminum Association, Inc., Washington, D.C., 1976.
‡ *Timber Construction Manual,* American Institute of Timber Construction, John Wiley and Sons, New York, 1974.

Figure 10-15

it can, of course, be designed analytically by solving the appropriate empirical formula for the necessary dimension.

The following examples and problems will demonstrate the use of empirical column formulas.

EXAMPLE 10-6

An aluminum alloy 6061-T6 column 8 ft long is to have a circular cross section. Find the required diameter if the column is to carry an axial load of 240 kips.

Solution:

First we must make a guess and assume that L/r lies between 9.5 and 66. We learn from Fig. 10-15 that

$$\sigma_{All} = 20.2 - 0.126 \left(\frac{L}{r} \right) = \frac{P_{All}}{A}$$

Next, we need the radius of gyration.

$$r = \sqrt{\frac{I}{A}} = \sqrt{\frac{\pi d^4/64}{\pi d^2/4}} = \sqrt{\frac{d^2}{4 \times 4}} = \frac{d}{4}$$

Remember that the r in the equation above refers to radius of gyration.

$$\sigma_{All} = 20.2 - 0.126 \left(\frac{96}{d/4} \right) = \frac{240}{\pi d^2/4}$$

Solving the quadratic yields

$$d = 3.24$$

$$r = \frac{d}{4} = 0.81$$

$$\frac{L}{r} = 120$$

Our initial assumption about slenderness ratio was wrong so we try again with the equation for $L/r > 66$

$$\sigma_{All} = \frac{51,000}{(L/r)^2}$$

Remembering that

$$\sigma_{All} = \frac{P}{A} = \frac{P}{\pi d^2/4} = \frac{240 \times 4}{\pi d^2} \quad \text{and} \quad r = \frac{d}{4}$$

and substituting, we find

$$\frac{960}{\pi d^2} = \frac{51,000}{\left(\dfrac{96}{d/4}\right)^2} = \frac{51,000 d^2}{(4 \times 96)^2}$$

$$\frac{960(4 \times 96)^2}{51,000\pi} = d^4$$

$$d = 5.45$$

$$r = \frac{d}{4} = 1.36$$

$$\frac{L}{r} = \frac{96}{1.36} = 70.4$$

Since $70.4 > 66$, we have used the correct formula and the required diameter is 5.45 in.

Naturally, it is not necessary to guess wrong first, although it does happen quite frequently. What is necessary is the final check on L/r to make sure the correct formula was used.

EXAMPLE 10-7

Using the AISC specifications for structural steel, determine the maximum allowable axial load for a W 8 $\times$ 31 used as a column with an effective length of 10 ft. The properties of the cross section may be found in the Appendix.

Solution:

For structural steel $\sigma_y = 35 \times 10^3$ ksi.

$$C_c = \sqrt{\frac{2\pi^2 \times 29 \times 10^6}{35 \times 10^3}} = 128$$

For the W 8 $\times$ 31 the radius of gyration is 2.01 so that $L/r \approx 60$. For $L/r < C_c$ the formula is

$$\sigma_{\text{All}} = \frac{\sigma_y}{F.S.}\left[1 - \frac{(L/r)^2}{2C_c^2}\right] = \frac{35}{F.S.}\left[1 - \frac{(60)^2}{2(128)^2}\right]$$

$$= \frac{35(0.89)}{F.S.}$$

$$F.S. = \frac{5}{3} + \frac{(3/8)(L/r)}{C_c} - \frac{1}{8}\left(\frac{L/r}{C_c}\right)^3$$

$$= 1.6667 + 0.1758 - 0.0129 = 1.8296 = 1.83$$

$$\sigma_{\text{All}} = \frac{35(0.89)}{1.83} = 17.03 \text{ ksi}$$

The cross-sectional area of a W 8 × 31 is 9.12 in.², so the allowable load is

$$P_{All} = A\sigma_{All} = (9.12)(17.03) = 155.3 \text{ kips}$$

PROBLEMS

10-55 (a) Plot the experimental column test data given below for pin-ended 2 × 2-in. wood columns. (b) Fit a straight line to these data. (c) Derive the equation of the straight line. (d) Apply a factor of safety of 3 to the derived equation to determine a working column formula for design.

Slenderness ratios	Buckling loads, lb
15	22,000
15	23,100
20	23,400
20	21,800
30	19,900
30	21,600
40	17,000
50	16,300
50	17,100
60	13,000
70	12,100
80	8,900
80	10,000
90	8,000
100	5,000
100	5,100

10-56 Plot the curve of unit buckling load (P/A) versus slenderness ratio (L/r) for pin-ended columns made from the magnesium alloy having the properties given in the illustration for the range of L/r from 20 to 200.

10-57 Plot the curve of unit buckling load (P/A) versus slenderness ratio (L/r) for pin-ended columns made from the 2024-T4 aluminum alloy having the properties given in the illustration for the range of L/r from 20 to 200.

10-58 A compression member is cast of AZ92A-T4 magnesium alloy (see the illustration for Prob. 10-56) with a box cross section having outside dimensions 4 × 2 in. and inside dimensions 3 × 1 in. If the member is pivoted at one end and fixed at the other, and

is 6 ft long, what is the maximum axial load it can withstand?

10-59 A pin-ended member is extruded from 2024-T4 aluminum alloy with a square cross section having sides $1\frac{1}{4}$ in. wide. If it must support an axial compressive load of 60,000 lb, what is the greatest length it can have? See the illustration for Prob. 10-57 for the compressive properties of the material.

10-60 Plot the curve of unit buckling load (P/A) versus slenderness ratio (L/r) for pin-ended columns made from the glass-resin laminate having the properties shown in the illustration for the range of L/r from 20 to 200. Assume that the warp of the fabric is parallel to the load axis.

10-61 A fixed-ended strut 20 in. long is made of the glass-resin laminate whose properties are given in the illustration for Prob. 10-60. The warp of the fabric is perpendicular to the long axis of the column. The cross section is a rectangle 1.5 × 1.0 in. What is the axial buckling load?

10-62 The side rod on a certain locomotive consists of a 6-ft-long structural-steel member having a rectangular cross section 4 in. deep and 2 in. wide. Determine the maximum axial compressive load to which the member may be subjected if it is assumed to be pivot ended. Use the AISC specifications.

10-63 Determine the diameter of (a) a structural-steel rod and (b) a rod of high-carbon steel SAE 1090 required to support an axial load of 100 kips in a pivot-ended column 18 ft long. Use AISC specifications.

10-64 Determine the maximum allowable compressive load for a red oak member 4 in. square and 8 ft long if it is (a) pivot ended, and (b) fixed ended. Refer to Fig. 10-15.

10-65 Solve Prob. 10-64 for a member composed of Douglas fir.

10-66 An extruded aluminum alloy (6061-T6) tube

Problem 10-56

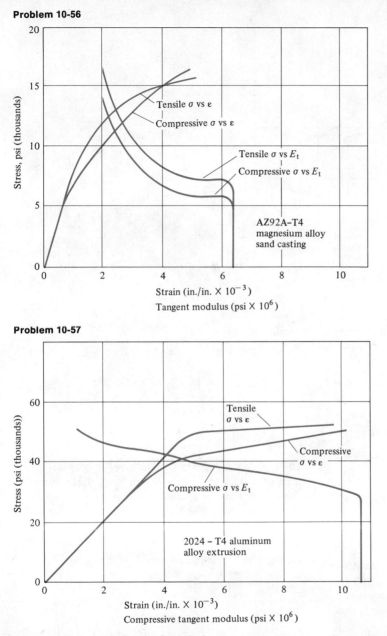

Tensile σ vs ε

Compressive σ vs ε

Tensile σ vs E_t

Compressive σ vs E_t

AZ92A–T4
magnesium alloy
sand casting

Stress, psi (thousands)

Strain (in./in. $\times 10^{-3}$)
Tangent modulus (psi $\times 10^6$)

Problem 10-57

Stress (psi (thousands))

Tensile
σ vs ε

Compressive
σ vs ε

Compressive σ vs E_t

2024 – T4 aluminum
alloy extrusion

Strain (in./in. $\times 10^{-3}$)
Compressive tangent modulus (psi $\times 10^6$)

3 ft long and pivoted at both ends is to carry a compressive load of 8000 lb. Determine the wall thickness required if the external diameter is (a) 1 in. and (b) $1\frac{1}{2}$ in. Use the formulas in Fig. 10-15.

10-67 A machine part 1 m long is to transmit an axial load of 7500 N and is to have a square cross section.

Determine the minimum size of (a) structural steel, (b) aluminum alloy 6061-T6 needed to meet the specifications in Fig. 10-15. The ends are assumed to be pivoted.

10-68 A 6061-T6 aluminum alloy member with fixed ends is to carry an axial compressive load of 3000 lb.

Problem 10-60

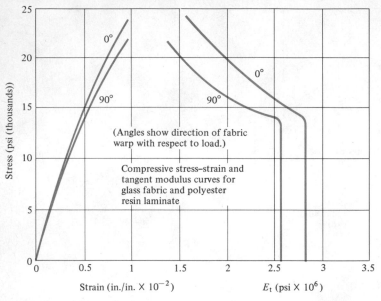

Stress (psi (thousands))

0°

90°

0°

90°

(Angles show direction of fabric warp with respect to load.)

Compressive stress–strain and tangent modulus curves for glass fabric and polyester resin laminate

Strain (in./in. × 10^{-2}) E_t (psi × 10^6)

The member is 10 ft long and is to have a rectangular cross section with one dimension twice the other. Determine the dimensions of the cross section.

10-69 to 10-74 Determine the maximum allowable load for the following structural steel sections used as axially loaded columns. Use the AISC specifications.

10-69 W 8 × 31   The effective length is 15 ft.

10-70 W 12 × 96   The effective length is 20 ft.

10-71 W 200 × 46.1   The effective length is 4 m.

10-72 W 8 × 24   The effective length is 12 ft.

10-73 S 12 × 31.8   The effective length is 12 ft.

10-74 S 200 × 34.2   The effective length is 10 ft.

10-75 A latticed steel column 24 ft long consists of two C 6 × 13 channels placed 4 in. back to back with the flanges projecting outward. Determine the maximum allowable load to which the column may be subjected.

10-76 A latticed steel column is made of two S 12 × 31.8 beams as shown in the illustration. (a) What should be the spacing center to center of the beams in order to provide a column of equal resistance each way? Give a dimensioned sketch. (b) If the column is 40 ft long, determine the maximum safe axial load it should carry.

Problem 10-76

10-77 Four ∠ 102 × 102 × 19 steel angles are to be incorporated into a column 8 m long. Determine the maximum allowable axial load that the four will carry if they are (a) loaded independently, (b) riveted as shown in the illustration, (c) welded to form a hollow 204 mm square.

Problems 10-77 and 10-78

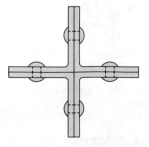

10-78 If the ∠ 102 × 102 × 19 angles shown in the

illustration for Prob. 10-77 are used in the cross section of a fixed-ended column 14 m long, determine the maximum axial load that may be supported.

10-79 A column 5 m long is made of two C 310 × 44.6 structural-steel channels latticed 100 mm back to back. Determine (a) the maximum safe axial compressive load the column can carry, (b) the maximum safe load that two such channels can carry if not latticed together but if each is loaded axially.

10-80 Select an economical rolled W-section of structural steel to carry an axial load of 72 kN as a pivot-ended column 8 m long. Do not allow the slenderness ratio to exceed C_c.

10-81 Determine a suitable rolled steel M-section for member U_1U_2 of the truss. Assume the member to be pivot ended. The slenderness ratio may not exceed C_c for any main member.

Problem 10-81

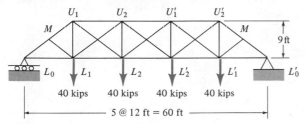

10-82 Select a suitable S-section for member L_0U_1 of the truss of Prob. 10-81, (a) with the collision strut ML_1, (b) without the strut. The slenderness ratio may not exceed C_c.

10-83 A column 5 ft long consists of a 6 × 2-in. flat plate of structural steel. Each of its ends is fastened by a connector like that in the diagram. What is the

maximum allowable axial load for the column according to the AISC specifications?

Problem 10-83

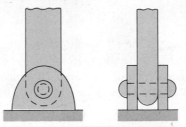

10-84 A W 200 × 46.1 is used as an axially loaded pivot-ended column 10 m long. It is braced laterally at its midpoint as shown. Determine the maximum safe load according to the AISC specifications.

Problem 10-84

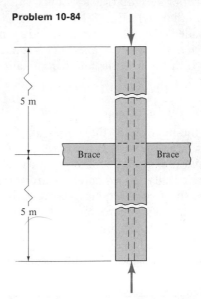

10-10 / Columns under eccentric loading

Our discussions to this point have been limited to columns subjected to centroidal axial loads. While such columns are much cleaner to work with from a theoretical standpoint, reality requires that we now address eccentrically loaded columns.

Eccentrically loaded columns are particularly common in structural applications where most wall columns and all corner columns are eccentrically loaded.

The most commonly used approach for examining eccentrically loaded columns is known as the interaction method and uses the relationship

$$\frac{P/A}{\sigma_{All}} + \frac{My/I}{\sigma'_{All}} \leq 1 \qquad (10\text{-}15)$$

In this equation σ_{All} is the allowable stress computed from a suitable column formula and σ'_{All} is the allowable stress in bending for the material.

Equation 10-15 can be modified into a simpler form for use in design by substituting S, the section modulus, for I/y. Thus, Eq. 10-15 becomes

$$\frac{P/A}{\sigma_{All}} + \frac{M/S}{\sigma'_{All}} \leq 1 \qquad (10\text{-}15a)$$

The loads P and M are as they were in section 7-11. In fact Eq. 7-19 can be used for column work if the stress is understood to be σ_{All} as computed from the appropriate empirical formula. This results in rather more conservative design than does Eq. 10-15.

Eccentrically loaded columns are often loaded so that there is a bending moment about each of the principal axes. (This is particularly common in corner columns.) If this is true, the following relation is used:

$$\frac{P/A}{\sigma_{All}} + \frac{(M_x c_x)/I_x}{\sigma'_{All}} + \frac{(M_y c_y)/I_y}{\sigma'_{All}} \leq 1 \qquad (10\text{-}16)$$

where M_x is the moment about the x axis, c_x is the extreme distance measured from the x axis, and I_x is the moment of inertia about it; M_y, c_y, and I_y are similar quantities based on the y axis.

Since most eccentrically loaded columns are studied by means of the empirical formulas, the process is quite tedious, particularly for design that must be done by some trial and error method. Consequently, eccentrically loaded columns are best handled by computers if the proper software is available.

The following examples will demonstrate the use of Eqs. 10-15 and 10-16.

EXAMPLE 10-8

A W 8 × 31 is to be used as a column with an effective length of 10 ft. A load of 100 kips is to be applied on the y axis at some distance e from the x axis. Determine the maximum allowable value of e. Use the AISC specifications for structural steel. Figure 10-16(a) shows a simplified top view of the column. Fig. 10-16(b) and (c) demonstrate the substitution of a force and a couple for an eccentric force.

Solution:

The first step is to replace the eccentric load in Fig. 10-16(b) with an axial load and a bending couple as shown in Fig. 10-16(c). Equation 10-16 becomes

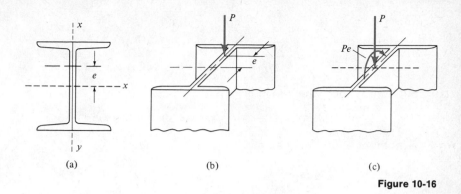

(a)   (b)   (c)

Figure 10-16

$$\frac{P/A}{\sigma_{\text{All}}} + \frac{Pec/I}{\sigma'_{\text{All}}} = \frac{P/A}{\sigma_{\text{All}}} + \frac{Pe/S_x}{\sigma'_{\text{All}}}$$

Using the value of σ_{All} obtained in Example 10-7, which employed the same column section, and $\sigma'_{\text{All}} = 20{,}000$, and finding A and S_x in the Appendix, we may make the following substitutions.

$$\frac{100/9.12}{17.03} + \frac{100e/27.5}{20} = 1$$

$$0.645 + 0.182e = 1$$

$$e = 1.95 \text{ in.}$$

EXAMPLE 10-9

A W 8 × 31 10 ft long is used as a corner column. It must carry a load of 25 kips acting on the x axis 1 in. from the y axis and a load of 50 kips acting on the y axis 5 in. from the x axis. (The loads result from beams supported on brackets.) Is the column safe? Use the allowables from the preceding example. Figure 10-17 shows the loads acting on the column.

Figure 10-17

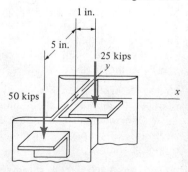

Solution:

The axial load is

$$P_1 + P_2 = 50 + 25 = 75 \text{ kips}$$

The moment about the x axis is

$$M_x = 50 \times 5 = 250 \text{ in.-kips}$$

The moment about the y axis is

$$M_y = 25 \times 1 = 25 \text{ in.-kips}$$

From the Appendix, $A = 9.12 \text{ in.}^2$, $S_x = 27.5 \text{ in.}^3$, and $S_y = 9.27 \text{ in.}^3$. From the preceding problem $\sigma_{\text{All}} = 17$ ksi and $\sigma'_{\text{All}} = 20$ ksi. Substituting all these values into Eq. 10-16 gives us

$$\frac{P/A}{\sigma_{\text{All}}} + \frac{M_x/S_x}{\sigma'_{\text{All}}} + \frac{M_y/S_y}{\sigma'_{\text{All}}} = \frac{75/9.12}{17} + \frac{250/27.5}{20} + \frac{25/9.27}{20}$$

$$= 0.483 + 0.455 + 0.135 = 1.07$$

Since the sum of the three terms is greater than 1, it is probably necessary to redesign. A safe design could be accomplished by using a larger column or by redesigning the brackets to reduce the eccentricity.

10-85 A red oak post 6 in. × 12 in. × 8 ft is used as a column. What is the maximum load that may be applied at a corner? Use a safety factor of 2 against bending.

10-86 An aluminum alloy 6061-T6 rod 4 in. in diameter and 4 ft long is used as a pivot-ended column to support a load of 100 kips. What is maximum allowable eccentricity? Use a safety factor of 2 against bending.

10-87 Solve Prob. 10-86 for a structural-steel rod of identical dimensions.

10-88 An S 8 × 23 steel section is used as a pivot-ended column 25 ft long. Determine the maximum allowable load that the column can carry if the load is applied at a point on the centerline of the web 2 in. from the centroid.

10-89 Determine the maximum allowable load that the 6061-T6 aluminum alloy column shown can carry as a 12-m column if the load is applied at (a) the centroid, (b) point A. Provide a safety factor of 2 against bending.

10-90 An S 200 × 27.4 steel section is used as a

Problem 10-89

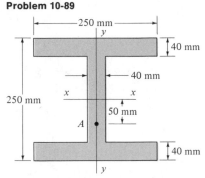

pivot-ended column 4 m long. Find the maximum distance from one axis of symmetry of the section that the load may be moved along the other axis of symmetry without reducing the permissible load on the column.

10-91 A pivot-ended aluminum alloy 6061-T6 column 25 ft long has the cross section shown. Determine the maximum allowable load the column can carry if the load is applied at (a) the centroid, (b) point A. Provide a safety factor of 2 against bending.

Problem 10-91

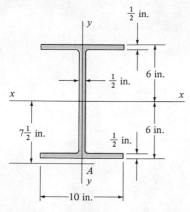

10-92   Two C 10 × 15.3 steel channels 35 ft in length are latticed together as shown to act as a pin-ended column. Determine the maximum safe working load if it is applied at (a) the centroid, (b) a point on the axis 3 in. from the centroid.

Problem 10-92

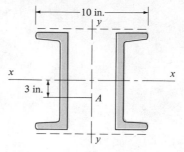

10-93   The length and end conditions of an S 380 × 74 steel section are such that it will carry a maximum allowable axial load of 800 kN acting as a column in the intermediate range. Determine the maximum permissible eccentricity along the major axis of symmetry for an eccentric load of 400 kN.

Chapter 11
Additional beam topics

11-1 / Objectives

Upon completion of this chapter you will be able to:

1 Explain the importance of the shear center in flexure members having thin cross sections.
2 Locate the shear center for a given cross section.
3 Locate the neutral axis in a curved beam.
4 Determine the normal stresses in a curved beam.
5 Find stresses and permissible loads for beams composed of two or more materials.

11-2 / Introduction

In Chapters 7, 8, and 9 we considered some of the basic ideas concerning the response of relatively long members subjected to transverse bending loads. In that presentation some rather restrictive assumptions were made in regard to the geometry of the member and the manner of loading. These assumptions will now be reviewed and discussed.

First, the beams in Chapters 7, 8, and 9 were initially assumed to be straight, to have a constant cross section, and to be made of a homogeneous and isotropic material. In the later parts of those chapters the restriction of constant cross section was removed, and beams with gradually varying cross sections (tapered beams) and beams with reduced cross sections were considered. Also, the deflections of beams composed of several sections of different materials joined end to end were investigated by use of the moment-area method. However, all the beams were *straight*.

Second, the beams were assumed to have a longitudinal plane of symmetry, and the loads were assumed to lie in this plane of symmetry. With these restrictions, the neutral surface of the beam was found to be perpendicular to this plane of symmetry. Also, the symmetry of loading and cross section prevented the possibility that the beam would twist. The only time the requirement of symmetry was removed was in Section 7-14. It was shown there that if the loading were pure bending (no shear forces), the bending couple could be resolved into component moments about the principal axes of the cross section. However, at no time did we consider unsymmetrical problems in which shear forces were present.

Third, it was assumed that bending, rather than shearing or buckling, was the primary consideration in determining the kinematic response of the member. This assumption will be carried over to the analyses presented in this chapter.

In addition to these first three major restrictions, other special assumptions were made at various times throughout Chapters 7, 8, and 9. It would be well for you to review briefly the important theory in those chapters before proceeding with the present chapter. The succeeding sections are devoted to types of members, such as beams with unsymmetrical shear loading, curved beams, and nonhomogeneous beams, which could not be discussed within the restrictions of those earlier chapters.

11-3 / Shear center

It was argued in Chapter 7 that if a beam had a longitudinal plane of symmetry and if the loads were applied in this plane of symmetry, then the beam would not twist and the resulting deformation would be caused primarily by the bending effects. On this basis the flexure formula

(Eq. 7-15a) was derived and applied to beams subjected to transverse shear forces as well as to pure bending couples. We might now pose the problem of a beam that is subjected to transverse loads and has no longitudinal plane of symmetry; or even if the beam does have a longitudinal plane of symmetry, we might suppose that the loads do not lie in this plane. How then do we determine the stresses in the beam? This problem is much too general to be presented in a text of this level. Instead, we shall concern ourselves with a problem that is solved much more easily but is still extremely useful and practical. In particular we shall consider a *beam whose cross section is relatively thin*, and shall determine where the transverse shear forces should be applied so that the beam does not twist. Since many standard structural beams have I-shaped, T-shaped, channel-shaped, and other types of thin cross sections, the solution of this problem has wide application.

The method of solution is conceptually simple. We merely try to locate the line of action of the resultant internal reaction shear forces assuming the beam bends *but does not twist*. Then for equilibrium the external resultant shear force must be colinear with the internal resultant shear force. Thus the problem is reduced to one of elementary statics.

Consider, for example, a thin channel-shaped section used as a cantilever beam and loaded as shown in Fig. 11-1(a). Since the bending moment varies throughout the length of the beam, we know that longitudinal shear force will be present (refer to section 7-12). One such force is ΔF_x

Figure 11-1

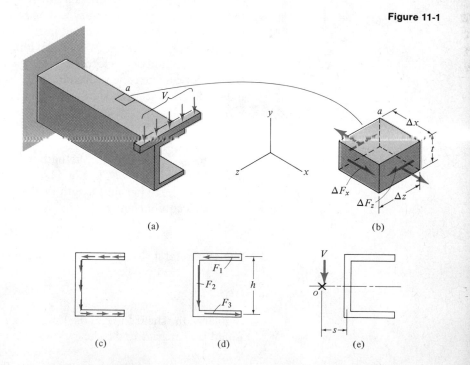

(a)

(b)

(c)

(d)

(e)

in Fig. 11-1(b). Rotational equilibrium then requires the existence of the horizontal shear force ΔF_z in the flange.

In a thin section of this type, it is assumed that the vertical web carries the entire vertical shear force and that the horizontal flanges do not carry any *vertical* shear forces. Thus, shear force "flows" through the cross section of the channel in a manner illustrated in Fig. 11-1(c). This condition is effectively a horizontal force in each of the flanges and a vertical force in the web, as in Fig. 11-1(d). From equilibrium requirements, this system of forces is statically equivalent to the single force shown in Fig. 11-1(e). The distance s is such that the force V must produce the same moment about any point on the centerline of the vertical web as does the couple formed by the horizontal forces F_1 and F_3. Thus,

$$s = \frac{F_1 h}{V} \qquad (11\text{-}1)$$

If the line of action of the resultant external vertical force is located at the distance s from the center of the web, there will be no twisting of the beam. If the line of action were in any other position, the beam would twist since the assumed internal reaction force distribution due to bending would not balance the external twisting effects and additional internal twisting reactions would be produced.

For this example, the distance s will completely locate the desired line of action, because the only external resultant force on the beam is a vertical one. In general, when there is a horizontal as well as a vertical resultant shear force, it may be necessary to determine both a vertical and a horizontal distance s. The point located in this manner, such as o in Fig. 11-1(e), is called the *shear center*. It locates a longitudinal axis through which the external resultant shear forces must pass in order that the beam will not twist. Notice, from the procedure used to determine the distance s, that if the thin cross section has an axis (or axes) of symmetry, the shear center will always lie on this axis. Thus, the intersection of any two axes of symmetry, if they exist, will locate the shear center.

Although Eq. 11-1 formally determines the distance s, the concept of shear flow is quite useful in actually evaluating s. (Refer to sections 6-9 and 7-12.) The force ΔF_x in Fig. 11-1(b) can be expressed as $q \Delta x$, where q is the shear force per unit length or the shear flow. Then by considering rotational equilibrium of an infinitesimal element $t\,dx\,dz$, we find that

$$dF_z = q\,dz$$

Hence, the total shear force in the flange is

$$F_z = \int_{\substack{\text{width of} \\ \text{flange}}} q\,dz \qquad (11\text{-}2)$$

Similarly, the shear force in the lower flange and in the vertical web can be found by integrating the shear flows over their respective lengths.

In many cases, such as in the example involving the channel section, the integrating operation indicated in Eq. 11-2 can be accomplished by rather simple methods. For example, in Fig. 11-1 we already know from equilibrium requirements that the vertical force F_2 in the web is equal to the external shear force V. Also, since the beam bends but does not twist, we know that for elastic behavior the shear flow is given by Eq. 7-24. Thus,

$$q = \frac{VQ}{I} \qquad Q = \int_{\text{area}} y \, da$$

The integral Q represents the first moment, about the centroidal bending axis, of the *area of that portion of the cross section that is considered to be separated from the remaining portion.* For instance, consider the channel section in Fig. 11-2(a). In evaluating the shear flow at c-c in the flange, Q would be the first moment of all the area to the right of c-c. By similar evaluation throughout the entire cross section of the channel, you will find that Q, and hence the shear flow q, varies linearly in the flanges and parabolically in the web, as indicated in Fig. 11-2(b). Thus, the horizontal force in either flange can be found by calculating the area under the shear-flow curve. The result is

$$F_1 = q_{\text{av}} d = \frac{VQ_{\text{av}}}{I} d$$

Substitution of this last result in Eq. 11-1 yields

$$s = \frac{VQ_{\text{av}}}{I} d \frac{h}{V} = \frac{Q_{\text{av}}(d)(h)}{I} \qquad\qquad (11\text{-}3)$$

where Q_{av} is the first moment, about the centroid, of half the area of one of the flanges.

Figure 11-2

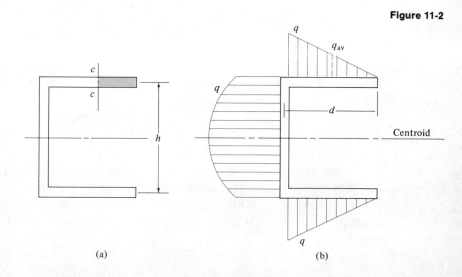

(a)   (b)

Although this last expression is strictly applicable only to a channel section, it illustrates the following important fact. *The location of the shear center depends only on the geometry of the cross section, and not on the loading.* The shear forces in the section and the shear flow used to calculate the location of the shear center were introduced merely to give a physical meaning to the shear center and to aid in finding its location. They did not influence the final result indicated by Eq. 11-3. This fact can be very useful in locating the shear center of an arbitrary cross section since we are free to assume that the external shear force is acting in any convenient direction, such as perpendicular to an axis of symmetry or a principal axis. The following example illustrates this idea.

EXAMPLE 11-1

Determine the location of the shear center for the $\angle\, 5 \times 5 \times \frac{7}{8}$ equal-angle section of Fig. 11-3(a).

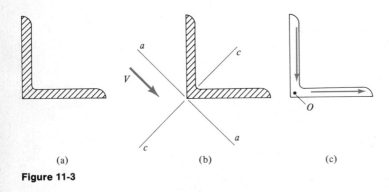

    (a)   (b)   (c)

Figure 11-3

Solution:

First, the axis c-c in Fig. 11-3(b) is an axis of symmetry, and thus the shear center must lie on this axis. We must now find where it is located on this axis. For this purpose, we shall find it convenient to assume that the applied external shear force V is parallel to the axis a-a or perpendicular to the axis c-c. If the *external* force is applied in this manner, the *internal* shear forces will flow as illustrated in Fig. 11-3(c), with their components parallel to the c axis canceling and their components parallel to the a axis adding to equal V. Since these two forces intersect at point O, the moment of these internal forces with respect to O will be zero. Hence, the external force V must have zero moment about O, and thus this external force must pass through O. Therefore, O is the shear center.

Note: It is unnecessary to locate O any more accurately than to say that it lies slightly inside the corner of the angle section, since our shear-flow theory is only an approximation based on the premise that the sections are very thin.

PROBLEMS

11-1 Place arrowheads on the two shear vectors shown for the element removed from the lower flange of the cantilever beam. The only load is vertically down through the shear center.

Problem 11-1

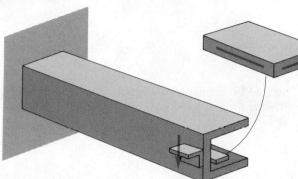

11-2 The cantilever beam in the illustration is made of a thin channel section and is loaded with a pure couple M as shown. For this loading condition, sketch the direction of the shear flow in the web and flanges.

Problem 11-2

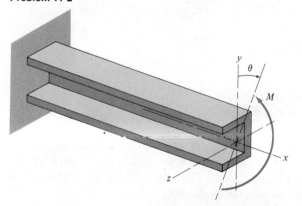

11-3 A thin-walled split tube whose cross section is shown in the illustration is used as a cantilever beam. Sketch the shear flow around this cross section if loaded as in Prob. 11-1. Estimate the location of the shear center.

11-4 to 11-6 For each of the thin cross sections shown, determine the location of the shear center.

Problem 11-3

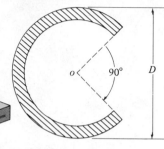

Problem 11-4

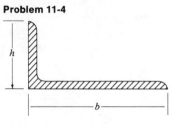

Problem 11-5

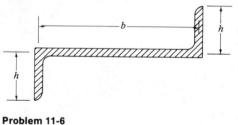

Problem 11-6

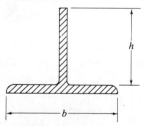

11-7 and 11-8 The section shown in the illustration has a moment of inertia about the horizontal centroidal axis of I in.⁴. Locate the shear center in terms of I and the dimensions given.

11-9 Locate the shear center for a C 10 × 30 (American standard channel).

Problem 11-7

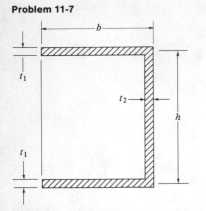

Problem 11-8

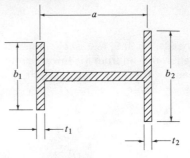

11-10 to 11-12   A cantilever beam with the cross section carries a vertical load of 10 tons. Sketch the direction of the shear flow throughout the cross section, and determine the location of the shear center. What is the value of the maximum longitudinal shear stress?

Problem 11-10

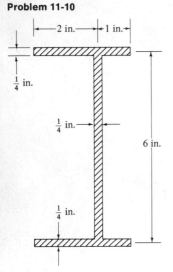

Problem 11-11

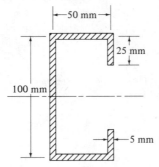

Problem 11-12

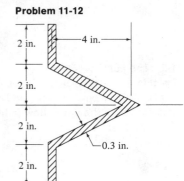

11-4 / Bending of curved beams

Many structural elements are long curved members that carry some type of flexural loading. Examples are crane hooks, punch-press frames, C clamps, and rocking-chair rockers. For the most part, this discussion will be concerned with determining the elastic longitudinal or circumferential flexural stresses* resulting from pure bending, although combined loading will be mentioned later.

Kinematic response   Essentially the same assumptions regarding material, symmetry, and loading will be made here as were made in Chapter 7 in deriving the elastic flexure formula (Eq. 7-15) for straight beams. The

---

* Although radial stresses do exist in curved flexural members, they are usually relatively small.

material is assumed to be homogeneous, isotropic, and elastic. The beam has a longitudinal plane of symmetry. The resultant pure bending couples lie in this plane of symmetry. When we make these assumptions and use arguments similar to those in section 7-7 for straight beams, we conclude that plane cross sections perpendicular to the longitudinal axis of the beam will remain plane after bending. Also, the beam will have a neutral surface throughout its length, and plane cross sections will remain perpendicular to this neutral surface after bending. A curved beam with a longitudinal plane of symmetry is shown in Fig. 11-4(a).

In Fig. 11-4(b) $\Delta\phi$ is the angle between two plane trapezoidal cross sections, *a-b* and *c-d*, *before* bending. During the application of the pure bending moments M, these two planes rotate relative to one another through a small angle $\Delta\theta$, as indicated by the dashed line *c'-d'*. The distances $\bar{r}$ and R are measured from the undeformed center of curvature to the centroidal axis and the neutral axis (wherever it may be), respectively. In the cross-sectional view in Fig. 11-4(c) a positive distance y is measured from the *neutral* surface *toward* the center of curvature. From the characteristics of the deformation and the geometry in Fig. 11-4(b) and (c), we see that the deformation at any distance y is given by the relationship

$$e = -y\,\Delta\theta$$

The minus sign indicates that those fibers on the concave side of the neutral surface are compressed and those fibers on the convex side are elongated.

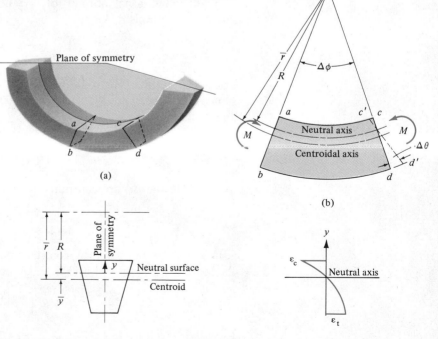

Figure 11-4

Although the *deformation varies linearly* throughout the depth of the beam, the *strain does not vary linearly* because all the longitudinal fibers between section *a-b* and section *c-d* do not have the same initial length. The initial length of a fiber at a distance y from the neutral surface is

$$\Delta L = (R - y)\,\Delta\phi$$

Then, by definition, the longitudinal strain in such a fiber is

$$\varepsilon = \frac{e}{\Delta L} = \frac{-y\,\Delta\theta}{(R - y)\Delta\phi} \tag{11-4}$$

This last equation shows that the strain varies hyperbolically, as illustrated in Fig. 11-4(d). The maximum compressive strain ε_c occurs at the innermost concave side, and the maximum tensile strain ε_t occurs at the outermost convex side.

Elastic flexural stresses Two questions still remain to be answered for a curved beam. Where is the location of the neutral surface, and what is the relation between the applied bending moment and the resulting flexural stresses? As was the case for a straight beam, these questions can be answered by utilizing the strain distribution given by Eq. 11-4 and the loading conditions for the curved beam.

First, since the loading is assumed to cause pure bending, there is no resultant longitudinal normal force. Thus, the stress distribution must satisfy the relationship

$$\int_{\substack{\text{cross} \\ \text{section}}} \sigma\,da = 0 \tag{11-5}$$

For *elastic* behavior, $\sigma = E\varepsilon$, where E is the elastic modulus and is assumed to be the same for tension and compression. Then by Eqs. 11-4 and 11-5, we have

$$\int_{\substack{\text{cross} \\ \text{section}}} E\left(-\frac{y\,\Delta\theta}{(R - y)\Delta\phi}\right) da = 0$$

Since E, $\Delta\theta$, and $\Delta\phi$ are constants insofar as this integral is concerned,

$$\int_{\text{area}} \frac{y}{R - y}\,da = 0 \tag{11-6}$$

If we let r be the distance from the center of curvature to any fiber of the beam, then $r = R - y$ and Eq. 11-6 becomes

$$\int_{\text{area}} \frac{R - r}{r}\,da = R\int_{\text{area}} \frac{da}{r} - \int_{\text{area}} da = 0$$

From this result we have

$$R = \frac{\int_{\text{area}} da}{\int_{\text{area}} \dfrac{da}{r}} \qquad (11\text{-}7)$$

Recalling the definition of R, we have now determined (at least formally) the location of the neutral surface, which does *not* coincide with the centroid of the cross section.

The stress distribution must also satisfy the relationship

$$M = -\int_{\substack{\text{cross} \\ \text{section}}} \sigma y \, da \qquad (11\text{-}8)$$

The minus sign results from the assumed conditions of loading and co-ordinates, just as it did in deriving Eq. 7-7. Substitution in Eq. 11-8 of the expression for the flexural stress used previously yields

$$M = -\int_{\text{area}} E\left(\frac{-y\,\Delta\theta}{(R-y)\,\Delta\phi}\right) y \, da$$

or

$$M = E \frac{\Delta\theta}{\Delta\phi} \int_{\text{area}} \frac{y^2}{R-y} \, da \qquad (11\text{-}9)$$

Now, with the aid of Eq. (11-6), this last integral can be written as

$$\int_{\text{area}} \frac{y^2}{R-y} \, da = \int_{\text{area}} \frac{Ry}{R-y} \, da - \int_{\text{area}} y \, da$$

$$= R(0) - a\bar{y}$$

or

$$\int_{\text{area}} \frac{y^2}{R-y} \, da = -A\bar{y} \qquad (11\text{-}10)$$

where A is the area of the cross section and $\bar{y}$ is the distance between the neutral axis and the centroidal axis.

Notice that the integral on the left-hand side of Eq. 11-10 is inherently positive. Since A is positive, $\bar{y}$ will necessarily be *negative*. From this we see that *the neutral axis will always lie closer to the center of curvature than will the centroidal axis.* Substitution of Eq. 11-10 in Eq. 11-9 yields

$$M = -E \frac{\Delta\theta}{\Delta\phi} A\bar{y}$$

from which

$$\frac{\Delta\theta}{\Delta\phi} = -\frac{M}{EA\bar{y}} \qquad (11\text{-}11)$$

Finally, from Eqs. 11-4 and 11-11, the relation between the external bending moment and the circumferential flexural stress is

$$\sigma = E\varepsilon$$

$$= E\frac{-y}{R-y}\frac{\Delta\theta}{\Delta\phi}$$

or

$$\sigma = \frac{My}{(R-y)A\bar{y}} = \frac{My}{rA\bar{y}} \tag{11-12}$$

Notice that a positive moment and a positive y (inside fibers) produce a negative flexural stress (recall that $\bar{y}$ is negative). This result is consistent with the assumed deformation response.

Although Eq. 11-12 was derived on the basis that the loading was pure bending with no resultant axial force, within limitations the result can be applied to more general loading conditions that are usually encountered in engineering structures. It is merely necessary to transfer the line of action of any resultant axial force to the *centroid* of the cross section by use of the theory of transformation of a force into a force and a couple. Then the stress caused by the centroidal axial load and the bending couple can first be evaluated separately and later be superposed to give the total longitudinal stress in the beam (provided, of course, that the total stress is within the elastic limit).

As a final comment, we mention that the major difficulty in using Eq. 11-12 is in determining the value of R. Even for a relatively simple cross section, such as a circular or triangular section, the evaluation of Eq. 11-7 can be a formidable problem. Also, considerable accuracy in determining R is required because the value of $\bar{y}$ greatly influences the stress given by Eq. 11-12.

EXAMPLE 11-2

The curved steel member in Fig. 11-5(a) has the cross section shown in Fig. 11-5(b). (a) Determine the maximum tensile and compressive stresses at section c-c. (b) What would be the error in the flexure stresses if the elastic flexure formula for straight beams were used instead of Eq. 11-12?

Solution:

(a) From Eq. 11-7,

$$R = \frac{\int da}{\int \frac{da}{r}} = \frac{2}{\int_5^7 \frac{(1)\,dr}{r}} = \frac{2}{\ln 7 - \ln 5}$$

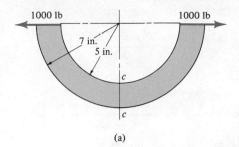

(a)

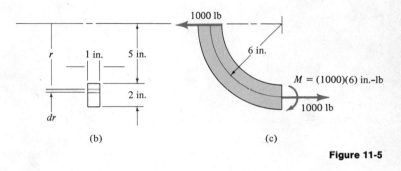

(b)   (c)

Figure 11-5

$$= \frac{2}{\ln \frac{7}{5}} = 5.944$$

Therefore, since the centroid is at $\bar{r} = 6$,

$$\bar{y} = 5.944 - 6 = -0.056 \text{ in.}$$

From Fig. 11-5(c) and Eq. 11-12 and the fact that the bending moment is negative, the stress on the innermost fiber is

$$\sigma_i = \frac{P}{A} + \frac{My}{rA\bar{y}}$$

$$= \frac{1000}{2} + \frac{(-6000)(0.944)}{(5)(2)(-0.056)} = 500 + 10{,}120 = 10{,}620 \text{ psi}$$

The stress in the outermost fiber is

$$\sigma_o = \frac{1000}{2} + \frac{(-6000)(-1.056)}{(7)(2)(-0.056)} = 500 - 8090 = -7590 \text{ psi}$$

(b) From the elastic flexure formula for straight beams, with the neutral axis at the centroid,

$$\sigma = -\frac{My}{I} = -\frac{(-6000)(1)}{(1)(2^3)/12} = 9000 \text{ psi}$$

The error in the tensile flexure stress is

$$\left(\frac{10{,}120 - 9000}{10{,}120}\right)(100) = \left(\frac{1120}{10{,}120}\right)(100) = 10.6 \text{ percent}$$

PROBLEMS

11-13 to 11-16 Rework Example 11-2 for the cross sections shown in the accompanying illustrations.

Problem 11-13

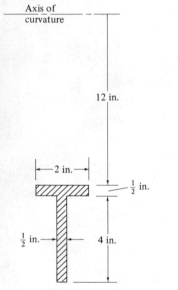

Problem 11-14

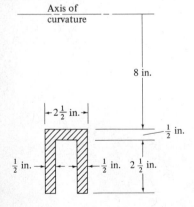

Problems 11-15 and 11-22

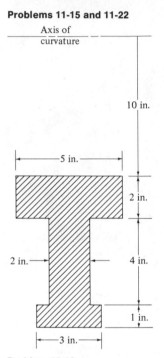

Problem 11-16

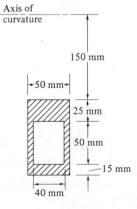

11-17 and 11-18 Determine the value of the dimension *b* of the cross section shown in each illustration that will make the maximum tensile and compressive stresses in the section numerically equal for a curved beam in pure bending.

Problem 11-17   **Problem 11-18**

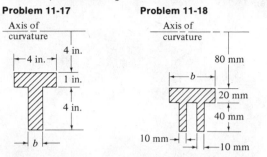

11-19 Determine the location of the neutral axis for a curved beam loaded in pure bending and having the cross section shown in the illustration.

11-20 Show that the distance *R* from the axis of curvature to the neutral axis for a curved beam with the cross section shown in the illustration is given by the relationship

$$R = \frac{\bar{r} + \sqrt{\bar{r}^2 - a^2}}{2}$$

(*Hint:* Use integral tables.)

Problem 11-21

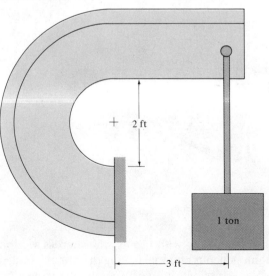

Problem 11-19

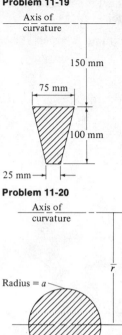

Problem 11-20

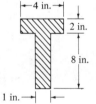

11-21 Find the maximum tensile and compressive stresses in the member in the illustration.

11-22 Same as Prob. 11-21 except the cross section is as shown in the illustration for Prob. 11-15.

11-23 The load applied to the frame is 300 lb. Find the stress at A and at B.

Problem 11-23

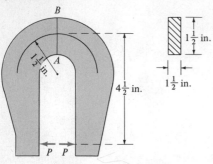

11-24 A ring is made of bar stock with a circular cross section 75 mm in diameter. The inside diameter of the ring is 100 mm. It is loaded as shown in the figure. Find the stresses at A and B.

Problem 11-24

$P = 16,000$ N

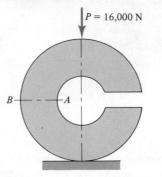

11-25 A load of 12,000 N is applied to the clamp shown. Determine the stresses at A and E.

Problem 11-25

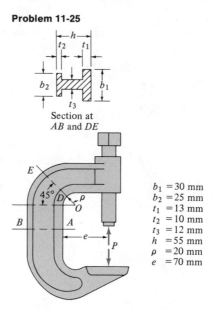

Section at
AB and DE

$b_1 = 30$ mm
$b_2 = 25$ mm
$t_1 = 13$ mm
$t_2 = 10$ mm
$t_3 = 12$ mm
$h = 55$ mm
$\rho = 20$ mm
$e = 70$ mm

11-5 / Nonhomogeneous beams

Thus far in this text we have dealt only with beams whose cross section was composed of a single homogeneous material having the same modulus of elasticity for tension and compression. It is fairly common engineering practice, however, to use beams in which a cross section is composed of two or more different materials. For example, steel plates are often used to reinforce timber beams, and steel rods are generally used to reinforce concrete beams. Also, several common structural materials, such as cast iron and concrete, do not have the same stiffness in compression as in tension and thus exhibit different elastic moduli in tension and compression.

One convenient method of analyzing the elastic stresses and deflections of a nonhomogeneous beam of the types mentioned above is to replace the actual beam by an *equivalent* beam composed of a single homogeneous material with the same modulus for tension and compression. By "equivalent" we mean that the new beam has the same kinematic response and internal *force* (not stress) distribution as does the original beam. Once the geometry of the equivalent beam has been determined, the elastic stresses and deflections can be found from the relations previously derived for beams made of a single homogeneous material.

Consider a straight, two-material beam whose symmetric cross section is shown in Fig. 11-6(a). The upper portion is made of a material with a modulus E_1, and the lower portion of a material with a modulus E_2. For convenience, we shall assume that E_1 is greater than E_2. If the loading causes pure bending and is applied in the plane of symmetry, we can repeat the arguments of section 7-7 for straight, homogeneous beams, and can conclude that plane sections remain plane after bending. The deformation, and hence the strain, will vary linearly from the neutral surface

Figure 11-6

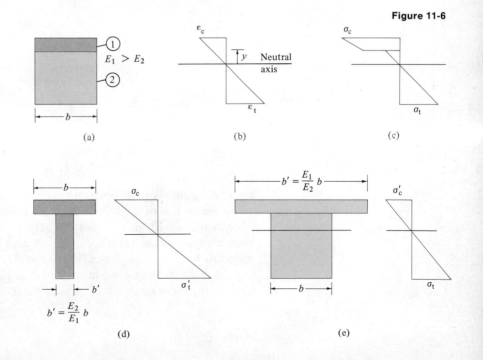

(a)   (b)   (c)

(d)   (e)

(wherever it may be), as indicated in Fig. 11-6(b), and the stress is given by the relationship $\sigma = E\varepsilon$. Because of the different moduli, however, the stress distribution will be as shown in Fig. 11-6(c), and an abrupt discontinuity will occur at the interface of the two materials.

As stated earlier, we wish to replace this beam having a nonhomogeneous cross section with an equivalent beam having a homogeneous cross section. First, we want the equivalent beam to have the same kinematic response. According to Eq. 7-6, the strain at any distance y from the neutral surface is

$$\varepsilon = -\frac{y}{\rho}$$

where ρ is the radius of curvature of the bent beam. Thus, for any given radius of curvature, if the strain in a fiber of the equivalent beam is to be the same as the strain in the corresponding fiber of the actual beam, we must not change the distance y from the neutral surface. Therefore, the *equivalent beam should have the same dimensions perpendicular to the neutral surface as the original beam.*

In addition to this kinematic equivalence, we also require static equivalence. That is, we require that the *force* on a fiber of the equivalent beam be the same as that on the corresponding fiber in the original beam. The force on any fiber at a distance y from the neutral surface is

$$dF = \sigma\,da = E\varepsilon\,da$$

$$= -E\frac{y}{\rho}\,da$$

Our requirement of static equivalence gives

$$dF' = dF$$

$$\sigma'\,da' = \sigma\,da$$

$$E'\frac{y}{\rho}\,da' = E\frac{y}{\rho}\,da$$

$$E'da' = E\,da \qquad\qquad (11\text{-}13)$$

where the primed quantities refer to the equivalent beam and the unprimed quantities to one of the materials of the actual beam.

Equation 11-13 shows that the elemental areas of the equivalent beam and the actual beam are inversely proportional to the moduli of the materials of the beams. Since we have said that we do not wish to change any height dimensions, we may only change the width of the elemental area. If the cross section has parallel straight sides,

$$da = b\,dy \qquad da' = b'\,dy$$

Hence, Eq. 11-13 yields

$$E'b' = Eb$$

from which

$$b' = \frac{E}{E'} b \qquad (11\text{-}14)$$

Thus, although the vertical dimensions of our equivalent beam will be the same as those of our original beam, the width of the equivalent beam is given by Eq. 11-14.

The equivalent beam is usually assumed to be made of one of the materials of the original beam. For example, if the beam of Fig. 11-6(a) is replaced by an equivalent beam made entirely of a material with a modulus E_1, the equivalent beam will have the dimensions shown in Fig. 11-6(d). On the other hand, if the equivalent beam is made of a material with a modulus E_2, its dimensions will be those of Fig. 11-6(e). The choice has no effect on the actual stresses in the original beam, but it is usually convenient to assume that the equivalent beam is made from the material of that portion of the original beam in which you are most interested. That is, if you wish to know the stresses in the upper fibers, you would probably choose the equivalent beam of Fig. 11-6(d). However, it is not necessary to do so, since

$$\sigma' \, da' = \sigma \, da$$

$$\sigma = \frac{da'}{da} \sigma' = \frac{E}{E'} \sigma' \qquad (11\text{-}15)$$

Thus, once an equivalent beam has been found, all the flexural stresses in the entire original beam (straight or curved) can be determined.

As mentioned in the first paragraph of this section, probably the most widely used type of composite beam is one made of concrete reinforced with steel. Such a beam can be analyzed by the method just presented. Concrete, however, is so relatively weak in tension that it is usually assumed that the portion of the concrete in tension *does not exist*. That is, it is assumed that the original beam is composed of sections of concrete in compression and the steel rods which are in tension. This beam is then transformed into an equivalent beam by the usual method.

EXAMPLE 11-3

A beam with the rectangular cross section of Fig. 11-7(a) is made of a material that has a tensile modulus of 10×10^6 psi and a compressive modulus of 15×10^6 psi. The beam is loaded in pure bending with a positive moment of 20,000 ft-lb. (a) Find an equivalent beam for this section. (b) Determine the maximum tensile and compressive flexural stresses in the original beam.

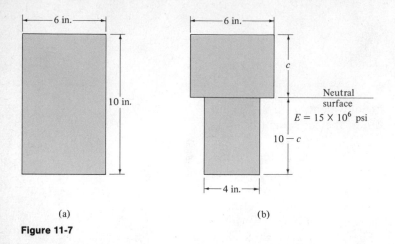

(a)   (b)

Figure 11-7

Solution:

(a) The fibers above the neutral surface are in compression ($E = 15 \times 10^6$), while those below the axis are in tension ($E = 10 \times 10^6$). However, we do not as yet know the location of the neutral surface. Assume that it is c inches from the top surface. Then by Eq. 11-14 an equivalent beam made of a material with a modulus of $E = 15 \times 10^6$ is shown in Fig. 11-7(b). For elastic bending the neutral surface coincides with the centroid. Hence,

$$(6)(c)\left(\frac{c}{2}\right) = (10 - c)(4)\left(\frac{10 - c}{2}\right)$$

$$c^2 + 40c - 200 = 0$$

$$c = 4.5 \text{ in.}$$

(b) The flexural stresses in the equivalent beam are given by the elastic flexure formula, which is

$$\sigma = -\frac{My}{I}$$

For the equivalent beam,

$$I = \frac{(6)(4.5^3)}{3} + \frac{(4)(5.5^3)}{3}$$

$$= 182 + 222 = 404 \text{ in.}^4$$

Then the stress in the uppermost fiber of the actual beam is

$$\sigma_u = -\frac{(20,000)(12)(4.5)}{404} = -2670 \text{ psi}$$

$$= 2670 \text{ psi (compressive)}$$

The stress in the lowermost fiber of the equivalent beam is

$$\sigma'_L = -\frac{(20,000)(12)(-5.5)}{404} = 3260 \text{ psi}$$

By Eq. 11-15 the stress in the lowermost fiber of the actual beam is

$$\sigma_L = \frac{E}{E'}\sigma'_L = \left(\frac{10 \times 10^6}{15 \times 10^6}\right)(3260) = 2175 \text{ psi (tensile)}$$

PROBLEMS

11-26 For the composite steel-reinforced timber beam shown in the illustration, find the maximum tensile and compressive flexural stresses in the wood when the beam is subjected to a bending moment sufficient to cause a maximum flexural tensile stress in the steel of 20,000 psi. The modulus of elasticity E for the steel is 30×10^6 and that for the wood is 1.5×10^6 psi.

Problem 11-26

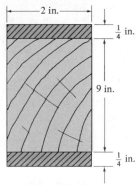

11-27 The requirements are the same as for Prob. 11-26, but the beam has a cross section with a rein-

Problem 11-27

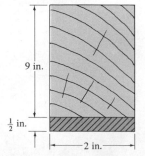

forcing plate only at the bottom as shown in the illustration.

11-28 The requirements are the same as for Prob. 11-40 except that the bending moment is about the vertical axis rather than the horizontal. Where is the shear center for this beam?

11-29 The requirements are the same as for Prob. 11-27 except that the bending moment is about the vertical axis rather than the horizontal. Where is the shear center for this beam?

11-30 A timber beam 6 in. high, 4 in. wide, and 16 ft long is to be simply supported at its ends and loaded with a uniformly distributed load of 250 lb/ft. The allowable bending stress in the wood is 1200 psi. (a) If the timber is to be reinforced with two steel plates 4 in. wide on the top and bottom, what should be the thickness of the plates? (b) What should be their width if steel plates $\frac{1}{2}$ in. thick are used? Use E for wood of 2×10^6 psi.

11-31 Select a timber beam approximately 4 in. wide to carry the loading in Prob. 11-30 over the same span. Compare the deflection of this timber beam to that of the reinforced beam in part (a) of Prob. 11-30. Discuss the relative merits of the two beams.

11-32 A positive bending moment of 20,000 ft-lb acts on a composite beam, the cross section of which is shown in the illustration. The steel and brass are bolted together. Determine the maximum flexural stresses in the materials if $E_{st} = 30 \times 10^6$ psi and $E_{br} = 15 \times 10^6$ psi.

11-33 A simply supported beam of reinforced concrete has the cross section shown in the illustration and carries a uniformly distributed load of 1500 lb/ft

Problem 11-32

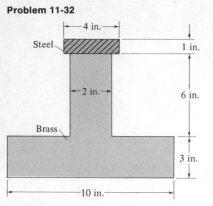

Steel

4 in.

1 in.

2 in.

6 in.

Brass

3 in.

10 in.

Problem 11-33

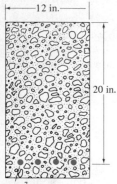

12 in.

20 in.

Four $\frac{7}{8}$ – in. diam. steel
reinforcing bars

over a span of 24 ft. Evaluate the maximum flexural
stress in the concrete and steel, assuming that $E_c =$
2.5×10^6 psi and $E_s = 30 \times 10^6$ psi.

11-34 A reinforced concrete beam whose cross
section is shown in the illustration is to carry a bending
moment of 20,000 ft-lb. If the ratio of the modulus of
elasticity of the reinforcing material to that of the

Problem 11-34

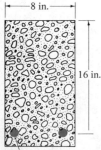

8 in.

16 in.

Two 1 in. diam.
reinforcing bars

concrete is 10, what are the maximum flexural stresses
in the concrete and in the reinforcing bars?

11-35 The beam in the illustration is made of concrete
reinforced with a material whose modulus of elasticity
is 12 times that of the concrete. If the allowable
bending stress in the concrete is 2000 psi and that
in the reinforcing bars is 30,000 psi, what is the allowable bending moment?

Problem 11-35

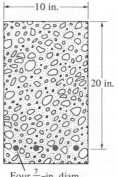

10 in.

20 in.

Four $\frac{7}{8}$-in. diam.
reinforcing bars

11-36 In designing a steel-reinforced concrete beam
it is possible to proportion the area of the steel to the
area of the concrete so that the maximum permissible
stresses in the steel and concrete occur simultaneously. Using the dimensions shown in the illustration, show that for balanced reinforcement

$$k = \frac{1}{1 + \dfrac{\sigma_s E_c}{\sigma_c E_s}}$$

where σ_c and σ_s are the permissible stresses in the
concrete and steel, respectively.

Problem 11-36

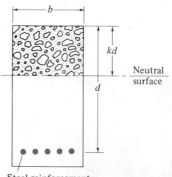

b

kd

Neutral
surface

d

Steel reinforcement

11-37 Refer to Prob. 11-36. Determine the required reinforcing area for balanced reinforcement if $\sigma_c -$ 14 MPa, $\sigma_s = 168$ MPa, $b = 200$ mm, $d = 500$ mm, $E_s = 209 \times 10^3$ MPa, and $E_c = 14 \times 10^3$ MPa.

11-38 The composite beam shown in the illustration is made of a timber reinforced with steel plates. The modulus of elasticity for the steel is 210×10^9 Pa, and that for the wood is 14×10^9 Pa. The allowable flexural stresses are 140 MPa for the steel and 10 MPa for the wood. What is the maximum permissible bending moment about the axis x-x?

Problem 11-38

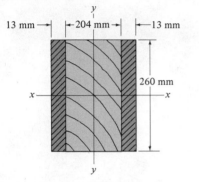

11-39 The cross section and the stresses are the the same as for Prob. 11-38. What bending moment can be applied about the axis y-y?

11-40 The composite beam shown in the illustration is a steel-reinforced timber beam. The properties of the steel and wood are the same as those in Prob. 11-38. Determine the maximum permissible bending moment about a horizontal axis.

Problem 11-40

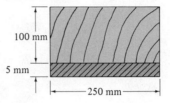

Chapter 12
Strain energy and theories of failure

Upon completion of this chapter you will be able to:

1 Calculate total strain energy and elastic strain energy.
2 Calculate volumetric strain energy and the energy of distortion.
3 Describe and use the following theories of failure:

> Maximum normal stress theory.
> Maximum shear stress theory.
> Maximum axial strain theory.
> Maximum total energy theory.
> Energy of distortion theory

12-2 / Strain energy and failure

Much of this text has been devoted to investigating the stresses, strains, and deformations in some of the more simple structural elements, such as shafts and beams. We have seen that while it is possible for the state of stress in a member to be uniaxial, it is much more likely that a state of combined or biaxial stress exists throughout a loaded member. Consequently, when we talk about the stress in a member, we must necessarily be concerned with the entire *state of stress* or *state of strain* in the member.

Why were we interested in finding stresses or strains? The main reason was so that we could predict or determine the load-carrying abilities of the various members. By comparing the stresses or strains developed in the member with some preset standard or specification, such as the yield-point stress or the proportional-limit strain, we tried to make a judgment as to the cause and type of failure.

Failures can be observed and thereby classified into certain categories, such as fracture, general yielding, buckling, and others which were discussed in some detail in earlier chapters. But, what *causes* a material to fracture? Or to yield? There is probably no simple answer to either of these questions, and usually the best we can do is to propose *theories* as to why fracture or yielding occurs. Such theories are called "theories of failure," and they are just theories, not true facts or laws. In later sections of this chapter, we shall discuss and compare some of the more commonly accepted theories of failure and try to point out why in certain cases some of the theories may seem more valid than the others. But remember that there is no one all-powerful, all-inclusive theory of failure, and therefore the engineer is called upon to exercise judgment as to which of the various theories is most applicable to a particular problem.

Why does a material fail? Because the largest stress exceeds some maximum permissible value? Because the largest strain exceeds some maximum permissible value? Both of these reasons seem plausible enough, but they do not provide the complete answer to the question of failure. What then? Perhaps the individual values of the stresses and strains are not so important as their combined effect; that is, maybe the combination of smaller stresses with their accompanying strains is a more critical factor than any individual larger stress or strain. How then do we measure their combined effect? The concept of strain energy provides a possible answer to this question.

Strain energy was first introduced and defined in section 3-8. However, that discussion was based mostly on the premise of a uniaxial state of stress that exists in a specimen during a uniaxial tension or compression test. The succeeding sections will be concerned with a more general development of strain energy in relation to biaxial and triaxial states of stress.

12-3 / Total strain energy

Strain-energy density (hereafter referred to simply as strain energy) is the energy per unit volume that a material absorbs while undergoing forced deformations. For example, consider a small elemental volume $\Delta x\, \Delta y\, \Delta z$ in Fig. 12-1. If a centroidal force is applied parallel to the x axis, deformations in the x, y, and z directions will result. Since, however, the only force is parallel to the x axis, the work done on the volume of material will depend only on this force and the deformation e_x in the x direction. Also, the magnitude of the force depends on the amount of deformation, and vice versa. Hence, the work per unit volume, or the strain-energy density is,

$$U = \frac{1}{V} \int_0^{e_x} P_x\, de_x = \int_0^{e_x} \frac{P_x}{\Delta z\, \Delta y} \frac{de_x}{\Delta x}$$

$$= \int_0^{\varepsilon_x} \sigma_x\, d\varepsilon_x \qquad (12\text{-}1)$$

In the last integral, the variable of integration is in terms of strain, and hence the limits are also in terms of strain.

The expression in Eq. 12-1 represents the strain energy (or stress work) for uniaxial loading only. Suppose, now, that the loading is not uniaxial, but that there are axial y and z forces as well as the x force. Then the total work done on the elemental volume of material will be the work done by all of the forces acting through their respective displacements. Thus,

$$U = \int_0^{\varepsilon_x} \sigma_x\, d\varepsilon_x + \int_0^{\varepsilon_y} \sigma_y\, d\varepsilon_y + \int_0^{\varepsilon_z} \sigma_z\, d\varepsilon_z \qquad (12\text{-}2)$$

where the strains ε_x, ε_y, and ε_z are the *total* strains in their respective directions. Remember from Chapters 3 and 4 that each of the individual total strains is influenced by each of the individual axial stresses, and vice versa. Hence, in order that the integrals in Eq. 12-2 may be evaluated, the generalized stress-strain law for the material must be known. In particular, if the material is homogeneous, elastic, and isotropic, so that the stress-strain law is the generalized Hooke's law, the evaluation of these integrals is considerably simplified as will be seen in the next section.

Figure 12-1

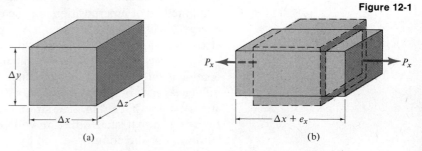

(a)   (b)

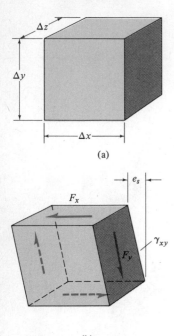

(a)

(b)

Figure 12-2

Suppose that, instead of axial loads, the forces acting on the elemental volume in Fig. 12-2(a) are shear forces that produce the deformation of Fig. 12-2(b). Then, from our definition of work, and assuming the deformations are small, the strain-energy density for this system becomes

$$U = \frac{1}{V} \int_0^{e_s} F_x \, de_s = \int_0^{e_s} \frac{F_x}{\Delta x \, \Delta z} \frac{de_s}{\Delta y}$$

$$= \int_0^{\gamma_{xy}} \tau_{yx} \, d\gamma_{xy} = \int_0^{\gamma_{xy}} \tau_{xy} d\gamma_{xy} \qquad (12\text{-}3)$$

Similarly, if all possible shear forces are taken into consideration, the strain energy due to these forces will be

$$U = \int_0^{\gamma_{xy}} \tau_{xy} \, d\gamma_{xy} + \int_0^{\gamma_{xz}} \tau_{xz} \, d\gamma_{xz} + \int_0^{\gamma_{yz}} \tau_{yz} \, d\gamma_{yz} \qquad (12\text{-}4)$$

Finally, in the most general case of a triaxial state of stress, the total strain energy will be that due to the three normal stresses and the three shear stresses, or the sum of Eqs. 12-2 and 12-4.

Before concluding this section, we shall make a very important observation. Work, and therefore strain energy, is a scalar quantity. That is, it is independent of the coordinate system used to identify the various forces, displacements, stresses, and strains. With this in mind, the calculation of the total strain energy for a combined state of stress can be considerably simplified if a principal coordinate system of stress and strain is used. For this case, there are no shear stresses or shear strains, and the total strain energy is then represented by Eq. 12-2 alone, where the stresses and strains are the principal values.

12-4 / Elastic strain energy

Of particular importance, insofar as theories of failure are concerned, is the *elastic* strain energy associated with a combined state of stress. For a linear, homogeneous, isotropic material, elastic behavior means that the stress-strain law is the generalized Hooke's law (refer to section 4-5). Also, since elastic strains and deformations are usually quite small, the convenient method of superposition may be used in calculating the elastic strain energy. When elastic stresses are superposed, the order in which the stresses are applied to the elemental volume is immaterial.

Before we proceed with the calculation of elastic strain energy, a comment on notation is worthwhile. Since each of the normal stresses influences each of the axial strains, a distinction will be made as to which stress causes what strain. The strain in a particular direction produced by the stress in that same direction will be denoted by ε_{xx}, ε_{yy}, and ε_{zz}. The strain in one direction produced by the Poisson effect of a normal stress in another direction will be denoted by ε_{xy} and ε_{xz}; ε_{yx} and ε_{yz}; and ε_{zx} and ε_{zy}. Thus, with this notation, the total strain in the x direction, for example, is

$$\varepsilon_{x(\text{total})} = \varepsilon_{xx} + \varepsilon_{xy} + \varepsilon_{xz}$$

$$= \frac{\sigma_x}{E} - \mu\frac{\sigma_y}{E} - \mu\frac{\sigma_z}{E} \tag{12-5}$$

Similar expressions can be obtained for the y and z directions as well.

In using the method of superposition to calculate elastic strain energy, we shall assume arbitrarily that the normal stresses are applied in the sequence σ_x, σ_y, and σ_z. You should satisfy yourself that the same result is obtained for any other sequence you may happen to select. Now, if the stress σ_x is the first one applied to an elemental volume, strains will be produced in the x, y, and z directions which are denoted by ε_{xx}, ε_{yx}, and ε_{zx}, respectively. However, since for this condition the loading is uniaxial, the strain energy is given by Eq. 12-1 and is

$$U = \int_0^{\varepsilon_{xx}} \sigma_x \, d\varepsilon_{xx}$$

For uniaxial loading, Hooke's law is simply

$$\varepsilon_{xx} = \frac{\sigma_x}{E}$$

and the elastic strain energy at this stage can be written as

$$U = \int_0^{\varepsilon_{xx}} E\varepsilon_{xx} \, d\varepsilon_{xx} = \frac{1}{2} E\varepsilon_{xx}^2 = \frac{1}{2}\sigma_x\varepsilon_{xx} = \frac{\sigma_x^2}{2E} \tag{a}$$

Next, the stress σ_y is applied. This stress also produces a strain in each of the directions; and these strains are denoted by ε_{yy}, ε_{xy}, and ε_{zy}. During this deformation process there are two stresses acting on the body, namely, σ_y and σ_x. The work done by the stress σ_y is

$$U = \int_0^{\varepsilon_{yy}} \sigma_y \, d\varepsilon_{yy} \tag{b}$$

since this stress varies from zero to its maximum value during the straining process. On the other hand, during the same process the stress σ_x retains a constant value. Thus, the work done by the stress σ_x during this deformation process is

$$U = \int_0^{\varepsilon_{xy}} \sigma_x \, d\varepsilon_{xy} = \sigma_x \int_0^{\varepsilon_{xy}} d\varepsilon_{xy} = \sigma_x\varepsilon_{xy} \tag{c}$$

Now, from Hooke's law,

$$\varepsilon_{yy} = \frac{\sigma_y}{E} \qquad \varepsilon_{xy} = -\mu\frac{\sigma_y}{E}$$

Substituting these values in equations (b) and (c) and adding, we have

$$U = \int_0^{\varepsilon_{yy}} \sigma_y \, d\varepsilon_{yy} + \sigma_x \varepsilon_{xy}$$

$$U = \frac{\sigma_y^2}{2E} - \mu \frac{\sigma_x \sigma_y}{E} \tag{d}$$

Finally, the stress σ_z is applied, causing the strains ε_{zz}, ε_{xz}, and ε_{yz}. During this process three stresses are acting on the element. The magnitudes of the work done by the stresses are, respectively,

$$\int_0^{\varepsilon_{zz}} \sigma_z \, d\varepsilon_{zz} \qquad \int_0^{\varepsilon_{xz}} \sigma_x \, d\varepsilon_{xz} \qquad \int_0^{\varepsilon_{yz}} \sigma_y \, d\varepsilon_{yz}$$

where σ_z varies while σ_x and σ_y are constant. By Hooke's law,

$$\varepsilon_{zz} = \frac{\sigma_z}{E} \qquad \varepsilon_{xz} = -\mu \frac{\sigma_z}{E} \qquad \varepsilon_{yz} = -\mu \frac{\sigma_z}{E}$$

So we find that the work during this final process is

$$U = \int_0^{\varepsilon_{zz}} \sigma_z \, d\varepsilon_{zz} + \sigma_x \int_0^{\varepsilon_{xz}} d\varepsilon_{xz} + \sigma_y \int_0^{\varepsilon_{yz}} d\varepsilon_{yz}$$

$$U = \frac{\sigma_z^2}{2E} - \mu \frac{\sigma_x \sigma_z}{E} - \mu \frac{\sigma_y \sigma_z}{E} \tag{e}$$

Thus, after the application of all three normal stresses, the final strain energy per unit volume is the algebraic sum of the work in equations (a), (d), and (e). This sum is

$$U = \frac{1}{2E} (\sigma_x^2 + \sigma_y^2 + \sigma_z^2) - \frac{\mu}{E} (\sigma_x \sigma_y + \sigma_x \sigma_z + \sigma_y \sigma_z) \tag{12-6}$$

This is a quadratic expression in the stresses σ_x, σ_y, and σ_z and represents the elastic strain energy due to the three normal stresses. By use of the generalized Hooke's law the expression for the strain energy (Eq. 12-6) can also be written in the bilinear form

$$U = \tfrac{1}{2}(\sigma_x \varepsilon_x + \sigma_y \varepsilon_y + \sigma_z \varepsilon_z) \tag{12-6a}$$

where ε_x, ε_y, and ε_z are the *total* strains in the x, y, and z directions.

If, in addition to the three normal stresses, shear stresses also act on the element, then the total elastic strain energy must include the energy due to the shear stresses. However, for an isotropic material a shear stress does not influence either the axial strains or the other shear strains. So we can compute the strain energy produced by each of the shear stresses individually and then add their contributions algebraically. By Hooke's law for elastic shear stresses,

$$\gamma_{xy} = \frac{\tau_{xy}}{G} \qquad \gamma_{xz} = \frac{\tau_{xz}}{G} \qquad \gamma_{yz} = \frac{\tau_{yz}}{G}$$

From Eq. 12-4 the elastic strain energy due to the shear stresses is

$$U = \int_0^{\gamma_{xy}} \tau_{xy}\, d\gamma_{xy} + \int_0^{\gamma_{xz}} \tau_{xz}\, d\gamma_{xz} + \int_0^{\gamma_{yz}} \tau_{yz}\, d\gamma_{yz}$$

$$U = \frac{1}{2G}(\tau_{xy}^2 + \tau_{xz}^2 + \tau_{yz}^2) \tag{12-7}$$

or

$$U = \tfrac{1}{2}(\tau_{xy}\gamma_{xy} + \tau_{xz}\gamma_{xz} + \tau_{yz}\gamma_{yz}) \tag{12-7a}$$

The *total* elastic strain energy per unit volume for a general state of triaxial stress is the sum of the energies due to the normal and shear stresses, or the sum of Eqs. 12-6 and 12-7. This sum is

$$U = \frac{1}{2E}(\sigma_x^2 + \sigma_y^2 + \sigma_z^2) - \frac{\mu}{E}(\sigma_x\sigma_y + \sigma_x\sigma_z + \sigma_y\sigma_z)$$

$$+ \frac{1}{2G}(\tau_{xy}^2 + \tau_{xz}^2 + \tau_{yz}^2) \tag{12-8}$$

or

$$U = \tfrac{1}{2}(\sigma_x\varepsilon_x + \sigma_y\varepsilon_y + \sigma_z\varepsilon_z + \tau_{xy}\gamma_{xy} + \tau_{xz}\gamma_{xz} + \tau_{yz}\gamma_{yz}) \tag{12-8a}$$

If the stresses and strains are measured in the principal directions, Eq. 12-8 reduces to Eq. 12-6, because the shear stresses and strains are zero for the principal directions. As a result, it is often more covenient to determine the strain energy after the principal stresses and strains have been found.

12-5 / Volumetric and distortion components of elastic energy

Although the total elastic strain energy may be an important criterion in determining the cause of failure of a material, an even more important consideration is *how* the strain energy input affects the material. One way of investigating the effect that the energy input has on the elemental volume of material is to divide the total energy into two components. One component is the strain energy associated with the change in volume of the element. The other component is the strain energy associated with the distortion that the element undergoes. Later we shall see what roles the two component energies play in the theories of failure.

First, we know that any generalized triaxial state of stress may be resolved into a system of normal stresses and a system of shear stresses, as in Fig. 12-3. Also, in section 4-6 it was shown that the change in volume per unit volume (cubical dilatation) ε_V was the sum of the axial strains, or $\varepsilon_x + \varepsilon_y + \varepsilon_z$. Hence, the shear stresses and strains do not influence the volume change that may occur during a deformation process, and only the normal stresses influence the volume change.

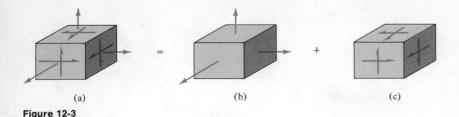

Figure 12-3

Let us now further decompose the system of normal stresses into two component systems of normal stresses. In one of these component stress systems, the value of the normal stress will be the same in all directions, and this stress will be equal to the average value of the original system of normal stresses; thus $\sigma_{av} = \frac{1}{3}(\sigma_x + \sigma_y + \sigma_z)$. The other component stress system will consist of the remainders of the original normal stresses, as illustrated in Fig. 12-4.

The axial strain of the element in Fig. 12-4(b) will be the same in each direction and denoted by ε_{av}. From Hooke's law, we have

$$\varepsilon_{av} = \frac{\sigma_{av}}{E} - \mu\frac{\sigma_{av}}{E} - \mu\frac{\sigma_{av}}{E} = \frac{\sigma_{av}}{E}(1 - 2\mu)$$

$$= \frac{\sigma_x + \sigma_y + \sigma_z}{3E}(1 - 2\mu) = \frac{1}{3}(\varepsilon_x + \varepsilon_y + \varepsilon_z) = \frac{1}{3}\varepsilon_V$$

Thus, the element in (b) undergoes a unit volume change of

$$3\varepsilon_{av} = \varepsilon_V = \varepsilon_x + \varepsilon_y + \varepsilon_z$$

This is also the unit volume change of the element in Fig. 12-4(a). Hence, the element in (c) actually undergoes no volume change, and the energy associated with the change of volume of the element in Fig. 12-4(a) is the energy produced by the system of stresses in (b). Using Eq. 12-6a we find this volumetric energy to be

$$U_V = 3\left(\tfrac{1}{2}\sigma_{av}\varepsilon_{av}\right) = 3\left(\tfrac{1}{2}\sigma_{av}\tfrac{1}{3}\varepsilon_V\right) = \tfrac{1}{2}\sigma_{av}\varepsilon_V \qquad (12\text{-}9)$$

Using the definition of bulk modulus (see section 4-6),

$$K = \frac{\sigma_{av}}{\varepsilon_V} = \frac{E}{3(1 - 2\mu)} \qquad [4\text{-}20]$$

the volumetric energy can also be written as

Figure 12-4

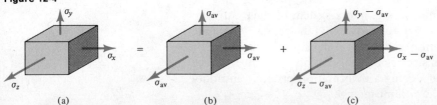

$$U_V = \frac{1}{2} K \varepsilon_V^2 = \frac{\sigma_{av}^2}{2K} \tag{12-9a}$$

The total energy produced by the original system of normal and shear stresses is given by Eq. 12-8. Thus,

$$U_{total} = \frac{1}{2E} (\sigma_x^2 + \sigma_y^2 + \sigma_z^2) - \frac{\mu}{E} (\sigma_x\sigma_y + \sigma_x\sigma_z + \sigma_y\sigma_z)$$

$$+ \frac{1}{2G} (\tau_{xy}^2 + \tau_{xz}^2 + \tau_{yz}^2) \tag{12-8}$$

If this expression represents the total elastic strain energy and that in Eq. 12-9a represents the volumetric strain energy, the remaining strain energy, or the energy of distortion, is

$$U_d = U_{total} - U_V$$

Using Eqs. 12-8, 12-9a, and 4-20, and simplifying, we obtain

$$U_d = \frac{(1 + \mu)}{6E} [(\sigma_x - \sigma_y)^2 + (\sigma_x - \sigma_z)^2 + (\sigma_y - \sigma_z)^2]$$

$$+ \frac{1}{2G} (\tau_{xy}^2 + \tau_{xz}^2 + \tau_{yz}^2)$$

Since $G = [E/2(1 + \mu)]$, the energy of distortion is

$$U_d = \frac{1}{12G} [(\sigma_x - \sigma_y)^2 + (\sigma_x - \sigma_z)^2 + (\sigma_y - \sigma_z)^2]$$

$$+ \frac{1}{2G} (\tau_{xy}^2 + \tau_{xz}^2 + \tau_{yz}^2) \tag{12-10}$$

This energy of distortion is the energy that causes the change in shape, or distortion, of the original elemental volume. It does not cause any volume change. As we shall soon see, energy of distortion plays an important role in the theories of failure. Note that Eq. 12-10 becomes simplified somewhat if the principal directions are used. In this case the shear stresses are zero and the normal stresses are the principal values.

EXAMPLE 12-1

Compute the total elastic strain energy, the energy of distortion, and the energy of volume change for a biaxial state of pure shear.

Solution:

For a biaxial state of pure shear with a maximum shear stress τ, the *principal* stresses are $\sigma_1 = \tau$, $\sigma_2 = -\tau$, and $\sigma_3 = 0$. Hence, if a principal coordinate system is used, the total strain energy is

$$U_{total} = \frac{1}{2E}(\sigma_1^2 + \sigma_2^2 + \sigma_3^2) - \frac{\mu}{E}(\sigma_1\sigma_2 + \sigma_1\sigma_3 + \sigma_2\sigma_3)$$

$$= \frac{1}{2E}[\tau^2 + (-\tau)^2 + 0^2] - \frac{\mu}{E}[\tau(-\tau) + \tau(0) + (-\tau)0]$$

or

$$U_{total} = \frac{1 + \mu}{E}\tau^2$$

The energy of distortion is

$$U_d = \frac{1}{12G}[(\sigma_1 - \sigma_2)^2 + (\sigma_1 - \sigma_3)^2 + (\sigma_2 - \sigma_3)^2]$$

$$= \frac{1}{12G}\{[\tau - (-\tau)]^2 + (\tau - 0)^2 + (-\tau - 0)^2\}$$

$$= \frac{1}{2G}\tau^2$$

The energy of volume change is

$$U_V = \frac{1}{2K}\sigma_{av}^2 = \frac{1}{2K}\left(\frac{\sigma_1 + \sigma_2 + \sigma_3}{3}\right)^2$$

$$= \frac{1}{2K}\left(\frac{\tau - \tau + 0}{3}\right) = 0$$

This last result could have been found from U_{total} and U_d by realizing that

$$G = \frac{E}{2(1 + \mu)}$$

Hence, for pure shear, $U_{total} = U_d$. This means that all the energy goes to distorting the element, and no volume change occurs.

PROBLEMS

12-1 Carry out the details to obtain Eq. 12-6a from Eq. 12-6 or vice versa.

12-2 Compute the total elastic energy, the energy of distortion, and the energy of volume change for a uniaxial state of stress.

12-3 Compute the total elastic energy, the energy of distortion, and the energy of volume change for a general state of plane stress for which $\sigma_z = \tau_{xz} = \tau_{yz} = 0$.

12-4 Compute the total elastic energy, the energy of distortion, and the energy of volume change for a state of plane strain for which $\varepsilon_z = \gamma_{xz} = \gamma_{yz} = 0$. Express your answer in terms of the stresses σ_x, σ_y, and τ_{xy}.

12-5 Carry out the details in the derivation of Eq. 12-10.

12-6 to 12-8 Calculate the elastic volumetric strain

energy and the energy of distortion for the element shown. Use $E = 30 \times 10^6$ psi and $\mu = \frac{1}{4}$. Express your answers in both the American and International units.

Problem 12-6 **Problem 12-7** **Problem 12-8**

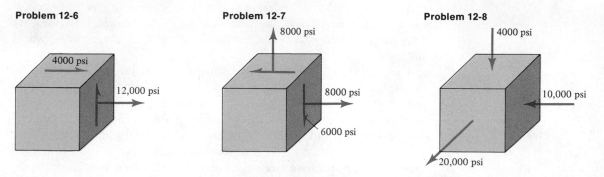

12-6 / Theories of failure

As noted in section 12-2, the phenomena of fracture and yielding can be observed and described, but the reasons (or reason) for their occurrence are as yet not completely understood. From a practical point of view, the designer or engineer would like to have some single criterion on which to base designs. That is, he or she would like to be able to say that failure in the form of fracture or yielding will occur when a certain critical quantity (whatever it may be) reaches a limiting value. But what is the critical quantity and what is its limiting value? These are the questions to which the theories of failure attempt to provide answers. In effect, a theory of failure establishes some quantity (or quantities) as a criterion for predicting failure.

As an example, consider a uniaxially loaded member. How can we establish a critical quantity for failure? In this case, comparison with experimental data could determine a fairly reliable criterion, such as the axial stress, axial strain, or strain energy. If plastic deformation were the mode of failure, the yield-point stress or yield strength could be used as the limiting value for the permissible axial stress. If fracture were the mode of failure, the ultimate strength would probably be used as the limiting value for the stress. Of course, the criterion need not be a stress. The elastic-limit strain might just as well be used as the criterion for plastic behavior, and the modulus of toughness could be used for fracture.

How is a critical quantity determined if the loading is not uniaxial and it produces some general biaxial or triaxial state of stress? Certainly it is not feasible to attempt to conduct experiments for all possible combinations of stresses. In fact, it is difficult, and in some cases impossible, to produce controlled states of triaxial stress. It is at this stage that the need for a *theory of failure* for any arbitrary state of stress arises. The following paragraphs describe some of the more commonly accepted theories of failure.

Maximum normal stress theory The maximum normal stress theory asserts that failure occurs at some point in a body only when the maximum principal normal stress at that point reaches some limiting value. According to this theory, only the magnitudes of the principal stresses are important, and the principal stress having the greatest magnitude governs the failure of the material.

This theory appears to give reasonable results for fracture of brittle materials. However, its limitations are rather self-evident, in that it does not take into account either the type of normal stress (tensile or compressive) or the orientation of the principal planes. Thus, this theory of failure is strictly applicable only to isotropic materials; and even for such materials it still does not take into account the effects of states of combined tensile and compressive stresses. For a ductile material, a biaxial combination of a tensile stress in one direction and a compressive stress in another direction can be more critical than the magnitude of either of the stresses individually.

Maximum shear stress theory The maximum shear stress theory predicts that failure will occur at some point in a loaded body when the maximum shear stress at that point reaches a certain limiting value. If the principal stresses are σ_1, σ_2, and σ_3, and if $\sigma_1 > \sigma_2 > \sigma_3$, then we recall from Chapter 1 that the maximum shear stress is

$$\tau_{max} = \frac{\sigma_1 - \sigma_3}{2}$$

Thus, if a limiting value of τ_{max} is assigned or known, this equation represents the mathematical formulation of the maximum shear stress theory.

This theory seems to be applicable to members made of ductile materials in which relatively large shear stresses are developed. However, as in the case of the maximum normal stress theory, this theory fails to take account of the orientation of the maximum shear stress. Hence, it is strictly applicable only to isotropic materials. Also, according to this theory, an element subjected to a state of triaxial hydrostatic stress (three equal principal stresses) would never fail; of course, such a conclusion is erroneous. Nevertheless, this theory is widely used in the design of steel members, since shear stresses play a significant role in the phenomenon of yielding, and is often referred to as the *Tresca yield condition*.

Maximum axial strain theory The maximum axial strain theory proposes that axial strain, rather than stress, is the criterion for failure of a material. In particular, it proposes that failure depends on the magnitude of the largest principal axial strain, regardless of the combination of · stresses causing the strains. If this theory is to be used, the generalized stress-strain law for the material must be known so that the principal strains can be evaluated.

As with the first two theories, this theory does not take into account the directions of the principal strains. Hence, this theory is strictly applicable only to isotropic materials. Also, since Hooke's law is valid only for elastic behavior, it becomes difficult to apply this theory if inelastic behavior is permitted to occur. For these and other reasons, this theory is rarely used in modern design.

Maximum total energy theory The maximum total energy theory states that failure occurs when the total energy per unit volume reaches some predetermined limiting value. This theory has one obvious advantage over the previous three theories in that the strain energy is independent of the orientations of the stresses and strains involved. For this reason an energy theory of this type can be more readily applied to nonisotropic materials. As indicated in section 12-3, the problem of evaluating the total strain energy for an arbitrary state of stress can be extremely difficult. This is especially true if inelastic behavior occurs, because the generalized stress-strain law is then quite complex and may possibly be unknown. For this reason, this theory is almost never used to predict failure by fracture for a ductile material.

If the total *elastic* strain energy is the criterion for failure, then the total elastic strain energy for an isotropic material can be evaluated by the equations derived in section 12-4, which require only that the elastic constants of the material be known. Thus, this theory can be used to predict fracture of a very brittle material or the initiation of yielding in a ductile material. The next theory, however, has been found to be more accurate in predicting the initiation of yielding, and the maximum normal stress theory is easier to use in predicting fracture of brittle materials. For these reasons, the total strain energy theory has been virtually abandoned.

Energy of distortion theory The energy of distortion theory, often called the Huber-Hencky-von Mises theory or the octahedral shear stress theory, predicts that failure will occur when an equivalent stress σ_e defined by

$$\sigma_e^2 = [(\sigma_1 - \sigma_2)^2 + (\sigma_1 - \sigma_3)^2 + (\sigma_2 - \sigma_3)^2] \qquad (12\text{-}11)$$

where σ_1, σ_2, σ_3 are the principal stresses, reaches a predetermined limiting value.

One development of this theory is based on the elastic energy of distortion, which was discussed in section 12-5. It was shown that any arbitrary state of stress can be decomposed into two systems of stresses. One system, consisting of equal normal stresses, produces only volume change, and the other system of stresses causes the distortion. Experimental evidence indicates that a state of stress due to hydrostatic compression has very little tendency to cause yielding in a ductile material. On this motivation, the energy of distortion theory states that yielding in a ductile material is independent of the energy of volume change and is *dependent* only on the energy of distortion. If the principal stress system is used, we see from Eq. 12-10 that, except for the elastic constant multiplier, the energy of distortion is proportional to the quantity indicated in Eq. 12-11. Hence, that quantity is used as the criterion for this theory of failure.

This theory of failure is distinctly different from the total energy theory. For example, if it were possible to construct a state of stress due to hydrostatic tension (three equal tensile principal stresses), there definitely would be strain energy but there would be no energy of distortion. Hence, according to the distortion theory of failure, yielding should never occur. However, except for the rare possibility of hydrostatic stress, the energy of distortion theory seems to be the most accurate for predicting failure by yielding in ductile materials.

12-7 / Use of the theories of failure

Each of the preceding theories of failure establishes a criterion for predicting failure of a material. Of course, by their very nature, some of the theories are more applicable to predicting failure by fracture, and others to predicting failure by yielding. Consequently, one might say that generally the maximum normal stress theory and the maximum axial strain theory are more valid for design of members made of brittle materials, and the maximum shear stress theory and the energy of distortion theory are more valid for design of ductile members. But, regardless of which theory of failure is employed, the question still remaining is, what number is to be used as the limiting value of the established criterion? From an engineering point of view, this value should be based on some experimental evidence. The most easily controlled states of stress are those associated with loading conditions causing uniaxial stress, torsion, or bending. Hence, the limiting value for the critical quantity is usually based on the value obtained from one or more of these simple loading conditions.

If all theories of failure (or even any one theory) were really correct, the value of the critical quantity would not depend on what experiment was used to determine it. Unfortunately, this is not true, as can be seen by a simple example. Suppose that the maximum shear stress theory is to be used to predict the initiation of yielding in a ductile material. Recalling Mohr's circle, for a uniaxial tensile test we see that the maximum shear stress is equal to one-half of the maximum tensile stress. Thus, by this test the shear yield point τ_{yp} would be one-half the tensile yield point σ_{yp}. However, it has been found that in a torsion test where the state of stress is pure shear, the value of the shear stress at yielding is approximately $0.57\sigma_{yp}$. Which of these values is to be used as the limiting value of the maximum shear stress? The answer depends on the particular problem and the experience of the engineer. The value obtained from a uniaxial

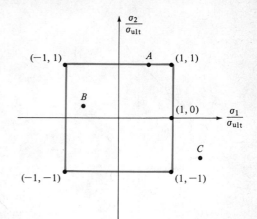

Figure 12-5

tension test is probably more conservative than that obtained from a torsion test. It is left as an exercise for the student (Prob. 12-16) to show that if the energy of distortion theory is used, and if this theory is correct for both uniaxial and torsion loading, the relation between σ_{yp} and τ_{yp} should be

$$\tau_{yp} = \frac{\sigma_{yp}}{\sqrt{3}} = 0.577\sigma_{yp}$$

which, as previously mentioned, agrees with experimental evidence for ductile materials in pure shear.

It is instructive to examine a graphical representation of a particular theory of failure. For example, consider the maximum normal stress theory, which for a state of plane stress ($\sigma_3 = 0$) takes the form

$$\sigma_1, \sigma_2 \leq \sigma_{ult} \tag{12-12}$$

where σ_{ult} is taken to be the limiting value of the normal stress. This inequality is represented by the diagram in Fig. 12-5.

Any point such as A or B representing a state of stress that lies inside, or on the boundary of, the rectangle satisfies the inequality (Eq. 12-12). On the other hand a state of stress represented by point C, which lies outside the boundary, does not satisfy the inequality. Thus, the rectangle is called the *failure envelope* for the maximum normal stress theory.

In a similar manner, for plane stress the maximum shear stress theory takes the form

$$\frac{\sigma_1 - \sigma_2}{2} \leq \tau_{yp} \tag{12-13}$$

when τ_{yp} is taken to be the limiting value. Note that when σ_1 and σ_2 have the same algebraic sign, this inequality becomes

$$\frac{\sigma_1}{2} \leq \tau_{yp} \quad \text{and} \quad \frac{\sigma_2}{2} \leq \tau_{yp} \tag{12-14}$$

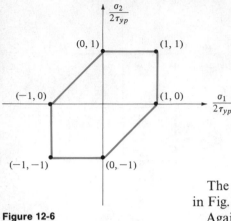

Figure 12-6

The inequalities (Eqs. 12-13 and 12-14) are represented by the diagram in Fig. 12-6.

Again, points lying within or on the hexagonal boundary, the failure envelope, satisfy the inequality (Eq. 12-13) while points outside do not.

Finally, the maximum energy of distortion theory for plane stress takes the form

$$\sigma_1^2 + \sigma_2^2 - \sigma_1\sigma_2 \le \frac{\sigma_e^2}{2} \tag{12-15}$$

This inequality is represented by the elliptical diagram in Fig. 12-7. This ellipse is the failure envelope for the energy of distortion theory.

Remember that nobody really knows what factor or combination of factors causes yielding or fracture to occur. A theory of failure does not try to explain why these phenomena occur, but merely attempts to establish some way of predicting when a certain phenomenon should occur. Thus, the theories of failure are really just reasonable guesses. Which theory should be used in a particular engineering problem? Here, again, an intelligent guess is involved. Before making this decision, the engineer or designer must assimilate the available information in regard to loading conditions, use of the structure, material, reliability of experimental data, and other influencing factors. Finally, the engineer must call on experience and good judgment.

Figure 12-7

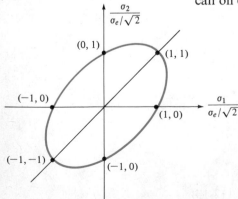

EXAMPLE 12-2

Data from a uniaxial tensile test are to be used to determine the limiting value for failure. If the mode of failure is the initiation of inelastic action, determine the limiting value for each of the theories of failure in terms of the available data.

Solution:

From the uniaxial test, the values of E, μ, and the proportional-limit stress σ_{PL} can be found. We can then determine the limiting value for each critical quantity in terms of these properties.

1 *Maximum normal stress theory.* The critical quantity is the largest principal stress. Thus

$$\sigma_{cr} = \sigma_{PL} \text{ psi}$$

2 *Maximum shear stress theory.* The critical quantity is the maximum shear stress. Thus

$$\tau_{cr} = \frac{\sigma_{PL}}{2} \text{ psi} \qquad \text{(since the test loading is uniaxial)}$$

3 *Maximum axial strain theory.* The critical quantity is the largest principal axial strain. Thus

$$\varepsilon_{cr} = \frac{\sigma_{PL}}{E} \text{ in./in.} \qquad \text{(since the test loading is uniaxial)}$$

4 *Maximum total energy theory.* The critical quantity is the total elastic strain energy. Thus

$$U_{cr} = \frac{1}{2} \frac{\sigma_{PL}^2}{E} \frac{\text{in.-lb}}{\text{in.}^3} \qquad \text{(from Eqs. 12-6 or 12-8)}$$

5 *Maximum energy of distortion theory.* The critical quantity is the effective stress given by Eq. 12-11. Thus

$$\sigma_{cr} = \sigma_e = \sqrt{\sigma_{PL}^2 + \sigma_{PL}^2} = \sqrt{2}\sigma_{PL} \text{ psi}$$

PROBLEMS

12-9 Data from a pure torsion test are to be used to determine the limiting value for failure. If the mode of failure is the initiation of inelastic action, determine the limiting value for each of the theories of failure in terms of the available data. The values of E, G, μ, and τ can be determined from an elastic torsion test.

12-10 A thin-walled spherical pressure vessel is made of a ductile material and the parts are riveted together. If the only significant loading is the internal pressure, discuss fully the relative merits of each of the theories of failure. Which is the most conservative? Which is the least conservative?

12-11 A thin-walled, riveted cylindrical pressure vessel is made of a ductile material. The longitudinal axis of the vessel is horizontal, and its ends are simply supported. (a) If the only significant loading is the internal pressure, discuss fully the relative merits of each of the theories of failure. (b) If the weight of the vessel causes significant bending effects, how will the applicability of each theory of failure be affected?

12-12 The member shown in the illustration is made of a relatively brittle material. Derive a design formula for the required radius of the member in terms of the dimensions L and b, the load P, and some limiting quantity. Also, specify the appropriate theory of failure and how the value of the limiting quantity should be determined.

Problem 12-12

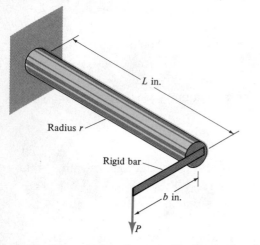

12-13 The member shown in the illustration is made of mild steel, and failure is due to the initiation of inelastic behavior. Derive an expression for the allowable value of P in terms of the radius of the member, some limiting quantity, and the torque T. Also, specify the appropriate theory of failure and how the value of the limiting quantity should be determined.

12-14 The member shown in the illustration for Prob. 12-13 is made of a linearly brittle material. The only available properties for the material are those from a uniaxial tensile test which indicated that

Problem 12-13

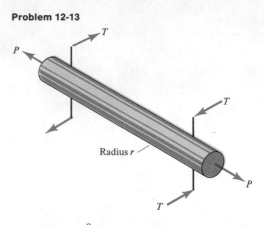

$E = 125 \times 10^9$ Pa, $\mu = 0.30$, and $\sigma_{ult} = 413$ MPa. If the radius of the member is 30 mm and $T = 5300$ N·m, determine the maximum permissible load P based on a safety factor of 2. Use the most conservative of the applicable theories of failure.

12-15 The member shown in the illustration for Prob. 12-12 is made of a ductile material. As determined from a uniaxial test, the yield strength (based on 0.2-percent offset) is $\sigma_{ys} = 40,000$ psi, $E = 24 \times 10^6$ psi, and $\mu = \frac{1}{4}$. Although data from a torsion test might be desirable, it is rather difficult to determine inelastic stresses in a torsion specimen. Hence, the uniaxial data must be used in the design of the member. If failure is due to excessive inelastic behavior, what theory or theories of failure are most applicable and why? If $L = 15$ in., $b = 4$ in., and $P = 10,000$ lb, what minimum radius is required based on a safety factor of $1\frac{1}{2}$?

12-16 In a ductile material failure is often based on the initiation of inelastic action. Compare the limiting values for each of the theories of failure obtained from a uniaxial test with those obtained from a torsion test. (*Hint:* See Example 12-2 and Prob. 12-9.) Plot and compare corresponding failure envelopes.

Chapter 13
Work and energy methods

13-1 / Objectives

Upon completion of this chapter you will be able to:

1   Determine the work done by forces and couples acting on elastic bodies.
2   Calculate the elastic strain energy of structural members.
3   Calculate elastic deformations of beams and shafts by energy methods.
4   Calculate displacements in trusses by energy methods.

13-2 / Introduction

We have seen in Chapter 3 how the concept of strain energy can be used as a material parameter, that is, the modulus of resilience and modulus of toughness. In Chapter 12 we showed how strain energy per unit volume can be used as a criterion for failure particularly in situations in which a multiaxial state of stress exists. We now turn to yet another, and perhaps the most widely used and practical, application of energy concepts, namely, the determination of loads and/or deformations from strain energy considerations.

13-3 / Work

Whenever external forces are applied to structural members such as beams shafts, etc., these members deform and the forces do work, the work of each force being equal to the product of the force and the displacement parallel to the force. In most deformation processes the applied forces vary from zero to their final values as the deformation progresses. Therefore, the work of each force would have to be measured continuously during the process in order to know the total amount of work done. Theoretically, if we knew how each force varied with the deformation, the work could be computed as

$$W_{\text{total}} = \sum_{i=1}^{n} \int_{0}^{s_i} P_i \, ds_i$$

where n is the number of forces on the body, P_i is the ith force, and s_i is the displacement of the ith force parallel to that force. In general, the obtainment of all the necessary information to compute the work would be a difficult and tedious task.

Recall from the discussion of section 4-3 that the essential characteristic of a *linearly elastic body* is that the deformation is directly proportional to the force producing that deformation, or vice versa. That is

$$P = ke$$

For this situation, the work done by a force as it is applied to an elastic body is

$$W = \int_{0}^{e} P \, de = \int_{0}^{e} ke \, de = k \frac{e^2}{2} = \frac{1}{2} Pe$$

or   (13-1)

$$= \int_{0}^{e} P \, de = \int_{0}^{P} P \frac{dP}{k} = \frac{1}{2} \frac{P^2}{k} = \frac{1}{2} Pe$$

In either expression we see that the work depends only on the proportionality constant k and the *final* value of the load P and/or the deformation e. Thus we need not know the entire history of the loading and deformation process in order to compute the work done by the force. The work done by forces producing linear elastic deformations can be computed by measuring the forces and the final deformation of the body at the points of application *without regard to the order in which the forces are applied.*

EXAMPLE 13-1

Compute the work done by a torque T applied to a solid circular shaft of length L and constant radius r loaded in pure torsion. Assume elastic behavior.

Solution:

The work done by a couple C is given by the integral

$$W = \int_0^\theta C \, d\theta$$

when θ is the angular displacement of the couple. From Chapter 6 we recall that the torque-angular deformation relationship for an elastic shaft is

$$\theta = \frac{TL}{JG} \quad \text{or} \quad T = \frac{JG\theta}{L}$$

Hence

$$W = \int_0^\theta \frac{JG\theta}{L} \, d\theta = \frac{JG}{L} \frac{1}{2} \theta^2 = \frac{1}{2} T\theta$$

or

$$W = \int_0^T T \frac{L \, dT}{JG} = \frac{L}{JG} \frac{1}{2} T^2 = \frac{1}{2} T\theta$$

EXAMPLE 13-2

A two-section bar shown in Fig. 13-1(a) is attached at the top to a rigid fixture. A load of 400 N is applied as shown in Fig. 13-1(b) causing section A to stretch 2.5 mm. Then a second 400-N load is applied as shown in Fig. 13-1(c) causing section A to stretch an additional 2.5 mm and section B to stretch 6 mm. What was the work done during this entire process if the bar is elastic?

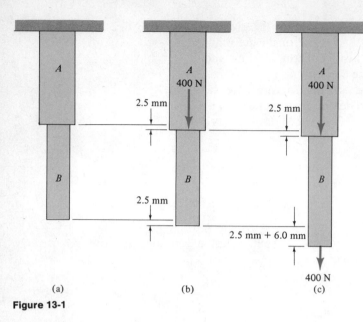

(a)   (b)   (c)

Figure 13-1

Solution:

During the application of the first force the work is by Eq. 13-1

$$W = \frac{1}{2} Pe = \frac{1}{2} (400) \left(\frac{2.5}{1000} \right) = 0.5 \text{ N} \cdot \text{m}$$

During the application of the second force the first force is constant while undergoing a displacement of an additional 2.5 mm. Hence the work of the first force is an additional

$$W = Ps = (400) \left(\frac{2.5}{1000} \right) = 1 \text{ N} \cdot \text{m}$$

while the work of the second force is

$$W = \frac{1}{2} Pe = \frac{1}{2} (400) \left(\frac{2.5 + 6.0}{1000} \right) = 1.7 \text{ N} \cdot \text{m}$$

since this second force produces a total elongation of 2.5 + 6.0 at the point of application. Thus the total work is

$$0.5 + 1.0 + 1.7 = 3.2 \text{ N} \cdot \text{m}$$

Alternative solution:

The total deformation of the point of application of the first force is

$$(2.5 + 2.5) = 5.0$$

and that of the point of application of the second force is

$$(2.5) + (2.5 + 6.0) = 11.0$$

Thus the total work is by Eq. 13-1

$$W = \frac{1}{2} P_1 e_1 + \frac{1}{2} P_2 e_2 = \frac{1}{2} (400) \left(\frac{5}{1000} \right) + \frac{1}{2} (400) \left(\frac{11}{1000} \right)$$

$$= 1.0 + 2.2 = 3.2 \text{ N} \cdot \text{m}$$

Restudy this example to be sure that you understand the ideas behind both procedures.

13-4 / Elastic strain energy of structural members

In Chapter 12 it was shown that the *work* done on a unit cube of *elastic* material by a general triaxial system of stresses during a quasi-static deformation process was given by the elastic strain energy density (Eq. 12-8)

$$U = \frac{1}{2E} (\sigma_x^2 + \sigma_y^2 + \sigma_z^2) - \frac{\mu}{E} (\sigma_x\sigma_y + \sigma_x\sigma_z + \sigma_y\sigma_z) + \frac{1}{2G} (\tau_{xy}^2 + \tau_{xz}^2 + \tau_{yz}^2)$$

$$[12\text{-}8]$$

Therefore, the total work done on an elastic structural member during a quasi-static process can be obtained by integrating the elastic strain energy per unit volume over the entire volume of the member.

$$W = \int_{\text{volume}} U \, dV = \text{total elastic strain energy of the body} \qquad (13\text{-}2)$$

For the more simple structural members, the stresses can be expressed in terms of the applied loads and the resulting integrals for W easily evaluated. Thus for such members Eq. 13-2 offers us a convenient method for determining the work done on an elastic member during quasi-static processes.

EXAMPLE 13-3

Using Eq. 13-2, rework Example 13-1.

Solution:

For a solid circular shaft subjected to pure torsion the shear stress is given by Eq. 6-3a

$$\tau_\rho = \frac{T\rho}{J}$$

and all other stresses are zero.

Hence from Eqs. 12-8 and 13-2

$$W = \int_V \frac{1}{2G} (\tau^2) \, dV = \frac{1}{2G} \int_V \frac{T^2 \rho^2}{J^2} \, dV = \frac{T^2}{2J^2 G} \int_V \rho^2 \, dV$$

Now an elemental volume dV can be written as

$$dV = da \, dL$$

when da is an elemental area and dL an elemental length parallel to the axis of the shaft. Then

$$\int_V \rho^2 \, dV = \int_{\text{length}} dL \int_{\text{area}} \rho^2 \, da$$

But

$$\int_{\text{area}} \rho^2 \, da = J \text{ polar moment of inertia}$$

Thus

$$\int_V \rho^2 \, dV = \int_{\text{length}} J \, dL = JL$$

provided the shaft is not tapered. Hence

$$W = \frac{T^2}{2J^2 G} JL = \frac{T^2 L}{2JG} = \frac{1}{2} T\theta \tag{13-3}$$

which is precisely that of Example 13-1, as expected.

In a manner similar to that employed in the previous examples, you can show (Probs. 13-1 and 13-2) that the work done during an elastic quasi-static process on an axially loaded member is given by

$$W = \frac{P^2 L}{2AE} \tag{13-4}$$

and for a straight beam loaded in *pure bending*

$$W = \frac{1}{2} \frac{M^2 L}{EI} \tag{13-5}$$

where M is the bending moment and L the length of the beam.

Note: In an arbitrary deformation process, the work done by external forces contributes to changes of the kinetic energy of the body of material as well as to the production of strain energy. However, for processes in which little or no changes in kinetic energy occur, the entire work of the external forces can be assumed to be equal to the stress work or strain energy as we have done here and in Chapter 12.

PROBLEMS

13-1 Carry out by two methods the details in the derivation of Eq. 13-4 for a straight elastic member loaded in pure tension.

13-2 Carry out by two methods the details in the derivation of Eq. 13-5 for a straight elastic beam loaded in pure bending.

13-3 Two elastic cylindrical bars of the same material and shown in the illustration are to absorb the same total amount of elastic strain energy delivered by uniaxial loads applied at their ends. Neglecting stress concentrations, compare the stresses in the two bars.

Problem 13-3

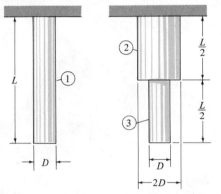

13-4 Two solid circular elastic shafts of the same material are to absorb the same amount of total elastic strain energy delivered by pure torques applied at their ends. Neglecting stress concentrations, compare the maximum shear stresses in the two shafts.

13-5 A 750-mm long piece of steel 10 mm thick has a width varying linearly from 50 mm at one end to 25 mm at the other end as shown. Determine the elastic

Problem 13-4

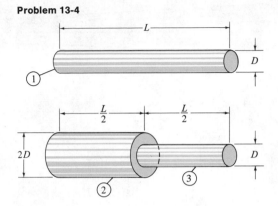

strain energy in the steel produced by an applied axial load P of 10,000 N.

13-6 Derive an expression for the total elastic strain energy of a solid homogeneous bar of constant cross-sectional area A, length L, and modulus E due to its own weight W when suspended vertically from one end.

13-7 What is the total elastic strain energy for a straight, circular, hollow steel shaft with outside diameter 40 mm, inside diameter 20 mm, and length 2 m subjected to a pure torque of 100 N · m?

13-8 A 40-in. long solid circular steel shaft is tapered linearly from a 2-in. diameter at one end to a 1-in. diameter at the other end as shown. Determine the elastic strain energy in the steel produced by an applied torque T of 800 in.-lb.

13-9 The elastic circular shaft in the illustration carries a uniformly distributed torque of T in.-lb/in. along its entire length. Derive an expression for the total elastic strain energy.

Problem 13-5

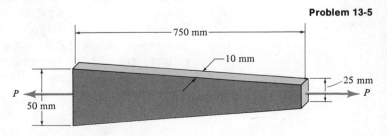

Problem 13-8

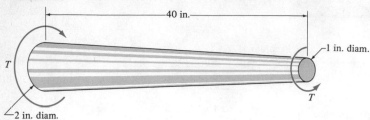

40 in.

1 in. diam.

T

T

2 in. diam.

Problem 13-9

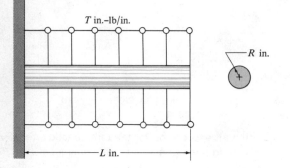

T in.–lb/in.

R in.

L in.

13-10   A simply supported straight beam carries a concentrated load P at the mid span. Determine the total elastic strain energy in the beam in terms of P, E, I, and L by integrating the strain energy per unit volume over the entire volume of the beam.

13-11   Same as Prob. 13-10 except that the beam is cantilevered with the load P applied at its free end.

13-5 / Castigliano's theorem

We stated in the introduction of this chapter that energy concepts can be used to determine loads and deformations of structural members. In the previous sections we have seen some samples of how the total elastic strain energy of a member can be related to the applied forces or vice versa. We now wish to make a more general study of the relation between the applied loads and the elastic strain energy.

For convenience, consider a uniaxially loaded member and suppose we record the force deformation data as it is loaded up to some load P as shown in Fig. 13-2. As we know, the work done by the force P would be the area *under* the load-deformation curve which we call W. The shaded area *above* the curve we shall call the complementary work W_c. Although no simple physical interpretation can be given to this quantity, we find it convenient to use in our discussions. Suppose now that P is given a very slight increase δP which results in a slight increase in deformation δe and accompanying increases in W and W_c of δW and δW_c, respectively, so that we may neglect second-order terms and write

$$\delta W = P\,\delta e \qquad \delta W_c = e\,\delta P$$

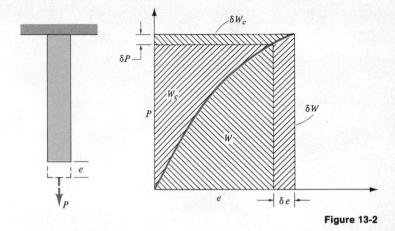

Figure 13-2

or
$$P = \frac{\delta W}{\delta e} \qquad e = \frac{\delta W_c}{\delta P} \qquad (13\text{-}6)$$

We can interpret these equations as stating that the rate of change of the work with respect to the deformation is equal to the load and that the rate of change of the complementary work with respect to the load is equal to the deformation. If the member had been *linearly elastic so that the load-deformation curve was a straight line*, then $W = W_c$ and we would have

$$\delta W = \delta W_c$$

and

$$e = \frac{\delta W}{\delta P} \qquad (13\text{-}7)$$

which is a simple example of Castigliano's theorem. A more general statement of this important and widely used theorem may be paraphrased as follows: *For an elastic body under equilibrated loads, the rate of change of work (or total elastic strain energy) with respect to any statically applied force P gives the deformation of the point of application of the force in the direction of the force.* We shall not attempt to derive the theorem in its full generality but shall demonstrate its derivation to point out its significant features.

Consider a simply supported beam with two concentrated loads P_1 and P_2 as shown in Fig. 13-3(b). We assume *elastic* behavior so that deflection of the beam is considerably exaggerated. From Eq. 13-1, the work done by these forces in producing the elastic deflection of the beam is

$$W(P_1, P_2) = \tfrac{1}{2}P_1 y_1 + \tfrac{1}{2}P_2 y_2 \qquad (13\text{-}8)$$

Now, apply an additional force at x_1, ΔP_1, which causes a change in

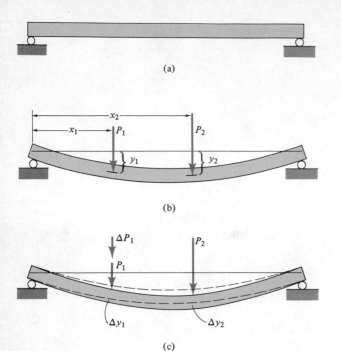

(a)

(b)

(c)

Figure 13-3

deflection Δy_1 at x_1 and Δy_2 at x_2 as indicated in Fig. 13-3(c). The work done during this change is

$$\Delta W = P_1 \Delta y_1 + P_2 \Delta y_2 + \tfrac{1}{2}\Delta P_1 \Delta y_1 \qquad (13\text{-}9)$$

since P_1 and P_2 were constant during the application of ΔP_1. The total work of the entire system is then

$$W = W(P_1, P_2) + \Delta W \qquad (13\text{-}10)$$

Suppose we had applied these same forces but in reverse order; that is, we apply ΔP_1 then P_1 and P_2. In this case the work during the application of ΔP_1 would be

$$\tfrac{1}{2}\Delta P_1 \Delta y_1 \qquad (13\text{-}8a)$$

while the work done during the application of P_1 and P_2 would be

$$\tfrac{1}{2}P_1 y_1 + \tfrac{1}{2}P_2 y_2 + \Delta P_1 y_1 \qquad (13\text{-}9a)$$

since ΔP_1 is constant during this application. According to our discussion in section 13-3, the work done on an elastic body is independent of the sequence of application of the forces and so the total work for the first sequence should be the same as that for the second sequence. Therefore, equating the sum of Eqs. 13-8 and 13-9 to the sum of Eqs. 13-8a and 13-9a, we obtain

$$P_1 \Delta y_1 + P_2 \Delta y_2 = \Delta P_1 y_1$$

and Eq. 13-9 becomes

$$\Delta W = \Delta P_1 y_1 + \tfrac{1}{2}\Delta P_1 \Delta y_1 \qquad (13\text{-}11)$$

Since the beam is linearly elastic $\Delta y_1 = k\,\Delta P_1$, and

$$\Delta W = \Delta P_1 y_1 + \tfrac{1}{2}k(\Delta P_1)^2$$

Dividing through by ΔP_1 and letting $\Delta P_1 \to 0$ we obtain

$$\frac{\partial W}{\partial P_1} = y_1 \qquad (13\text{-}12)$$

The partial notation must be employed since W, you recall, was dependent upon P_1 and P_2. Equation 13-12 is the mathematical equivalent of Castigliano's theorem stated earlier. In a similar manner, by considering couples C and their resulting angular deformations θ, we can obtain the result

$$\frac{\partial W}{\partial C} = \theta \qquad (13\text{-}13)$$

You must realize that in order to use Eq. 13-12 or 13-13, you must express the work (or total elastic strain energy) as a function of the applied loads. The practical usefulness of this theorem depends upon how easy or difficult it is to obtain the necessary differentiable expression for the work.

We now have a new method for computing the deformations at points of application of concentrated loads or concentrated couples. However, suppose we wish to determine the deformation at a point where no load is applied? We get around this dilemma by realizing that a zero load is the same as no load at all. That is, we can place a *fictitious load* at the point in question, then carry out the formal calculations as if that load were actually there, and finally *after differentiation* set this load equal to zero. We demonstrate these ideas in the following examples.

EXAMPLE 13-4

What is the maximum deflection of a straight, simply supported elastic beam loaded with a single load P at its mid span? (See Fig. 13-4.)

Figure 13-4

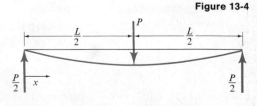

Solution:

From symmetry, the maximum deflection will occur directly under the load. Hence

$$y_{max} = \frac{\partial W}{\partial P} \tag{a}$$

For an elastic beam the work (strain energy) is given by

$$W = \int_{volume} U\, dV = \int_V \frac{\sigma^2}{2E} dV = \int_V \left(\frac{My}{I}\right)^2 \frac{dV}{2E}$$

$$= \int_{length} dx \int_{area} \frac{M^2 y^2\, da}{2EI^2} = \int_{length} \frac{M^2\, dx}{2EI}$$

Before this expression can be integrated, we must obtain an expression for the bending moment M in terms of the loads on the beam. In this case

$$M = \frac{P}{2} x \qquad 0 \le x \le \frac{L}{2}$$

$$= \frac{P}{2} x - P\left(x - \frac{L}{2}\right) \qquad \frac{L}{2} \le x \le L$$

Thus, since E and I are constant along the length of the beam

$$W = \frac{1}{2EI} \int_0^{L/2} \left(\frac{P}{2} x\right)^2 dx + \frac{1}{2EI} \int_{L/2}^L \left(\frac{-Px}{2} + \frac{PL}{2}\right)^2 dx \tag{b}$$

Expression (b) is representative of the type that is usually encountered in using energy methods to solve beam deflection problems. Since P is constant insofar as the variable of integration x is concerned we may, if we wish, perform the differentiation required in (a) *before* integrating (b). This procedure is often economical and saves some unnecessary integration operations. In this case

$$\frac{\partial W}{\partial P} = \frac{1}{2EI} \int_0^{L/2} \frac{P}{2} x^2\, dx + \frac{1}{2EI} \int_{L/2}^L 2P\left(\frac{-x}{2} + \frac{L}{2}\right)^2 dx$$

$$= \frac{1}{2EI} \frac{P}{6} \left(\frac{L}{2}\right)^3 + \frac{2P}{2EI} \frac{L^3}{96}$$

$$= \frac{PL^3}{48EI}$$

Note: Because of the symmetry in this problem, the strain energy in the left half of the beam will be one-half of the total strain energy. Thus, we could have written

$$W = 2 \int_0^{L/2} \frac{M^2}{2EI} dx = \frac{1}{EI} \int_0^{L/2} \left(\frac{P}{2} x\right)^2 dx$$

instead of (b), which would have simplified the subsequent differentiation and integration operations.

EXAMPLE 13-5

Assuming elastic behavior for the beam shown in Fig. 13-5(a), determine the deflection of the free end.

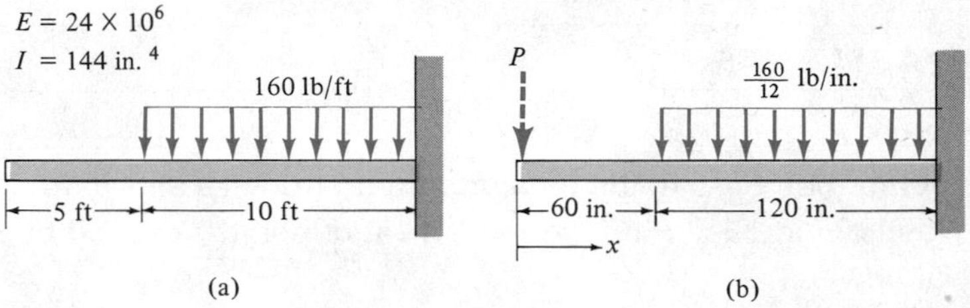

$E = 24 \times 10^6$
$I = 144$ in.^{4}
160 lb/ft
$\frac{160}{12}$ lb/in.
P

5 ft — 10 ft — 60 in. — 120 in.

x

(a)   (b)

Figure 13-5

Solution:

Although the free end has no load applied there, we shall put a fictitious load of P at the left end as shown in Fig. 13-5(b). Then the moment equations are

$$M = -Px \qquad 0 \leq x \leq 60$$

$$= -Px - \frac{160}{12}\frac{(x - 60)^2}{2} \qquad 60 \leq x \leq 180$$

From the previous example, the elastic strain energy for the beam is given by

$$W = \int_0^L \frac{M^2}{2EI}\, dx$$

$$= \frac{1}{2EI}\int_0^{60} (-Px)^2\, dx + \frac{1}{2EI}\int_{60}^{180}\left[-Px - \frac{160}{24}(x - 60)^2\right]^2 dx$$

Since we want the deflection at the left end

$$y_L = \frac{\partial W}{\partial P} = \frac{1}{2EI}\int_0^{60} 2Px^2\, dx + \frac{1}{2EI}\int_{60}^{180} 2\left[-Px - \frac{160}{24}(x - 60)^2\right](-x)\, dx$$

Since P is actually equal to zero, we need only evaluate the second term in the latter integral and thus

$$y_L = \frac{1}{EI}\frac{160}{24}\left[\frac{x^4}{4} - 40x^3 + (60)^2\frac{x^2}{2}\right]\Bigg|_{60}^{180}$$

$$= \frac{1}{(24)(10^6)(144)}\frac{160}{24}(400)(60)^3 = \frac{1}{6}\text{ in.}$$

Compare this example with Example 8-3.

In both of the previous examples singularity functions (sections 9-3 and 9-4) could have been employed in writing the moment equations, although in these examples there was no particular advantage in doing so.

We close this chapter with a final example illustrating how the energy method can be used in composite structures.

EXAMPLE 13-6

All the members in the pin-connected structure shown in Fig. 13-6 have the same cross section and elastic modulus. Find the movement of point D caused by the application of a horizontal force P at point D.

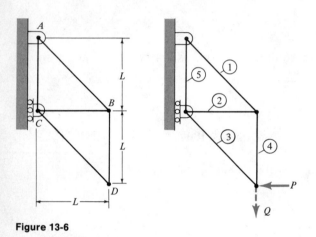

Figure 13-6

Solution:

Even though only a horizontal force is applied at D, chances are that point D will move vertically and horizontally. Therefore, we shall also apply a fictitious vertical force Q at D.

The total elastic strain energy in the structure will be the sum of the energy in each of the members. Since each member is an axially loaded one, by Eq. 13-4

$$W = \sum_{i=1}^{5} \frac{P_i^2 L_i}{2 A_i E_i}$$

Now

$$L_1 = \sqrt{2}L \qquad L_2 = L \qquad L_3 = \sqrt{2}L \qquad L_4 = L \qquad L_5 = L$$

By the method of joints (or sections)

$$P_1 = \sqrt{2}(P + Q) \text{ tension} \qquad P_2 = (P + Q) \text{ compression}$$

$$P_3 = \sqrt{2}P \text{ compression} \qquad P_4 = (P + Q) \text{ tension}$$

$$P_5 = P \text{ compression}$$

Therefore

$$W = \frac{1}{2AE} [2(P + Q)^2 \sqrt{2}L + (P + Q)^2 L + 2P^2 \sqrt{2}L + (P + Q)^2 L + P^2 L]$$

From Castigliano's theorem, the horizontal displacement will be

$$\delta_H = \frac{\partial W}{\partial P} = \frac{1}{AE} [2(P + Q)\sqrt{2}L + (P + Q)L + 2P\sqrt{2}L + (P + Q)L + PL]$$

For $Q = 0$,

$$\delta_H = \frac{PL}{AE} (2\sqrt{2} + 1 + 2\sqrt{2} + 1 + 1) = 8.6 \frac{PL}{AE} \leftarrow$$

Similarly, the vertical displacement will be

$$\delta_V = \frac{\partial W}{\partial Q} = \frac{1}{AE} [2(P + Q)\sqrt{2}L + (P + Q)L + (P + Q)L]$$

For $Q = 0$,

$$\delta_V = \frac{PL}{AE} (2\sqrt{2} + 1 + 1) = 4.8 \frac{PL}{AE} \downarrow$$

PROBLEMS

13-12 Using a fictitious axial load at the lower end of the bar of Prob. 13-6, determine the total elastic deformation of the bar due to its own weight.

13-13 Using a fictitious torque at the right end of the shaft of Prob. 13-9, determine the total elastic angular twist of the shaft due to the distributed torque.

13-14 A cantilevered beam of length L, second moment I, and modulus E carries a uniformly distributed load of w lb/ft over its entire length. Using a fictitious load at its free end, determine the deflection of the free end due to the distributed load.

13-15 A simply supported elastic beam of length L, second moment I, and modulus E is loaded with a uniformly distributed load of w lb/in. along its entire length. Derive an expression for the maximum deflection of the beam using a fictitious load at its midspan.

13-16 Work Prob. 8-45 using the energy methods of this chapter.

13-17 Work Prob. 8-46 using the energy methods of this chapter.

13-18 Work Prob. 8-47 using the energy methods of this chapter.

13-19 Work Prob. 8-55 using the energy methods of this chapter.

13-20 Work Prob. 8-56 using the energy methods of this chapter.

13-21 Work Prob. 8-57 using the energy methods of this chapter.

13-22 Work Prob. 8-58 using the energy methods of this chapter.

13-23 Work Prob. 6-56 using the energy methods of this chapter.

13-24 and 13-25 Using Castigliano's theorem, determine the wall reactions on the beam shown in the illustration.

Problem 13-24

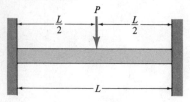

Problem 13-25

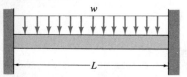

13-26 Work Prob. 8-67 using the energy methods of this chapter.

13-27 Work Prob. 8-68 using the energy methods of this chapter.

13-28 Work Prob. 8-69 using the energy methods of this chapter.

13-29 Work Prob. 8-71 using the energy methods of this chapter.

13-30 to 13-35 The pin-connected trusses shown in the accompanying illustrations are made of mild structural steel. The cross-sectional area of each member is 2 in.^{2} Determine the displacement of point B in each case.

Problem 13-30

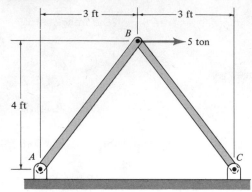

Problem 13-31

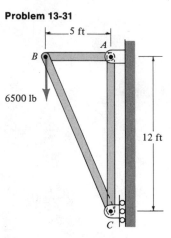

Problem 13-32

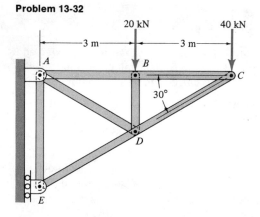

Problem 13-33

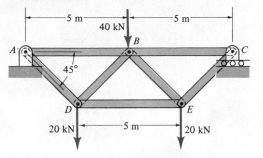

Problem 13-34

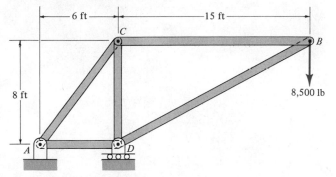

Problem 13-35

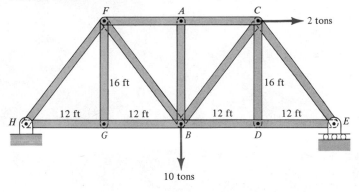

Chapter 14
Nonstatic loads, strain concentrations, and time-dependent properties

14-1 / Objectives

Upon completion of this chapter you will be able to:

1   Describe the causes and effects of stress concentrations.
2   Use stress concentration factors to compute stresses in, or allowable loads on, members with abrupt changes in cross section.
3   Predict when, where, and how a fatigue failure is likely to occur.
4   Design and check members subject to repeated or fluctuating loads.
5   Describe the effects of creep and relaxation.
6   Include the effects of creep and relaxation in design calculations.
7   Use work and energy concepts to determine the effects of low-velocity impact loadings.

14-2 / Introduction

Thus far we have been concerned primarily with the response of engineering members to static loads. We have not considered in any detail the influence of time effects in loading, nor have we considered abrupt geometry changes in members causing concentrations of strain and stress. Concentrations resulting from abrupt geometry changes and material imperfections become increasingly important under repeated loading conditions (fatigue). It is important and appropriate to discuss these two topics, concentrations and repeated loading, in the same chapter so that their relations with each other can be emphasized. These and several other topics, including impact and time-dependent properties, are discussed in the following sections.

14-3 / Stress concentrations

The "flow" of force through a member will be disrupted by an abrupt change in the geometry of the member, and a concentration of the lines of force will result. For example, the lines of force in a simple plate loaded uniaxially are shown in Fig. 14-1(a), and Fig. 14-1(b) suggests how the lines of force become concentrated around a hole in the plate.

The lengths of the vectors in Fig. 14-1(c) and (d) indicate the magnitudes of the stresses at various points along plane *A-A*. In (c) the stress is uniform and undisturbed in the portion away from the end effects of the loads. In (d), where a hole is introduced, the strains and therefore the stresses are more concentrated at the hole. The peak stress on plane *A-A* is adjacent to the hole and has a much larger value than the stress in a cross section without a hole. The peak stresses in structures at points of abrupt changes in geometry are extremely important in the design of members of any material subjected to repeated loads and in the design of members made of brittle material subjected to loading of any type.

It is often difficult to determine the maximum value of the stress at a concentration by theoretical analysis, but designers can usually determine the maximum stress by applying a factor to an average stress obtained from the elementary load-stress relationship. Such factors are called *stress concentration factors* and have been determined experimentally for many common geometry changes, such as holes, grooves, notches, and fillets in plates and shafts subjected to axial, bending, and torsion stresses.

The stress concentration factor, denoted here by k, is defined as the ratio of the maximum stress to the average stress. That is,

$$k = \frac{\sigma_{\max}}{\sigma_{\mathrm{av}}}$$

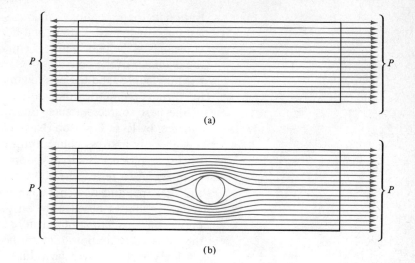

(a)

(b)

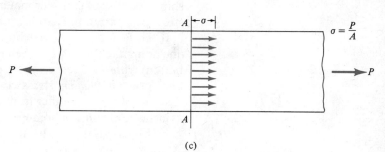

$$\sigma = \frac{P}{A}$$

(c)

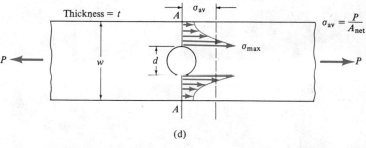

$$\sigma_{av} = \frac{P}{A_{net}}$$

(d)

Figure 14-1

The average stress is usually determined by using the net area of the cross section. For example, in Fig. 14-1(d),

$$\sigma_{max} = k \frac{P}{A_{net}}$$

where A_{net} is $(w - d)t$, t being the plate thickness. The stress concentration factor, except in very unusual instances, is greater than unity. For a uniaxially loaded plate with a hole, the factor approaches a theoretical value of 3 as the plate width becomes extremely large in comparison with the diameter of the hole.

Many curves and tables showing values of stress concentration factors for various geometries and loadings are given in technical references. One of the best sources of such information is *Stress Concentration Design Factors*, by R. E. Peterson (New York: Wiley, 1953).

Generally a stress concentration factor is determined experimentally by using models (such as photoelastic models discussed in the next chapter). As a result, the factor is based only on geometrical considerations (shape) and does not take into account the effect of the material. A factor that is based only on geometry is called a *theoretical* stress concentration factor and is denoted by k_t. Factors obtained by comparing the stress distributions determined from the mathematical theory of elasticity with the average values determined from elementary relationships also do not consider material effects and are considered theoretical stress concentration factors.

If the factor is obtained by comparing the actual stress distribution (including the effects of the behavior of the specific material on the distribution) with the average stress determined from elementary theory, it is known as an *effective* stress concentration factor and is denoted by k_e. Effective factors can differ from theoretical factors because of such things as inelastic action, nonhomogeneity due to large grain size, and anisotropy.

In Fig. 14-2(a) and (b) are shown the effects of a typical geometry

Figure 14-2

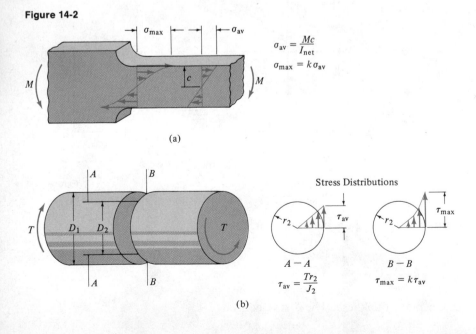

$$\sigma_{av} = \frac{Mc}{I_{net}}$$

$$\sigma_{max} = k\sigma_{av}$$

(a)

Stress Distributions

$$\tau_{av} = \frac{Tr_2}{J_2}$$

$$\tau_{max} = k\tau_{av}$$

$A - A$

$B - B$

(b)

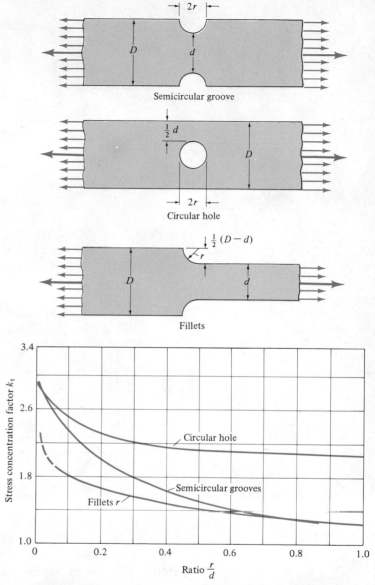

Figure 14-3 Theoretical stress concentration factors for axial loading obtained by photoelasticity. (After Frocht.)

change on the stress distributions in bending and torsion members, respectively. The curves in Fig. 14-3 and Fig. 14-4 give some values of the theoretical stress concentration factors for plates with axial and bending loads, respectively, and show their variation with various dimension parameters. The values were determined by photoelastic methods.

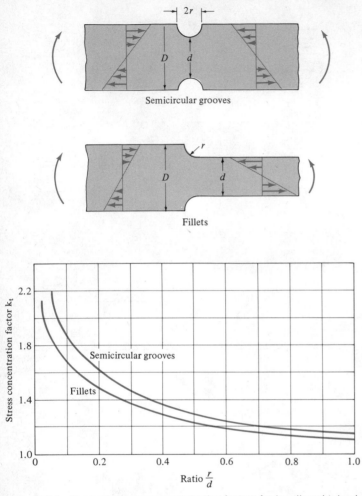

Figure 14-4 Theoretical stress concentration factors for bending obtained by photoelasticity. (After Frocht.)

For a member made of a very *brittle* material, the "peaked" stress distribution will be present at any load level, and a rupture will ordinarily start at the point of maximum stress when that stress reaches the value of the ultimate stress. For this reason, abrupt geometry changes that cause stress concentrations must always be carefully considered, and avoided if possible, in the design of members of brittle materials.

In a member made of a *ductile* material the "peaked" stress distribution, or stress concentration, will be reduced as yielding occurs. The yielding will start at the point of maximum stress. If the material has a large inelastic range, as do several common structural materials (for example, mild steel), the material can yield in the region of the stress concentration

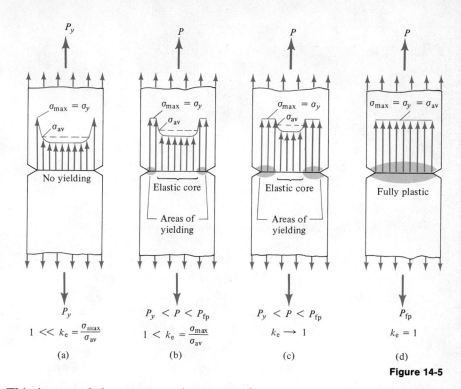

Figure 14-5

without danger of rupture. This is one of the greatest advantages of ductility in a material. In the designing of a member made of a ductile material for static loads, stress concentration factors are seldom considered. If, however, repeated loads are expected, even a ductile material may fail as a result of the propagation of cracks which start at concentrations of stress. Repeated loads are discussed in more detail in the next section.

Figure 14-5 illustrates the reduction of the stress concentration factor and the redistribution of stress for a notched plate of a very ductile material. It is assumed that the material has a flat-topped stress-strain diagram. Notice the progression of yielding and how it is accompanied by a redistribution of stress to the extent that the stress finally becomes uniform over the cross section of the plate. The effect of the abrupt change in geometry is nullified and the effective stress concentration factor is reduced to unity.

EXAMPLE 14-1

The member in Fig. 14-6 is made from a brittle ceramic whose stress-strain relationship is linear to rupture at 8000 psi. Using a factor of safety of 3, find the allowable load.

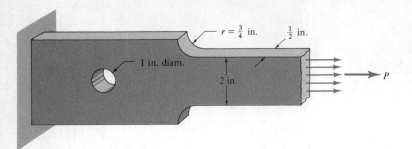

Figure 14-6

Solution:

For a brittle material the stress concentrations must be considered. For the section at the hole,

$$\sigma_{max} = k_t\sigma_{av} = k_t \frac{P_f}{A_{net}}$$

The failure load is $P_f = NP_a$, where P_a is the allowable load and N is the factor of safety. Therefore

$$P_a = \frac{\sigma_{max}A_{net}}{Nk_t}$$

Referring to Fig. 14-3, where $D = 3\frac{1}{2}$ in., $d = 2\frac{1}{2}$ in., $r = \frac{1}{2}$ in., and $r/d = 0.20$, we get $k_t = 2.3$. Therefore, the allowable load based on the stress at the hole is

$$P_a = \frac{(8000)(2\frac{1}{2})(\frac{1}{2})}{(3)(2.3)} = 1449 \text{ lb}$$

For the section at the fillet, $r/d = 0.375$. From Fig. 14-3, $k_t = 1.5$. Hence, the allowable load based on the stress here is

$$P_a = \frac{(8000)(2)(\frac{1}{2})}{(3)(1.5)} = 1778 \text{ lb}$$

The allowable load for the member is the smaller of the two, or 1449 lb.

PROBLEMS

14-1 Find the maximum stress that occurs in an axially loaded plate whose cross section is $3\frac{1}{2}$ by $\frac{1}{4}$ in. with a hole $\frac{7}{8}$ in. in diameter in the center, if the plate is made of a linearly elastic material and the axial load is 12 kips.

14-2 Find the maximum stress that occurs in an axially loaded plate whose cross section is 80 by 6 mm with a hole 20 mm in diameter in the center, if the plate

is made of a linearly elastic material and the axial load is 50,000 N.

14-3 Find the maximum stress that occurs in an axially loaded plate whose cross section is 100 by 10 mm with semicircular grooves opposite each other cut across each of the thinner edges of the plate. The radius of each groove is 12 mm. The plate is made of a linearly elastic material and the axial load is 25,000 N.

Problem 14-5

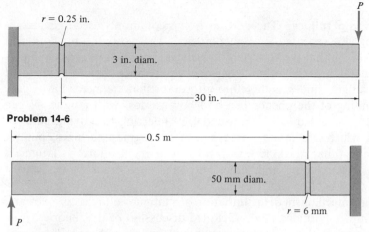

Problem 14-6

14-4 Find the maximum stress that occurs in an axially loaded plate whose cross section is 4 by $\frac{3}{8}$ in. with semicircular grooves opposite each other cut across each of the thinner edges of the plate. The radius of each groove is $\frac{1}{2}$ in. The plate is made of a linearly elastic material and the axial load is 6.25 kips.

14-5 How much load can be applied to the end of the cantilever beam with a circular cross section shown in the illustration, if a flexural stress of 18,000 psi must not be exceeded? Assume elastic action.

14-6 Find the load that can be applied to the end of the cantilever beam with a circular cross section shown in the illustration, if a flexural stress of 125 MPa must not be exceeded? Assume elastic action.

14-7 A long plate whose cross section is 1.5 by $\frac{3}{16}$ in. is loaded with an axial static load of 3500 lb. A hole

$\frac{1}{4}$ in. in diameter is drilled in the center of the plate. (a) Find the maximum stress in the plate if the material is mild steel with a yield-point stress of 30,000 psi and an ultimate stress of 65,000 psi. (b) Find the maximum stress if the plate is made from a brittle material with a linear stress-strain curve to its ultimate stress of 40,000 psi.

14-8 The 120 by 30 mm bar shown in the illustration is made from a brittle material with an ultimate stress of 60×10^6 N/m². How narrow can it be cut down (find d) and still carry the 40,000-N load? Assume a factor of safety of 2.4 based on failure by rupture. (*Hint:* Use trial-and-error solution.)

14-9 Where does the maximum stress occur, and what is its magnitude, for the beam shown in the illustration? Assume elastic action.

Problem 14-8

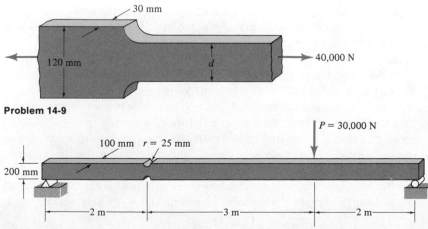

Problem 14-9

14-4 / Fatigue

Description of failure The word *fatigue* is commonly used in engineering to describe repeated-load phenomena. The word is a misnomer in the lay sense, in that the materials do not get "tired." Upwards of 75 percent of the fractures that occur in engineering members are due to repeated loads or fatigue, and it is therefore important that engineers have some understanding of the phenomena and how to deal with fatigue. When we speak of repeated loads we usually are referring to moving parts or members which are found, for example, in engines, turbines, pumps, motors, and other machinery with moving parts. Failure by fatigue is a progressive cracking and, unless detected, this cracking leads to a rupture which is often catastrophic.

The exact mechanism of the initiation of a fatigue failure is complex and is not completely understood. Detailed discussion of the theories of its initiation and initial propagation will not be given here. It is more important at this stage to be able to determine where, when, and under what conditions such a fatigue crack is likely to start. If a repeated load is large enough to cause a fatigue crack, the crack will start at a point of maximum stress. This maximum stress is usually due to a stress concentration (often called a stress raiser). Stress concentrations can occur in the interior of a member as a result of the inclusion of foreign matter or voids in the material. They can occur on the exterior surface of the member because of scratches, rust pits, machining marks, or the more obvious sharp corners or other abrupt geometry changes in the designs.

If a stress raiser occurs in a region of high overall stress (for example, at a section of maximum bending moment) and if the effect of the stress raiser is superimposed on this already high stress, then the chances of a fatigue crack developing under repeated loading are greatly increased. Since a fatigue crack or failure is undesirable, it is important for the engineer to take great care to eliminate or reduce the possibility of conditions that might lead to the initiation of a fatigue crack. High quality control may reduce chances of interior imperfections. Polishing the surface in critical areas may be necessary. Careful designs can also reduce stress concentrations. The idea is not to give a fatigue crack a place to start!

In the previous sections we mentioned that stress concentrations are sometimes ignored in ductile materials subjected to *static* loads. Under repeated loading ductile materials, as well as brittle materials, are susceptible to fatigue failures. In a fatigue failure of a ductile material there is no large amount of plastic deformation, or "necking," as in a static-load rupture of the same material. The fatigue failure of a ductile material appears somewhat as would the failure in a brittle material.

After a fatigue crack is initiated at some microscopic or macroscopic stress raiser, the crack itself acts as an additional stress raiser. The crack grows with each repetition of the load until the effective cross section is reduced to such an extent that the remaining portion will fail with the next application of the load. It may take many thousands, or even millions, of stress repetitions for a fatigue crack to grow to such an extent that it causes rupture. The required number of repetitions depends on the magnitude of the load (and stress) and other related factors.

The appearance of most of the fatigue-ruptured surface of a metal is rather rough and crystalline, and it is often erroneously thought by laymen that the metal has crystallized in the process of breaking. Since the solid metal is crystalline, and has been since it was transformed from a liquid, the appearance is perfectly understandable. What is interesting about the appearance is the fact that you can usually detect the point of initiation of the crack. The progression of the crack from this point causes the appearance of rings about the point somewhat like growth rings in timber. This portion of the fracture is rather smooth because of the rubbing action in the development of the crack.

The *S-N* curve   Now that we have some idea where and how a fatigue failure might occur, we must try to determine *when* it is likely to occur. Since susceptibility of a member to fatigue failure is greatly influenced by the material from which it is made, we must depend on experimental tests of the material in order to design against failure by fatigue. One common test used to evaluate fatigue properties of a material is a rotating-beam test, in which the number of completely reversed cycles of bending stress required to cause failure (this is called the fatigue life) is measured at various stress levels. By a completely reversed cycle of stress we mean that in one complete cycle the stress goes from some maximum tensile stress, to zero, to a maximum compressive stress of the same magnitude as the maximum tensile stress, and then back to zero and on to the original tensile stress again. These test data are usually plotted on semilog paper, and the resulting plot is referred to as an *S-N* curve. Some typical curves are shown in Fig. 14-7.

An individual test in which the specimen is fractured is required for each experimental point on an *S-N* curve. The inherent nature of fatigue tests gives rise to a great deal of scatter in the data. That is, if several specimens that have been carefully machined and polished (as all fatigue-test specimens should be) are tested at the same stress level, it is not unusual to have a variation of 10 to 20 percent in their fatigue life measured in terms of the number of loading cycles at which the specimen ruptures. It therefore requires quite a few tests to correctly identify an *S-N* curve for a material.

The property called *fatigue strength* refers to the stress level that can be applied a specific number of times before failure. For any certain life or number of stress repetitions, there is a corresponding fatigue strength. This strength can be taken from the *S-N* curve.

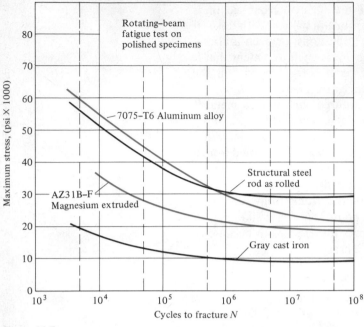

Figure 14-7

Endurance limit As specimens are tested at lower stress levels, more cycles or repetitions of this stress are required to cause failure. Notice that each *S-N* curve of Fig. 14-7 has an approximately horizontal line as a lower limit. When the stress level for a specimen is at or below this limit, which is called the *endurance limit* or *fatigue limit*, the specimen does not fail. The machine is stopped and the unbroken specimen is removed usually after about 10^8 (one hundred million) cycles. Such a test is called a "run out," and the corresponding point is usually plotted on the diagram with an arrow extending from it to the right. The endurance limit for a material is defined as the stress below which an indefinite number of stress repetitions may be applied without a fatigue failure.

It is important for the designer to know what type of fatigue test was used to determine the fatigue properties of a material. Although data from rotating-beam fatigue tests are perhaps the most common, other types of fatigue testing are in common use. Some of them involve bending of specimens of thin plates, uniaxial repeated loading, and torsion loading.

The endurance limit for most engineering materials is less than the yield strength, elastic limit, or proportional limit. It is important for you to realize that a member can fracture when loaded at an average stress level in the elastic range if the stress is high enough and is repeated enough times. It has been proved that a member is damaged to some degree after very few repetitions of stress. This damage is permanent and remains after the member is unloaded and left stressfree. If at any future time the member is again stressed repeatedly, the damage progresses from where it left off. Hence, we see that stress history is important in materials. Generally, tensile stresses are more damaging in fatigue than are compressive stresses.

Effective stress concentration factors for use in the design of members subjected to repeated loads are commonly determined from fatigue tests. To determine the factor in such a case, the endurance limit for a material with the stress concentration in question (for example, a grooved or notched specimen may be used) is divided into the endurance limit for the same material obtained from specimens free of the stress concentration.

Fatigue properties for materials are often determined at elevated temperatures and also in various corrosive environments. Temperature and/or environment can drastically influence the fatigue properties. Fatigue machines can be equipped for such special tests as these. Also, many fatigue machines are designed to subject the specimen to a varying deflection of some preset amplitude, whereas others impose a varying load with some preset maximum.

EXAMPLE 14-2

A part of the landing gear of a military airplane is made from 7075-T6 aluminum alloy. The landing gear is to be designed for a life of 10,000 landings. A factor of safety of 1.5 is to be used. The part is cylindrical and is subjected to a reversed bending moment of 800 ft-lb as the plane lands. Find the required diameter.

Solution:

From the *S-N* curve of Fig. 14-7 the fatigue strength for a life of 10,000 cycles is 54,000 psi. For bending we assume that the flexure formula is applicable. The 800 ft-lb, or (800)(12) in.-lb, is the working moment, and the failure moment is (800)(12)(1.5) in.-lb. Hence,

$$\sigma = \frac{M_f y}{I}$$

where $y = d/2$ and $I = \pi d^4/64$,

$$54,000 = \frac{(800)(12)(1.5)(d/2)}{\pi d^4/64}$$

$$d = 1.4 \text{ in.}$$

EXAMPLE 14-3

An engine part made from gray cast iron must be designed for an indefinite fatigue life. A factor of safety of 4 is needed because of the quality of the material, the consequences of a failure, and the fact that weight is not important. If the cross section is 12 mm square, how much bending moment may be applied to the part?

Solution:

Since the life is indefinite, the endurance-limit stress will be used. As obtained from Fig. 14-7, it is about 9000 psi or 63 MPa. Assuming that the flexure formula is applicable, we have

$$M_f = \frac{\sigma I}{y} \quad \text{and} \quad M_w = \frac{\sigma I}{Ny}$$

$$M_w = \frac{(63 \times 10^6)(\frac{1}{12})(0.012)^4}{(4)(0.006)} = 4.54 \text{ N·m}$$

Nonreversed cycles of loading Up to this point we have discussed only completely reversed cycles of stress in relation to fatigue failures. In many cases fluctuating loads do not approximate complete reversals. They do, however, in many instances approximate complete reversal superimposed on a uniform mean stress. This is often true where a load varies from some high limit to some low limit and both limits are tensile or both are compressive. To determine the allowable limits for such a range of stress, it is desirable to have experimental data from fatigue tests that cover the range of interest. In the absence of such test data, there are several approximate and conservative criteria that have been proposed for design under conditions of fluctuating stress.

Two approximate criteria are shown in Fig. 14-8(a), where the initial mean stress is plotted against the alternating stress. If the criterion of failure is yielding (slip), the so-called Soderberg straight line is assumed. Its limits are the endurance limit for zero mean stress and the yield point or yield strength for the case of no alternating load. If the criterion for failure is fracture, the so-called Goodman diagram is commonly used. It

Figure 14-8

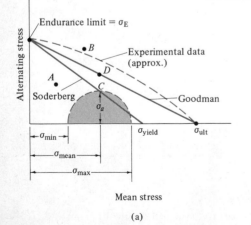

(a)

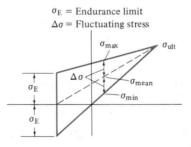

(b)

allows a higher value of alternating stress for a given value of mean stress, or vice versa.

For any combination of alternating stress and mean stress that plots as a point below the line, such as point A in Fig. 14-8(a), there would be no failure. For a combination of stresses that plots outside the line, as at point B, it is supposed that failure will occur.

For a mean stress σ_{mean} when the criterion of failure is yielding, the maximum alternating stress that can be superposed on the mean stress is σ_a, up to point C. If the fracture criterion is assumed, an alternating stress up to the value corresponding to point D can be superposed on the mean stress. Actual test data have verified the fact that the Soderberg and Goodman diagrams are on the safe or conservative side. For the fracture criterion, for example, points plotted from the experimental test data fall outside the straight Goodman line about as indicated by the dashed line. Thus, if adequate test data were available, we would not need the Soderberg or the Goodman diagram.

Another common way of presenting the Goodman criterion is shown in Fig. 14-8(b). The fluctuating stress $\Delta\sigma$ that can be superposed on a steady or mean stress is the distance between either sloping solid line and the sloping dashed line indicating the mean stress. If the mean stress is equal to the ultimate stress (right-hand end of diagram), no fluctuating stress may be added. If the mean stress is zero (left-hand end of diagram), the fluctuating stress may be equal to the endurance limit σ_E.

EXAMPLE 14-4

What amount of fluctuating bending stress would cause rupture if superposed on a maximum steady tensile stress of 8000 psi in a machine part of AZ31B-F magnesium whose ultimate static tensile strength is 35,000 psi and whose S-N curve is given in Fig. 14-7?

Solution:

Since the mode of failure is rupture, we shall use the Goodman diagram. The endurance limit from Fig. 14-7 is 18,000 psi, and the ultimate stress is given as 35,000 psi. The Goodman diagram is plotted in Fig. 14-9. The equation of the inclined line is

$$\sigma_a = 18,000 - \frac{18,000}{35,000}\sigma_m$$

For the mean stress of 8000 psi, the superimposed alternating stress that would cause rupture is

$$\sigma_a = 18,000 - \frac{18,000}{35,000}\,8000$$

$$= 13,880 \text{ psi}$$

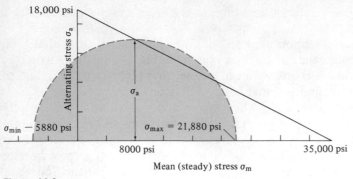

Figure 14-9

The maximum and minimum values of stress that would lead to rupture, based on a steady tensile stress of 8000 psi, are

$$\sigma_{max} = 8000 + 13{,}880 = 21{,}880 \text{ psi, tension}$$

$$\sigma_{min} = 8000 - 13{,}880 = 5880 \text{ psi, compression}$$

EXAMPLE 14-5

A machine part of AZ31B-F magnesium is subjected to a varying load that causes the maximum bending stress to vary from 8000 psi, tension, to 20,000 psi, tension. The ultimate tensile strength of the material is 35,000 psi. What is the ratio of the fluctuating stress that will cause rupture (failure stress) to the actual fluctuating stress on the member?

Solution:

Since the mode of failure is rupture, we shall use the Goodman diagram. The endurance limit from Fig. 14-7 is 18,000 psi, and the ultimate stress is given as 35,000 psi. From the Goodman diagram in Fig. 14-10, it is seen by similar triangles that the failure stress σ_f is

$$\frac{\sigma_f}{35{,}000 - \sigma_m} = \frac{18{,}000}{35{,}000}$$

where

$$\sigma_m = \frac{20{,}000 + 8000}{2} = 14{,}000 \text{ psi}$$

$$\sigma_f = 10{,}800 \text{ psi}$$

The actual fluctuating stress is

$$\frac{20{,}000 - 8000}{2} = 6000 \text{ psi}$$

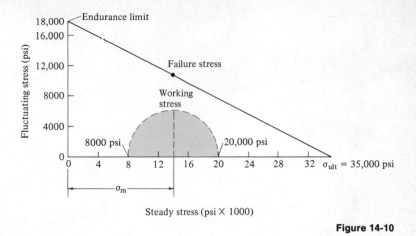

Figure 14-10

Hence, the ratio of failure stress to actual stress is 10,800/6000 = 1.77. This ratio might be interpreted as a margin of safety.

PROBLEMS

14-10 It is estimated that the service life of a machine part made from extruded magnesium alloy AZ31B-F is one million cycles. Determine a suitable working stress, using a factor of safety of 2.

14-11 What is the magnitude of a completely reversed load in newtons that can be applied to the free end of a cantilever beam of structural steel that is 1 m long and has a cross section 50 by 100 mm? It bends about the axis of least moment of inertia, and the load is to be applied an indefinite number of times. A factor of safety of 3 is required.

14-12 The beam shown in the illustration is made of 7075-T6 aluminum alloy. How many times can the load be completely reversed before you can expect a failure? Neglect stress concentrations at the connections.

Problem 14-12

14-13 At a certain engine speed a thin plate of 7075-T6 aluminum alloy in an airplane vibrates to the extent that a completely reversed stress of 25,000 psi is caused. After how many hours of flying time at this engine speed would you expect a failure, if the frequency of vibration is 20 Hz? Needless to say, the pilot will avoid this critical engine speed!

14-14 The vertical cantilever beam shown in the illustration is made of 7075-T6 aluminum alloy and is to have a life of 10^6 cycles of completely reversed bending under a load P of 40,000 N. What radius r of fillet should be used if a factor of safety of 1.5 is required? Assume that the theoretical stress concentration factors of Fig. 14-4 are applicable here.

14-15 What maximum value can P have if the material is gray cast iron? The load is to be repeated only 5000 times during the life of the part. A factor of safety of 2.5 is to be used. Refer to the illustration.

14-16 An automobile torsion bar is subjected to completely reversed cycles of torsional load. The torsional fatigue strength at the design life is 140 MPa, and the diameter of the shaft is 20 mm. Determine the torsional load, using a factor of safety of 2. Assume elastic behavior.

Problem 14-14

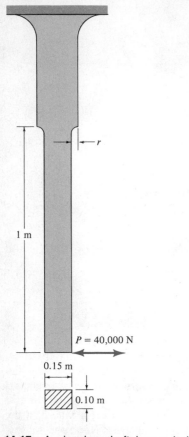

Problem 14-15

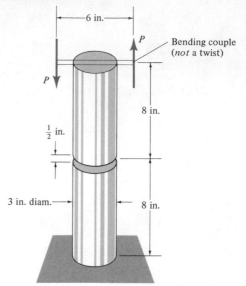

Bending couple
(*not* a twist)

14-17 A circular shaft has a hollow cross section, and the outside diameter is twice the inside diameter. It is subjected to a completely reversed bending moment of 12,000 in.-lb. If the material is gray cast iron and the shaft must have an indefinite fatigue life, determine the outside diameter. Use a factor of safety of 2.5.

14-18 Determine the factor of safety with respect to failure by yielding for a structural-steel machine part for which the stress varies repeatedly from 18,000 psi tension to 6000 psi compression.

14-19 A simple beam 16 ft long has a downward concentrated load at the center that varies from 500 to 4000 lb. The ultimate strength and the endurance limit of the material are 55,000 psi and 28,000 psi, respectively. If the depth of the beam is twice the width, what are the minimum dimensions needed to prevent rupture?

14-20 A machine part of AZ31B-F magnesium is subjected to a varying load that causes the maximum bending stress to vary from 55 MPa, tension, to 125 MPa, tension. The yield strength of the material is 155 MPa. What is the margin of safety based on failure by yielding?

14-21 A circular cantilever beam 45 mm in diameter is subjected to a transverse load (on its free end) that varies from 3200 N upward to 1200 N downward. The material is 7075-T6 aluminum alloy (see Fig. 14-7 and the Appendix for properties). Assuming a yielding mode of failure, determine the permissible length of the beam.

14-5 / Creep and relaxation

Creep   As defined in sections 3-9 and 4-3, creep is the behavior wherein the strain in a member undergoes changes while the loading remains essentially unchanged. Most engineering materials, particularly metals, do not creep appreciably at ordinary temperatures. Creep is a function of stress and time, as well as temperature, and a material is more likely to creep at a higher stress level or after a long period of time even if the temperature is not elevated. However, a great majority of creep problems in engineering materials arise from conditions of high temperatures. For example, the chemical engineer must often design containers or vessels for a process that takes place at very high pressures and temperatures. Mechanical engineers must consider creep problems in tubes of high-pressure boilers, in steam-turbine blades, in jet-engine blades, and in other applications where high stresses and high temperatures must be sustained for long periods of time.

The engineer must design the part so that the stress is low enough to limit creep to an *allowable* value. This maximum permissible stress is often called the *creep strength* or *creep limit*. The creep strength is defined as the highest stress that a material can stand for a specified time without excessive deformation. The *creep-rupture strength*, sometimes called the *rupture strength*, is defined as the highest stress a material can stand for a specified time without rupture. Creep properties are determined primarily from experimental tests on the materials of interest. For the most common type of creep test, a tensile specimen is loaded uniaxially with a dead load (constant stress) at a constant temperature while the data for strain versus time are recorded. The general shape of the creep curve for many materials is illustrated in Fig. 14-11.

Creep behavior for most materials exhibits the three stages of creep shown in the curve of Fig. 14-11 (compare with Fig. 4-3). After the "instantaneous" strain at the time of application of the load, the *rate* of strain in most cases (depending on test conditions) gradually decreases to

Figure 14-11

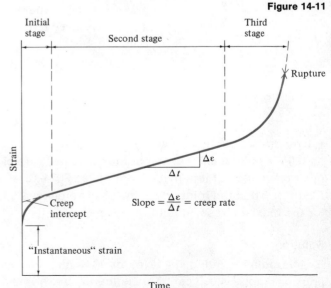

Creep curve (constant stress)

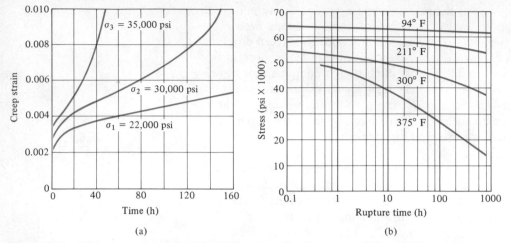

Figure 14-12 Creep curves for Alclad 2024-T3 aluminum alloy sheet (a) at 400°F and (b) at a thickness of 0.040 in. (After Flanagan, Tedsen, and Dorn.)

a more or less constant value. In this second stage the material is almost entirely viscous. The constant slope, in this stage, defined as the creep rate, is the minimum strain rate encountered in the test. In the third stage, called the *tertiary creep* range, the strain rate increases abruptly primarily because of the necking of the specimen, and rupture follows. In Fig. 14-12(a) is shown a family of actual creep curves for a constant temperature and three stress levels. In Fig. 14-12(b) is shown another common way of presenting creep data. Here the time to rupture is plotted against stress for various temperatures. Notice that only one creep test is required to obtain each curve in Fig. 14-12(a), whereas a separate creep test is required for each experimental point of each curve in Fig. 14-12(b).

EXAMPLE 14-6

A cylindrical space capsule 48 in. in diameter must resist a pressure differential of 8 psi for 36 hr at 400°F. To prevent leakage at the joints, a strain of 0.006 in./in. must not be exceeded. How thick must the wall be if it is made from Alclad 2024-T3 aluminum alloy? Assume a stress concentration factor of 2.0 at the joint and a factor of safety of 1.5. Assume that the curves of Fig. 14-12(a) apply for the biaxial stress condition of this wall.

Solution:

By interpolating in Fig. 14-12(a) for 36 hr and a strain of 0.006 in./in., we obtain a constant stress of approximately 34,000 psi. This is the failure, or maximum, stress. The average permissible stress is the maximum peak stress of 34,000 psi divided by the stress concentration factor of 2. Thus,

$$\sigma_{av} = \frac{34,000}{2} = 17,000 \text{ psi}$$

The design pressure is 8 psi, and the failure pressure is therefore $8N = (8)(1.5) = 12$ psi. For a thin-walled pressure vessel,

$$\sigma = \frac{pD}{2t}$$

$$t = \frac{pD}{2\sigma} = \frac{(12)(48)}{(2)(17,000)} = 0.017 \text{ in.}$$

Relaxation Relaxation is defined as the relief of stress by *internal* creep. In other words, it is the time-dependent decrease in stress in a member that is held in a relatively fixed external configuration. A material that is susceptible to creep is also susceptible to relaxation. The same mechanisms within the material are involved in both behaviors. In creep, external or overall deformation occurs while the member is subjected to a constant load; whereas in relaxation, internal creep causes a decrease and redistribution of the stresses *without* significant deformation of the boundaries.

Bolts and other connectors used at joints of such elements as pipes, valves, pressure vessels, turbine casings, and cylinder heads are often subjected to high temperatures and initial stresses. In these instances the boundary deformation remains essentially constant but, as internal creep occurs, the bolt relaxes. To analyze this relaxation behavior for constant strain, a plot of the reduction of stress with time can be made as in Fig. 14-13. Compare this with Fig. 4-4.

Figure 14-13

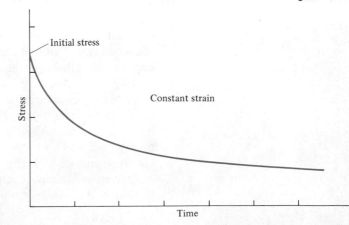

Data for plots of stress versus time to determine relaxation properties are usually obtained by holding tensile specimens at a constant temperature while the load (stress) is reduced with time so that the strain remains essentially constant. As with creep, relaxation is more prevalent and occurs at a faster rate at elevated temperatures.

PROBLEMS

14-22 From Fig. 14-12(a) determine the minimum creep rate in inches per inch per hour for the material at the 35,000 psi stress level.

14-23 Determine the rupture time for a uniaxial tensile specimen of Alclad 2024-T3 aluminum alloy 2 in. wide and 0.025 in. thick, if it is subjected to a constant load of 2200 lb and a constant temperature of 375°F.

14-24 The service life of an oven part is 400 hr at 300°F. If it is made of Alclad 2024-T3 aluminum alloy,

sustains a constant bending moment of 200 in.-lb, and is 0.5 in. wide, what must be the height of its cross section? Assume a factor of safety of 2.

14-25 A cantilever beam of Alclad 2024-T3 aluminum alloy is 0.063 in. deep, 0.750 in. wide, and 12 in. long. It supports a dead load of 2 lb on the free end. If the temperature of the beam is raised from 70 to 300°F, will the beam fail? If so, how long will it take for failure to occur after the temperature is raised? Neglect stress concentrations.

14-6 / Impact loads

Creep and relaxation concern time-dependent behavior from the standpoint of what happens over relatively long periods of time. Now let us consider briefly material and structural behavior when loads are applied suddenly or are varied rapidly for very short periods of time.

Loads that are applied for extremely short periods of time are called impact loads, or shock loads, and a detailed analysis of their effects is very complex. Strain waves are set up in the material, and the magnitudes and distributions of the resulting stresses and strains depend on the velocity of propagation of these strain waves, as well as on the usual material properties, dimensions, and loading. If there is only elastic action, the member will vibrate until it comes back to equilibrium. If the time of application of the load is short compared to the natural period of vibration of the member, the load is usually said to be *high-velocity impact*.

Since theoretical analyses are difficult for impact loading, empirical test results are commonly used to get an index of a material's ability to resist high-velocity impact loads. A pendulum-type impact test machine is commonly used for this purpose. The pendulum, which has a known kinetic energy at impact, strikes a $10 \times 10 \times 50$-mm standard notched specimen supported as a simple beam (Charpy test) or as a cantilever beam (Izod test), and the energy required to rupture the specimen is recorded by

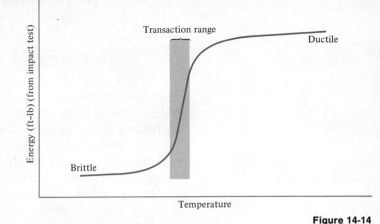

Figure 14-14

measuring the loss of energy of the pendulum. Such tests are good for comparing materials, but this test does not give quantitative data that can be used directly in design. Also, the results do not correlate well with those of tests for other mechanical properties.

One interesting and important phenomenon illustrated by impact tests is the sudden ductile–brittle transition that is exhibited by some materials as the temperature is lowered. Recall from Chapter 3 that the amount of energy absorbed to rupture can be used as a measure of the ductility (or brittleness) of a material. As illustrated in Fig. 14-14, some materials (particularly steels) lose their ductility (as measured by the impact test) and become brittle very abruptly at a low temperature. The center of this transition range can vary for common engineering steels from about $+50°F$ to $-100°F$, the exact value for a particular material depending on the specific metallurgical structure.

14-7 / Low-velocity impact

When the velocity of impact of loads is not too large, that is, when the time of application of the load is greater than several times the lowest natural frequency of the loaded member, and when the load causes only elastic behavior, the engineer can usually make simplifying assumptions that will permit development of a reasonably simple rational analysis of the problem. Under these conditions, the rate of loading is low enough to enable the resisting body to behave in the same way as it would under a static load. The inertia of the resisting member or system is usually neglected, and the deformation of the member is directly proportional to the magnitude of the applied force as is the case in elastic static loading.

The energy imparted to the member by the low-velocity impact load is equal to the work done by some equivalent static load causing the same deformation. Thus,

$$\text{energy imparted} = \text{work of equivalent static load}$$

This approach is often called the equivalent static load method.

The energy imparted may be, for example, the potential energy of a weight W to be dropped from a height h onto a member; or the imparted energy may be kinetic energy of a translating mass imparted to the resisting member. In the first case, the energy imparted would be $W(h + \Delta)$, where Δ is the vertical deformation of the member. In the second case, the imparted energy would be $Wv^2/2g$, where W and v are the weight and velocity of the translating mass, respectively.

Recall that for a quasi-static deformation process the work done by a static load is

$$\text{work} = \int_0^e P \, de \quad \text{or} \quad \int_0^\theta T \, d\theta$$

The first integral represents the work done by a force P acting through a collinear deformation e, and the second integral represents the work done by a moment of torque T acting through an angular deformation θ (measured in radians). For linear elastic behavior, the deformations are directly proportional to the loads causing them. Hence

$$\text{work} = \frac{P_{\text{max}}}{2} e \quad \text{or} \quad \frac{T_{\text{max}}}{2} \theta$$

Now, by equating the imparted energy to the work done by an equivalent static load, we can determine the equivalent static load. We can then use this equivalent load to investigate the stresses and deformations in the member.

EXAMPLE 14-7

The cantilever beam shown in Fig. 14-15 is made of mild steel. Determine the maximum stress that occurs in the beam when the 5-lb weight is dropped on it.

Figure 14-15

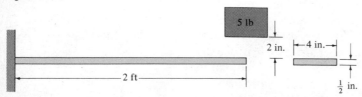

Solution:

The 5-lb weight will move a distance of 2 in. plus the beam deflection Δ, and the energy imparted to the resisting beam will be equal to $W(2 + \Delta)$. The static deflection Δ of a cantilever beam with a concentrated load at the free end is $PL^3/(3EI)$. In this case $E = 30 \times 10^6$ psi and $I = (4)(0.5)^3/12 = 1/24$, and $L = 24$ in. Therefore,

$$\text{energy imparted} = \text{work of equivalent load } P$$

$$5(2 + \Delta) = \frac{P}{2}\Delta$$

$$5\left(2 + \frac{PL^3}{3EI}\right) = \left(\frac{P}{2}\right)\left(\frac{PL^3}{3EI}\right)$$

$$(10) + \frac{(5)(P)(24^3)}{(3)(30)(10^6)(\frac{1}{24})} = \frac{P^2(24^3)}{(6)(30)(10^6)(\frac{1}{24})}$$

$$P = 5 \pm \sqrt{25 + 5420}$$

$$P = +79 \text{ lb} \quad \text{or} \quad -69 \text{ lb}$$

The positive root is the value of the force required to cause the maximum downward deflection Δ (the negative root would be the force required to cause the maximum upward deflection that would result if the beam vibrated back upward).

The maximum stress would occur in the outer fibers at the left end where the maximum moment occurs. Neglecting stress concentrations and using the elastic flexure formula, we find that the maximum stress σ is

$$\sigma = \frac{My}{I} = \frac{(79)(24)(\frac{1}{4})}{(\frac{1}{24})} = 11,370 \text{ psi}$$

PROBLEMS

14-26 A load of 120 N is dropped from a height of 50 mm onto a helical spring. The modulus of the spring is 200 N/mm. Find the maximum deflection of the spring.

14-27 A 160-lb student jumps from a height of 3 ft. onto the diving board shown in the illustration. The board is hickory with a modulus of 1.75 million psi. How much will the board deflect, and what maximum

Problem 14-27

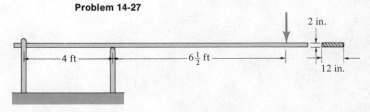

stress will it be subjected to? How much would the board deflect if the student walked out slowly and stood still on the end?

14-28 A simply supported beam 12 ft long is 2 in. wide and 1 in. deep and is made of an aluminum alloy that has a modulus of 10×10^6 psi. A weight of 60 lb is dropped a distance of 3 ft onto the center of the beam. Will any yielding occur in the beam, if the elastic limit stress is 30,000 psi?

14-29 When the 200 N weight shown in the illustration is dropped and hits the stop, all of its energy is transferred into the steel bar. What is the maximum height through which the weight can be dropped without the stress exceeding 140 MPa in the bar?

14-30 When the block shown in the illustration strikes the square steel bar, which is 1 in. on a side, assume that all of the energy of the block is transferred to the bar. Do you think the bar will buckle? If so, why? If not, why not?

Problem 14-29

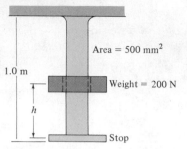

Area = 500 mm²

1.0 m

Weight = 200 N

h

Stop

14-31 For the conditions shown in the illustration, from what height h must the weight be dropped to cause the beam to deflect three times what it would if the weight were applied very slowly? The spring is force free before the weight hits.

14-32 A hollow aluminum-alloy shaft 1 m long is made from a tube with a wall thickness of 3 mm and an average diameter of 50 mm. How much shear stress will result in the shaft from a pure torsional energy load of $4/\pi$ N·m?

Problem 14-30

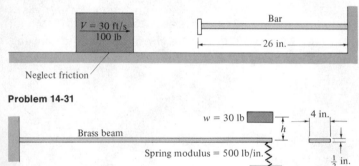

Bar

V = 30 ft/s
100 lb

26 in.

Neglect friction

Problem 14-31

w = 30 lb

4 in.

Brass beam

h

Spring modulus = 500 lb/in.

½ in.

48 in.

Chapter 15
Experimental mechanics

15-1 / Objectives

Upon completion of this chapter you will be able to:

1   Compute the relationship between strain and change in resistance
    for an electrical resistance strain gage.
2   Write the appropriate balance (imbalance) equation for a
    Wheatstone-bridge circuit composed of one or more active gages.
3   Sketch and explain the design and circuitry of simple strain-gage
    transducers.
4   Determine the photoelastic constant for a material given appropriate
    information on size, load, and fringe order.

15-2 / Introduction

Suppose it is imperative for an aerospace engineer to know the magnitude and direction of the maximum stress that occurs at a critical point in a rocket structure for less than a second during the acceleration of the rocket when it is several miles from the earth. Perhaps it is necessary to know the stress in a blade of a jet-engine turbine while it is turning at 10,000 rpm at a temperature of 1800°F and an altitude of 60,000 ft. Or it may be necessary to know the forces acting on a drill bit rotating in an oil well several miles below the earth's surface, or to know the complete state of stress in the wall of a nuclear reactor in a highly radioactive environment. Can the answers to such difficult problems be obtained with experimental techniques available today? Generally, the answer is yes. The answer 40 years ago was no. Great advances have been made in the field of *experimental* mechanics in recent years. This field of study is commonly called experimental stress analysis, although experimental strain analysis is perhaps a better name. We compromise with "experimental mechanics."

This chapter is a brief introduction to experimental mechanics. It outlines some of the more important principles, techniques, equipment, and limitations of the state of the art today. Because of the tremendous advancements now being made in this field, such a presentation may be somewhat obsolete before this text is available to the student, particularly in regard to the equipment used. The underlying principles, however, remain basically the same, regardless of new refinements in techniques and equipment.

The methods of experimental mechanics to be discussed in the following sections are based on the use of electrical resistance strain gages, photoelasticity (including birefringent coatings), brittle lacquer coatings, and Moiré analysis.

Experimental analyses of stresses and strains, and/or their distribution, by actual measurements are important in *supplementing* theoretical analyses, as well as substituting for them. In many cases the geometry or loading of a member or structure may be so complex that a theoretical solution is too cumbersome or uneconomical, or impossible. In such a case, the engineer may have no choice but to resort to experimental methods. However, experimental analyses are often used to establish sufficient boundary conditions (usually surface conditions) to enable the completion of a theoretical analysis. Experimental measurements are also used many times to verify theoretical predictions. Also it is possible to use an experimental approach to determine the stresses and strains, or their distributions, when the loading condition or load distribution is unknown.

Experimental stress or strain analysis is based on *measurements*. Measurement of what? We learned in Chapters 1 and 2 that stress and strain are mathematical quantities. *We cannot measure stress or strain directly.* We must measure quantities that are related to stress or strain. We can measure force, length, and time.

Perhaps the most common measurement in experimental mechanics is the *surface* deformation in a small gage length. Generally, any device used to measure surface deformations is called a *strain gage.* A suitable mechanical device, such as a tensile extensometer on a test specimen, can be used to translate this deformation into some other form of language, such as a dial reading, an electrical signal by the proportional movement of a miniature transformer core, or a light-spot movement greatly magnified by a proportional mirror rotation. Much sophistication and precision can be built into these essentially mechanical devices, but they are inherently rather clumsy compared to electrical resistance strain gages.

15-3 / Electrical resistance strain gages

Strictly speaking, any strain gage—whether it be mechanical, electrical, optical, or of some other type—does not measure strain. Strain gages are deformation-sensitive; that is, they are able to sense and respond to a deformation in the form of a *change* of a finite length. Therefore, the "measured" quantity obtained from a strain gage is *proportional* to an *average* strain in the gage length of the gage. In a region of a high strain gradient, an average reading of a strain gage may be appreciably less than the maximum occurring at some specific point within the gage length. This difficulty may be overcome to a great degree by using a strain gage with a very small gage length, but there is a physical limitation on the size of the smallest gage that can be made. At present, electrical resistance gages can be made with a gage length as small as 0.015 in., although the most common convenient gage length for such a gage is about $\frac{1}{4}$ to $\frac{1}{2}$ in. This size is quite satisfactory except in areas of high strain gradient.

The most common electrical resistance strain gages used universally are *bonded* gages. That is, the gage is intimately bonded to the surface on which the strain is desired, and it is therefore deformed along with the surface. Essentially, a wire gage consists of a length of very fine wire (about 1 mil in diameter) which is looped into a pattern, as in Fig. 15-1(a) or (b). A foil gage is made by etching a pattern on very thin metal foil (about 0.0001 in. thick), as in Fig. 15-1(b) or (c). The foil or wire is usually bonded to a thin base of plastic or paper. When in use, the bonded gage is cemented firmly to the member under investigation, with the foil or wire side out. The foil-plastic gage is much more widely used today.

An electric current is passed through the foil or wire. The resistance of the element (foil or wire) changes as the surface under it (and therefore the gage) is strained. The basic principle involved is simply Lord Kelvin's

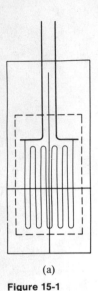

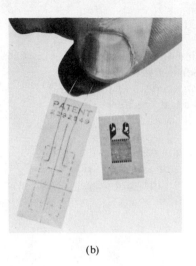

(a)

(b)

(c)

Figure 15-1

discovery that a wire changes its electrical resistance when deformed. Since the foil or wire is bonded throughout its length, the gage is able to sense a compressive strain as well as a tensile strain. The resistance change, which is accurately proportional to the strain, is measured by appropriate instruments.

Resistance gages are made in a great variety of shapes, sizes, and types. Figure 15-2 is a photograph of only a few types of gages.

A resistance gage can be used on the surface of almost any solid material important in engineering, such as metal, plastic, concrete, wood, glass, and paper. In special cases the gage can be embedded in the interior of a cast material such as concrete. The weight it adds to the member is usually negligible, and it can be used for static or dynamic strains. Also, it does not have to be read "on the spot." That is, a remote indicating device can be used to record the output information from the gage. Gages are sensitive to strains as small as 5 millionths of an inch per inch and usually have a range up to strains of 1 to 2 percent (10,000 to 20,000 μin./in.). Special gages have ranges up to a strain of 10 percent. If properly protected, typical gages can be used in various environments, such as underwater and exposed to temperatures of about 300 to 400°F. At the present time, special high-temperature gages can be used up to 2000°F, and this limit is sure to go higher with the tremendous interest and research in this direction. The cost of a gage varies from about $2.50 for a common general-purpose foil gage with a $\frac{1}{2}$-in. gage length to about $10.00 to $15.00 for a three-element foil rosette gage.

Previously we stated that the resistance change of the gage was proportional to strain. How are these two quantities related? Each gage

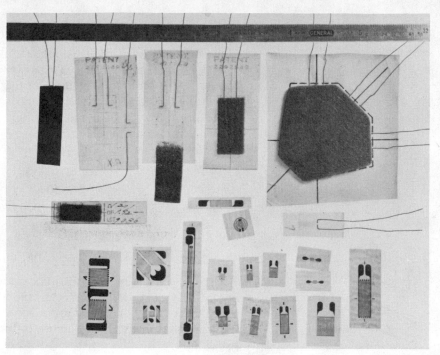

Figure 15-2

has a gage factor, which we shall denote by GF. The gage factor is defined as the ratio of the unit change in resistance to the unit change in length. That is,

$$GF = \frac{\Delta R/R}{\Delta L/L} = \frac{\Delta R/R}{\varepsilon_{av}} \tag{15-1}$$

where ΔR is the total change in resistance and ΔL is the total change in length.

The gage factor is an index of the strain sensitivity of a gage and is a constant for the small range of resistance changes and strains usually encountered. The higher the gage factor, the more sensitive is the strain gage. The gage factor is a function of the material of the foil or wire. The most typical gages have a gage factor of about 2. Although it is possible to use materials that will give higher gage factors, such materials usually have other undesirable properties that offset their high gage factor.

The success of a bonded strain gage depends to a great extent on the quality of the bond between it and the member on which it is used. It is of the utmost importance to use great care in cleaning and preparing the surface, applying the adhesive, and curing the adhesive to make sure that an intimate bond is obtained between the gage and the test piece.

15-4 / Strain-gage circuitry

How is the resistance change due to strain in a gage measured? Can we use an ordinary ohmmeter? To answer these questions, we must first determine the order of magnitude of the resistance change. A standard gage has a resistance of 120 ohms, and its gage factor is 2.0. If we bond such a gage to a bar of aluminum with a modulus of 10 million psi, and subject it to a uniaxial elastic stress of 500 psi in the direction of the gage axis, we can determine the strain and resistance change from Hooke's law and Eq. 15-1. Thus,

$$\varepsilon = \frac{\sigma}{E} = \frac{500}{10,000,000} = 0.00005 \text{ in./in.}$$

$$\Delta R = \varepsilon(GF)R$$

$$\Delta R = (0.00005)(2.0)(120) = 0.0120 \text{ ohms}$$

Resistance changes of this magnitude cannot be measured accurately with an ordinary ohmmeter. A Wheatstone-bridge circuit is usually used, in which one or more of the four arms of the bridge are strain gages. Following is a brief discussion of the use of the Wheatstone bridge for strain gages.

The basic Wheatstone-bridge circuit is shown in Fig. 15-3(a), and the basic relationship for a balanced bridge with zero output is

$$R_A \times R_D = R_B \times R_C \tag{15-2}$$

If a strain gage is placed in one arm of the bridge to replace the fixed resistance R_A, as in Fig. 15-3(b), when the gage is applied to a member and subjected to a strain, say tensile, the magnitude of R_A in the basic equation will be increased, causing the equation to become unbalanced. *The*

Figure 15-3

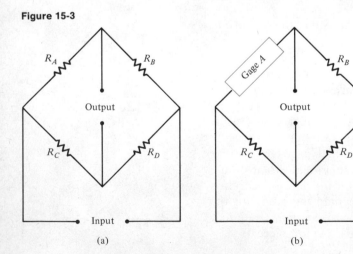

(a) (b)

magnitude of this unbalance is measured as output of the bridge and is *proportional to the strain* in the member. Thus, because of unbalance due to ΔR_A,

$$(R_A + \Delta R_A) \times R_D \neq R_B \times R_C \qquad (15\text{-}2a)$$

We are here and hereafter assuming that initially all the resistances R_A, R_B, R_C, and R_D are equal and that they change only as indicated.

The fact that the strain gage is very sensitive to temperature prevents the circuit in Fig. 15-3(b) from being entirely satisfactory. If the temperature of the member changes even slightly, this change will manifest itself as a "strain" or resistance change in the gage A. Therefore, the output or equation unbalance measured in such a circuit would actually be due to temperature strain as well as strain due to the loading. The correct equation, including unbalance due to ΔR_A and ΔR_T, is

$$(R_A + \Delta R_A + \Delta R_T) \times R_D \neq R_B \times R_C \qquad (15\text{-}2b)$$

If R_B is replaced with another gage B, identical to gage A, mounted on a similar but unstressed material, and placed in the immediate vicinity of gage A, then gage B will be subjected to the same temperature influences as gage A. This circuit is shown in Fig. 15-4(a), and the basic equation for it would be

$$(R_A + \Delta R_A + \Delta R_T) \times R_D \neq (R_B + \Delta R_T) \times R_C \qquad (15\text{-}2c)$$

Notice that since R_B (gage B) is on the opposite side of the equation from R_A (gage A), the terms ΔR_T due to temperature are canceled out, and the only unbalance of the equation (output of the bridge) is due to the applied strain in the gage A which is desired. For this reason the gage B is often referred to as a *temperature-compensating* gage or *dummy* gage.

Figure 15-4

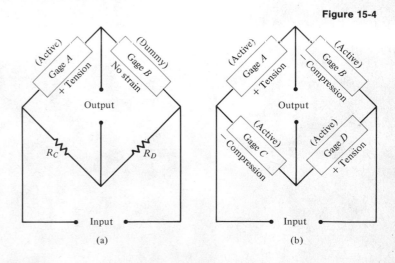

(a)   (b)

,15-5 / Measuring instruments

Static measurement For static measurement the null balance is commonly used, various types of which are shown in Fig. 15-5. After the active and compensating gages are connected to the indicator and the gage-factor adjustment is made, the indicator is adjusted until the circuit is balanced, as indicated by a zero deflection of the meter dial. A reading is then taken. After the active gage undergoes a strain, the circuit is again balanced and a new reading is taken. The difference in readings is the change in strain directly in microinches per inch.

This type of indicator is usually battery operated and is therefore completely portable. Power for the bridge and amplification of the signal is provided within the indicator. Specifically, such an indicator usually consists of an audiofrequency oscillator for the bridge power, an amplifier, a detection circuit, and an indicating meter. Such an indicator can also be used as an amplifier in connection with a cathode-ray oscilloscope for dynamic work up to frequencies of about 100 Hz. With a slight external adjustment, most null indicators can readily be used with four external active gages (full external bridge). More will be said about the value of four active gages in section 15-6 on transducers.

Frequently it is necessary to have many gages on a single structure subjected to static loads. Since each of these gages must be read individually, various types of switching units, or switching-and-balancing units, are available to handle such situations with a minimum of time and maximum convenience. In Fig. 15-6 is shown a commonly used 20-channel switching-and-balancing unit linked with a null indicator. Each channel has a balance potentiometer which makes it possible, in most cases, for a convenient zero reading on the indicator to be the same for all gages. For any load on the structure the gages are read, one by one, as each gage is selected on the switch box and balanced on the null indicator. If there is not too much difference in the strain readings between channels, each gage can be read in about 15 to 30 sec with a little practice.

Figure 15-5

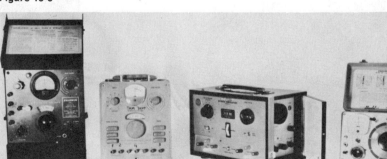

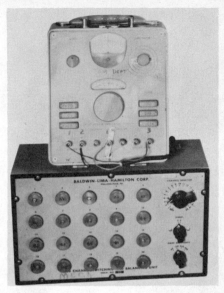

Figure 15-6

For higher speed and less chance of human reading error, many types of automatic recording systems are available. Where the switching is automatic, the balancing is accomplished with a servo or electronically (rather than by hand), and the readings and recordings are made by an automatic printer. After such a system has been properly adjusted for initial balance, it is only necessary to push one button to obtain a complete and permanent record of strain in each gage from the printer in a few seconds. Or the data can be stored on a magnetic tape or in a computer memory and/or used directly in calculations or feedback control circuits depending on how the automatic system or computer is programmed.

Dynamic measurement   For applications where a strain gage is applied to a member in which the strain is changing rapidly, many types of recording instruments are available. In practically every case it is necessary to have means for powering the bridge and amplifying the output signal so that this signal can drive some type of readout device.

Several commonly used types of simple readout recording instruments are shown in Fig. 15-7. The portable instrument in Fig. 15-7(a) is a two-channel recording oscillograph which plots output versus time. The plotting is done with a mechanical pen. The inertia of the pen mechanism limits this common type of device to frequencies below approximately 100 Hz. For higher frequency response a recorder such as the "light beam" or "mirror galvanometer" type of oscillograph shown in Fig. 15-7(b) is commonly used. Here a low-inertia galvanometer with an attached mirror deflects a light beam in proportion to the signal, and the deflection is

(a)

(b)

(c)

Figure 15-7

recorded as a trace on photosensitive paper. Since the paper is moving with a constant speed, a plot of output versus time is obtained. The particular model shown is 14 channels with eight amplifiers just to its left and six in the rack above. Galvanometers of various frequency response (up to approximately 12,000 Hz) are available for these types of instruments.

A cathode-ray oscilloscope (CRO) may also be used as the readout instrument for strain signals. One type used for strain-gage work is shown in Fig. 15-7(c), along with a camera attachment at the top which clamps over the tube face. This instrument can be used to photograph a steady-state dynamic signal or a transient signal. Since the CRO uses an essentially inertialess electron beam for indication, it is suitable for any range of frequency up to the limit of the amplifier.

This section is a rather condensed survey of strain-gage instrumentation. It is highly recommended that the student read Chapter 5 in *The Strain Gage Primer*, 2d ed., by Perry and Lissner (New York: McGraw-Hill, 1962) or other reference on instrumentation of strain gages from manufacturers of current equipment. Other chapters of *The Strain Gage Primer* are also of great value for the beginning student in strain-gage work.

15-6 / Strain-gage transducers

A strain-gage transducer is a device that uses a strain gage to produce an electrical signal that is proportional to some mechanical phenomenon. This is done by making the strain in some part of the device proportional to the phenomenon to be measured. Strain-gage transducers are used to measure force (tension or compression), displacement, pressure, torque, or acceleration. These devices can usually be made very small, rugged, and simple, and they have the advantages that are inherent in electrical strain gages; that is, they are very accurate, are reliable, and can be used for static or dynamic phenomena, and the readout instrument can be remote from the sensing device. Many types of very elegant and sophisticated strain-gage transducers can be purchased commercially. The prices are quite high but reliable transducers, such as a torque pickup with slip rings or an accurately calibrated and damped accelerometer, are worth a high price. A simple (but reliable) "homemade" load, displacement, or pressure transducer can be constructed from a few dollars' worth of gages and materials. A few basic transducers will be discussed later in this section.

Generally, when a transducer is used, it is highly desirable to increase the output from the Wheatstone bridge as much as possible, in order to increase the sensitivity of the transducer. To increase the output, we must increase the unbalance of the basic bridge equation (Eq. 15-2). If tension in gage A in one arm of the bridge caused a certain unbalanced resistance $(+\Delta R_A)$ on the *left-hand* side of the equation, as in Eq. 15-2a, then the unbalance and hence the output could be *doubled* if at the same time the gage B in another arm could be subjected to an equal compression which would cause an equal but opposite change $(-\Delta R_B)$ on the *right-hand* side of the equation. This condition is often referred to as having two "active" gages or arms of the bridge, or an "active half bridge." The equation for the circuit would then be

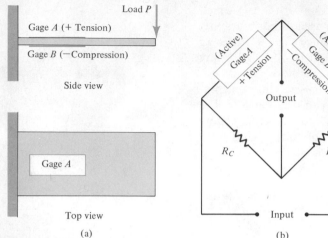

Load P

Gage A (+ Tension)

Gage B (−Compression)

Side view

Gage A

Top view

(a)

Figure 15-8

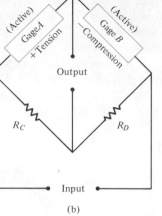

(Active)
Gage A
+ Tension

(Active)
Gage B
−Compression

Output

R_C R_D

Input

(b)

$$(R_A + \Delta R_A + \Delta R_T) \times R_D \neq (R_B - \Delta R_B + \Delta R_T) \times R_C \quad (15\text{-}2d)$$

In this case, the output (unbalance due to $+\Delta R_A$ and $-\Delta R_B$) is doubled, while the temperature effects are still canceled out. The most common example of doubling the output by the use of two active gages is the application of putting gages "back to back in bending," as in the cantilever beam in Fig. 15-8(a). The circuit is represented in Fig. 15-8(b).

By the same reasoning as above, to *quadruple* (multiply four times) the output, it is possible also to replace the other two fixed resistances R_C and R_D in Fig. 15-8(b) with gages, as shown in Fig. 15-4(b) or Fig. 15-9(b). Then if gage D is subjected to tension since it is on the left-hand side of Eq. 15-2, and if gage C is subjected to compression since it is on the right-hand side of the equation, the unbalance of the equation (the output of the bridge) will be four times as great as it was in Eq. 15-2a or 15-2c. This condition is often referred to as having four active gages or a "full bridge." The equation for a full bridge is

$$(R_A + \Delta R_A + \Delta R_T) \times (R_D + \Delta R_D + \Delta R_T)$$
$$\neq (R_B - \Delta R_B + \Delta R_T) \times (R_C - \Delta R_C + \Delta R_T) \quad (15\text{-}2e)$$

It can be seen that the output (unbalance due to $+\Delta R_A$, $+\Delta R_D$, $-\Delta R_B$, and $-\Delta R_C$) is quadrupled while the significant (measurable) temperature effects are still canceled out. The determination of the unbalance for a full bridge does not consider the unbalance due to the products of the Δ terms (higher-order terms), which are usually negligible.

An example of a full bridge arrangement, which is often used in the sensing elements of transducers, is shown in Fig. 15-9(a). The circuit is represented in Fig. 15-9(b). This is the most desirable arrangement for transducers because it yields the greatest output from the bridge and therefore the greatest sensitivity to the transducer.

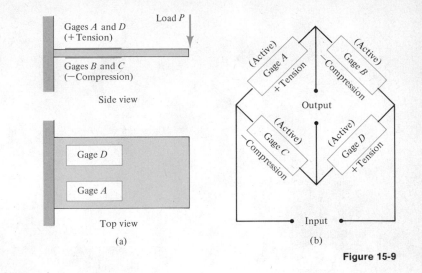

Figure 15-9

Sketches of several types of simple transducers, which can be "home-made," are shown in Fig. 15-10. Notice how the strain gages are used in a full bridge. They are placed back to back in bending wherever possible. In Fig. 15-10(b), (c), (e), (f), and (g), four fully active gages give the maximum sensitivity possible. In (a), the lateral gages B and C increase the output somewhat as a result of the Poisson effect. They, of course, also compensate for temperature. Only in the pressure transducer of Fig. 15-10(d) is it impossible to find places for gages B and C where they can read strains that are opposite to those read by gages A and D. Here the only function of gages B and C is to compensate for temperature.

The load and torque transducers may easily be calibrated with dead weights or with a reliable test machine. A displacement transducer may usually be calibrated with a good micrometer, and a pressure transducer with a good pressure source and a reliable pressure gage or with a dead-weight pressure tester. Can you recommend a way to calibrate the accelerometer?

Electrical resistance gages have revolutionized experimental mechanics in the last 40 years. They have tremendous advantages over their pre-decessors, but they are not without some limitations. A resistance strain gage, except in a few special applications, measures only the *average surface* strain over the finite length of the gage *in the direction of the gage axis*. Rosette gages are useful in that they enable the determination of the maximum strain at a "point" (spot or small area is a better word). The important thing strain gages cannot do is to tell you *where* the maximum strain occurs. At exactly what point should the gage be bonded, and if a single gage is used, how should it be oriented? This question has worried many an engineer. The methods discussed in the next four sections help, in many cases, to answer the question.

Figure 15-10

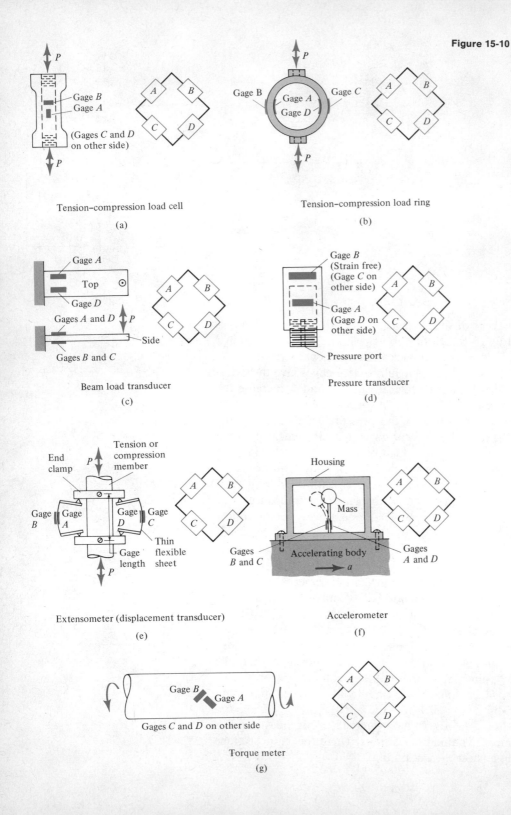

Tension–compression load cell

(a)

Tension–compression load ring

(b)

Beam load transducer

(c)

Pressure transducer

(d)

Extensometer (displacement transducer)

(e)

Accelerometer

(f)

Torque meter

(g)

15-7 / Photoelasticity

Photoelasticity is a method of stress analysis in which a *model* of the member or structure of interest is used. The model is made of special plastic that possesses desirable strain-optical properties. Since the model is geometrically similar to the actual structure, the stress distribution in the model indicates the effects of the geometry on the stress distribution in the actual structure. The mechanical properties of the material of the model and the material of the structure may differ considerably, however, and the model may therefore give a false impression of the actual behavior of the structure.

This method differs from strain-gage analysis in that it provides an overall determination of the stresses at all points, both surface and interior, and not just a measure of the stress at one specific point. Photoelasticity is primarily a two-dimensional method in that the models used have a constant thickness. In advanced techniques, stresses are "frozen" in three-dimensional models and then two-dimensional slices are analyzed.

A photoelastic model is analyzed by use of an optical apparatus called a polariscope. It consists of a source of polarized light that passes through the model, a means of loading the model, and some means of analyzing the resulting phenomena (viewing screen or camera). A typical polariscope is shown in Fig. 15-11. Photoelasticity is based on the phenomenon that occurs when polarized light passes through the special plastic of the model. The plastic is *birefringent* when loaded. This means that it has unique properties that cause a ray of polarized light passing through the

Figure 15-11

thickness at any point to be resolved into two components along the two principal stress directions at the point. One of these components is retarded, relative to the other, as they pass through the thickness of the stressed plastic. In other words, one component needs more time to go through than the other. This relative retardation (time difference) is proportional to the difference of principal stresses at the point. Thus,

$$\sigma_1 - \sigma_2 = \frac{C}{t} f \tag{15-3}$$

where C is the photoelastic constant for the material, t is the model thickness, and f is the number of full wavelengths of relative retardation and is called the fringe order. In the usual polariscope setup, if the relative retardation at some point of the model is exactly one full wavelength or any multiple thereof, the light is extinguished in the optical system and a black spot results on the viewing screen or camera plate for that point of the model. If the retardation is a half wavelength or any multiple thereof, then a maximum amount of light is passed by the optical system and a bright spot on the screen represents that point in the model.

Since the model represents an infinite number of points, an infinite number of light rays pass through it, and the image seen on the screen is an infinite number of light, dark, and shady spots which merge to form a pattern of light and dark bands. Light very close to one frequency (monochromatic light) is usually used, so that the bands are sharper and more nearly black and white, rather than colored. A band, also called a *fringe* or *isochromatic*, represents the locus of points having the same *difference* of principal stresses. A qualitative study of the viewing screen will give an immediate and overall "picture" of the stress distribution in the model. Points of maximum stress can be detected, and the effect of the model geometry can readily be seen.

A photograph of the isochromatic fringes is usually made for future quantitative study. Figure 15-12 is such a photograph. Special devices or techniques of compensation are available that make possible the determination of the stress differences at points of partial fringe orders. By use of white light, removal of certain elements of the optical system, and manipulation of other elements, the locus of points having the same *directions* of principal stress can be sketched on the screen. Several such lines can be drawn. These lines, called *isoclinics*, permit the determination of the directions of the principal stresses at any point.

The isochromatic fringe patterns represent the *differences* of the principal stresses at a point. Since $\sigma_1 - \sigma_2 = 2\tau$, we see that the fringe patterns actually give an overall picture of the *shear-stress* distributions. Along the edges of the model, where the maximum stress usually occurs, one principal stress is zero. There the difference of the principal stresses, or $\sigma_1 - \sigma_2$, reduces directly to one stress. Elsewhere, the principal stresses can usually be separated only by tedious numerical computation.

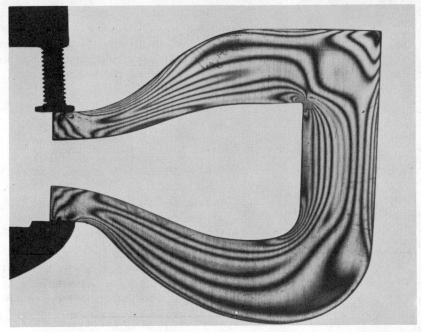

Figure 15-12

A photoelastic material is usually chosen for its high value of C (see Eq. 15-3), ease of machining, high proportional limit, optical transparency, good creep properties, and reasonable price. The most commonly used materials are a type of Bakelite, usually referred to by the number BT-61-893, a Columbia Resin called CR-39, and several epoxies.

Before it is used, a material must be "calibrated." That is, the constant C must be determined. This is usually done by loading a simple tensile-test specimen in a polariscope and plotting the axial load versus the fringe order. The slope of this curve is proportional to C. The other common method is to photograph the isochromatic pattern for a simple beam with a constant bending-moment loading. The slope of the plotted stress distribution is proportional to C.

15-8 / Birefringent coatings

In the past few years, with the development of excellent adhesives such as those of the epoxy family, photoelastic coatings (sometimes called birefringent coatings) have come into wide use. Techniques and instrumentation for their use have become highly developed. In this method of analysis the birefringent plastic is bonded directly to the surface of the actual structure to be stress analyzed. When the surface of the structure undergoes a strain, the bond causes the plastic coating to undergo the same strain. The resulting isochromatic fringes and isoclinics of the plastic can be seen, sketched, or photographed, and they can be interpreted by using the principles of ordinary photoelasticity. Provision must be made to assure that the light directed through the plastic will be properly reflected for analysis. In this case, the light passes through the plastic twice. Instruments called reflective polariscopes are required for this technique. Some are extremely simple, while others are rather complex and expensive. For example, extremely valuable qualitative analyses are often made with a small piece of Polaroid plastic used as a reflective polariscope.

Special techniques and equipment are available for dynamic work, for application to curved surfaces, for easy determination of separate principal stresses, for taking account of the stiffening effect of the plastic on thin sections, for use on nonmetallic surfaces, and for many other special applications. Although conventional photoelastic analysis of models is considered the more quantitatively accurate method, the photoelastic-coating method has the one big advantage of allowing an analysis of the actual member or structure of interest, rather than requiring the analysis of a model.

By necessity this text can include only a very brief survey of the photoelastic method. You are encouraged to refer to a text, such as *Photoelasticity*, Vol. I, by M. Frocht (New York: Wiley, 1941) or *Introduction to Photomechanics*, by Durelli and Riley (Englewood Cliffs, New Jersey: Prentice-Hall, 1965) for a detailed coverage of the basic principles of photoelasticity and the techniques of classical photoelastic model analysis. For more information about photoelastic-coating techniques, refer to the more recent issues of the *Proceedings of the Society for Experimental Stress Analysis*.

15-9 / Brittle coatings

Another valuable tool of experimental mechanics is the use of brittle coatings for stress analysis. A coating is sprayed onto a surface to be analyzed. The coating is designed to crack at a strain level within the elastic range of the material of the structure being analyzed. As a coated structure is being loaded, the first cracks occur at the points at which the "threshold strain" of the coating is reached first. These first cracks indicate the points of maximum tensile strain. By watching the progression of the crack patterns, the engineer can "see" the development and distribution of the strains as loading of the structure is continued. The crack at any point will be perpendicular to the direction of the maximum principal tensile strain at that point. Therefore, the directions of the principal stresses can usually also be determined.

The coating is calibrated by coating both the structure or member to be analyzed and several calibration bars at the same time and in exactly the same manner. After curing, and at the time of test, the calibration bars are subjected to known variations of strain and their crack patterns are analyzed so that crack size and appearance may be associated with the corresponding known strain. The threshold strain is also determined by this calibration. The threshold strain or strains at the initiation of cracking is usually in the range from 500 to 1000 μin./in. The calibration is usually done with a fixture that deflects the calibration bar, loaded as a cantilever beam, to a known deflection by means of an eccentric roller at the free end.

This method sounds good in principle, but it also has its drawbacks. The coatings used and their threshold strains are somewhat sensitive to the influences of temperature, humidity, and creep. There is a premium on technique in that it takes an experienced technician working under optimal conditions to apply a uniform and reliable coating. Under the best conditions, fairly good quantitative results are possible. Plus or minus 10 to 15 percent for strain magnitudes is about the best accuracy that can usually be expected. The method perhaps has its main value in allowing the engineer to get a rather inexpensive and quick look at the overall qualitative picture of the strains and their distribution. This analysis can tell the engineer where and in what directions to place strain gages for further and more refined analyses.

One of the better references describing outstanding work done with this method is *Analysis of Stress and Strain*, by Durelli, Phillips, and Tsao (New York: McGraw-Hill, 1958).

15-10 / Moiré analysis

Moiré analysis is another "field" method that allows an overall view for investigation as do photoelasticity and brittle coatings. These methods have this advantage over strain gages that give a "point by point" analysis. The Moiré technique is usually applied to models (but sometimes to the actual piece) and involves application of a fine-pitched grid (or grating) being intimately bonded to the surface to be investigated such that the grid undergoes the same displacements as the surface. When another identical unbonded and undistorted grid is placed over the bonded grid, fringe patterns appear. These Moiré fringes are loci of points that have the same value of displacements along directions perpendicular to undistorted grid lines. The difference in the displacement components of adjacent fringes is equal to the distance between the lines of the grid, or pitch.

Strains can be determined by plotting the displacements and determining slopes. For example

$$\varepsilon_x = \frac{\text{grating pitch}}{\text{distance between fringes}}$$

The differentiation of the Moiré patterns is usually difficult and often lacking in precision, and this is a major limitation of the method. This is particularly true when rotations are involved. The method does, however, lend itself well for determination of displacements and its use is relatively simple in this respect. It is important to note that this method gives the surface geometry changes directly without going through any intermediate related property as do photoelasticity and strain gages and, therefore, is not influenced by any changes in these intermediate properties.

The sensitivity of the Moiré method is dependent on the grating pitch, and the present state of the art is such that it is very difficult to make a conveniently usable grating with more than about 1000 lines per inch. Therefore, the sensitivity of the method is limited for most engineering materials. The method is better suited for low-modulus materials.

PROBLEMS

15-1 A 120-ohm electrical resistance strain gage with a gage factor of 2.04 undergoes a change of resistance of 0.085 ohm. What is the strain in the direction of the gage axis?

15-2 A 120-ohm electrical resistance strain gage is mounted on a solid steel shaft 15 mm in radius with the gage axis 20° from the longitudinal axis of the shaft. What will be the resistance change of the gage if the shaft is subjected to 400 N · m of pure torque? The gage factor is 1.98. Assume elastic behavior.

Problem 15-3

15-3 For the beam shown in the illustration, what is the strain on the surface of the beam where the gages are? The beam is made of a high-strength alloy steel and has a rectangular cross section 3 mm high and 6 mm wide. Each gage has a resistance of 120 ohms and a gage factor of 2.03. Gages C and D are mounted on similar unstrained material. What is the voltage drop across the voltmeter V, assuming the voltmeter to have a very high resistance?

15-4 Same as Prob. 15-3 except gage C is mounted beside gage B, and gage D is mounted beside gage A,

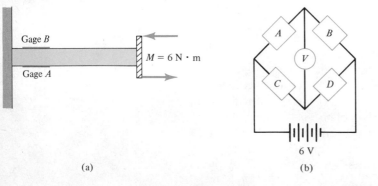

(a)

(b)

all gages being parallel to the longitudinal axis of the beam.

15-5 The conditions are the same as in Prob. 15-3, except that gage C is mounted beside gage A with its axis perpendicular to the beam axis and perpendicular to the axis of gage A, and gage D is mounted on the top surface beside gage B and its axis is perpendicular to the beam axis and to the axis of gage B.

15-6 The conditions are the same as in Prob. 15-5, except that gage C is mounted on the top surface and gage D is mounted on the bottom.

15-7 Sketch the design and circuitry for a simple strain-gage transducer that will measure the pull on a drawbar between a farm tractor and a plow. Use a readout instrument that will give a permanent record of pull versus time. Explain carefully how you would calibrate the device.

15-8 Sketch the design and circuitry for a strain-gage transducer to measure the impact force of a golf club on a golf ball. A complete record of force versus time is needed. How would you calibrate the device?

15-9 Sketch the design and circuity for a strain-gage transducer that could be screwed into a spark-plug hole in an automobile engine to measure the pressure in the cylinder. A record of pressure versus time is desired. How would you calibrate the device?

15-10 A simple uniaxial tensile specimen of a photo-elastic material is tested in a polariscope. The resulting values recorded for load in pounds versus fringe order are given below. What is the photoelastic constant for the material? The cross section of the specimen is 0.335 in. wide by 0.275 in. thick.

Load (P)	0	8	14	20	28	35	42
Fringe order (f)	0	1	2	3	4	5	6

Appendix

Table A-1 Average mechanical properties of selected engineering materials

Material	E 10⁹ Pa	E 10⁶ psi	G 10⁶ psi	μ	0.2% Yield strength (tension), 10³ psi	Ultimate strength 10⁶ Pa	Ultimate strength 10³ psi		Elongation at rupture in 2 in., %	Weight, lb/in.³	Coefficient of thermal expansion, 10⁻⁶ per °F
Hot rolled steel (SAE 1020)	207	30	12	0.27	36	450	65	(ten)	30	0.283	6.5
Structural steel (A-7)	207	30	12	0.27	35	410–500	60–72	(ten)	30	0.283	6.5
High-carbon steel (SAE 1090)	207	30	12	0.27	67	840	122	(ten)	10	0.283	6.5
Alloy steel (SAE 4130) (heat treated)	207	30	12	0.30	100	860	125	(ten)	10	0.283	6.5
Stainless steel (18-8)	193	28	9.5	0.30	80	830	120	(ten)		0.284	9.6
Gray cast iron (ASTM Class 30)	101	14.7	5.9	0.20		210 / 850	31 / 124	(ten) / (comp)	<1	0.260	6.7
Cast iron (pearlitic malleable)	182	26.4	10		80	690 / 2070	100 / 300	(ten) / (comp)	7	0.266	6.6
Aluminum 1100-0 (annealed)	69	10.0	3.8	0.33	3.5	76	11	(ten)	25	0.098	13.1
Aluminum alloy 2024-T3 (sheet and plate)	73	10.6	4.0	0.33	50	480	70	(ten)	18	0.100	12.6
Aluminum alloy 6061-T6 (extruded)	69	10.0	3.8	0.33	35	260	38	(ten)	10	0.098	13.1
Aluminum alloy 7075-T6 (sheet and plate)	72	10.4	3.9	0.33	70	550	80	(ten)	5	0.101	12.9
Magnesium alloy (H K31A-H24 sheet)	45	6.5	2.4	0.35	23	230	34	(ten)	4	0.0647	15
Titanium alloy (6Al-4V sheet)	110	15.9	6.2	0.34	120	900	130	(ten)	10	0.160	4.6
Brass, hard yellow	103	15	5.6	0.35	60	510	74	(ten)	10	0.306	10.5
Copper DHP (Hard Temper) (Pipe)	117	17	6.4		45	340	50	(ten)	10	0.323	18
Douglas fir timber (air dry; parallel to grain)	12	1.7				56 / 51	8.1 / 7.4	(ten) / (comp)		0.020	3.0
Red oak timber (air dry; parallel to grain)	12	1.8				48	6.9	(comp)		0.025	1.9
Lead (rolled)	14	2	0.7	0.43	2	17	2.5	(ten)	50	0.410	16.4
Tungsten carbide (Carboloy, Grade 999)	690	100		0.24		4140 / 9	600 / 1.3	(comp) / (ten)			2.2
Glass (fused silica)	69	10.0		0.17		90	13	(comp)	Nil	0.15	4.0
Concrete (low strength)	14	2		0.15		14	2	(comp)		0.087	6.0
Concrete (high strength)	21	3		0.15		34	5	(comp)		0.087	6.0
Polystyrene (average)	3.4	0.5				96	14	(comp)	2		70
Polyethylene (average)	12	1.8		0.45		14	2	(ten)	350	0.033	150
Epoxy (cast; average)	4.5	0.65				48 / 210	7 / 30	(ten) / (comp)	4		33
Rubber (natural; molded)				0.50		21	3	(ten)	800		90

Table A-2   Properties of plane areas

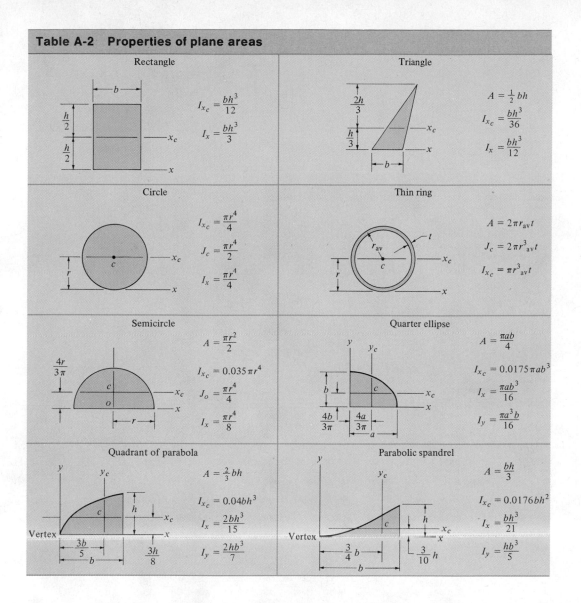

Rectangle

$$I_{x_c} = \frac{bh^3}{12}$$

$$I_x = \frac{bh^3}{3}$$

Triangle

$$A = \tfrac{1}{2} bh$$

$$I_{x_c} = \frac{bh^3}{36}$$

$$I_x = \frac{bh^3}{12}$$

Circle

$$I_{x_c} = \frac{\pi r^4}{4}$$

$$J_c = \frac{\pi r^4}{2}$$

$$I_x = \frac{\pi r^4}{4}$$

Thin ring

$$A = 2\pi r_{av} t$$

$$J_c = 2\pi r^3_{av} t$$

$$I_{x_c} = \pi r^3_{av} t$$

Semicircle

$$A = \frac{\pi r^2}{2}$$

$$I_{x_c} = 0.035\,\pi r^4$$

$$J_o = \frac{\pi r^4}{4}$$

$$I_x = \frac{\pi r^4}{8}$$

Quarter ellipse

$$A = \frac{\pi ab}{4}$$

$$I_{x_c} = 0.0175\,\pi ab^3$$

$$I_x = \frac{\pi ab^3}{16}$$

$$I_y = \frac{\pi a^3 b}{16}$$

Quadrant of parabola

$$A = \tfrac{2}{3} bh$$

$$I_{x_c} = 0.04bh^3$$

$$I_x = \frac{2bh^3}{15}$$

$$I_y = \frac{2hb^3}{7}$$

Parabolic spandrel

$$A = \frac{bh}{3}$$

$$I_{x_c} = 0.0176bh^2$$

$$I_x = \frac{bh^3}{21}$$

$$I_y = \frac{hb^3}{5}$$

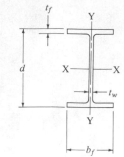

Table A-3 W shapes: properties and dimensions for design (English customary units)[a]

Section and weight, lb/ft	A, in.^{2}	d, in.	b_f, in.	t_f, in.	t_w, in.	I_x, in.^{4}	S_x, in.^{3}	r_x, in.	I_y, in.^{4}	S_y, in.^{3}	r_y, in.
W 36 × 300.0	88.30	36.74	16.655	1.680	0.945	20300.0	1110.00	15.20	1300.00	156.00	3.830
× 280.0	82.40	36.52	16.595	1.570	0.885	18900.0	1030.00	15.10	1200.00	144.00	3.810
× 260.0	76.50	36.26	16.551	1.440	0.840	17300.0	953.00	15.00	1090.00	132.00	3.780
× 245.0	72.10	36.08	16.510	1.350	0.800	16100.0	895.00	15.00	1010.00	123.00	3.750
× 230.0	67.60	35.90	16.470	1.260	0.760	15000.0	837.00	14.90	940.00	114.00	3.730
× 210.0	61.80	36.69	12.180	1.360	0.830	13200.0	719.00	14.60	411.00	67.50	2.580
W 36 × 194.0	57.00	36.49	12.115	1.260	0.765	12100.0	664.00	14.60	375.00	61.90	2.560
× 182.0	53.60	36.33	12.075	1.180	0.725	11300.0	623.00	14.50	347.00	57.60	2.550
× 170.0	50.00	36.17	12.030	1.100	0.680	10500.0	580.00	14.50	320.00	53.20	2.530
× 160.0	47.00	36.01	12.000	1.020	0.650	9750.0	542.00	14.40	295.00	49.10	2.500
× 150.0	44.20	35.85	11.975	0.940	0.625	9040.0	504.00	14.30	270.00	45.10	2.470
× 135.0	39.70	35.55	11.950	0.790	0.600	7800.0	439.00	14.00	225.00	37.70	2.380
W 33 × 241.0	70.90	34.18	15.860	1.400	0.830	14200.0	829.00	14.10	933.00	118.00	3.630
× 221.0	65.00	33.93	15.805	1.275	0.775	12800.0	757.00	14.10	840.00	106.00	3.590
× 201.0	59.10	33.68	15.745	1.150	0.715	11500.0	684.00	14.00	749.00	95.20	3.560
W 33 × 152.0	44.70	33.49	11.565	1.055	0.635	8160.0	487.00	13.50	273.00	47.20	2.470
× 141.0	41.60	33.30	11.535	0.960	0.605	7450.0	448.00	13.40	246.00	42.70	2.430
× 130.0	38.30	33.09	11.510	0.855	0.580	6710.0	406.00	13.20	218.00	37.90	2.390
× 118.0	34.70	32.86	11.480	0.740	0.550	5900.0	359.00	13.00	187.00	32.60	2.320
W 30 × 211.0	62.00	30.94	15.105	1.315	0.775	10300.0	663.00	12.90	757.00	100.00	3.490
× 191.0	56.10	30.68	15.040	1.185	0.710	9170.0	598.00	12.80	673.00	89.50	3.460
× 173.0	50.80	30.44	14.985	1.065	0.655	8200.0	539.00	12.70	598.00	79.80	3.430
W 30 × 132.0	38.90	30.31	10.545	1.000	0.615	5770.0	380.00	12.20	196.00	37.20	2.250
× 124.0	36.50	30.17	10.515	0.930	0.585	5360.0	355.00	12.10	181.00	34.40	2.230
× 116.0	34.20	30.01	10.495	0.850	0.565	4930.0	329.00	12.00	164.00	31.30	2.190
× 108.0	31.70	29.83	10.475	0.760	0.545	4470.0	299.00	11.90	146.00	27.90	2.150
× 99.0	29.10	29.65	10.450	0.670	0.520	3990.0	269.00	11.70	128.00	24.50	2.100
W 27 × 178.0	52.30	27.81	14.085	1.190	0.725	6990.0	502.00	11.60	555.00	78.80	3.260
× 161.0	47.40	27.59	14.020	1.080	0.660	6280.0	455.00	11.50	497.00	70.90	3.240
× 146.0	42.90	27.38	13.965	0.975	0.605	5630.0	411.00	11.40	443.00	63.50	3.210
W 27 × 114.0	33.50	27.29	10.070	0.930	0.570	4090.0	299.00	11.00	159.00	31.50	2.180
× 102.0	30.00	27.09	10.015	0.830	0.515	3620.0	267.00	11.00	139.00	27.80	2.150
× 94.0	27.70	26.92	9.990	0.745	0.490	3270.0	243.00	10.90	124.00	24.80	2.120
× 84.0	24.80	26.71	9.960	0.640	0.460	2850.0	213.00	10.70	106.00	21.20	2.070
W 24 × 162.0	47.70	25.00	12.955	1.220	0.705	5170.0	414.00	10.40	443.00	68.40	3.050
× 146.0	43.00	24.74	12.900	1.090	0.650	4580.0	371.00	10.30	391.00	60.50	3.010
× 131.0	38.50	24.48	12.855	0.960	0.605	4020.0	329.00	10.20	340.00	53.00	2.970
× 117.0	34.40	24.26	12.800	0.850	0.550	3540.0	291.00	10.10	297.00	46.50	2.940
× 104.0	30.60	24.06	12.750	0.750	0.500	3100.0	258.00	10.10	259.00	40.70	2.910
W 24 × 94.0	27.70	24.31	9.065	0.875	0.515	2700.0	222.00	9.87	109.00	24.00	1.980
× 84.0	24.70	24.10	9.020	0.770	0.470	2370.0	196.00	9.79	94.40	20.90	1.950
× 76.0	22.40	23.92	8.990	0.680	0.440	2100.0	176.00	9.69	82.50	18.40	1.920
× 68.0	20.10	23.73	8.965	0.585	0.415	1830.0	154.00	9.55	70.40	15.70	1.870
W 24 × 62.0	18.20	23.74	7.040	0.590	0.430	1550.0	131.00	9.23	34.50	9.80	1.380
× 55.0	16.20	23.57	7.005	0.505	0.395	1350.0	114.00	9.11	29.10	8.30	1.340

[a] From *The Structural Steel Design Manual*, by Joseph E. Bowles (New York: McGraw-Hill, © 1980).

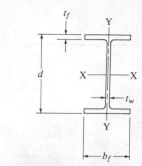

Table A-3 W shapes: properties and dimensions for design (SI units)

Section and mass, kg/m	Wt, kN/m	A, m² × 10⁻³	d, mm	b_f, mm	t_f, mm	t_w, mm	I_x, m⁴ × 10⁻⁶	S_x, m³ × 10⁻³	r_x, mm	I_y, m⁴ × 10⁻⁶	S_y, m³ × 10⁻³	r_y, mm
W 920 × 446.4	4.38	56.97	933.	423.	42.7	24.0	8449.5	18.19	386.1	541.10	2.556	97.3
× 416.7	4.09	53.16	928.	422.	39.9	22.5	7866.8	16.88	383.5	499.48	2.360	96.8
× 386.9	3.79	49.35	921.	420.	36.6	21.3	7200.8	15.62	381.0	453.69	2.163	96.0
× 364.6	3.58	46.52	916.	419.	34.3	20.3	6701.3	14.67	381.0	420.39	2.016	95.2
× 342.3	3.36	43.61	912.	418.	32.0	19.3	6243.5	13.72	378.5	391.26	1.868	94.7
× 312.5	3.06	39.87	932.	309.	34.5	21.1	5494.2	11.78	370.8	171.07	1.106	65.5
W 920 × 288.7	2.83	36.77	927.	308.	32.0	19.4	5036.4	10.88	370.8	156.09	1.014	65.0
× 270.8	2.66	34.58	923.	307.	30.0	18.4	4703.4	10.21	368.3	144.43	0.944	64.8
× 253.0	2.48	32.26	919.	306.	27.9	17.3	4370.4	9.50	368.3	133.19	0.872	64.3
× 238.1	2.34	30.32	915.	305.	25.9	16.5	4058.2	8.88	365.8	122.79	0.805	63.5
× 223.2	2.19	28.52	911.	304.	23.9	15.9	3762.7	8.26	363.2	112.38	0.739	62.7
× 200.9	1.97	25.61	903.	304.	20.1	15.2	3246.6	7.19	355.6	93.65	0.618	60.5
W 840 × 358.6	3.52	45.74	868.	403.	35.6	21.1	5910.5	13.58	358.1	388.34	1.934	92.2
× 328.9	3.23	41.94	862.	401.	32.4	19.7	5327.8	12.40	358.1	349.63	1.737	91.2
× 299.1	2.93	38.13	855.	400.	29.2	18.2	4786.7	11.21	355.6	311.76	1.560	90.4
W 840 × 226.2	2.22	28.84	851.	294.	26.8	16.1	3396.4	7.98	342.9	113.63	0.773	62.7
× 209.8	2.06	26.84	846.	293.	24.4	15.4	3100.9	7.34	340.4	102.39	0.700	61.7
× 193.5	1.90	24.71	840.	292.	21.7	14.7	2792.9	6.65	335.3	90.74	0.621	60.7
× 175.6	1.72	22.39	835.	292.	18.8	14.0	2455.8	5.88	330.2	77.84	0.534	58.9
W 760 × 314.0	3.08	40.00	786.	384.	33.4	19.7	4287.2	10.86	327.7	315.09	1.639	88.6
× 284.2	2.79	36.19	779.	382.	30.1	18.0	3816.8	9.80	325.1	280.12	1.467	87.9
× 257.5	2.52	32.77	773.	381.	27.1	16.6	3413.1	8.83	322.6	248.91	1.308	87.1
W 760 × 196.4	1.93	25.10	770.	268.	25.4	15.6	2401.7	6.23	309.9	81.58	0.610	57.1
× 184.5	1.81	23.55	766.	267.	23.6	14.9	2231.0	5.82	307.3	75.34	0.564	56.6
× 172.6	1.69	22.06	762.	267.	21.6	14.4	2052.0	5.39	304.8	68.26	0.513	55.6
× 160.7	1.58	20.45	758.	266.	19.3	13.8	1860.6	4.90	302.3	60.77	0.457	54.6
× 147.3	1.44	18.77	753.	265.	17.0	13.2	1660.8	4.41	297.2	53.20	0.401	53.3
W 690 × 264.9	2.60	33.74	706.	358.	30.2	18.4	2909.5	8.23	294.6	231.01	1.291	82.8
× 239.6	2.35	30.58	701.	356.	27.4	16.8	2613.9	7.46	292.1	206.87	1.162	82.3
× 217.3	2.13	27.68	695.	355.	24.8	15.4	2343.4	6.74	289.6	184.39	1.041	81.5
W 690 × 169.7	1.66	21.61	693.	256.	23.6	14.5	1702.4	4.90	279.4	66.18	0.516	55.4
× 151.8	1.49	19.35	688.	254.	21.1	13.1	1506.8	4.38	279.4	57.86	0.456	54.6
× 139.9	1.37	17.87	684.	254.	18.9	12.4	1361.1	3.98	276.9	51.61	0.406	53.8
× 125.0	1.23	16.00	678.	253.	16.3	11.7	1186.3	3.49	271.8	44.12	0.347	52.6
W 610 × 241.1	2.36	30.77	635.	329.	31.0	17.9	2151.9	6.78	264.2	184.39	1.121	77.5
× 217.3	2.13	27.74	628.	328.	27.7	16.5	1906.3	6.08	261.6	162.75	0.991	76.5
× 194.9	1.91	24.84	622.	327.	24.4	15.4	1673.2	5.39	259.1	141.52	0.869	75.4
× 174.1	1.71	22.19	616.	325.	21.6	14.0	1473.5	4.77	256.5	123.62	0.762	74.7
× 154.8	1.52	19.74	611.	324.	19.0	12.7	1290.3	4.23	256.5	107.80	0.667	73.9
W 610 × 139.9	1.37	17.87	617.	230.	22.2	13.1	1123.8	3.64	250.7	45.37	0.393	50.3
× 125.0	1.23	15.94	612.	229.	19.6	11.9	986.5	3.21	248.7	39.29	0.342	49.5
× 113.1	1.11	14.45	608.	228.	17.3	11.2	874.1	2.88	246.1	34.34	0.302	48.8
× 101.2	0.99	12.97	603.	228.	14.9	10.5	761.7	2.52	242.6	29.30	0.257	47.5
W 610 × 92.3	0.90	11.74	603.	179.	15.0	10.9	645.2	2.15	234.4	14.36	0.161	35.1
× 81.8	0.80	10.45	599.	178.	12.8	10.0	561.9	1.87	231.4	12.11	0.136	34.0

(continued)

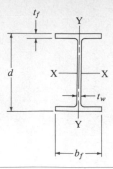

Table A-3 W shapes: properties and dimensions for design, continued (English customary units)

Second and weight, lb/ft	A, in.^{2}	d, in.	b_f, in.	t_f, in.	t_w, in.	I_x, in.^{4}	S_x, in.^{3}	r_x, in.	I_y, in.^{4}	S_y, in.^{3}	r_y, in.
W 21 × 147.0	43.20	22.06	12.510	1.150	0.720	3630.0	329.00	9.17	376.00	60.10	2.950
× 132.0	38.80	21.83	12.440	1.035	0.650	3220.0	295.00	9.12	333.00	53.50	2.930
× 122.0	35.90	21.68	12.390	0.960	0.600	2960.0	273.00	9.09	305.00	49.20	2.920
× 111.0	32.70	21.51	12.340	0.875	0.550	2670.0	249.00	9.05	274.00	44.50	2.900
× 101.0	29.80	21.36	12.290	0.800	0.500	2420.0	227.00	9.02	248.00	40.30	2.890
W 21 × 93.0	27.30	21.62	8.420	0.930	0.580	2070.0	192.00	8.70	92.90	22.10	1.840
× 83.0	24.30	21.43	8.355	0.835	0.515	1830.0	171.00	8.67	81.40	19.50	1.830
× 73.0	21.50	21.24	8.295	0.740	0.455	1600.0	151.00	8.64	70.60	17.00	1.810
× 68.0	20.00	21.13	8.270	0.685	0.430	1480.0	140.00	8.60	64.70	15.70	1.800
× 62.0	18.30	20.99	8.240	0.615	0.400	1330.0	127.00	8.54	57.50	14.00	1.770
W 21 × 57.0	16.70	21.06	6.555	0.650	0.405	1170.0	111.00	8.36	30.60	9.35	1.350
× 50.0	14.70	20.83	6.530	0.535	0.380	984.0	94.50	8.18	24.90	7.64	1.300
× 44.0	13.00	20.66	6.500	0.450	0.350	843.0	81.60	8.06	20.70	6.37	1.260
W 18 × 119.0	35.10	18.97	11.265	1.060	0.655	2190.0	231.00	7.90	253.00	44.90	2.690
× 106.0	31.10	18.73	11.200	0.940	0.590	1910.0	204.00	7.84	220.00	39.40	2.660
× 97.0	28.50	18.59	11.145	0.870	0.535	1750.0	188.00	7.82	201.00	36.10	2.650
× 86.0	25.30	18.39	11.090	0.770	0.480	1530.0	166.00	7.77	175.00	31.60	2.630
× 76.0	22.30	18.21	11.035	0.680	0.425	1330.0	146.00	7.73	152.00	27.60	2.610
W 18 × 71.0	20.80	18.47	7.635	0.810	0.495	1170.0	127.00	7.50	60.30	15.80	1.700
× 65.0	19.10	18.35	7.590	0.750	0.450	1070.0	117.00	7.49	54.80	14.40	1.690
× 60.0	17.60	18.24	7.555	0.695	0.415	984.0	108.00	7.47	50.10	13.30	1.690
× 55.0	16.20	18.11	7.530	0.630	0.390	890.0	98.30	7.41	44.90	11.90	1.670
× 50.0	14.70	17.99	7.495	0.570	0.355	800.0	88.90	7.38	40.10	10.70	1.650
W 18 × 46.0	13.50	18.06	6.060	0.605	0.360	712.0	78.80	7.25	22.50	7.43	1.290
× 40.0	11.80	17.90	6.015	0.525	0.315	612.0	68.40	7.21	19.10	6.35	1.270
× 35.0	10.30	17.70	6.000	0.425	0.300	510.0	57.60	7.04	15.30	5.12	1.220
W 16 × 100.0	29.40	16.97	10.425	0.985	0.585	1490.0	175.00	7.10	186.00	35.70	2.510
× 89.0	26.20	16.75	10.365	0.875	0.525	1300.0	155.00	7.05	163.00	31.40	2.490
× 77.0	22.60	16.52	10.295	0.760	0.455	1110.0	134.00	7.00	138.00	26.90	2.470
× 67.0	19.70	16.33	10.235	0.665	0.395	954.0	117.00	6.96	119.00	23.20	2.460
W 16 × 57.0	16.80	16.43	7.120	0.715	0.430	758.0	92.20	6.72	43.10	12.10	1.600
× 50.0	14.70	16.26	7.070	0.630	0.380	659.0	81.00	6.68	37.20	10.50	1.590
× 45.0	13.30	16.13	7.035	0.565	0.345	586.0	72.70	6.65	32.80	9.34	1.570
× 40.0	11.80	16.01	6.995	0.505	0.305	518.0	64.70	6.63	28.90	8.25	1.570
W 16 × 36.0	10.60	15.86	6.985	0.430	0.295	448.0	56.50	6.51	24.50	7.00	1.520
× 31.0	9.12	15.88	5.525	0.440	0.275	375.0	47.20	6.41	12.40	4.49	1.170
× 26.0	7.68	15.69	5.500	0.345	0.250	301.0	38.40	6.26	9.59	3.49	1.120
W 14 × 730.0	215.00	22.42	17.890	4.910	3.070	14300.0	1280.00	8.17	4720.00	527.00	4.690
× 665.0	196.00	21.64	17.650	4.520	2.830	12400.0	1150.00	7.98	4170.00	472.00	4.620
× 605.0	178.00	20.92	17.415	4.160	2.595	10800.0	1040.00	7.80	3680.00	423.00	4.550
× 550.0	162.00	20.24	17.200	3.820	2.380	9430.0	931.00	7.63	3250.00	378.00	4.490
× 500.0	147.00	19.60	17.010	3.500	2.190	8210.0	838.00	7.48	2880.00	339.00	4.430
× 455.0	134.00	19.02	16.835	3.210	2.015	7190.0	756.00	7.33	2560.00	304.00	4.380
W 14 × 426.0	125.00	18.67	16.695	3.035	1.875	6600.0	707.00	7.26	2360.00	283.00	4.340
× 398.0	117.00	18.29	16.590	2.845	1.770	6000.0	656.00	7.16	2170.00	262.00	4.310
× 370.0	109.00	17.92	16.475	2.660	1.655	5440.0	607.00	7.07	1990.00	241.00	4.270
× 342.0	101.00	17.54	16.360	2.470	1.540	4900.0	559.00	6.98	1810.00	221.00	4.240

Table A-3 W shapes: properties and dimensions for design, continued (SI units)

Section and mass, kg/m	Wt, kN/m	A, m² ×10⁻³	d, mm	b_f, mm	t_f, mm	t_w, mm	I_x, m⁴ ×10⁻⁶	S_x, m³ ×10⁻³	r_x, mm	I_y, m⁴ ×10⁻⁶	S_y, m³ ×10⁻³	r_y, mm
W 530 × 218.8	2.15	27.87	560.	318.	29.2	18.3	1510.9	5.39	232.9	156.50	0.985	74.9
× 196.4	1.93	25.03	554.	316.	26.3	16.5	1340.3	4.83	231.6	138.60	0.877	74.4
× 181.6	1.78	23.16	551.	315.	24.4	15.2	1232.0	4.47	230.9	126.95	0.806	74.2
× 165.2	1.62	21.10	546.	313.	22.2	14.0	1111.3	4.08	229.9	114.05	0.729	73.7
× 150.3	1.47	19.23	543.	312.	20.3	12.7	1007.3	3.72	229.1	103.23	0.660	73.4
W 530 × 138.4	1.36	17.61	549.	214.	23.6	14.7	861.6	3.15	221.0	38.67	0.362	46.7
× 123.5	1.21	15.68	544.	212.	21.2	13.1	761.7	2.80	220.2	33.88	0.320	46.5
× 108.6	1.07	13.87	539.	211.	18.8	11.6	666.0	2.47	219.5	29.39	0.279	46.0
× 101.2	0.99	12.90	537.	210.	17.4	10.9	616.0	2.29	218.4	26.93	0.257	45.7
× 92.3	0.90	11.81	533.	209.	15.6	10.2	553.6	2.08	216.9	23.93	0.229	45.0
W 530 × 84.8	0.83	10.77	535.	166.	16.5	10.3	487.0	1.82	212.3	12.74	0.153	34.3
× 74.4	0.73	9.48	529.	166.	13.6	9.7	409.6	1.55	207.8	10.36	0.125	33.0
× 65.5	0.64	8.39	525.	165.	11.4	8.9	350.9	1.34	204.7	8.62	0.104	32.0
W 460 × 177.1	1.74	22.65	482.	286.	26.9	16.6	911.5	3.79	200.7	105.31	0.736	68.3
× 157.7	1.55	20.06	476.	284.	23.9	15.0	795.0	3.34	199.1	91.57	0.646	67.6
× 144.4	1.42	18.39	472.	283.	22.1	13.6	728.4	3.08	198.6	83.66	0.592	67.3
× 128.0	1.26	16.32	467.	282.	19.6	12.2	636.8	2.72	197.4	72.84	0.518	66.8
× 113.1	1.11	14.39	463.	280.	17.3	10.8	553.6	2.39	196.3	63.27	0.452	66.3
W 460 × 105.7	1.04	13.42	469.	194.	20.6	12.6	487.0	2.08	190.5	25.10	0.259	43.2
× 96.7	0.95	12.32	466.	193.	19.0	11.4	445.4	1.92	190.2	22.81	0.236	42.9
× 89.3	0.88	11.35	463.	192.	17.7	10.5	409.6	1.77	189.7	20.85	0.218	42.9
× 81.8	0.80	10.45	460.	191.	16.0	9.9	370.4	1.61	188.2	18.69	0.195	42.4
× 74.4	0.73	9.48	457.	190.	14.5	9.0	333.0	1.46	187.5	16.69	0.175	41.9
W 460 × 68.5	0.67	8.71	459.	154.	15.4	9.1	296.4	1.29	184.1	9.37	0.122	32.8
× 59.5	0.58	7.61	455.	153.	13.3	8.0	254.7	1.12	183.1	7.95	0.104	32.3
× 52.1	0.51	6.65	450.	152.	10.8	7.6	212.3	0.94	178.8	6.37	0.084	31.0
W 410 × 148.8	1.46	18.97	431.	265.	25.0	14.9	620.2	2.87	180.3	77.42	0.585	63.8
× 132.4	1.30	16.90	425.	263.	22.2	13.3	541.1	2.54	179.1	67.85	0.515	63.2
× 114.6	1.12	14.58	420.	261.	19.3	11.6	462.0	2.20	177.8	57.44	0.441	62.7
× 99.7	0.98	12.71	415.	260.	16.9	10.0	397.1	1.92	176.8	49.53	0.380	62.5
W 410 × 84.8	0.83	10.84	417.	181.	18.2	10.9	315.5	1.51	170.7	17.94	0.198	40.6
× 74.4	0.73	9.48	413.	180.	16.0	9.7	274.3	1.33	169.7	15.48	0.172	40.4
× 67.0	0.66	8.58	410.	179.	14.4	8.8	243.9	1.19	168.9	13.65	0.153	39.9
× 59.5	0.58	7.61	407.	178.	12.8	7.7	215.6	1.06	168.4	12.03	0.135	39.9
W 410 × 53.6	0.53	6.84	403.	177.	10.9	7.5	186.5	0.93	165.4	10.20	0.115	38.6
× 46.1	0.45	5.88	403.	140.	11.2	7.0	156.1	0.77	162.8	5.16	0.074	29.7
× 38.7	0.38	4.95	399.	140.	8.8	6.3	125.3	0.63	159.0	3.99	0.057	28.4
W 360 × 1086.4	10.65	138.71	569.	454.	124.7	78.0	5952.1	20.98	207.5	1964.6	8.636	119.1
× 989.6	9.70	126.45	550.	448.	114.8	71.9	5161.3	18.85	202.7	1735.7	7.735	117.3
× 900.3	8.83	114.84	531.	442.	105.7	65.9	4495.3	17.04	198.1	1531.7	6.932	115.6
× 818.5	8.03	104.52	514.	437.	97.0	60.5	3925.1	15.26	193.8	1352.8	6.194	114.0
× 744.1	7.30	94.84	498.	432.	88.9	55.6	3417.3	13.73	190.0	1198.7	5.555	112.5
× 677.1	6.64	86.45	483.	428.	81.5	51.2	2992.7	12.39	186.2	1065.6	4.982	111.3
W 360 × 634.0	6.22	80.64	474.	424.	77.1	47.6	2747.1	11.59	184.4	982.3	4.638	110.2
× 592.3	5.81	75.48	465.	421.	72.3	45.0	2497.4	10.75	181.9	903.2	4.293	109.5
× 550.6	5.40	70.32	455.	418.	67.6	42.0	2264.3	9.95	179.6	828.30	3.949	108.5
× 509.0	4.99	65.16	446.	416.	62.7	39.1	2039.5	9.16	177.3	753.38	3.622	107.7

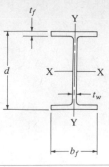

Table A-3 W shapes: properties and dimensions for design, continued (English customary units)

Section and weight, lb/ft	A, in.^{2}	d, in.	b_f, in.	t_f, in.	t_w, in.	I_x, in.^{4}	S_x, in.^{3}	r_x, in.	I_y, in.^{4}	S_y, in.^{3}	r_y, in.
W 14 × 311.0	91.40	17.12	16.230	2.260	1.410	4330.0	506.00	6.88	1610.00	199.00	4.200
× 283.0	83.30	16.74	16.110	2.070	1.290	3840.0	459.00	6.79	1440.00	179.00	4.170
× 257.0	75.60	16.38	15.995	1.890	1.175	3400.0	415.00	6.71	1290.00	161.00	4.130
× 233.0	68.50	16.04	15.890	1.720	1.070	3010.0	375.00	6.63	1150.00	145.00	4.100
× 211.0	62.00	15.72	15.800	1.560	0.980	2660.0	338.00	6.55	1030.00	130.00	4.070
× 193.0	56.80	15.48	15.710	1.440	0.890	2400.0	310.00	6.50	931.00	119.00	4.050
× 176.0	51.80	15.22	15.650	1.310	0.830	2140.0	281.00	6.43	838.00	107.00	4.020
× 159.0	46.70	14.98	15.565	1.190	0.745	1900.0	254.00	6.38	748.00	96.20	4.000
× 145.0	42.70	14.78	15.500	1.090	0.680	1710.0	232.00	6.33	677.00	87.30	3.980
W 14 × 132.0	38.80	14.66	14.725	1.030	0.645	1530.0	209.00	6.28	548.00	74.50	3.760
× 120.0	35.30	14.48	14.670	0.940	0.590	1380.0	190.00	6.24	495.00	67.50	3.740
× 109.0	32.00	14.32	14.605	0.860	0.525	1240.0	173.00	6.22	447.00	61.20	3.730
× 99.0	29.10	14.16	14.565	0.780	0.485	1110.0	157.00	6.17	402.00	55.20	3.710
× 90.0	26.50	14.02	14.520	0.710	0.440	999.0	143.00	6.14	362.00	49.90	3.700
× 82.0	24.10	14.31	10.130	0.855	0.510	882.0	123.00	6.05	148.00	29.30	2.480
× 74.0	21.80	14.17	10.070	0.785	0.450	796.0	112.00	6.04	134.00	26.60	2.480
W 14 × 68.0	20.00	14.04	10.035	0.720	0.415	723.0	103.00	6.01	121.00	24.20	2.460
× 61.0	17.90	13.89	9.995	0.645	0.375	640.0	92.20	5.98	107.00	21.50	2.450
× 53.0	15.60	13.92	8.060	0.660	0.370	541.0	77.80	5.89	57.70	14.30	1.920
× 48.0	14.10	13.79	8.030	0.595	0.340	485.0	70.30	5.85	51.40	12.80	1.910
× 43.0	12.60	13.66	7.995	0.530	0.305	428.0	62.70	5.82	45.20	11.30	1.890
× 38.0	11.20	14.10	6.770	0.515	0.310	385.0	54.60	5.87	26.70	7.88	1.550
× 34.0	10.00	13.98	6.745	0.455	0.285	340.0	48.60	5.83	23.30	6.91	1.530
W 14 × 30.0	8.85	13.84	6.730	0.385	0.270	291.0	42.00	5.73	19.60	5.82	1.490
× 26.0	7.69	13.91	5.025	0.420	0.255	245.0	35.30	5.65	8.91	3.55	1.080
× 22.0	6.49	13.74	5.000	0.335	0.230	199.0	9.00	5.54	7.00	0.80	1.040
W 12 × 336.0	98.80	16.82	13.385	2.955	1.775	4060.0	483.00	6.41	1190.00	177.00	3.470
× 305.0	89.60	16.32	13.235	2.705	1.625	3550.0	435.00	6.29	1050.00	159.00	3.420
× 279.0	81.90	15.85	13.140	2.470	1.530	3110.0	393.00	6.16	937.00	143.00	3.380
× 252.0	74.10	15.41	13.005	2.250	1.395	2720.0	353.00	6.06	828.00	127.00	3.340
× 230.0	67.70	15.05	12.895	2.070	1.285	2420.0	321.00	5.97	742.00	115.00	3.310
× 210.0	61.80	14.71	12.790	1.900	1.180	2140.0	292.00	5.89	664.00	104.00	3.280
W 12 × 190.0	55.80	14.38	12.670	1.735	1.060	1890.0	263.00	5.82	589.00	93.00	3.250
× 170.0	50.00	14.03	12.570	1.560	0.960	1650.0	235.00	5.74	517.00	82.30	3.220
× 152.0	44.70	13.71	12.480	1.400	0.870	1430.0	209.00	5.66	454.00	72.80	3.190
× 136.0	39.90	13.41	12.400	1.250	0.790	1240.0	186.00	5.58	398.00	64.20	3.160
× 120.0	35.30	13.12	12.320	1.105	0.710	1070.0	163.00	5.51	345.00	56.00	3.130
× 106.0	31.20	12.89	12.220	0.990	0.610	933.0	145.00	5.47	301.00	49.30	3.110
W 12 × 96.0	28.20	12.71	12.160	0.900	0.550	833.0	131.00	5.44	270.00	44.40	3.090
× 87.0	25.60	12.53	12.125	0.810	0.515	740.0	118.00	5.38	241.00	39.70	3.070
× 79.0	23.20	12.38	12.080	0.735	0.470	662.0	107.00	5.34	216.00	35.80	3.050
× 72.0	21.10	12.25	12.040	0.670	0.430	597.0	97.40	5.31	195.00	32.40	3.040
W 12 × 65.0	19.10	12.12	12.000	0.605	0.390	533.0	87.90	5.28	174.00	29.10	3.020
× 58.0	17.00	12.19	10.010	0.640	0.360	475.0	78.00	5.28	107.00	21.40	2.510
× 53.0	15.60	12.06	9.995	0.575	0.345	425.0	70.60	5.23	95.80	19.20	2.480
× 50.0	14.70	12.19	8.080	0.640	0.370	394.0	64.70	5.18	56.30	13.90	1.960
× 45.0	13.20	12.06	8.045	0.575	0.335	350.0	58.10	5.15	50.00	12.40	1.940
× 40.0	11.80	11.94	8.005	0.515	0.295	310.0	51.90	5.13	44.10	11.00	1.940
W 12 × 35.0	10.30	12.50	6.560	0.520	0.300	285.0	45.60	5.25	24.50	7.47	1.544
× 30.0	8.79	12.34	6.520	0.440	0.260	238.0	38.60	5.01	20.30	6.04	1.500

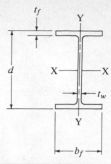

Table A-3 W shapes: properties and dimensions for design, continued (SI units)

Section and mass, kg/m	Wt, kN/m	A, m² × 10⁻³	d, mm	b_f, mm	t_f, mm	t_w, mm	I_x, m⁴ × 10⁻⁶	S_x, m³ × 10⁻³	r_x, mm	I_y, m⁴ × 10⁻⁶	S_y, m³ × 10⁻³	r_y, mm
W 360 × 462.8	4.54	58.97	435.	412.	57.4	35.8	1802.3	8.29	174.8	670.13	3.261	106.7
× 421.2	4.13	53.74	425.	409.	52.6	32.8	1598.3	7.52	172.5	599.37	2.933	105.9
× 382.5	3.75	48.77	416.	406.	48.0	29.8	1415.2	6.80	170.4	536.94	2.638	104.9
× 346.7	3.40	44.19	407.	404.	43.7	27.2	1252.9	6.15	168.4	478.67	2.376	104.1
× 314.0	3.08	40.00	399.	401.	39.6	24.9	1107.2	5.54	166.4	428.72	2.130	103.4
× 287.2	2.82	36.65	393.	399.	36.6	22.6	999.0	5.08	165.1	387.51	1.950	102.9
× 261.9	2.57	33.42	387.	398.	33.3	21.1	890.7	4.60	163.3	348.80	1.753	102.1
× 236.6	2.32	30.13	380.	395.	30.2	18.9	790.8	4.16	162.1	311.34	1.576	101.6
× 215.8	2.12	27.55	375.	394.	27.7	17.3	711.8	3.80	160.8	281.79	1.431	101.1
W 360 × 196.4	1.93	25.03	372.	374.	26.2	16.4	636.8	3.42	159.5	228.09	1.221	95.5
× 178.6	1.75	22.77	368.	373.	23.9	15.0	574.4	3.11	158.5	206.03	1.106	95.0
× 162.2	1.59	20.65	364.	371.	21.8	13.3	516.1	2.83	158.0	186.06	1.003	94.7
× 147.3	1.44	18.77	360.	370.	19.8	12.3	462.0	2.57	156.7	167.32	0.905	94.2
× 133.9	1.31	17.10	356.	369.	18.0	11.2	415.8	2.34	156.0	150.68	0.818	94.0
× 122.0	1.20	15.55	363.	257.	21.7	13.0	367.1	2.02	153.7	61.60	0.480	63.0
× 110.1	1.08	14.06	360.	256.	19.9	11.4	331.3	1.84	153.4	55.77	0.436	63.0
W 360 × 101.2	0.99	12.90	357.	255.	18.3	10.5	300.9	1.69	152.7	50.36	0.397	62.5
× 90.8	0.89	11.55	353.	254.	16.4	9.5	266.4	1.51	151.9	44.54	0.352	62.2
× 78.9	0.77	10.06	354.	205.	16.8	9.4	225.2	1.27	149.6	24.02	0.234	48.8
× 71.4	0.70	9.10	350.	204.	15.1	8.6	201.9	1.15	148.6	21.39	0.210	48.5
× 64.0	0.63	8.13	347.	203.	13.5	7.7	178.1	1.03	147.8	18.81	0.185	48.0
× 56.6	0.55	7.23	358.	172.	13.1	7.9	160.2	0.89	149.1	11.11	0.129	39.4
× 50.6	0.50	6.45	355.	171.	11.6	7.2	141.5	0.80	148.1	9.70	0.113	38.9
W 360 × 44.6	0.44	5.71	352.	171.	9.8	6.9	121.1	0.69	145.5	8.16	0.095	37.8
× 38.7	0.38	4.96	353.	128.	10.7	6.5	102.0	0.58	143.5	3.71	0.058	27.4
× 32.7	0.32	4.19	349.	127.	8.5	5.8	82.8	0.15	140.7	2.91	0.013	26.4
W 310 × 500.0	4.90	63.74	427.	340.	75.1	45.1	1689.9	7.91	162.8	495.31	2.901	88.1
× 453.9	4.45	57.81	415.	336.	68.7	41.3	1477.6	7.13	159.8	437.04	2.606	86.9
× 415.2	4.07	52.84	403.	334.	62.7	38.9	1294.5	6.44	156.5	390.01	2.343	85.9
× 375.0	3.68	47.81	391.	330.	57.1	35.4	1132.1	5.78	153.9	344.64	2.081	84.8
× 342.3	3.36	43.68	382.	328.	52.6	32.6	1007.3	5.26	151.6	308.84	1.885	84.1
× 312.5	3.06	39.87	374.	325.	48.3	30.0	890.7	4.79	149.6	276.38	1.704	83.3
W 310 × 282.8	2.77	36.00	365.	322.	44.1	26.9	786.7	4.31	147.8	245.16	1.524	82.5
× 253.0	2.48	32.26	356.	319.	39.6	24.4	686.8	3.85	145.8	215.19	1.349	81.8
× 226.2	2.22	28.84	348.	317.	35.6	22.1	595.2	3.42	143.8	188.97	1.193	81.0
× 202.4	1.98	25.74	341.	315.	31.7	20.1	516.1	3.05	141.7	165.66	1.052	80.3
× 178.6	1.75	22.77	333.	313.	28.1	18.0	445.4	2.67	140.0	143.60	0.918	79.5
× 157.7	1.55	20.13	327.	310.	25.1	15.5	388.3	2.38	138.9	125.29	0.808	79.0
W 310 × 142.9	1.40	18.19	323.	309.	22.9	14.0	346.7	2.15	138.2	112.38	0.728	78.5
× 129.5	1.27	16.52	318.	308.	20.6	13.1	308.0	1.93	136.7	100.31	0.651	78.0
× 117.6	1.15	14.97	314.	307.	18.7	11.9	275.5	1.75	135.6	89.91	0.587	77.5
× 107.1	1.05	13.61	311.	306.	17.0	10.9	248.5	1.60	134.9	81.17	0.531	77.2
W 310 × 96.7	0.95	12.32	308.	305.	15.4	9.9	221.9	1.44	134.1	72.42	0.477	76.7
× 86.3	0.85	10.97	310.	254.	16.3	9.1	197.7	1.28	134.1	44.54	0.351	63.8
× 78.9	0.77	10.06	306.	254.	14.6	8.8	176.9	1.16	132.8	39.87	0.315	63.0
× 74.4	0.73	9.48	310.	205.	16.3	9.4	164.0	1.06	131.6	23.43	0.228	49.8
× 67.0	0.66	8.52	306.	204.	14.6	8.5	145.7	0.95	130.8	20.81	0.203	49.3
× 59.5	0.58	7.61	303.	203.	13.1	7.5	129.0	0.85	130.3	18.36	0.180	49.3
W 310 × 52.1	0.51	6.65	317.	167.	13.2	7.6	118.6	0.75	133.3	10.20	0.122	39.2
× 44.6	0.44	5.67	313.	166.	11.2	6.6	99.1	0.63	127.3	8.45	0.099	38.1

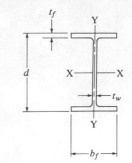

only for lateral loads

Section and weight, lb/ft	A, in.²	d, in.	b_f, in.	t_f, in.	t_w, in.	I_x, in.⁴	S_x, in.³	r_x, in.	I_y, in.⁴	S_y, in.³	r_y, in.
W 12 × 26.0	7.65	12.22	6.490	0.380	0.230	204.0	33.40	5.17	17.30	5.34	1.510
× 22.0	6.48	12.31	4.030	0.425	0.260	156.0	25.40	4.91	4.66	2.31	0.848
W 12 × 19.0	5.57	12.16	4.005	0.350	0.235	130.0	21.30	4.82	3.76	1.88	0.822
× 16.0	4.71	11.99	3.990	0.265	0.220	103.0	17.10	4.67	2.82	1.41	0.773
× 14.0	4.16	11.91	3.970	0.225	0.200	88.60	14.90	4.62	2.36	1.19	0.753
W 10 × 112.0	32.90	11.36	10.415	1.250	0.755	716.0	126.00	4.66	236.00	45.30	2.680
× 100.0	29.40	11.10	10.340	1.120	0.680	623.0	112.00	4.60	207.00	40.00	2.650
× 88.0	25.90	10.84	10.265	0.990	0.605	534.0	98.40	4.54	179.00	34.80	2.630
× 77.0	22.60	10.60	10.190	0.870	0.530	455.0	85.90	4.49	154.00	30.10	2.600
× 68.0	20.00	10.40	10.130	0.770	0.470	394.0	75.70	4.44	134.00	26.40	2.590
× 60.0	17.60	10.22	10.080	0.680	0.420	341.0	66.70	4.39	116.00	23.00	2.570
× 54.0	15.80	10.09	10.030	0.615	0.370	303.0	60.00	4.37	103.00	20.60	2.560
× 49.0	14.40	9.98	10.000	0.560	0.340	272.0	54.60	4.35	93.40	18.70	2.540
W 10 × 45.0	13.30	10.10	8.020	0.620	0.350	248.0	49.10	4.32	53.40	13.30	2.010
× 39.0	11.50	9.92	7.985	0.530	0.315	209.0	42.10	4.27	45.00	11.30	1.980
× 33.0	9.71	9.73	7.960	0.435	0.290	171.0	35.00	4.19	36.60	9.20	1.940
× 30.0	8.84	10.47	5.810	0.510	0.300	170.0	32.40	4.38	16.70	5.75	1.370
× 26.0	7.61	10.33	5.770	0.440	0.260	144.0	27.90	4.35	14.10	4.89	1.360
× 22.0	6.49	10.17	5.750	0.360	0.240	118.0	23.20	4.27	11.40	3.97	1.330
W 10 × 19.0	5.62	10.24	4.020	0.395	0.250	96.30	18.80	4.14	4.29	2.14	0.874
× 17.0	4.99	10.11	4.010	0.330	0.240	81.90	16.20	4.05	3.56	1.78	0.845
× 15.0	4.41	9.99	4.000	0.270	0.230	68.90	13.80	3.95	2.89	1.45	0.810
× 12.0	3.54	9.87	3.960	0.210	0.190	53.80	10.90	3.90	2.18	1.10	0.785
W 8 × 67.0	19.70	9.00	8.280	0.933	0.570	272.0	60.40	3.72	88.60	21.40	2.120
× 58.0	17.10	8.75	8.220	0.810	0.510	228.0	52.00	3.65	75.10	18.30	2.100
× 48.0	14.10	8.50	8.110	0.685	0.400	184.0	43.30	3.61	60.90	15.00	2.080
× 40.0	11.70	8.25	8.070	0.560	0.360	146.0	35.50	3.53	49.10	12.20	2.040
× 35.0	10.30	8.12	8.020	0.495	0.310	127.0	31.20	3.51	42.60	10.60	2.030
× 31.0	9.13	8.00	7.995	0.435	0.285	110.0	27.50	3.47	37.10	9.27	2.020
W 8 × 28.0	8.25	8.06	6.535	0.465	0.285	98.00	24.30	3.45	21.70	6.63	1.620
× 24.0	7.08	7.93	6.495	0.400	0.245	82.80	20.90	3.42	18.30	5.63	1.610
× 21.0	6.16	8.28	5.270	0.400	0.250	75.30	18.20	3.49	9.77	3.71	1.260
W 8 × 18.0	5.26	8.14	5.250	0.330	0.230	61.90	15.20	3.43	7.97	3.04	1.230
× 15.0	4.44	8.11	4.015	0.315	0.245	48.00	11.80	3.29	3.41	1.70	0.876
× 13.0	3.84	7.99	4.000	0.255	0.230	39.60	9.91	3.21	2.73	1.37	0.843
× 10.0	2.96	7.89	3.940	0.205	0.170	30.80	7.81	3.22	2.09	1.06	0.841
W 6 × 25.0	7.34	6.38	6.080	0.455	0.320	53.40	16.70	2.70	17.10	5.61	1.520
× 20.0	5.87	6.20	6.020	0.365	0.260	41.40	13.40	2.66	13.30	4.41	1.500
× 15.0	4.43	5.99	5.990	0.260	0.230	29.10	9.72	2.56	9.32	3.11	1.450
W 6 × 16.0	4.74	6.28	4.030	0.405	0.260	32.10	10.20	2.60	4.43	2.20	0.967
× 12.0	3.55	6.03	4.000	0.280	0.230	22.10	7.31	2.49	2.99	1.50	0.918
× 9.0	2.68	5.90	3.940	0.215	0.170	16.40	5.56	2.47	2.20	1.11	0.905
W 5 × 19.0	5.56	5.15	5.030	0.430	0.270	26.30	10.20	2.17	9.13	3.63	1.280
× 16.0	4.71	5.01	5.000	0.360	0.240	21.40	8.55	2.13	7.51	3.00	1.260
W 4 × 13.0	3.83	4.16	4.060	0.345	0.280	11.30	5.46	1.72	3.86	1.90	1.000

Table A-3 W shapes: properties and dimensions for design, continued (English customary units)

Section Modulus · *Cgyration* (radius of gyration)

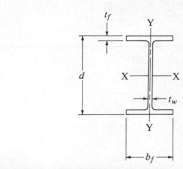

Table A-3   W shapes: properties and dimensions for design, continued (SI units)

Section and mass, kg/m	Wt, kN/m	A, m² × 10⁻³	d, mm	b_f, mm	t_f, mm	t_w, mm	I_x, m⁴ × 10⁻⁶	S_x, m³ × 10⁻³	r_x, mm	I_y, m⁴ × 10⁻⁶	S_y, m³ × 10⁻³	r_y, mm
W 310 × 38.7	0.38	4.94	310.	165.	9.7	5.8	84.9	0.55	131.3	7.20	0.088	38.4
× 32.7	0.32	4.18	313.	102.	10.8	6.6	64.9	0.42	124.7	1.94	0.038	21.5
W 310 × 28.3	0.28	3.59	309.	102.	8.9	6.0	54.1	0.35	122.4	1.57	0.031	20.9
× 23.8	0.23	3.04	305.	101.	6.7	5.6	42.9	0.28	118.6	1.17	0.023	19.6
× 20.8	0.20	2.68	303.	101.	5.7	5.1	36.9	0.24	117.3	0.98	0.020	19.1
W 250 × 166.7	1.63	21.23	289.	265.	31.7	19.2	298.0	2.06	118.4	98.23	0.742	68.1
× 148.8	1.46	18.97	282.	263.	28.4	17.3	259.3	1.84	116.8	86.16	0.655	67.3
× 131.0	1.28	16.71	275.	261.	25.1	15.4	222.3	1.61	115.3	74.51	0.570	66.8
× 114.6	1.12	14.58	269.	259.	22.1	13.5	189.4	1.41	114.0	64.10	0.493	66.0
× 101.2	0.99	12.90	264.	257.	19.6	11.9	164.0	1.24	112.8	55.77	0.433	65.8
× 89.3	0.88	11.35	260.	256.	17.3	10.7	141.9	1.09	111.5	48.28	0.377	65.3
× 80.4	0.79	10.19	256.	255.	15.6	9.4	126.1	0.98	111.0	42.87	0.338	65.0
× 72.9	0.72	9.29	253.	254.	14.2	8.6	113.2	0.89	110.5	38.88	0.306	64.5
W 250 × 67.0	0.66	8.58	257.	204.	15.7	8.9	103.2	0.80	109.7	22.23	0.218	51.1
× 58.0	0.57	7.42	252.	203.	13.5	8.0	87.0	0.69	108.5	18.73	0.185	50.3
× 49.1	0.48	6.26	247.	202.	11.0	7.4	71.2	0.57	106.4	15.23	0.151	49.3
× 44.6	0.44	5.70	266.	148.	13.0	7.6	70.8	0.53	111.3	6.95	0.094	34.8
× 38.7	0.38	4.91	262.	147.	11.2	6.6	59.9	0.46	110.5	5.87	0.080	34.5
× 32.7	0.32	4.19	258.	146.	9.1	6.1	49.1	0.38	108.5	4.75	0.065	33.8
W 250 × 28.3	0.28	3.63	260.	102.	10.0	6.3	40.1	0.31	105.2	1.79	0.035	22.2
× 25.3	0.25	3.22	257.	102.	8.4	6.1	34.1	0.27	102.9	1.48	0.029	21.5
× 22.3	0.22	2.85	254.	102.	6.9	5.8	28.7	0.23	100.3	1.20	0.024	20.6
× 17.9	0.18	2.28	251.	101.	5.3	4.8	22.4	0.18	99.1	0.91	0.018	19.9
W 200 × 99.7	0.98	12.71	229.	210.	23.7	14.5	113.2	0.99	94.5	36.88	0.351	53.8
× 86.3	0.85	11.03	222.	209.	20.6	13.0	94.9	0.85	92.7	31.26	0.300	53.3
× 71.4	0.70	9.10	216.	206.	17.4	10.2	76.6	0.71	91.7	25.35	0.246	52.8
× 59.5	0.58	7.55	210.	205.	14.2	9.1	60.8	0.58	89.7	20.44	0.200	51.8
× 52.1	0.51	6.65	206.	204.	12.6	7.9	52.9	0.51	89.2	17.73	0.174	51.6
× 46.1	0.45	5.89	203.	203.	11.0	7.2	45.8	0.45	88.1	15.44	0.152	51.3
W 200 × 41.7	0.41	5.32	205.	166.	11.8	7.2	40.8	0.40	87.6	9.03	0.109	41.1
× 35.7	0.35	4.57	201.	165.	10.2	6.2	34.5	0.34	86.9	7.62	0.092	40.9
× 31.3	0.31	3.97	210.	134.	10.2	6.3	31.3	0.30	88.6	4.07	0.061	32.0
W 200 × 26.8	0.26	3.39	207.	133.	8.4	5.8	25.8	0.25	87.1	3.32	0.050	31.2
× 22.3	0.22	2.86	206.	102.	8.0	6.2	20.0	0.19	83.6	1.42	0.028	22.3
× 19.3	0.19	2.48	203.	102.	6.5	5.8	16.5	0.16	81.5	1.14	0.022	21.4
× 14.9	0.15	1.91	200.	100.	5.2	4.3	12.8	0.13	81.8	0.87	0.017	21.4
W 150 × 37.2	0.36	4.74	162.	154.	11.6	8.1	22.2	0.27	68.6	7.12	0.092	38.6
× 29.8	0.29	3.79	157.	153.	9.3	6.6	17.2	0.22	67.6	5.54	0.072	38.1
× 22.3	0.22	2.86	152.	152.	6.6	5.8	12.1	0.16	65.0	3.88	0.051	36.8
W 150 × 23.8	0.23	3.06	160.	102.	10.3	6.6	13.4	0.17	66.0	1.84	0.036	24.6
× 17.9	0.18	2.29	153.	102.	7.1	5.8	9.2	0.12	63.2	1.24	0.025	23.3
× 13.4	0.13	1.73	150.	100.	5.5	4.3	6.8	0.09	62.7	0.92	0.018	23.0
W 130 × 28.3	0.28	3.59	131.	128.	10.9	6.9	10.9	0.17	55.1	3.80	0.059	32.5
× 23.8	0.23	3.04	127.	127.	9.1	6.1	8.9	0.14	54.1	3.13	0.049	32.0
W 100 × 19.3	0.19	2.47	106.	103.	8.8	7.1	4.7	0.09	43.7	1.61	0.031	25.4

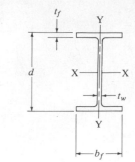

Table A-4 M shapes: properties and dimensions for design (English customary units)[a]

Section and weight, lb/ft	A, in.^{2}	d, in.	b_f, in.	t_f, in.	t_w, in.	I_x, in.^{4}	S_x, in.^{3}	r_x, in.	I_y, in.^{4}	S_y, in.^{3}	r_y, in.
M 14 × 17.2	5.05	14.00	4.000	0.272	0.210	147.0	21.10	5.40	2.65	1.33	0.725
M 12 × 11.8	3.47	12.00	3.065	0.225	0.177	71.90	12.00	4.55	0.98	0.64	0.532
M 10 × 29.1	8.56	9.88	5.937	0.389	0.427	131.0	26.60	3.92	11.20	3.76	1.140
× 22.9	6.73	9.88	5.752	0.389	0.242	117.0	23.60	4.16	10.00	3.48	1.220
M 10 × 9.0	2.65	10.00	2.690	0.206	0.157	38.80	7.76	3.83	0.61	0.45	0.480
M 8 × 37.7	11.10	8.12	8.002	0.521	0.377	132.0	32.60	3.46	40.40	10.10	1.910
× 34.3	10.10	8.00	8.003	0.459	0.378	116.0	29.10	3.40	34.90	8.73	1.860
× 32.6	9.58	8.00	7.940	0.459	0.315	114.0	28.40	3.44	34.10	8.58	1.890
M 8 × 22.5	6.60	8.00	5.395	0.353	0.375	68.20	17.10	3.22	7.48	2.77	1.060
× 18.5	5.44	8.00	5.250	0.353	0.230	62.00	15.50	3.38	6.82	2.60	1.120
M 8 × 6.5	1.92	8.00	2.281	0.189	0.135	18.50	4.62	3.10	0.34	0.30	0.423
M 7 × 5.5	1.62	7.00	2.080	0.180	0.128	12.00	3.44	2.73	0.25	0.24	0.392
M 6 × 33.8	9.93	6.25	6.114	0.605	0.488	64.70	20.70	2.55	21.40	6.99	1.470
× 22.5	6.62	6.00	6.060	0.379	0.372	41.20	13.70	2.49	12.40	4.08	1.370
× 20.0	5.89	6.00	5.938	0.379	0.250	39.00	13.00	2.57	11.60	3.90	1.400
M 6 × 4.4	1.29	6.00	1.844	0.171	0.114	7.20	2.40	2.36	0.16	0.18	0.358
M 5 × 18.9	5.55	5.00	5.003	0.416	0.316	24.10	9.63	2.08	7.86	3.14	1.190
M 4 × 16.3	4.80	4.20	3.938	0.472	0.312	14.00	6.67	1.71	4.44	2.25	0.962
× 13.8	4.06	4.00	4.000	0.371	0.313	10.80	5.42	1.63	3.58	1.79	0.939
× 13.0	3.81	4.00	3.940	0.371	0.254	10.50	5.24	1.66	3.36	1.71	0.939

[a] From *The Structural Steel Design Manual*, by Joseph E. Bowles (New York: McGraw-Hill, 1980).

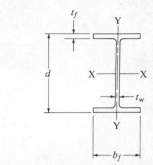

Table A-4 M shapes: properties and dimensions for design (SI units)

Section and mass, kg/m	Wt, kN/m	A, m^2 $\times 10^{-3}$	d, mm	b_f, mm	t_f, mm	t_w, mm	I_x, m^4 $\times 10^{-6}$	S_x, m^3 $\times 10^{-3}$	r_x, mm	I_y, m^4 $\times 10^{-6}$	S_y, m^3 $\times 10^{-3}$	r_y, mm
M 360 × 25.6	0.25	3.26	356.	102.	6.9	5.3	61.2	0.35	137.2	1.10	0.022	18.4
M 310 × 17.6	0.17	2.24	305.	78.	5.7	4.5	29.9	0.20	115.6	0.41	0.010	13.5
M 250 × 43.3	0.42	5.52	251.	151.	9.9	10.8	54.5	0.44	99.6	4.66	0.062	29.0
× 34.1	0.33	4.34	251.	146.	9.9	6.1	48.7	0.39	105.7	4.16	0.057	31.0
M 250 × 13.4	0.13	1.71	254.	68.	5.2	4.0	16.1	0.13	97.3	0.25	0.007	12.2
M 200 × 56.1	0.55	7.16	206.	203.	13.2	9.6	54.9	0.53	87.9	16.82	0.166	48.5
× 51.0	0.50	6.52	203.	203.	11.7	9.6	48.3	0.48	86.4	14.53	0.143	47.2
× 48.5	0.48	6.18	203.	202.	11.7	8.0	47.5	0.47	87.4	14.19	0.141	48.0
M 200 × 33.5	0.33	4.26	203.	137.	9.0	9.5	28.4	0.28	81.8	3.11	0.045	26.9
× 27.5	0.27	3.51	203.	133.	9.0	5.8	25.8	0.25	85.9	2.84	0.043	28.4
M 200 × 9.7	0.09	1.24	203.	58.	4.8	3.4	7.7	0.08	78.7	0.14	0.005	10.7
M 180 × 8.2	0.08	1.05	178.	53.	4.6	3.3	5.0	0.06	69.3	0.10	0.004	10.0
M 150 × 50.2	0.49	6.41	159.	155.	15.4	12.4	26.9	0.34	64.8	8.91	0.115	37.3
× 33.5	0.33	4.27	152.	154.	9.6	9.4	17.1	0.22	63.2	5.16	0.067	34.8
× 29.8	0.29	3.80	152.	151.	9.6	6.3	16.2	0.21	65.3	4.83	0.064	35.6
M 150 × 6.5	0.06	0.83	152.	47.	4.3	2.9	3.0	0.04	59.9	0.07	0.003	9.1
M 130 × 28.1	0.28	3.58	127.	127.	10.6	8.0	10.0	0.16	52.8	3.27	0.051	30.2
M 100 × 24.3	0.24	3.10	107.	100.	12.0	7.9	5.8	0.11	43.4	1.85	0.037	24.4
× 20.5	0.20	2.62	102.	102.	9.4	8.0	4.5	0.09	41.4	1.49	0.029	23.9
× 19.3	0.19	2.46	102.	100.	9.4	6.5	4.4	0.09	42.2	1.40	0.028	23.9

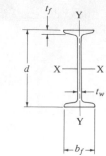

Table A-5 S shapes: properties and dimensions for design (English customary units)[a]

Section and weight, lb/ft	A, in.[2]	d, in.	b_f, in.	t_f, in.	t_w, in.	I_x, in.[4]	S_x, in.[3]	r_x, in.	I_y, in.[4]	S_y, in.[3]	r_y, in.
S 24 × 121.0	35.60	24.00	8.050	1.090	0.800	3160.0	258.00	9.43	84.20	20.90	1.540
× 106.0	31.20	24.50	7.870	1.090	0.620	2940.0	240.00	9.71	78.20	19.80	1.580
S 24 × 100.0	29.30	24.00	7.245	0.870	0.745	2390.0	199.00	9.02	47.80	13.20	1.270
S 24 × 90.00	26.50	24.00	7.125	0.870	0.625	2250.0	187.00	9.21	44.90	12.60	1.300
× 80.00	23.50	24.00	7.000	0.870	0.500	2100.0	175.00	9.47	42.30	12.10	1.340
S 20 × 96.00	28.20	20.30	7.200	0.920	0.800	1670.0	165.00	7.71	49.70	13.80	1.330
× 85.00	25.30	20.30	7.060	0.920	0.660	1580.0	155.00	7.89	46.20	13.10	1.360
S 20 × 75.00	22.00	20.00	6.385	0.795	0.635	1280.0	128.00	7.62	29.60	9.28	1.160
× 66.00	19.40	20.00	6.255	0.795	0.505	1190.0	118.00	7.83	27.40	8.77	1.190
S 18 × 70.00	20.60	18.00	6.251	0.691	0.711	926.0	103.00	6.71	24.10	7.72	1.080
× 54.70	16.10	18.00	6.001	0.691	0.461	804.0	89.40	7.07	20.80	6.94	1.140
S 15 × 50.00	14.70	15.00	5.640	0.622	0.550	486.0	64.80	5.75	15.70	5.57	1.030
× 42.90	12.60	15.00	5.501	0.622	0.411	447.0	59.60	5.95	14.40	5.23	1.070
S 12 × 50.00	14.70	12.00	5.477	0.659	0.687	305.0	50.80	4.55	15.70	5.74	1.030
× 40.80	12.00	12.00	5.252	0.659	0.462	272.0	45.40	4.77	13.60	5.16	1.060
S 12 × 35.00	10.30	12.00	5.078	0.544	0.423	229.0	38.20	4.72	9.87	3.89	0.980
× 31.80	9.35	12.00	5.000	0.544	0.350	218.0	36.40	4.83	9.36	3.74	1.000
S 10 × 35.00	10.30	10.00	4.944	0.491	0.594	147.0	29.40	3.78	8.36	3.38	0.901
× 25.40	7.46	10.00	4.661	0.491	0.311	124.0	24.70	4.07	6.79	2.91	0.954
S 8 × 23.00	6.77	8.00	4.171	0.425	0.441	64.9	16.20	3.10	4.31	2.07	0.798
× 18.40	5.41	8.00	4.001	0.425	0.271	57.6	14.40	3.26	3.73	1.86	0.831
S 7 × 20.00	5.88	7.00	3.860	0.392	0.450	42.4	12.10	2.69	3.17	1.64	0.734
× 15.30	4.50	7.00	3.662	0.392	0.252	36.7	10.50	2.86	2.64	1.44	0.766
S 6 × 17.25	5.07	6.00	3.565	0.359	0.465	26.3	8.77	2.28	2.31	1.30	0.675
× 12.50	3.67	6.00	3.332	0.359	0.232	22.1	7.37	2.45	1.82	1.09	0.705
S 5 × 14.75	4.34	5.00	3.284	0.326	0.494	15.2	6.09	1.87	1.67	1.01	0.620
× 10.00	2.94	5.00	3.004	0.326	0.214	12.3	4.92	2.05	1.22	0.81	0.643
S 4 × 9.50	2.79	4.00	2.796	0.293	0.326	6.8	3.39	1.56	0.90	0.65	0.569
× 7.70	2.26	4.00	2.663	0.293	0.193	6.1	3.04	1.64	0.76	0.57	0.581
S 3 × 7.50	2.21	3.00	2.509	0.260	0.349	2.9	1.95	1.15	0.59	0.47	0.516
× 5.70	1.67	3.00	2.330	0.260	0.170	2.5	1.68	1.23	0.45	0.39	0.522

[a] From *The Structural Steel Design Manual*, by Joseph E. Bowles (New York: McGraw-Hill, 1980).

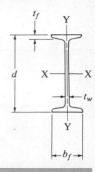

Section and mass, kg/m	Wt, kN/m	A, m² ×10⁻³	d, mm	b_f, mm	t_f, mm	t_w, mm	I_x, m⁴ ×10⁻⁶	S_x, m³ ×10⁻³	r_x, mm	I_y, m⁴ ×10⁻⁶	S_y, m³ ×10⁻³	r_y, mm
Table A-5 S shapes: properties and dimensions for design (SI units)												
S 610 × 180.1	1.77	22.97	610.	204.	27.7	20.3	1315.3	4.23	239.5	35.05	0.342	39.1
× 157.7	1.55	20.13	622.	200.	27.7	15.7	1223.7	3.93	246.6	32.55	0.324	40.1
S 610 × 148.8	1.46	18.90	610.	184.	22.1	18.9	994.8	3.26	229.1	19.90	0.216	32.3
S 610 × 133.9	1.31	17.10	610.	181.	22.1	15.9	936.5	3.06	233.9	18.69	0.206	33.0
× 119.1	1.17	15.16	610.	178.	22.1	12.7	874.1	2.87	240.5	17.61	0.198	34.0
S 510 × 142.9	1.40	18.19	516.	183.	23.4	20.3	695.1	2.70	195.8	20.69	0.226	33.8
× 126.5	1.24	16.32	516.	179.	23.4	16.8	657.6	2.54	200.4	19.23	0.215	34.5
S 510 × 111.6	1.09	14.19	508.	162.	20.2	16.1	532.8	2.10	193.5	12.32	0.152	29.5
× 98.2	0.96	12.52	508.	159.	20.2	12.8	495.3	1.93	198.9	11.40	0.144	30.2
S 460 × 104.2	1.02	13.29	457.	159.	17.6	18.1	385.4	1.69	170.4	10.03	0.127	27.4
× 81.4	0.80	10.39	457.	152.	17.6	11.7	334.6	1.47	179.6	8.66	0.114	29.0
S 380 × 74.4	0.73	9.48	381.	143.	15.3	14.0	202.3	1.06	146.0	6.53	0.091	26.2
× 63.8	0.63	8.13	381.	140.	15.3	10.4	186.1	0.98	151.1	5.99	0.086	27.2
S 310 × 74.4	0.73	9.48	305.	139.	16.7	17.4	127.0	0.83	115.6	6.53	0.094	26.2
× 60.7	0.60	7.74	305.	133.	16.7	11.7	113.2	0.74	121.2	5.66	0.085	26.9
S 310 × 52.1	0.51	6.65	305.	129.	13.8	10.9	95.3	0.63	119.9	4.11	0.064	24.9
× 47.3	0.46	6.03	305.	127.	13.8	8.9	90.7	0.60	122.7	3.90	0.061	25.4
S 250 × 52.1	0.51	6.65	254.	126.	12.5	15.1	61.2	0.48	96.0	3.48	0.055	22.9
× 37.8	0.37	4.81	254.	118.	12.5	7.9	51.6	0.40	103.4	2.83	0.048	24.2
S 200 × 34.2	0.34	4.37	203.	106.	10.8	11.2	27.0	0.27	78.7	1.79	0.034	20.3
× 27.4	0.27	3.49	203.	102.	10.8	6.9	24.0	0.24	82.8	1.55	0.030	21.1
S 180 × 29.8	0.29	3.79	178.	98.	10.0	11.4	17.6	0.20	68.3	1.32	0.027	18.6
× 22.8	0.22	2.90	178.	93.	10.0	6.4	15.3	0.17	72.6	1.10	0.024	19.5
S 150 × 25.7	0.25	3.27	152.	91.	9.1	11.8	10.9	0.14	57.9	0.96	0.021	17.1
× 18.6	0.18	2.37	152.	85.	9.1	5.9	9.2	0.12	62.2	0.76	0.018	17.9
S 130 × 22.0	0.22	2.80	127.	83.	8.3	12.5	6.3	0.10	47.5	0.70	0.017	15.7
× 14.9	0.15	1.90	127.	76.	8.3	5.4	5.1	0.08	52.1	0.51	0.013	16.3
S 100 × 14.1	0.14	1.80	102.	71.	7.4	8.3	2.8	0.06	39.6	0.38	0.011	14.5
× 11.5	0.11	1.46	102.	68.	7.4	4.9	2.5	0.05	41.7	0.32	0.009	14.8
S 75 × 11.2	0.11	1.43	76.	64.	6.6	8.9	1.2	0.03	29.2	0.24	0.008	13.1
× 8.5	0.08	1.08	76.	59.	6.6	4.3	1.0	0.03	31.2	0.19	0.006	13.3

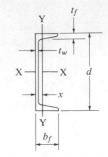

Table A-6   Channels, American standard: properties and dimensions for design (English customary units)[a]

Section and weight, lb/ft	A, in.^{2}	d, in.	b_f, in.	t_f, in.	t_w, in.	I_x, in.^{4}	S_x, in.^{3}	r_x, in.	I_y, in.^{4}	S_y, in.^{3}	r_y, in.
C 15 × 50.00	14.70	15.00	3.716	0.650	0.716	404.0	53.80	5.24	11.00	3.78	0.867
× 40.00	11.80	15.00	3.520	0.650	0.520	349.0	46.50	5.44	9.23	3.36	0.886
× 33.90	9.96	15.00	3.400	0.650	0.400	315.0	42.00	5.62	8.13	3.11	0.904
C 12 × 30.00	8.82	12.00	3.170	0.501	0.510	162.0	27.00	4.29	5.14	2.06	0.763
× 25.00	7.35	12.00	3.047	0.501	0.387	144.0	24.10	4.43	4.47	1.88	0.780
× 20.70	6.09	12.00	2.942	0.501	0.282	129.0	21.50	4.61	3.88	1.73	0.799
C 10 × 30.00	8.82	10.00	3.033	0.436	0.673	103.0	20.70	3.42	3.94	1.65	0.669
× 25.00	7.35	10.00	2.886	0.436	0.526	91.2	18.20	3.52	3.36	1.48	0.676
× 20.00	5.88	10.00	2.739	0.436	0.379	78.9	15.80	3.66	2.81	1.32	0.691
× 15.30	4.49	10.00	2.600	0.436	0.240	67.4	13.50	3.87	2.28	1.16	0.713
C 9 × 20.00	5.88	9.00	2.648	0.413	0.448	60.9	13.50	3.22	2.42	1.17	0.642
× 15.00	4.41	9.00	2.485	0.413	0.285	51.0	11.30	3.40	1.93	1.01	0.661
× 13.40	3.94	9.00	2.433	0.413	0.233	47.9	10.60	3.48	1.76	0.96	0.668
C 8 × 18.75	5.51	8.00	2.527	0.390	0.487	44.0	11.00	2.82	1.98	1.01	0.599
× 13.75	4.04	8.00	2.343	0.390	0.303	36.1	9.03	2.99	1.53	0.85	0.615
× 11.50	3.38	8.00	2.260	0.390	0.220	32.6	8.14	3.11	1.32	0.78	0.625
C 7 × 14.75	4.33	7.00	2.299	0.366	0.419	27.2	7.78	2.51	1.38	0.78	0.564
× 12.25	3.60	7.00	2.194	0.366	0.314	24.2	6.93	2.60	1.17	0.70	0.571
× 9.80	2.87	7.00	2.090	0.366	0.210	21.3	6.08	2.72	0.97	0.63	0.581
C 6 × 13.00	3.83	6.00	2.157	0.343	0.437	17.4	5.80	2.13	1.05	0.64	0.525
× 10.50	3.09	6.00	2.034	0.343	0.314	15.2	5.06	2.22	0.86	0.56	0.529
× 8.20	2.40	6.00	1.920	0.343	0.200	13.1	4.38	2.34	0.69	0.49	0.537
C 5 × 9.00	2.64	5.00	1.885	0.320	0.325	8.9	3.56	1.83	0.63	0.45	0.489
× 6.70	1.97	5.00	1.750	0.320	0.190	7.5	3.00	1.95	0.48	0.38	0.493
C 4 × 7.25	2.13	4.00	1.721	0.296	0.321	4.6	2.29	1.47	0.43	0.34	0.450
× 5.40	1.59	4.00	1.584	0.296	0.184	3.8	1.93	1.56	0.32	0.28	0.449
C 3 × 6.00	1.76	3.00	1.596	0.273	0.356	2.1	1.38	1.08	0.30	0.27	0.416
× 5.00	1.47	3.00	1.498	0.273	0.258	1.8	1.24	1.12	0.25	0.23	0.410
× 4.10	1.21	3.00	1.410	0.273	0.170	1.7	1.10	1.17	0.20	0.20	0.404

[a] From *The Structural Steel Design Manual*, by Joseph E. Bowles (New York: McGraw-Hill, 1980).

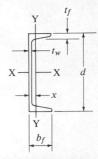

Table A-6 Channels, American standard: properties and dimensions for design (SI units)

Section and mass, kg/m	Wt, kN/m	A, m^2 $\times 10^{-3}$	d, mm	b_f, mm	t_f, mm	t_w, mm	I_x, m^4 $\times 10^{-6}$	S_x, m^3 $\times 10^{-3}$	r_x, mm	I_y, m^4 $\times 10^{-6}$	S_y, m^3 $\times 10^{-3}$	r_y, mm
C 380 × 74.4	0.730	9.48	381.0	94.4	16.5	18.2	168.16	0.882	133.1	4.579	0.062	22.0
× 59.5	0.584	7.61	381.0	89.4	16.5	13.2	145.26	0.762	138.2	3.842	0.055	22.5
× 50.4	0.495	6.43	381.0	86.4	16.5	10.2	131.11	0.688	142.7	3.384	0.051	23.0
C 310 × 44.6	0.438	5.69	304.8	80.5	12.7	13.0	67.43	0.442	109.0	2.139	0.034	19.4
× 37.2	0.365	4.74	304.8	77.4	12.7	9.8	59.94	0.395	112.5	1.861	0.031	19.8
× 30.8	0.302	3.93	304.8	74.7	12.7	7.2	53.69	0.352	117.1	1.615	0.028	20.3
C 250 × 44.6	0.438	5.69	254.0	77.0	11.1	17.1	42.87	0.339	86.9	1.640	0.027	17.0
× 37.2	0.365	4.74	254.0	73.3	11.1	13.4	37.96	0.298	89.4	1.399	0.024	17.2
× 29.8	0.292	3.79	254.0	69.6	11.1	9.6	32.84	0.259	93.0	1.170	0.022	17.6
× 22.8	0.223	2.90	254.0	66.0	11.1	6.1	28.05	0.221	98.3	0.949	0.019	18.1
C 230 × 29.8	0.292	3.79	228.6	67.3	10.5	11.4	25.35	0.221	81.8	1.007	0.019	16.3
× 22.3	0.219	2.85	228.6	63.1	10.5	7.2	21.23	0.185	86.4	0.803	0.017	16.8
× 19.9	0.196	2.54	228.6	61.8	10.5	5.9	19.94	0.174	88.4	0.733	0.016	17.0
C 200 × 27.9	0.274	3.55	203.2	64.2	9.9	12.4	18.31	0.180	71.6	0.824	0.017	15.2
× 20.5	0.201	2.61	203.2	59.5	9.9	7.7	15.03	0.148	75.9	0.637	0.014	15.6
× 17.1	0.168	2.18	203.2	57.4	9.9	5.6	13.57	0.133	79.0	0.549	0.013	15.9
C 180 × 22.0	0.215	2.79	177.8	58.4	9.3	10.6	11.32	0.127	63.8	0.574	0.013	14.3
× 18.2	0.179	2.32	177.8	55.7	9.3	8.0	10.07	0.114	66.0	0.487	0.012	14.5
× 14.6	0.143	1.85	177.8	53.1	9.3	5.3	8.87	0.100	69.1	0.403	0.010	14.8
C 150 × 19.3	0.190	2.47	152.4	54.8	8.7	11.1	7.24	0.095	54.1	0.437	0.011	13.3
× 15.6	0.153	1.99	152.4	51.7	8.7	8.0	6.33	0.083	56.4	0.360	0.009	13.4
× 12.2	0.120	1.55	152.4	48.8	8.7	5.1	5.45	0.072	59.4	0.288	0.008	13.6
C 130 × 13.4	0.131	1.70	127.0	47.9	8.1	8.3	3.70	0.058	46.5	0.263	0.007	12.4
× 10.0	0.098	1.27	127.0	44.4	8.1	4.8	3.12	0.049	49.5	0.199	0.006	12.5
C 100 × 10.8	0.106	1.37	101.6	43.7	7.5	8.2	1.91	0.038	37.3	0.180	0.006	11.4
× 8.0	0.079	1.03	101.6	40.2	7.5	4.7	1.60	0.032	39.6	0.133	0.005	11.4
C 75 × 8.9	0.088	1.14	76.2	40.5	6.9	9.0	0.86	0.023	27.4	0.127	0.004	10.6
× 7.4	0.073	0.95	76.2	38.0	6.9	6.6	0.77	0.020	28.4	0.103	0.004	10.4
× 6.1	0.060	0.78	76.2	35.8	6.9	4.3	0.69	0.018	29.7	0.082	0.003	10.3

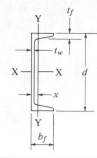

Table A-7 Channels, miscellaneous: properties and dimensions for design (English customary units)[a]

Section and weight, lb/ft	A, in.²	d, in.	b_f, in.	t_f, in.	t_w, in.	I_x, in.⁴	S_x, in.³	r_x, in.	I_y, in.⁴	S_y, in.³	r_y, in.
MC 18 × 58.00	17.10	18.00	4.200	0.625	0.700	676.0	75.10	6.29	17.80	5.32	1.020
× 51.90	15.30	18.00	4.100	0.625	0.600	627.0	69.70	6.41	16.40	5.07	1.040
× 45.80	13.50	18.00	4.000	0.625	0.500	578.0	64.30	6.56	15.10	4.82	1.060
× 42.70	12.60	18.00	3.950	0.625	0.450	554.0	61.60	6.64	14.40	4.69	1.070
MC 13 × 50.00	14.70	13.00	4.412	0.610	0.787	314.0	48.40	4.62	16.50	4.79	1.060
× 40.00	11.80	13.00	4.185	0.610	0.560	273.0	42.00	4.82	13.70	4.26	1.080
× 35.00	10.30	13.00	4.072	0.610	0.447	252.0	38.80	4.95	12.30	3.99	1.100
× 31.80	9.35	13.00	4.000	0.610	0.375	239.0	36.80	5.06	11.40	3.81	1.110
MC 12 × 50.00	14.70	12.00	4.135	0.700	0.835	269.0	44.90	4.28	17.40	5.65	1.090
× 45.00	13.20	12.00	4.012	0.700	0.712	252.0	42.00	4.36	15.80	5.33	1.090
× 40.00	11.80	12.00	3.890	0.700	0.590	234.0	39.00	4.46	14.30	5.00	1.100
× 35.00	10.30	12.00	3.767	0.700	0.467	216.0	36.10	4.59	12.70	4.67	1.110
MC 12 × 37.00	10.90	12.00	3.600	0.600	0.600	205.0	34.20	4.34	9.81	3.59	0.950
× 32.90	9.67	12.00	3.500	0.600	0.500	191.0	31.80	4.44	8.91	3.39	0.960
× 30.90	9.07	12.00	3.450	0.600	0.450	183.0	30.60	4.50	8.46	3.28	0.966
MC 12 × 10.60	3.10	12.00	1.500	0.309	0.190	55.4	9.23	4.22	0.38	0.31	0.351
MC 10 × 41.10	12.10	10.00	4.321	0.575	0.796	158.0	31.50	3.61	15.80	4.88	1.140
× 33.60	9.87	10.00	4.100	0.575	0.575	139.0	27.80	3.75	13.20	4.38	1.160
× 28.50	8.37	10.00	3.950	0.575	0.425	127.0	25.30	3.89	11.40	4.02	1.170
MC 10 × 28.30	8.32	10.00	3.502	0.575	0.477	118.0	23.60	3.77	8.21	3.20	0.993
× 25.30	7.43	10.00	3.550	0.500	0.425	107.0	21.40	3.79	7.61	2.89	1.010
× 24.90	7.32	10.00	3.402	0.575	0.377	110.0	22.00	3.87	7.32	2.99	1.000
× 21.90	6.43	10.00	3.450	0.500	0.325	98.5	19.70	3.91	6.74	2.70	1.020
MC 10 × 8.40	2.46	10.00	1.500	0.280	0.170	32.0	6.40	3.61	0.33	0.27	0.365
× 6.50	1.91	10.00	1.127	0.202	0.152	22.1	4.42	3.40	0.11	0.12	0.242
MC 9 × 25.40	7.47	9.00	3.500	0.550	0.450	88.0	19.60	3.43	7.65	3.02	1.010
× 23.90	7.02	9.00	3.450	0.550	0.400	85.0	18.90	3.48	7.22	2.93	1.010
MC 8 × 22.80	6.70	8.00	3.502	0.525	0.427	63.8	16.00	3.09	7.07	2.84	1.030
× 21.40	6.28	8.00	3.450	0.525	0.375	61.6	15.40	3.13	6.64	2.74	1.030
MC 8 × 20.00	5.88	8.00	3.025	0.500	0.400	54.5	13.60	3.05	4.47	2.05	0.872
× 18.70	5.50	8.00	2.978	0.500	0.353	52.5	13.10	3.09	4.20	1.97	0.874
MC 8 × 8.50	2.50	8.00	1.874	0.311	0.179	23.3	5.83	3.05	0.63	0.43	0.501
MC 7 × 22.70	6.67	7.00	3.603	0.500	0.503	47.5	13.60	2.67	7.29	2.85	1.050
× 19.10	5.61	7.00	3.452	0.500	0.352	43.2	12.30	2.77	6.11	2.57	1.040
× 17.60	5.17	7.00	3.000	0.475	0.375	37.6	10.80	2.70	4.01	1.89	0.881
MC 6 × 18.00	5.29	6.00	3.504	0.475	0.379	29.7	9.91	2.37	5.93	2.48	1.060
× 15.30	4.50	6.00	3.500	0.385	0.340	25.4	8.47	2.38	4.97	2.03	1.050
MC 6 × 16.30	4.79	6.00	3.000	0.475	0.375	26.0	8.68	2.33	3.82	1.84	0.892
× 15.10	4.44	6.00	2.941	0.475	0.316	25.0	8.32	2.37	3.51	1.75	0.889
MC 6 × 12.00	3.53	6.00	2.497	0.375	0.310	18.7	6.24	2.30	1.87	1.04	0.728
MC 3 × 9.00	2.65	3.00	2.122	0.351	0.497	3.1	2.10	1.09	0.97	0.68	0.604
× 7.10	2.09	3.00	1.938	0.351	0.312	2.7	1.82	1.14	0.71	0.56	0.583

[a] From *The Structural Steel Design Manual*, by Joseph E. Bowles (New York: McGraw-Hill, 1980).

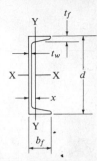

Table A-7   Channels, miscellaneous: properties and dimensions for design (SI units)

Section and mass, kg/m	Wt, kN/m	A, m^2 $\times 10^{-3}$	d, mm	b_f, mm	t_f, mm	t_w, mm	I_x, m^4 $\times 10^{-6}$	S_x, m^3 $\times 10^{-3}$	r_x, mm	I_y, m^4 $\times 10^{-6}$	S_y, m^3 $\times 10^{-3}$	r_y, mm
MC 460 × 86.3	0.846	11.03	457.2	106.7	15.9	17.8	281.37	1.231	159.8	7.409	0.087	25.9
× 77.2	0.757	9.87	457.2	104.1	15.9	15.2	260.98	1.142	162.8	6.826	0.083	26.4
× 68.2	0.668	8.71	457.2	101.6	15.9	12.7	240.58	1.054	166.6	6.285	0.079	26.9
× 63.5	0.623	8.13	457.2	100.3	15.9	11.4	230.59	1.009	168.7	5.994	0.077	27.2
MC 330 × 74.4	0.730	9.48	330.2	112.1	15.5	20.0	130.70	0.793	117.3	6.868	0.078	26.9
× 59.5	0.584	7.61	330.2	106.3	15.5	14.2	113.63	0.688	122.4	5.702	0.070	27.4
× 52.1	0.511	6.65	330.2	103.4	15.5	11.4	104.89	0.636	125.7	5.120	0.065	27.9
× 47.3	0.464	6.03	330.2	101.6	15.5	9.5	99.48	0.603	128.5	4.745	0.062	28.2
MC 310 × 74.4	0.730	9.48	304.8	105.0	17.8	21.2	111.97	0.736	108.7	7.242	0.093	27.7
× 67.0	0.657	8.52	304.8	101.9	17.8	18.1	104.89	0.688	110.7	6.576	0.087	27.7
× 59.5	0.584	7.61	304.8	98.8	17.8	15.0	97.40	0.639	113.3	5.952	0.082	27.9
× 52.1	0.511	6.65	304.8	95.7	17.8	11.9	89.91	0.592	116.6	5.286	0.077	28.2
MC 310 × 55.1	0.540	7.03	304.8	91.4	15.2	15.2	85.33	0.560	110.2	4.083	0.059	24.1
× 49.0	0.480	6.24	304.8	88.9	15.2	12.7	79.50	0.521	112.8	3.709	0.056	24.4
× 46.0	0.451	5.85	304.8	87.6	15.2	11.4	76.17	0.501	114.3	3.521	0.054	24.5
MC 310 × 15.8	0.155	2.00	304.8	38.1	7.8	4.8	23.06	0.151	107.2	0.158	0.005	8.9
MC 250 × 61.2	0.600	7.81	254.0	109.8	14.6	20.2	65.76	0.516	91.7	6.576	0.080	29.0
× 50.0	0.490	6.37	254.0	104.1	14.6	14.6	57.86	0.456	95.2	5.494	0.072	29.5
× 42.4	0.416	5.40	254.0	100.3	14.6	10.8	52.86	0.415	98.8	4.745	0.066	29.7
MC 250 × 42.1	0.413	5.37	254.0	89.0	14.6	12.1	49.12	0.387	95.8	3.417	0.052	25.2
× 37.7	0.369	4.79	254.0	90.2	12.7	10.8	44.54	0.351	96.3	3.168	0.047	25.7
× 37.1	0.363	4.72	254.0	86.4	14.6	9.6	45.79	0.361	98.3	3.047	0.049	25.4
× 32.6	0.320	4.15	254.0	87.6	12.7	8.3	41.00	0.323	99.3	2.805	0.044	25.9
MC 250 × 12.5	0.123	1.59	254.0	38.1	7.1	4.3	13.32	0.105	91.7	0.137	0.004	9.3
× 9.7	0.095	1.23	254.0	28.6	5.1	3.9	9.20	0.072	86.4	0.046	0.002	6.1
MC 230 × 37.8	0.371	4.82	228.6	88.9	14.0	11.4	36.63	0.321	87.1	3.184	0.049	25.7
× 35.6	0.349	4.53	228.6	87.6	14.0	10.2	35.38	0.310	88.4	3.005	0.048	25.7
MC 200 × 33.9	0.333	4.32	203.2	89.0	13.3	10.8	26.56	0.262	78.5	2.943	0.047	26.2
× 31.8	0.312	4.05	203.2	87.6	13.3	9.5	25.64	0.252	79.5	2.764	0.045	26.2
MC 200 × 29.8	0.292	3.79	203.2	76.8	12.7	10.2	22.68	0.223	77.5	1.861	0.034	22.1
× 27.8	0.273	3.55	203.2	75.6	12.7	9.0	21.85	0.215	78.5	1.748	0.032	22.2
MC 200 × 12.6	0.124	1.61	203.2	47.6	7.9	4.5	9.70	0.096	77.5	0.262	0.007	12.7
MC 180 × 33.8	0.331	4.30	177.8	91.5	12.7	12.8	19.77	0.223	67.8	3.034	0.047	26.7
× 28.4	0.279	3.62	177.8	87.7	12.7	8.9	17.98	0.202	70.4	2.543	0.042	26.4
× 26.2	0.257	3.34	177.8	76.2	12.1	9.5	15.65	0.177	68.6	1.669	0.031	22.4
MC 150 × 26.8	0.263	3.41	152.4	89.0	12.1	9.6	12.36	0.162	60.2	2.468	0.041	26.9
× 22.8	0.223	2.90	152.4	88.9	9.8	8.6	10.57	0.139	60.5	2.069	0.033	26.7
MC 150 × 24.3	0.238	3.09	152.4	76.2	12.1	9.5	10.82	0.142	59.2	1.590	0.030	22.7
× 22.5	0.220	2.86	152.4	74.7	12.1	8.0	10.41	0.136	60.2	1.461	0.029	22.6
MC 150 × 17.9	0.175	2.28	152.4	63.4	9.5	7.9	7.78	0.102	58.4	0.778	0.017	18.5
MC 75 × 13.4	0.131	1.71	76.2	53.9	8.9	12.6	1.31	0.034	27.7	0.402	0.011	15.3
× 10.6	0.104	1.35	76.2	49.2	8.9	7.9	1.14	0.030	29.0	0.296	0.009	14.8

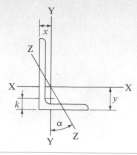

Table A-8 Angles, unequal leg: properties and dimensions for design (English customary units)[a]

Angle size and thickness, in.	Weight per foot, lb	Area, in.²	X axis I, in.⁴	X axis S, in.³	X axis r, in.	X axis y, in.	Y axis I, in.⁴	Y axis S, in.³	Y axis r, in.	Y axis x, in.	Z axis r, in.	Z axis Tan α	k, in.
$9 \times 4 \times 1$	40.80	12.000	97.000	17.600	2.840	3.500	12.000	4.000	1.000	1.000	0.834	0.203	1.500
$9 \times 4 \times \frac{7}{8}$	36.10	10.600	86.800	15.700	2.860	3.450	10.800	3.560	1.010	0.953	0.836	0.208	1.375
$9 \times 4 \times \frac{3}{4}$	31.30	9.190	76.100	13.600	2.880	3.410	9.630	3.110	1.020	0.906	0.841	0.212	1.250
$9 \times 4 \times \frac{5}{8}$	26.30	7.730	64.900	11.500	2.900	3.360	8.320	2.650	1.040	0.858	0.847	0.216	1.125
$9 \times 4 \times \frac{9}{16}$	23.80	7.000	59.100	10.400	2.910	3.330	7.630	2.410	1.040	0.834	0.850	0.218	1.063
$9 \times 4 \times \frac{1}{2}$	21.30	6.250	53.200	9.340	2.920	3.310	6.920	2.170	1.050	0.810	0.854	0.220	1.000
$8 \times 6 \times 1$	44.20	13.000	80.800	15.100	2.490	2.650	38.800	8.920	1.730	1.650	1.280	0.543	1.500
$8 \times 6 \times \frac{7}{8}$	39.10	11.500	72.300	13.400	2.510	2.610	34.900	7.940	1.740	1.610	1.280	0.547	1.375
$8 \times 6 \times \frac{3}{4}$	33.80	9.940	63.400	11.700	2.530	2.560	30.700	6.620	1.760	1.560	1.290	0.551	1.250
$8 \times 6 \times \frac{5}{8}$	28.50	8.360	54.100	9.870	2.540	2.520	26.300	5.880	1.770	1.520	1.290	0.554	1.125
$8 \times 6 \times \frac{9}{16}$	25.70	7.560	49.300	8.950	2.550	2.500	24.000	5.340	1.780	1.500	1.300	0.556	1.063
$8 \times 6 \times \frac{1}{2}$	23.00	6.750	44.300	8.020	2.560	2.470	21.700	4.790	1.790	1.470	1.300	0.558	1.000
$8 \times 6 \times \frac{7}{16}$	20.20	5.930	39.200	7.070	2.570	2.450	19.300	4.230	1.800	1.450	1.310	0.560	0.938
$8 \times 4 \times 1$	37.40	11.000	69.600	14.100	2.520	3.050	11.600	3.940	1.030	1.050	0.846	0.247	1.500
$8 \times 4 \times \frac{7}{8}$	33.10	9.730	62.500	12.500	2.530	3.000	10.500	3.510	1.040	0.999	0.848	0.253	1.375
$8 \times 4 \times \frac{3}{4}$	28.70	8.440	54.900	10.900	2.550	2.950	9.360	3.070	1.050	0.953	0.852	0.258	1.250
$8 \times 4 \times \frac{5}{8}$	24.20	7.110	46.900	9.210	2.570	2.910	8.100	2.620	1.070	0.906	0.857	0.262	1.125
$8 \times 4 \times \frac{9}{16}$	21.90	6.430	42.800	8.350	2.580	2.880	7.430	2.380	1.070	0.882	0.861	0.265	1.063
$8 \times 4 \times \frac{1}{2}$	19.60	5.750	38.500	7.490	2.590	2.860	6.740	2.150	1.080	0.859	0.865	0.267	1.000
$8 \times 4 \times \frac{7}{16}$	17.20	5.060	34.100	6.600	2.600	2.830	6.020	1.900	1.090	0.835	0.869	0.269	0.938
$7 \times 4 \times \frac{7}{8}$	30.20	8.860	42.900	9.650	2.200	2.550	10.200	3.460	1.070	1.050	0.856	0.318	1.375
$7 \times 4 \times \frac{3}{4}$	26.20	7.690	37.800	8.420	2.220	2.510	9.050	3.030	1.090	1.010	0.860	0.324	1.250
$7 \times 4 \times \frac{5}{8}$	22.10	6.480	32.400	7.140	2.240	2.460	7.840	2.580	1.100	0.963	0.865	0.329	1.125
$7 \times 4 \times \frac{9}{16}$	20.00	5.870	29.600	6.480	2.240	2.440	7.190	2.350	1.110	0.940	0.868	0.332	1.063
$7 \times 4 \times \frac{1}{2}$	17.90	5.250	26.700	5.810	2.250	2.420	6.530	2.120	1.110	0.917	0.872	0.335	1.000
$7 \times 4 \times \frac{7}{16}$	15.80	4.620	23.700	5.130	2.260	2.390	5.830	1.880	1.120	0.893	0.876	0.337	0.938
$7 \times 4 \times \frac{3}{8}$	13.60	3.980	20.600	4.440	2.270	2.370	5.100	1.630	1.130	0.870	0.880	0.340	0.875
$6 \times 4 \times \frac{7}{8}$	27.20	7.980	27.700	7.150	1.860	2.120	9.750	3.390	1.110	1.120	0.857	0.421	1.375
$6 \times 4 \times \frac{3}{4}$	23.60	6.940	24.500	6.250	1.880	2.080	8.680	2.970	1.120	1.080	0.860	0.428	1.250
$6 \times 4 \times \frac{5}{8}$	20.00	5.860	21.100	5.310	1.900	2.030	7.520	2.540	1.130	1.030	0.864	0.435	1.125
$6 \times 4 \times \frac{9}{16}$	18.10	5.310	19.300	4.830	1.900	2.010	6.910	2.310	1.140	1.010	0.866	0.438	1.063
$6 \times 4 \times \frac{1}{2}$	16.20	4.750	17.400	4.330	1.910	1.990	6.270	2.080	1.150	0.987	0.870	0.440	1.000
$6 \times 4 \times \frac{7}{16}$	14.30	4.180	15.500	3.830	1.920	1.960	5.600	1.850	1.160	0.964	0.873	0.443	0.938
$6 \times 4 \times \frac{3}{8}$	12.30	3.610	13.500	3.320	1.930	1.940	4.900	1.600	1.170	0.941	0.877	0.446	0.875
$6 \times 4 \times \frac{5}{16}$	10.30	3.030	11.400	2.790	1.940	1.920	4.180	1.350	1.170	0.918	0.882	0.448	0.813
$6 \times 4 \times \frac{1}{4}$	8.30	2.440	9.270	2.260	1.950	1.890	3.410	1.100	1.180	0.894	0.887	0.451	0.750
$6 \times 3\frac{1}{2} \times \frac{1}{2}$	15.30	4.500	16.600	4.240	1.920	2.080	4.250	1.590	0.972	0.833	0.759	0.344	1.000
$6 \times 3\frac{1}{2} \times \frac{3}{8}$	11.70	3.420	12.900	3.240	1.940	2.040	3.340	1.230	0.988	0.787	0.767	0.350	0.875
$6 \times 3\frac{1}{2} \times \frac{5}{16}$	9.80	2.870	10.900	2.730	1.950	2.010	2.850	1.040	0.996	0.763	0.772	0.532	0.813
$5 \times 3\frac{1}{2} \times \frac{3}{4}$	19.80	5.810	13.900	4.280	1.550	1.750	5.550	2.220	0.977	0.996	0.748	0.464	1.250
$5 \times 3\frac{1}{2} \times \frac{5}{8}$	16.80	4.920	12.000	3.650	1.560	1.700	4.830	1.900	0.991	0.951	0.751	0.472	1.125
$5 \times 3\frac{1}{2} \times \frac{1}{2}$	13.60	4.000	9.990	2.990	1.580	1.660	4.050	1.560	1.010	0.906	0.755	0.479	1.000
$5 \times 3\frac{1}{2} \times \frac{7}{16}$	12.00	3.530	8.900	2.640	1.590	1.630	3.630	1.390	1.010	0.883	0.758	0.482	0.938
$5 \times 3\frac{1}{2} \times \frac{3}{8}$	10.40	3.050	7.780	2.290	1.600	1.610	3.180	1.210	1.020	0.861	0.762	0.486	0.875
$5 \times 3\frac{1}{2} \times \frac{5}{16}$	8.70	2.560	6.600	1.940	1.610	1.590	2.720	1.020	1.030	0.838	0.766	0.489	0.813
$5 \times 3\frac{1}{2} \times \frac{1}{4}$	7.00	2.060	5.390	1.570	1.620	1.560	2.230	0.830	1.040	0.814	0.770	0.492	0.750
$5 \times 3 \times \frac{1}{2}$	12.80	3.750	9.450	2.910	1.590	1.750	2.580	1.150	0.829	0.750	0.648	0.357	1.000
$5 \times 3 \times \frac{7}{16}$	11.30	3.310	8.430	2.580	1.600	1.730	2.320	1.020	0.837	0.727	0.651	0.361	0.938
$5 \times 3 \times \frac{3}{8}$	9.80	2.860	7.370	2.240	1.610	1.700	2.040	0.888	0.845	0.704	0.654	0.364	0.875
$5 \times 3 \times \frac{5}{16}$	8.20	2.400	6.260	1.890	1.610	1.680	1.750	0.753	0.853	0.681	0.658	0.368	0.813
$5 \times 3 \times \frac{1}{4}$	6.60	1.940	5.110	1.530	1.620	1.660	1.440	0.614	0.861	0.657	0.663	0.371	0.750

[a] From *The Structural Steel Design Manual*, by Joseph E. Bowles (New York: McGraw-Hill, 1980).

512

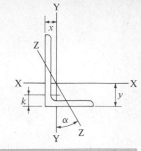

Table A-8 Angles, unequal leg: properties and dimensions for design (SI units)

Angle size and thickness, mm	Weight per meter, kN/m	Area, m² × 10⁻³	X axis I, m⁴ × 10⁻⁶	S, m³ × 10⁻³	r, mm	y, mm	Y axis I, m⁴ × 10⁻⁶	S, m³ × 10⁻³	r, mm	x, mm	Z axis r, mm	Tan α	k, mm
229 × 102 × 25.4	0.595	7.742	40.374	0.288	72.1	88.9	5.0	0.066	25.4	25.4	21.2	0.203	38.1
229 × 102 × 22.2	0.527	6.839	36.129	0.257	72.6	87.6	4.5	0.058	25.7	24.2	21.2	0.208	34.9
229 × 102 × 19.0	0.457	5.929	31.675	0.223	73.2	86.6	4.0	0.051	25.9	23.0	21.4	0.212	31.7
229 × 102 × 15.9	0.384	4.987	27.013	0.188	73.7	85.3	3.5	0.043	26.4	21.8	21.5	0.216	28.6
229 × 102 × 14.3	0.347	4.516	24.599	0.170	73.9	84.6	3.2	0.039	26.4	21.2	21.6	0.218	27.0
229 × 102 × 12.7	0.311	4.032	22.143	0.153	74.2	84.1	2.9	0.036	26.7	20.6	21.7	0.220	25.4
203 × 152 × 25.4	0.645	8.387	33.631	0.247	63.2	67.3	16.1	0.146	43.9	41.9	32.5	0.543	38.1
203 × 152 × 22.2	0.571	7.419	30.093	0.220	63.8	66.3	14.5	0.130	44.2	40.9	32.5	0.547	34.9
203 × 152 × 19.0	0.493	6.413	26.389	0.192	64.2	65.0	12.8	0.108	44.7	39.6	32.8	0.551	31.7
203 × 152 × 15.9	0.416	5.394	22.518	0.162	64.5	64.0	10.9	0.096	45.0	38.6	32.8	0.554	28.6
203 × 152 × 14.3	0.375	4.877	20.520	0.147	64.8	63.5	10.0	0.088	45.2	38.1	33.0	0.556	27.0
203 × 152 × 12.7	0.336	4.355	18.439	0.131	65.0	62.7	9.0	0.078	45.5	37.3	33.0	0.558	25.4
203 × 152 × 11.1	0.295	3.826	16.316	0.116	65.3	62.2	8.0	0.069	45.7	36.8	33.3	0.560	23.8
203 × 102 × 25.4	0.546	7.097	28.970	0.231	64.0	77.5	4.8	0.065	26.2	26.7	21.5	0.247	38.1
203 × 102 × 22.2	0.483	6.277	26.014	0.205	64.3	76.2	4.4	0.058	26.4	25.4	21.5	0.253	34.9
203 × 102 × 19.0	0.419	5.445	22.851	0.179	64.8	74.9	3.9	0.050	26.7	24.2	21.6	0.258	31.7
203 × 102 × 15.9	0.353	4.587	19.521	0.151	65.3	73.9	3.4	0.043	27.2	23.0	21.8	0.262	28.6
203 × 102 × 14.3	0.320	4.148	17.815	0.137	65.5	73.2	3.1	0.039	27.2	22.4	21.9	0.265	27.0
203 × 102 × 12.7	0.286	3.710	16.025	0.123	65.8	72.6	2.8	0.035	27.4	21.8	22.0	0.267	25.4
203 × 102 × 11.1	0.251	3.265	14.193	0.108	66.0	71.9	2.5	0.031	27.7	21.2	22.1	0.269	23.8
178 × 102 × 22.2	0.441	5.716	17.856	0.158	55.9	64.8	4.2	0.057	27.2	26.7	21.7	0.318	34.9
178 × 102 × 19.0	0.382	4.961	15.734	0.138	56.4	63.8	3.8	0.050	27.7	25.7	21.8	0.324	31.7
178 × 102 × 15.9	0.322	4.181	13.486	0.117	56.9	62.5	3.3	0.042	27.9	24.5	22.0	0.329	28.6
178 × 102 × 14.3	0.292	3.787	12.320	0.106	56.9	62.0	3.0	0.039	28.2	23.9	22.0	0.332	27.0
178 × 102 × 12.7	0.261	3.387	11.113	0.095	57.1	61.5	2.7	0.035	28.2	23.3	22.1	0.335	25.4
178 × 102 × 11.1	0.231	2.981	9.865	0.084	57.4	60.7	2.4	0.031	28.4	22.7	22.3	0.337	23.8
178 × 102 × 9.5	0.198	2.568	8.574	0.073	57.7	60.2	2.1	0.027	28.7	22.1	22.4	0.340	22.2
152 × 102 × 22.2	0.397	5.148	11.530	0.117	47.2	53.8	4.1	0.056	28.2	28.4	21.8	0.421	34.9
152 × 102 × 19.0	0.344	4.477	10.198	0.102	47.8	52.8	3.6	0.049	28.4	27.4	21.8	0.428	31.7
152 × 102 × 15.9	0.292	3.781	8.782	0.087	48.3	51.6	3.1	0.042	28.7	26.2	21.9	0.435	28.6
152 × 102 × 14.3	0.264	3.426	8.033	0.079	48.3	51.1	2.9	0.038	29.0	25.7	22.0	0.438	27.0
152 × 102 × 12.7	0.236	3.065	7.242	0.071	48.5	50.5	2.6	0.034	29.2	25.1	22.1	0.440	25.4
152 × 102 × 11.1	0.209	2.697	6.452	0.063	48.8	49.8	2.3	0.030	29.5	24.5	22.2	0.443	23.8
152 × 102 × 9.5	0.179	2.329	5.619	0.054	49.0	49.3	2.0	0.026	29.7	23.9	22.3	0.446	22.2
152 × 102 × 7.9	0.150	1.955	4.745	0.046	49.3	48.8	1.7	0.022	29.7	23.3	22.4	0.448	20.7
152 × 102 × 6.4	0.121	1.574	3.858	0.037	49.5	48.0	1.4	0.018	30.0	22.7	22.5	0.451	19.0
152 × 89 × 12.7	0.223	2.903	6.909	0.069	48.8	52.8	1.8	0.026	24.7	21.2	19.3	0.344	25.4
152 × 89 × 9.5	0.171	2.206	5.369	0.053	49.3	51.8	1.4	0.020	25.1	20.0	19.5	0.350	22.2
152 × 89 × 7.9	0.143	1.852	4.537	0.045	49.5	51.1	1.2	0.017	25.3	19.4	19.6	0.532	20.7
152 × 89 × 6.4	0.115	1.490	3.688	0.036	49.8	50.5	1.0	0.014	25.7	18.8	19.7	0.355	19.0
127 × 89 × 19.0	0.289	3.748	5.786	0.070	39.4	44.4	2.3	0.036	24.8	25.3	19.0	0.464	31.7
127 × 89 × 15.9	0.245	3.174	4.995	0.060	39.6	43.2	2.0	0.031	25.2	24.2	19.1	0.472	28.6
127 × 89 × 12.7	0.198	2.581	4.158	0.049	40.1	42.2	1.7	0.026	25.7	23.0	19.2	0.479	25.4
127 × 89 × 11.1	0.175	2.277	3.704	0.043	40.4	41.4	1.5	0.023	25.7	22.4	19.3	0.482	23.8
127 × 89 × 9.5	0.152	1.968	3.238	0.038	40.6	40.9	1.3	0.020	25.9	21.9	19.4	0.486	22.2
127 × 89 × 7.9	0.127	1.652	2.747	0.032	40.9	40.4	1.1	0.017	26.2	21.3	19.5	0.489	20.7
127 × 89 × 6.4	0.102	1.329	2.243	0.026	41.1	39.6	0.9	0.014	26.4	20.7	19.6	0.492	19.0
127 × 76 × 12.7	0.187	2.419	3.933	0.048	40.4	44.4	1.1	0.019	21.1	19.0	16.5	0.357	25.4
127 × 76 × 11.1	0.165	2.135	3.509	0.042	40.6	43.9	1.0	0.017	21.3	18.5	16.5	0.361	23.8
127 × 76 × 9.5	0.143	1.845	3.068	0.037	40.9	43.2	0.8	0.015	21.5	17.9	16.6	0.364	22.2
127 × 76 × 7.9	0.120	1.548	2.606	0.031	40.9	42.7	0.7	0.012	21.7	17.3	16.7	0.368	20.7
127 × 76 × 6.4	0.096	1.252	2.127	0.025	41.1	42.2	0.6	0.010	21.9	16.7	16.8	0.371	19.0

513

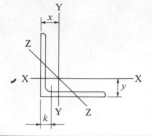

Table A-9 Angles, equal leg: properties and dimensions for design. (Very small sections not included.) (English customary units)[a]

Angle size and thickness, in.	Weight per foot, lb	Area, in.²	Axis XX or axis YY				Axis ZZ r, in.	k, in.
			I, in.⁴	S, in.³	r, in.	x or y, in.		
$8 \times 8 \times 1\frac{1}{8}$	56.90	16.700	98.000	17.500	2.420	2.410	1.560	1.75
$8 \times 8 \times 1$	51.00	15.000	89.000	15.800	2.440	2.370	1.560	1.63
$8 \times 8 \times \frac{7}{8}$	45.00	13.200	79.600	14.000	2.450	2.320	1.570	1.50
$8 \times 8 \times \frac{3}{4}$	38.90	11.400	69.700	12.200	2.470	2.280	1.580	1.38
$8 \times 8 \times \frac{5}{8}$	32.70	9.610	59.400	10.300	2.490	2.230	1.580	1.25
$8 \times 8 \times \frac{9}{16}$	29.60	8.680	54.100	9.340	2.500	2.210	1.590	1.19
$8 \times 8 \times \frac{1}{2}$	26.40	7.750	48.600	8.360	2.500	2.190	1.590	1.13
$6 \times 6 \times 1$	37.40	11.000	35.500	8.570	1.800	1.860	1.170	1.50
$6 \times 6 \times \frac{7}{8}$	33.10	9.730	31.900	7.630	1.810	1.820	1.170	1.38
$6 \times 6 \times \frac{3}{4}$	28.70	8.440	28.200	6.660	1.830	1.780	1.170	1.25
$6 \times 6 \times \frac{5}{8}$	24.20	7.110	24.200	5.660	1.840	1.730	1.180	1.13
$6 \times 6 \times \frac{9}{16}$	21.90	6.430	22.100	5.140	1.850	1.710	1.180	1.19
$6 \times 6 \times \frac{1}{2}$	19.60	5.750	19.900	4.610	1.860	1.680	1.180	1.00
$6 \times 6 \times \frac{7}{16}$	17.20	5.060	17.700	4.080	1.870	1.660	1.190	0.94
$6 \times 6 \times \frac{3}{8}$	14.90	4.360	15.400	3.530	1.880	1.640	1.190	0.88
$6 \times 6 \times \frac{5}{16}$	12.40	3.650	13.000	2.970	1.890	1.620	1.200	0.81
$5 \times 5 \times \frac{7}{8}$	27.20	7.980	17.800	5.170	1.490	1.570	0.973	1.38
$5 \times 5 \times \frac{3}{4}$	23.60	6.940	15.700	4.530	1.510	1.520	0.975	1.25
$5 \times 5 \times \frac{5}{8}$	20.00	5.860	13.600	3.860	1.520	1.480	0.978	1.13
$5 \times 5 \times \frac{1}{2}$	16.20	4.750	11.300	3.160	1.540	1.430	0.983	1.00
$5 \times 5 \times \frac{7}{16}$	14.30	4.180	10.000	2.790	1.550	1.410	0.986	0.94
$5 \times 5 \times \frac{3}{8}$	12.30	3.610	8.740	2.420	1.560	1.390	0.990	0.88
$5 \times 5 \times \frac{5}{16}$	10.30	3.030	7.420	2.040	1.570	1.370	0.994	0.81
$4 \times 4 \times \frac{3}{4}$	18.50	5.440	7.670	2.810	1.190	1.270	0.778	1.13
$4 \times 4 \times \frac{5}{8}$	15.70	4.610	6.660	2.400	1.200	1.230	0.779	1.00
$4 \times 4 \times \frac{1}{2}$	12.80	3.750	5.560	1.970	1.220	1.180	0.782	0.88
$4 \times 4 \times \frac{7}{16}$	11.30	3.310	4.970	1.750	1.230	1.160	0.785	0.81
$4 \times 4 \times \frac{3}{8}$	9.80	2.860	4.360	1.520	1.230	1.140	0.788	0.75
$4 \times 4 \times \frac{5}{16}$	8.20	2.400	3.710	1.290	1.240	1.120	0.791	0.69
$4 \times 4 \times \frac{1}{4}$	6.60	1.940	3.040	1.050	1.250	1.090	0.795	0.63
$3\frac{1}{2} \times 3\frac{1}{2} \times \frac{1}{2}$	11.10	3.250	3.640	1.490	1.060	1.060	0.683	0.88
$3\frac{1}{2} \times 3\frac{1}{2} \times \frac{7}{16}$	9.80	2.870	3.260	1.320	1.070	1.040	0.684	0.81
$3\frac{1}{2} \times 3\frac{1}{2} \times \frac{3}{8}$	8.50	2.480	2.870	1.150	1.070	1.010	0.687	0.75
$3\frac{1}{2} \times 3\frac{1}{2} \times \frac{5}{16}$	7.20	2.090	2.450	0.976	1.080	0.990	0.690	0.69
$3\frac{1}{2} \times 3\frac{1}{2} \times \frac{1}{4}$	5.80	1.690	2.010	0.794	1.090	0.968	0.694	0.63
$3 \times 3 \times \frac{1}{2}$	9.40	2.750	2.220	1.070	0.898	0.932	0.584	0.81
$3 \times 3 \times \frac{7}{16}$	8.30	2.430	1.990	0.954	0.905	0.910	0.585	0.75
$3 \times 3 \times \frac{3}{8}$	7.20	2.110	1.760	0.833	0.913	0.888	0.587	0.69
$3 \times 3 \times \frac{5}{16}$	6.10	1.780	1.510	0.707	0.922	0.865	0.589	0.63
$3 \times 3 \times \frac{1}{4}$	4.90	1.440	1.240	0.577	0.930	0.842	0.592	0.56
$3 \times 3 \times \frac{3}{16}$	3.71	1.090	0.962	0.441	0.939	0.820	0.596	0.50
$2\frac{1}{2} \times 2\frac{1}{2} \times \frac{1}{2}$	7.70	2.250	1.230	0.724	0.739	0.806	0.487	0.81
$2\frac{1}{2} \times 2\frac{1}{2} \times \frac{3}{8}$	5.90	1.730	0.984	0.566	0.753	0.762	0.487	0.69
$2\frac{1}{2} \times 2\frac{1}{2} \times \frac{5}{16}$	5.00	1.460	0.849	0.482	0.761	0.740	0.489	0.63
$2\frac{1}{2} \times 2\frac{1}{2} \times \frac{1}{4}$	4.10	1.190	0.703	0.394	0.769	0.717	0.491	0.56
$2\frac{1}{2} \times 2\frac{1}{2} \times \frac{3}{16}$	3.07	0.902	0.547	0.303	0.778	0.694	0.495	0.50
$2 \times 2 \times \frac{3}{8}$	4.70	1.360	0.479	0.351	0.594	0.636	0.389	0.69
$2 \times 2 \times \frac{5}{16}$	3.92	1.150	0.416	0.300	0.601	0.614	0.390	0.63
$2 \times 2 \times \frac{1}{4}$	3.19	0.938	0.348	0.247	0.609	0.592	0.391	0.56
$2 \times 2 \times \frac{3}{16}$	2.44	0.715	0.272	0.190	0.617	0.569	0.394	0.50
$2 \times 2 \times \frac{1}{8}$	1.65	0.484	0.190	0.131	0.626	0.546	0.398	0.44

[a] From *The Structural Steel Design Manual*, by Joseph E. Bowles (New York: McGraw-Hill, 1980).

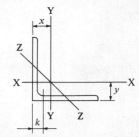

Table A-9 Angles, equal leg: properties and dimensions for design (SI units)

Angle size and thickness, mm	Weight per meter, kN/m	Axis XX or axis YY						k, mm
		Area, $m^2 \times 10^{-3}$	I, $m^4 \times 10^{-6}$	S, $m^3 \times 10^{-3}$	r, mm	x or y, mm	r, mm	
203 × 203 × 28.6	0.830	10.774	40.791	0.287	61.5	61.2	39.6	44.4
203 × 203 × 25.4	0.744	9.677	37.045	0.259	62.0	60.2	39.6	41.3
203 × 203 × 22.2	0.657	8.516	33.132	0.229	62.2	58.9	39.9	38.1
203 × 203 × 19.0	0.568	7.355	29.011	0.200	62.7	57.9	40.1	34.9
203 × 203 × 15.9	0.477	6.200	24.724	0.169	63.2	56.6	40.1	31.7
203 × 203 × 14.3	0.432	5.600	22.518	0.153	63.5	56.1	40.4	30.2
203 × 203 × 12.7	0.385	5.000	20.229	0.137	63.5	55.6	40.4	28.6
152 × 152 × 25.4	0.546	7.097	14.776	0.140	45.7	47.2	29.7	38.1
152 × 152 × 22.2	0.483	6.277	13.278	0.125	46.0	46.2	29.7	34.9
152 × 152 × 19.0	0.419	5.445	11.738	0.109	46.5	45.2	29.7	31.7
152 × 152 × 15.9	0.353	4.587	10.073	0.093	46.7	43.9	30.0	28.6
152 × 152 × 14.3	0.320	4.148	9.199	0.084	47.0	43.4	30.0	30.2
152 × 152 × 12.7	0.286	3.710	8.283	0.076	47.2	42.7	30.0	25.4
152 × 152 × 11.1	0.251	3.265	7.367	0.067	47.5	42.2	30.2	23.8
152 × 152 × 9.5	0.217	2.813	6.410	0.058	47.8	41.7	30.2	22.2
152 × 152 × 7.9	0.181	2.355	5.411	0.049	48.0	41.1	30.5	20.7
127 × 127 × 22.2	0.397	5.148	7.409	0.085	37.8	39.9	24.7	34.9
127 × 127 × 19.0	0.344	4.477	6.535	0.074	38.4	38.6	24.8	31.7
127 × 127 × 15.9	0.292	3.781	5.661	0.063	38.6	37.6	24.8	28.6
127 × 127 × 12.7	0.236	3.065	4.703	0.052	39.1	36.3	25.0	25.4
127 × 127 × 11.1	0.209	2.697	4.162	0.046	39.4	35.8	25.0	23.8
127 × 127 × 9.5	0.179	2.329	3.638	0.040	39.6	35.3	25.1	22.2
127 × 127 × 7.9	0.150	1.955	3.088	0.033	39.9	34.8	25.2	20.7
102 × 102 × 19.0	0.270	3.510	3.192	0.046	30.2	32.3	19.8	28.6
102 × 102 × 15.9	0.229	2.974	2.772	0.039	30.5	31.2	19.8	25.4
102 × 102 × 12.7	0.187	2.419	2.314	0.032	31.0	30.0	19.9	22.2
102 × 102 × 11.1	0.165	2.135	2.069	0.029	31.2	29.5	19.9	20.7
102 × 102 × 9.5	0.143	1.845	1.815	0.025	31.2	29.0	20.0	19.0
102 × 102 × 7.9	0.120	1.548	1.544	0.021	31.5	28.4	20.1	17.3
102 × 102 × 6.4	0.096	1.252	1.265	0.017	31.7	27.7	20.2	15.9
89 × 89 × 12.7	0.162	2.097	1.515	0.024	26.9	26.9	17.3	22.2
89 × 89 × 11.1	0.143	1.852	1.357	0.022	27.2	26.4	17.4	20.7
89 × 89 × 9.5	0.124	1.600	1.195	0.019	27.2	25.7	17.4	19.0
89 × 89 × 7.9	0.105	1.348	1.020	0.016	27.4	25.1	17.5	17.5
89 × 89 × 6.4	0.085	1.090	0.837	0.013	27.7	24.6	17.6	15.9
76 × 76 × 12.7	0.137	1.774	0.924	0.018	22.8	23.7	14.8	20.7
76 × 76 × 11.1	0.121	1.568	0.828	0.016	23.0	23.1	14.9	19.0
76 × 76 × 9.5	0.105	1.361	0.733	0.014	23.2	22.6	14.9	17.5
76 × 76 × 7.9	0.089	1.148	0.629	0.012	23.4	22.0	15.0	15.9
76 × 76 × 6.4	0.071	0.929	0.516	0.009	23.6	21.4	15.0	14.3
76 × 76 × 4.8	0.054	0.703	0.400	0.007	23.9	20.8	15.1	12.7
64 × 64 × 12.7	0.112	1.452	0.512	0.012	18.8	20.5	12.4	20.7
64 × 64 × 9.5	0.086	1.116	0.410	0.009	19.1	19.4	12.4	17.5
64 × 64 × 7.9	0.073	0.942	0.353	0.008	19.3	18.8	12.4	15.9
64 × 64 × 6.4	0.060	0.768	0.293	0.006	19.5	18.2	12.5	14.3
64 × 64 × 4.8	0.045	0.582	0.228	0.005	19.8	17.6	12.6	12.7
51 × 51 × 9.5	0.069	0.877	0.199	0.006	15.1	16.2	9.9	17.5
51 × 51 × 7.9	0.057	0.742	0.173	0.005	15.3	15.6	9.9	15.9
51 × 51 × 6.4	0.047	0.605	0.145	0.004	15.5	15.0	9.9	14.3
51 × 51 × 4.8	0.036	0.461	0.113	0.003	15.7	14.5	10.0	12.7
51 × 51 × 3.2	0.024	0.312	0.079	0.002	15.9	13.9	10.1	11.1

Table A-10 Design data for steel pipe commonly used as columns[a] (English customary units)

Nominal diameter: standard weight	Weight, lb/ft	Outside diameter, in.	Wall thickness, in.	Area, in.^{2}	I_x or I_y, in.^{4}	S, in.^{3}	r, in.
12	49.56	12.750	0.375	14.6	279.	43.8	4.38
10	40.48	10.750	0.365	11.9	161.	29.9	3.67
8	28.55	8.625	0.322	8.40	72.5	16.8	2.94
6	18.97	6.625	0.280	5.58	28.1	8.50	2.25
5	14.62	5.563	0.258	4.30	15.2	5.45	1.88
4	10.79	4.500	0.237	3.17	7.23	3.21	1.51
$3\frac{1}{2}$	9.11	4.000	0.226	2.68	4.79	2.39	1.34
3	7.58	3.500	0.216	2.23	3.02	1.72	1.16
Extra strong							
12	65.42	12.750	0.500	19.2	362.	56.7	4.33
10	54.74	10.750	0.500	16.1	212.	39.4	3.63
8	43.39	8.625	0.500	12.8	106.	24.5	2.88
6	28.57	6.625	0.432	8.40	40.5	12.2	2.19
5	20.78	5.563	0.375	6.11	20.7	7.43	1.84
4	14.98	4.500	0.337	4.41	9.61	4.27	1.48
$3\frac{1}{2}$	12.50	4.000	0.318	3.68	6.28	3.14	1.31
3	10.25	3.500	0.300	3.02	3.89	2.23	1.14

Table A-10 Design data for steel pipe commonly used as columns (SI units)

Nominal diameter, mm: standard weight	Weight, kN/m	Outside diameter, mm	Wall thickness, mm	Area, m$^2 \times 10^{-3}$	I_x or I_y, m$^4 \times 10^{-6}$	S, m$^3 \times 10^{-3}$	r, mm
300	0.723	324.	9.5	9.42	116.1	0.718	111.
250	0.591	273.	9.3	7.68	67.0	0.490	93.
200	0.417	219.	8.2	5.42	30.2	0.275	75.
150	0.277	168.	7.1	3.60	11.7	0.139	57.
130	0.213	141.	6.6	2.77	6.33	0.089	47.8
100	0.157	114.	6.0	2.05	3.01	0.053	38.4
90	0.133	102.	5.7	1.73	1.994	0.039	34.0
75	0.111	89.	5.5	1.44	1.257	0.028	29.5
Extra strong							
300[a]	0.955	324.	12.7	12.39	150.7	0.929	110.
250	0.799	273.	12.7	10.39	88.2	0.646	92.
200	0.633	219.	12.7	8.26	44.1	0.401	73.
150	0.417	168.	11.0	5.42	16.9	0.200	56.
130	0.303	141.	9.5	3.94	8.61	0.122	46.7
100	0.219	114.	8.6	2.85	4.00	0.070	37.6
90	0.182	102.	8.1	2.37	2.614	0.051	33.3
75	0.150	89.	7.6	1.95	1.619	0.037	29.0

[a] From *The Structural Steel Design Manual*, by Joseph E. Bowles (New York: McGraw-Hill, 1980).

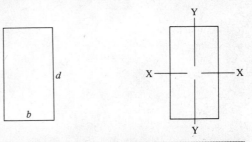

Table A-11 Selected timber standard sizes[a] properties of sections

Nominal size, in. b d	Actual size, in. b d	Area, in.²	Axis XX S, in.³	Axis XX I, in.⁴	Axis YY S, in.³	Axis YY I, in.⁴	Board measure per lineal foot	Weight per lineal foot (lb)
2 × 2	1½ × 1½	2.25	0.56	0.42	0.56	0.42	0.33	0.63
3	2$\frac{9}{16}$	3.84	1.64	2.10	0.96	0.72	0.50	1.07
4	3$\frac{9}{16}$	5.34	3.17	5.65	1.36	1.00	0.67	1.48
6	5½	8.25	7.56	20.80	2.06	1.54	1	2.29
8	7½	11.25	14.06	52.73	2.81	2.11	1.33	3.12
10	9½	14.25	22.56	107.17	3.56	2.67	1.67	3.96
12	11½	17.25	33.06	190.11	4.31	3.23	2	4.79
14	13½	20.25	45.56	307.55	5.06	3.80	2.33	5.62
4 × 4	3$\frac{5}{8}$ × 3$\frac{5}{8}$	13.14	7.94	14.39	7.94	14.39	1.33	3.64
6	5½	19.94	18.27	50.26	12.04	21.83	2	5.54
8	7½	27.19	33.98	127.45	16.42	29.77	2.67	7.55
10	9½	34.44	54.53	259.02	20.80	37.71	3.33	9.57
12	11½	41.69	79.90	459.13	25.18	45.65	4	11.58

[a] Southern Pine Association Technical Bulletin No. 2.

Table A-12 Slopes and deflections of selected beams

	Beam	Slope	Deflection (+ upward)
1		$\theta_1 = -\dfrac{PL^2}{16EI}$ at $x = 0$ $\theta_2 = +\dfrac{PL^2}{16EI}$ at $x = L$	$y_{max} = -\dfrac{PL^3}{48EI}$ at $x = L/2$
2		$\theta_1 = -\dfrac{wL^3}{24EI}$ at $x = 0$ $\theta_2 = +\dfrac{wL^3}{24EI}$ at $x = L$	$y_{max} = -\dfrac{5wL^4}{384EI}$ at $x = L/2$
3		$\theta_1 = -\dfrac{Pb(L^2 - b^2)}{6LEI}$ at $x = 0$ $\theta_2 = +\dfrac{Pa(L^2 - a^2)}{6LEI}$ at $x = L$	$y_{max} = -\dfrac{Pb(L^2 - b^2)^{3/2}}{9\sqrt{3}LEI}$ at $x = \sqrt{(L^2 - b^2)/3}$ $y_{center} = -\dfrac{Pb(3L^2 - 4b^2)}{48EI}$
4		$\theta = -\dfrac{PL^2}{2EI}$ at $x = L$	$y_{max} = -\dfrac{PL^3}{3EI}$ at $x = L$
5		$\theta = -\dfrac{wL^3}{6EI}$ at $x = L$	$y_{max} = -\dfrac{wL^4}{8EI}$ at $x = L$
6		Zero at ends Zero at center	y_{max} (at center) $= -\dfrac{wL^4}{384EI}$ Moment at ends is max. $M_{max} = -\dfrac{wL^2}{12}$
7		Zero at ends Zero at center	y_{max} (at center) $= -\dfrac{PL^3}{192EI}$ Moment at ends is max. $M_{max} = -\dfrac{PL}{8}$

Table A-13   Unit conversion factors

Length	1 ft $= 0.3048$ m 1 in. $= 25.40$ mm	1 m $= 3.281$ ft 1 mm $= 0.03937$ in.
Area	1 ft^2 $= 0.0929$ m^2 1 in.^{2} $= 645.2$ mm^2	1 m^2 $= 10.764$ ft^2 1 mm^2 $= 0.001550$ in.^{2}
Force	1 lb $= 4.448$ N $= 444{,}823$ dyn	1 N $= 0.2248$ lb 1 dyn $= 2.248 \times 10^{-6}$ lb
Stress, pressure	1 lb/in.^{2} $= 6895$ N/m^2 $= 68{,}950$ dyn/cm^2	1 N/m^2 $= 145 \times 10^{-6}$ lb/in.^{2} 1 dyn/cm^2 $= 14.5 \times 10^{-6}$ lb/in.^{2}
Energy, work	1 ft-lb $= 1.356$ Nm (joule) $= 1.356 \times 10^7$ dyn-cm (erg)	1 Nm $= 0.7376$ ft-lb 1 dyn-cm $= 8.850 \times 10^{-7}$ in.-lb
Mass	1 slug (lb-sec^2/ft) $= 14.59$ kg (N-sec^2/m)	1 kilogram (N-sec^2/m) $= 0.0685$ slug (lb-sec^2/ft)

Appendix I A FORTRAN Program for the Calculation of Principal Stresses or Principal Strains[*]

This program calculates the extreme values of stress (or strain) and the orientation, for given state at a point. A flowchart is given in Fig. A-1. This program can calculate the stresses (or strains) on a plane passing through a point, given the state of stress (or strain) at that point and the angle between the normal of the plane and the vertical axis. All the calculations are done for only planar situation (plane stress/strain). The following is the explanation for input or output:

```
C       VARIABLE                      EXPLANATION                    COMMENT
C
C       XVALUE       THE VALUE (STRESS OR STRAIN)                    INPUT
C                    PARALLEL TO X-AXIS
C       YVALUE       THE VALUE (STRESS OR STRAIN)                    INPUT
C                    PARALLEL TO Y-AXIS
C       SHEAR        SHEAR (STRESS OR STRAIN)                        INPUT
C       NSTRAN       =0  PLANE STRESS (STRESS VALUES)                INPUT
C                    =1  PLANE STRAIN (STRAIN VALUES)                INPUT
C       GTHETA       ANGLE MADE BY THE NORMAL OF THE PLANE TO Y-     INPUT
C                    AXIS (DEGREES)
C                    =0  IF (NTHETA=0)
C       NTHETA       =0  IF ONLY EXTREME VALUES ARE SOUGHT           INPUT
C                    =1  OTHERWISE
C       NMAX         =0  IF ONLY STRESSES (OR STRAINS) ON A PLANE    INPUT
C                    ARE TO BE CALCULATED
C                    (NO MAX. VALUE)
C                    =1  OTHERWISE
C       AMAXX        =MAXIMUM NORMAL VALUE                           OUTPUT
C       AMIN         =MINIMUM NORMAL VALUE                           OUTPUT
C       SHEARM       =MAXIMUM SHEAR VALUE (NOT ABSOLUTE)             OUTPUT
C       THETAN       =ANGLE OF MAX. NORMAL                           OUTPUT
C       THETAS       =ANGLE OF MAX. SHEAR                            OUTPUT
C       ANORMAL      =NORMAL VALUE ON THE PLANE                      OUTPUT
C       ASHEAR       =SHEAR VALUES ALONG THE PLANE                   OUTPUT
C
C
```

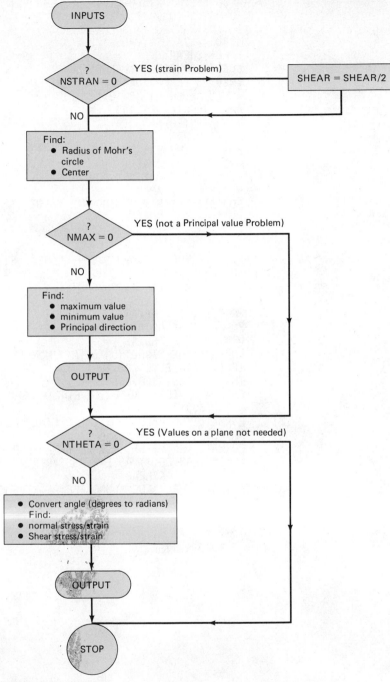

Figure A-1 Flowchart.

```
          READ (5,10) XVALUE,YVALUE,SHEAR,NSTRAN,NTHETA,NMAX,GTHETA
10        FORMAT(3F10.4,3I1,F10.4)
          IF (NSTRAN.EQ.1) SHEAR=SHEAR/2.
4RADX=(XVALUE-YVALUE)/2.
5RADY=SHEAR.
6CENTER=(XVALUE+YVALUE)/2.
7RADIUS=SQRT(RADX**2+RADY**2)
          IF (NMAX.EQ.0) GO TO 1
          AMAXX=CENTER+RADIUS
          AMIN=CENTER-RADIUS
          WRITE(6,21)
21        FORMAT('THE EXTREMUM VALUES OF THE PROGRAM ARE:',//.)
          WRITE(6,22) AMAXX
22        FORMAT('THE MAXIMUM NORMAL VALUE IS=',E14.7,//.)
          WRITE(6,23) AMIN
23        FORMAT('THE MINIMUM NORMAL VALUE IS=',E14.7,//.)
          WRITE(6,24) SHEARM


24        FORMAT('THE MAXIMUM SHEAR VALUE=',E14.7,//)
          ANG=RADX/RADY
          THETAS=(ATAN(ANG))*28.648
          THETAN=45.+THETAS
          WRITE(6,25) THETAN
25        FORMAT('THE PRINCIPAL DIRECTION(W.R.T.X-AXIS)',F6.2,'DEGREES',//)
          WRITE(6,26) THETAS
26        FORMAT('THE DIRECTION OF MAXIMUM SHEAR',F6.2,5X,'DEGREES',//)
1         IF (NTHETA.EQ.0) GO TO 2
          THETA2=GTHETA/28.648
          ANORMA=CENTER+RAD*COS(THETA2)-SHEAR*SIN(THETA2)
          WRITE(6,31)
31        FORMAT('THE VALUES ON THE DEFINED PLANE',//)
          WRITE(6,32) ANORMAL
32        FORMAT('THE VALUE IN THE NORMAL DIRECTION',E14.7,//)
          IF (NSTRAN.EQ.1) ASHEAR=ASHEAR*2.
          WRITE(6,33) ASHEAR
33        FORMAT('THE VALUE ALONG THE PLANE(SHEAR)',E14.7,//)
2         STOP
          END
```

* Courtesy of Subhotash Khan

Input parameters are shown in the picture and rests are explained in the program (Fig. A-2 for stress and Fig. A-3 for strain). Output parameters are explained and properly labeled in the program.

The strength of materials convention is followed (i.e., tension positive, compression negative, clockwise moment = positive shear). Check the format statements if large numbers are involved (greater than 10^5). Answers are given in exponent form (power of 10). If SHEAR = 0, NMAX *must* be equal to zero, otherwise the program will terminate with error.

In stress or strain, GTHETA is the angle of the normal of the plane with respect to the x-axis.

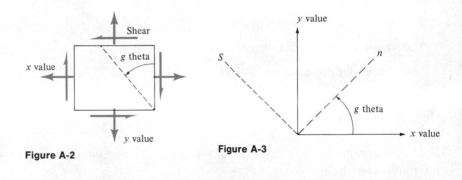

Figure A-2   Figure A-3

EXAMPLE A-1

A timber beam is to be loaded in such a manner that the resulting state of stress near the load will be that shown in Fig. A-4. It is feared that the timber will split if the shear stress parallel to the grain reaches or exceeds 4×10^6 N/m². Is the beam safe?

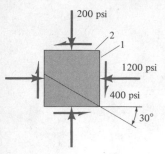

Figure A-4

Solution:

Input

XVALUE = −12000.0

YVALUE = −200.0

SHEAR = 400.0

NSTRAN = 0

NTHETA = 1

NMAX = 0

GTHETA = 60.0

Comments

Positive shear.

Stress problem.

Need 10 final values on a plane. Need not find extremum according to convention.

Output

The values on the defined plane.

The value in the normal direction is −0.7964102 E 03.

The value along the plane (SHEAR) is −0.6330127 E 03.

Interpretation

$$4 \times 10^6 \text{ N/m}^2 = 580 \text{ psi}$$

The maximum shear calculated is −633 psi. Hence, the structure is unsafe. The negative value only indicates the direction.

EXAMPLE A-2

Some point on the surface of a loaded body was found to have a tensile strain of 0.0001 in./in. in the *x* direction and a compressive strain of 0.0007 in./in. in the *y* direction, and the angular distortion

γ_{xy} was -0.0006 rad. (a) Determine the values and directions of the principal strains. (b) Determine the magnitude of the maximum shear strain. (c) Determine the axial and shear strains associated with the n axis which is 30° counterclockwise from the x axis, as indicated in Fig. 2-11(a).

Solution:

Input	Comments
XVALUE = 100.0	Converted to μ strain.
YVALUE = 700.0	Strain problem.
SHEAR = 600.0	Follow the convention.
NSTRAN = 1	
NTHETA = 1	
NMAX = 1	
GTHETA = 30.0	

Output

The extremum values of the problem are:

The maximum normal value is 0.2000000 E 03.

The minimum normal value is -0.8000000 E 03.

The principal direction (w.r.t. x axis) is 18.43°.

The direction of maximum shear is $-26.57°$.

The values in the defined plane:

The value in the normal direction is 0.1598076 E 03.

The value in the shear direction is 0.3928203 E 03.

Answers to selected problems

1-1 $F_a = 12$ kips tension
$F_b = F_c = 38$ kips tension

1-3 $F_a = 2500$ N tension
$M_a = 50$ N·m bending
$F_b = F_c = 2500$ N tension

1-5 Section a-a $F_N = 183.9$ N compression
                $F_T = 154.3$ N shear
     Section b-b $F_b = 240$ N tension
                $M_b = 38.4$ N·m bending
     Section c-c $F_c = 240$ N shear
                $M_c = 19.2$ N·m bending

1-7 Section a-a $F_a = 0$
                $M_a = 200$ N·m bending

     Section b-b $F_N = 0$
                $F_T = 800$ N shear
                $M_b = 1000$ N·m bending
     Section c-c $F_x = 0$
                $F_y = 800$ N shear
                $M_c = 1800$ N·m bending

1-9 Section a-a $F_a = 200$ N shear
                $M_a = 4$ N·m bending
     Section b-b $F_b = 200$ N shear
                $M_b = 16$ N·m bending
                $C_b = 12$ N·m twisting
     Section c-c $F_c = 200$ N shear
                $M_c = 24$ N·m bending
                $C_c = 12$ N·m twisting

1-11 Section *a-a* $F_x = 1067$ N tension
                       $F_y = 1000$ N shear
                       $M_a = 3000$ N·m bending
        Section *b-b* $F_x = 267$ N shear
                       $F_y = 400$ N compression
                       $M_b = 800$ N·m bending
        Section *c-c* $F_x = 800$ N shear
                       $F_y = 600$ N tension
                       $M_c = 800$ N·m bending

1-13 Section *a-a* $F_x = 600$ N shear
                       $F_y = 800$ N tension
                       $M_a = 2100$ N·m bending
        Section *b-b* $F_x = 400$ N tension
                       $F_y = 150$ N shear
                       $M_b = 225$ N·m bending
        Section *c-c* $F_x = 400$ N shear
                       $F_y = 150$ N tension
                       $M_c = 200$ N·m bending

1-15 $U_2L_2 = 0$
      $L_1L_2 = 2250$ lb tension
      $U_1'L_2 = 3750$ lb tension

1-17 Section *a-a* $F_T = 70.7$ N shear
                       $F_N = 70.7$ N tension
        Section *b-b* $F_x = 0$
                       $F_y = 100$ N tension
                       $M_b = 10$ N·m bending
        Section *c-c* $F_x = 0$
                       $F_y = 100$ N tension

1-19 Section *a-a* $F_N = 1732$ lb tension
                       $F_T = 1000$ lb shear
                       $M_a = 505$ ft-lb bending
        Section *b-b* $F_b = 2000$ lb tension
        Section *c-c* $F_N = 2000$ lb tension
                       $F_T = 0$
                       $M_c = 583$ ft-lb bending

1-21 Section *a-a* $C_a = 0.50$ ft-lb twist
                       $F_a = 15.7$ lb shear
                       $M_a = 23$ ft-lb bending
        Section *b-b* $F_b = 15.3$ lb shear
                       $C_b = 0.25$ ft-lb twist
                       $M_b = 27$ ft-lb bending
        Section *c-c* $F_c = 43.33$ lb shear
                       $M_c = 7.22$ ft-lb bending

1-23 $\sigma_a = 125 \times 10^3$ Pa compression
      $\sigma_c = 50 \times 10^3$ Pa tension

1-25 $\sigma = 23.5 \times 10^3$ Pa compression
      $\tau = 19.7 \times 10^3$ Pa shear

1-27 $\sigma = 350$ psi compression

1-29 $\sigma = 0.72$ ksi tension
      $\tau = 0.96$ ksi shear

1-31 $\sigma = 0$
      $\tau = 8$ MPa

1-33 $\sigma = 144.6$ MPa compression
      $\tau = 516.5$ MPa shear

1-35 $\sigma = 4213$ psi compression
      $\tau = 1089$ psi shear

1-39 $\sigma_{max} = \sigma_y$
      $\sigma_{min} = \sigma_y$
      $\tau = 0$

1-41 $\sigma_{max} = \sigma_y$
      $\sigma_{min} = \sigma_x$
      $\tau_{max} = \sigma_y$

1-43 $\sigma_{max} = \tau$
      $\sigma_{min} = -\tau$
      $\tau_{max} = \tau$

1-45 $\sigma_{max} = 1.33\sigma_x$
      $\sigma_{min} = 0.667\sigma_x$
      $\tau_{max} = \tau$

1-47 $\sigma_{max} = 4.14$ ksi
      $\sigma_{min} = -24.1$ ksi
      $\tau_{max} = 14.1$ ksi

1-49 $\sigma_{max} = 212$ MPa
      $\sigma_{min} = -212$ MPa
      $\tau_{max} = 212$ MPa

1-51 $\sigma_\theta = -6000$ psi
      $\tau_\theta = 0$

1-53 $\sigma_\theta = -12.4$ MPa
      $\tau_\theta = -11.5$ MPa

1-55 $\sigma_\theta = 2.83$ MPa
      $\tau_\theta = -6.83$ MPa

1-57 $\tau_\theta = 4.36$ MPa, $\Theta = 64.8°$ CW
                  or
      $\tau_\theta = -4.36$ MPa, $\Theta = 38.2°$ CCW

1-59 $\sigma_x = -3.55$ to 19.55 MPa

1-61 $\sigma_x < -100$ psi

1-63 $\sigma_{max} = 7.66$ ksi
$\sigma_{min} = 3.66$ ksi
$\tau_{max} = 5.66$ ksi

1-65 $\sigma_{max} = \sigma_z = 0$
$\sigma_{min} = -26.2$ ksi
$\tau_{max} = 13.1$ ksi

1-67 $\sigma_{max} = 8$ ksi
$\sigma_{min} = -8$ ksi

$\sigma_z = 0$
$\tau_{max} = 8$ ksi

1-69 $\sigma_{max} = 13.0$ MPa
$\sigma_{min} = -23.0$ MPa
$\sigma_z = 0$
$\tau_{max} = 18.0$ MPa

1-71 $\sigma_{max} = 1340$ psi
$\sigma_{min} = \sigma_z = 0$
$\tau_{max} = 670$ psi

CHAPTER 2

2-1 $\varepsilon = 0.33$ in./in.

2-3 $\varepsilon = 0.26$ m/m

2-5 $\varepsilon - 0.00625$ in./in.

2-7 $e_B = e_A + 0.01$ m
$\varepsilon_B = 0.6\varepsilon_A + 0.004$
$e_C = 0.015 + e_A$ m

2-9 $\varepsilon = 0.0025$ in./in.

2-11 $\varepsilon = 0.0047$ mm/mm

2-13 $\varepsilon_{AC} = 0.0469$ in./in.
$\varepsilon_{BD} = -0.0492$ in./in.

2-15 $\varepsilon_{AP} = +\dfrac{1}{4}\dfrac{x}{L}$

$\varepsilon_{DP} = -\dfrac{1}{4}\dfrac{x}{L}$

2-17 $\varepsilon_L = \dfrac{L^2\theta^2}{8b^2}$

2-19 $\gamma = 0.0488$ rad

2-23 $\varepsilon_{max} = 500$ μin./in.
$\varepsilon_{min} = -2100$ μin./in.
$\gamma_{max} = 2600$ μin./in.
$\theta = -11.3°$

2-25 $\varepsilon_{max} = 2400$ μin./in.
$\varepsilon_{min} = 400$ μin./in.
$\gamma_{max} = 2000$ μin./in.
$\theta = 63.4°$

2-27 $\varepsilon_{max} = 1697$ microstrain
$\varepsilon_{min} = -1697$ microstrain
$\gamma_{max} = 3394$ microstrain
$\theta = -22.5°$

2-29 $\varepsilon_{max} = 1530$ microstrain
$\varepsilon_{min} = 680$ microstrain
$\gamma_{max} = 850$ microstrain
$\theta = -76°$

2-31 $AC' = 147.1$ mm
$B'D = 135.8$ mm
Angle change is 0 because there is no shear strain for $\theta = 45°$.

2-41 $\varepsilon_{max} = 939$ mm/mm $\times 10^{-6}$
$\varepsilon_{min} = -139$ mm/mm $\times 10^{-6}$
$\gamma_{max} = 1077$ mm/mm $\times 10^{-6}$

2-43 $\varepsilon_{max} = 441$ mm/mm $\times 10^{-6}$
$\varepsilon_{min} = 159$ mm/mm $\times 10^{-6}$
$\gamma_{max} = 282$ mm/mm $\times 10^{-6}$

2-45 $\varepsilon_{max} = 1273$ microstrain
$\varepsilon_{min} = -639$ microstrain
$\gamma_{max} = 1912$ microstrain

2-47 $\varepsilon_{max} = 1216$ microstrain
$\varepsilon_{min} = -748$ microstrain
$\gamma_{max} = 1964$ microstrain

2-49 $\varepsilon_{max} = 1103$ microstrain
$\varepsilon_{min} = -603$ microstrain
$\gamma_{max} = 1706$ microstrain

CHAPTER 3

3-1 $E = 6.25 \times 10^6$ psi
$\sigma_{PL} = 25{,}000$ psi

3-4 $E_t = 10.44 \times 10^6$ psi
$\sigma_{PL} = 48{,}000$ psi

3-6 $E = 1.6 \times 10^6$ psi
$\sigma_{PL} = 8500$ psi

3-8 $\mu = 0.3$

3-10 $e_1 = -125 \times 10^{-6}$ in.
$e_2 = -31.5 \times 10^{-6}$ in.

3-12 loaded dimensions: 100.12 mm $\times$ 39.984 mm $\times$ 9.996 mm

3-14 $\sigma_{YS} = 42{,}000$ psi
$R = 50$ in.-lb/in.^{3}
$U = 414$ in.-lb/in.^{3}

3-16 $\sigma_{YS} = 53.4$ ksi
$R = 110.4$ in.-lb/in.^{3}
$U = 367.2$ in.-lb/in.^{3}

3-18 $\sigma_{YS} = 21.21$ ksi
$R = 0.0734$ ksi
$T = 0.242$ ksi

3-20 $E = 110.3$ GPa
$\sigma_{PL} = 661.9$ MPa
$\sigma_{YS} = 882.6$ MPa
$R = 1.99$ MPa

3-22

	tension	compression
	$E_t = 11.11 \times 10^6$ psi	$E_c = 11.11 \times 10^6$ psi
	$\sigma_{PL} = 52$ ksi	$\sigma_{PL} = 34$ ksi
	$\sigma_Y = 58$ ksi	$\sigma_Y = 52$ ksi
	$R = 121.7$ in.-lb/in.^{3}	$R = 52.03$ in.-lb/in.^{3}

CHAPTER 4

4-3 For relaxation $e =$ constant:

a) $F = C \exp\left(\dfrac{-kt}{\eta}\right)$

b) $F = ke$ (constant)

c) $F = C \exp\left(\dfrac{-k_2 t}{\eta}\right) + k_1 e$

4-5 $F = \eta\dot{e}$ for $F < F_c$
$F = F_c$ otherwise

4-7 $F < F_c \qquad F = ke$

$F > F_c \qquad \dot{e} = \dfrac{\dot{F}}{k} + \dfrac{F - F_c}{\eta}$

4-9 $E_t = 0.333 \times 10^6$ psi $\qquad E_s = 1.0 \times 10^6$ psi

4-11 $E_t = 2500$ ksi $\qquad E_s = 5400$ ksi

4-13 $E_t = 1.125$ GPa $\qquad E_s = 2.91$ GPa

4-15 $E_t = 490$ ksi $\qquad E_s = 1380$ ksi

4-17 $E_t = 10.3 \times 10^6$ psi $\qquad E_s = 15.2 \times 10^6$ psi

4-19 strain energy $= 22.95$ in.-lb/in.^{3}

4-21 $\varepsilon_x = \dfrac{1 - \mu^2}{E}\sigma_x - \dfrac{\mu(1 + \mu)}{E}\sigma_y$

$\sigma_x = \dfrac{E}{1 + \mu}\varepsilon_x + \dfrac{E\mu(\varepsilon_x + \varepsilon_y)}{(1 - 2\mu)(1 + \mu)}$

$\tau_{xy} = G\gamma_{xy}$

4-23 For G to be positive $\mu > -1$.
For K to be positive, $\mu < \frac{1}{2}$

4-27 $C_1 = \dfrac{E}{1 + \mu}, \; C_2 = \dfrac{E\mu}{(1 + \mu)(1 - 2\mu)}$

4-29 $\sigma_1 = -4725$ psi
$\sigma_2 = -45{,}675$ psi
$\sigma_3 = 0$
$\tau_{max} = 22{,}837$ psi

4-31 $\sigma_1 = 27{,}625$ psi
$\sigma_2 = 14{,}823$ psi
$\sigma_3 = 0$
$\tau_{max} = 13{,}813$ psi

4-33 $\sigma_1 = 119$ MPa
$\sigma_2 = -119$ MPa
$\sigma_3 = 0$
$\tau_{max} = 119$ MPa

4-35 $\sigma_1 = 380.8$ MPa
$\sigma_2 = 238.0$ MPa
$\sigma_3 = 0$
$\tau_{max} = 190.4$ MPa

4-37 $\Delta V = 7.1 \times 10^{-3}$ in.^{3}

CHAPTER 5

5-1 360,000 lb, 2.67 in. from top edge

5-3 173.4 kN, 118.5 mm from bottom edge

5-5 630,000 lb, 1.43 in. from bottom edge

5-7 1536 kips, 3.0 in. from bottom edge

5-9 1728 kips, 3.56 in. from bottom edge

5-11 28.52 kips, 1.2 in. from bottom edge

5-13 8000 lb, $\varepsilon_A = -133 \ \mu\varepsilon$

5-15 $\varepsilon_A = -147\mu\varepsilon$, $P = 9800$ lb

5-17 $\varepsilon_B = 773 \times 10^{-6}$
$\varepsilon_A = 193 \times 10^{-6}$

5-19 $F = 6400$ N

5-21 $F = 500$ lb

5-23 $F_{CB} = 14{,}900$ N compression

5-25 $\tau = 5.69$ MPa

5-27 $a = 5$ in. $c = 2.39$ in. $d = 1.07$ in.
$e = 1.875$ in. b may be any convenient value

5-29 $\sigma_{max} = 11.67$ ksi
$\tau_{max} = 5.83$ ksi

5-31 $\tau_{max} = 500$ psi

5-35 $\sigma_\theta = 530$ psi
$\sigma_\theta = 918$ psi

5-37 $\tau = 4.16$ MPa

5-39 $\sigma_\theta = 18.75$ MPa
$\tau_\theta = 10.82$ MPa

5-41 $d_A = 6.38$ mm
$d_C = d_B = 8.71$ mm

5-43 5 rivets

5-45 $D = 94$ mm

5-47 $\sigma_t = 37.7$ MPa
$\sigma_b = 370$ MPa

5-49 $\tau = 40.7$ MPa
$\sigma_b = 80$ MPa
$\sigma_t = 100$ MPa

5-51 $\tau = 5820$ psi
$\sigma_b = 13{,}700$ psi
$\sigma_t = 7380$ psi

5-53 $\tau = 4.97$ MPa
$\sigma_b = 15.6$ MPa
$\sigma_t = 14.7$ MPa

5-55 $\tau = 12{,}300$ psi
$\sigma_b = 19{,}400$ psi
$\sigma_t = 12{,}500$ psi

5-59 $P = 1.10$ MPa

5-61 $\tau = -450$ lb/in.
$\sigma = 3479$ lb/in.

5-63 6 bolts

5-65 $D = \frac{3}{4}$ in.

5-67 $\sigma_1 = 91.1$ MPa
$\sigma_2 = 70.9$ MPa
$\sigma_L = 54.0$ MPa
$\sigma_C = 108.0$ MPa
$p = 0.864$ MPa

5-69 $\Delta R = \dfrac{PR^2}{2tE}(1 - \mu)$,

$\Delta V = \dfrac{2\pi PR^4(1 - \mu)}{tE}$

5-71 $\sigma_{max} = 1.132 \, t - 68$

5-73 7 bolts on each side, 14 total

5-75 $e = 2.04$ in.

5-77 $e = 0.03$ in.

5-79 $P = 1473$ lb

5-81 $e = 0.0217$ in.
$\Delta d = -0.000713$ in.

5-83 $e = 0.232$ in.

5-85 $e = 3.2$ mm $\Delta V = 2 \times 10^{-6}$ m^3

5-87 $L_B = 1.78$ m $L_S = 0.22$ m

5-89 $\tau = 1875$ psi
$\varepsilon = 62.5 \times 10^{-6}$ in./in.

5-91 $P = 235$ N

5-93 $e = 4.5 \times 10^{-3}$ in.

5-95 a) $e = 0.00146$ m
b) $e = 0.0212$ m

5-97 $\Delta L = \dfrac{\rho\omega^2 L^3}{3E}$ $\qquad \sigma = \dfrac{\rho\omega^2}{2}(L^2 - l^2)$

5-99 $e = \dfrac{wb}{2E}\left[\dfrac{a(a + 2b)}{2b} - b\ln\left(\dfrac{a + b}{b}\right)\right]$

5-101 $A = 1.94$ in.^{2}

5-103 $P = 1056$ lb

5-105 $P = 10,000$ lb

5-107 $D = 1.23$ $\qquad$ Use $1\frac{1}{4}$ in. bolts

5-109 $T_B = 6618$ lb
$\qquad T_C = 11,030$ lb

5-111 $P = 40,000$ lb

5-113 $\sigma_s = 3690$ psi
$\qquad \tau_{ci} = 2422$ psi

5-115 $P_B = 395$ kN
$\qquad P_T = 525$ kN

5-117 $\sigma_B = 210$ MPa $\qquad e_B = 3.25$ mm
$\qquad \sigma_C = 1207$ MPa

5-119 $R_T = 137,000$ N
$\qquad R_B = 112,000$ N

5-121 Scale reads 0.0225 in.

CHAPTER 6

6-1 $T = \dfrac{\tau_{max}\pi r^3}{2}$

6-3 $T = \dfrac{\tau_{max}\pi}{2}r^3(1 - c^4)$

6-5 $T = \dfrac{\tau_{max}\pi r^3}{6}(4 - c^3)$

6-7 $T = \tau_{max}\pi r^3\dfrac{55}{99}$

6-9 $T = \dfrac{\pi r^3}{6}Gc(4 - c^3)\gamma_{max}$

$\qquad T = \tau_{max}\dfrac{\pi r^3}{6}(4 - c^3)$

6-11 $J = 25.1$ cm^4 for both elastic and plastic behavior

6-13 percent decrease in $J = -100c^4$

$\qquad$ percent increase in $\tau = \dfrac{c^4}{1 - c^4} \times 100$

6-15 percent increase in torque $= 700\%$ for both cases

6-17 44%

6-19 $T = 796,000$ in.-lb

6-21 $\tau_0 = \dfrac{PR}{\pi r_0^2 L}$, $\tau_i = \dfrac{PR}{\pi r_i^2 L}$

6-23 **(a)** $T = 125,700$ in.-lb
$\qquad$ **(b)** $T = 990,000$ in.-lb

6-25 $\sigma_{max} = \tau_{max} = 238$ MPa for fully plastic condition if shaft does not rupture

6-27 48.3 hp

6-29 **(a)** 1870 hp
$\qquad$ **(b)** 873 hp

6-31 $d = 16$ in.

6-33 $T_w = 36,300$ N·m
$\qquad \tau_s = 11.9$ MPa

6-35 $J_h > J_s$
$\qquad T_h > T_s$

6-37 9349 hp
$\qquad$ 0.03 rad

6-39 $1.076°$

6-41 14.9×10^3 N·m/rad

6-43 27,490 N·m, $\theta = 5.72°$

6-45 $0.458°$

6-47 $0.99°$

6-49 $6.7°$

6-51 95,500 in.-lb

6-53 $13.87°$

6-55 $\tau = \dfrac{4TL}{3R^3}$

$\qquad \theta = \dfrac{TL^2}{\pi R^4 G}$

6-57 7095 N·m, $\theta = 0.286°$

6-59 40.3 hp

6-61 $r = 2.26$ in.

6-63 50-mm-diameter shaft cannot meet twist requirement

6-65 1842 hp

6-67 1047 hp

6-69 $\sigma_{max} = 9.67$ MPa
$\tau_{max} = 5.43$ MPa

6-71 $\sigma_{max} = 68.8 \times 10^6$ Pa
$\tau_{max} = 38.2 \times 10^6$ Pa

6-73 1498 μin./in. tension or 1462 μin./in. compression

6-75 11.3 MPa

6-77 78,897 in.-lb

6-79 $\tau_S = 12{,}732$ psi
$\tau_A = 8488$ psi

6-81 $\Delta T = 5300$ in.-lb

6-83 $D = 44$ mm

6-85 15,175 in.-lb

6-87 83 MPa, $\theta = 27.5°$

CHAPTER 7

7-1

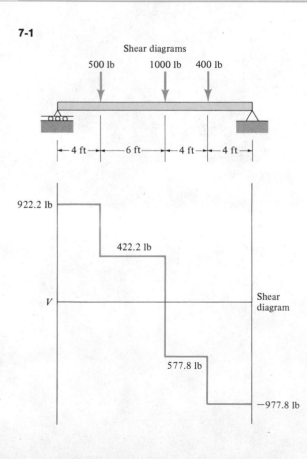

7-3

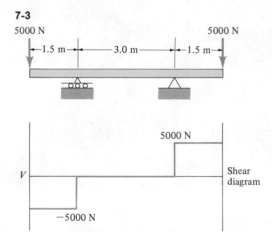

7-5

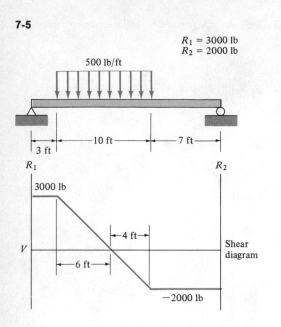

$R_1 = 3000$ lb
$R_2 = 2000$ lb

500 lb/ft

10 ft

7 ft

3 ft

R_1

R_2

3000 lb

4 ft

V

6 ft

Shear diagram

-2000 lb

7-7

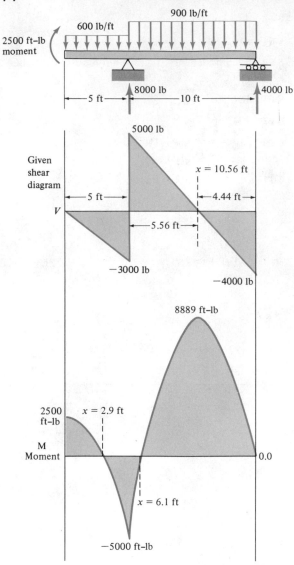

900 lb/ft

600 lb/ft

2500 ft–lb moment

8000 lb

4000 lb

5 ft

10 ft

5000 lb

Given shear diagram

$x = 10.56$ ft

5 ft

4.44 ft

V

5.56 ft

-3000 lb

-4000 lb

8889 ft–lb

2500 ft–lb

$x = 2.9$ ft

M Moment

0.0

$x = 6.1$ ft

-5000 ft–lb

7-9

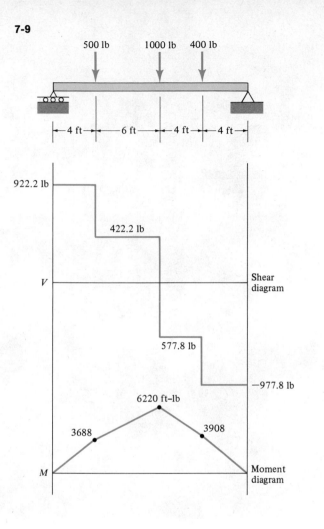

7-11

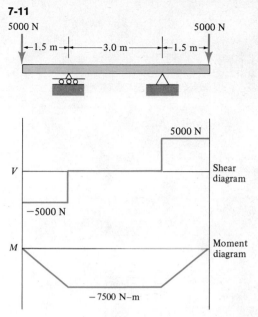

7-13

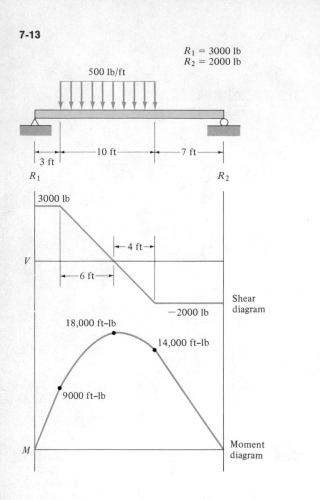

$R_1 = 3000$ lb
$R_2 = 2000$ lb

500 lb/ft

10 ft

7 ft

3 ft

R_1

R_2

3000 lb

4 ft

V

6 ft

Shear
diagram

−2000 lb

18,000 ft–lb

14,000 ft–lb

9000 ft–lb

M

Moment
diagram

7-15

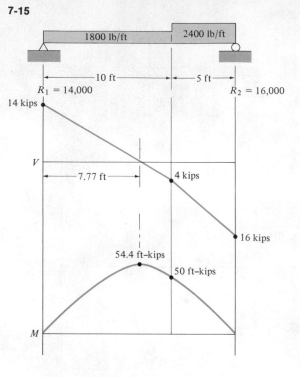

1800 lb/ft

2400 lb/ft

10 ft

5 ft

$R_1 = 14,000$

$R_2 = 16,000$

14 kips

V

7.77 ft

4 kips

16 kips

54.4 ft–kips

50 ft–kips

M

7-17

7-19

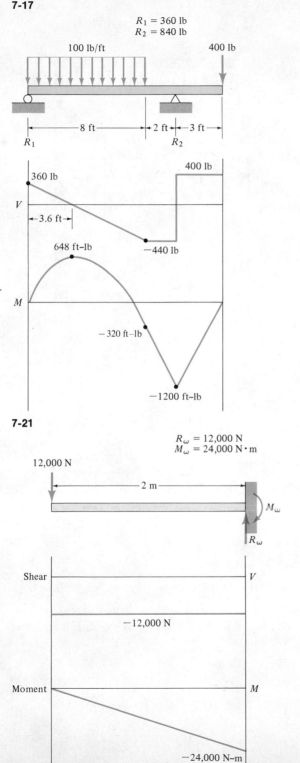

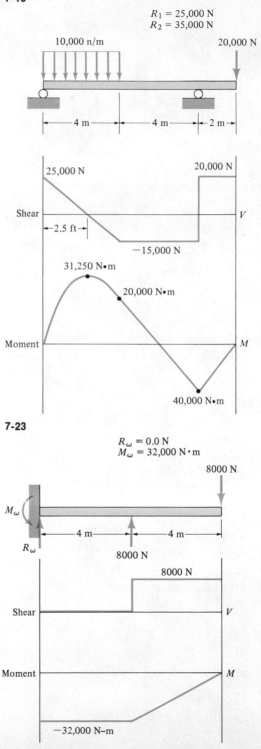

7-21

7-23

7-25

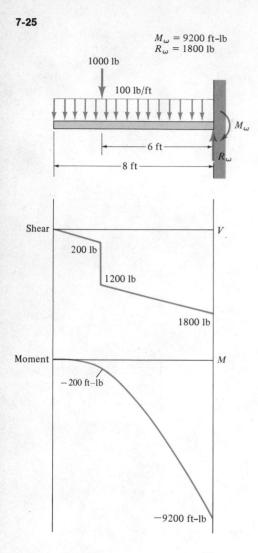

7-27

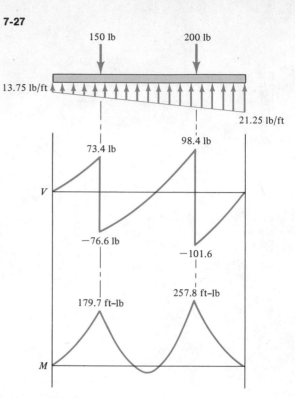

7-29   186 mm from bottom
225 mm from bottom

7-31   3.29 in. from bottom
3.50 in. from bottom

6-33   (b) is better

7-35   1.40

7-37   1.50

7-39   2.343

7-41 9.12×10^6 in.-lb

7-43 $\sigma_{yp} = 20{,}000$ psi

7-45   14,000 lb

7-47 $K = 10/12$

7-49 $M = 306{,}000$ in.-lb

7-51 $\sigma_{max} = 882$ psi in the bottom fibre at midspan

7-53 $\sigma_{max} = 4.22$ MPa compression at 1.72 m from left end

7-55 $\sigma = 272$ psi

7-57 $P = 523.6$ lb

7-59 $\sigma_t = 147$ MPa, $\sigma_c = 44.3$ MPa

7-61   21.33 ft

7-63 40.4 ft

7-65 $\sigma_T = 34.2$ MPa
$\sigma_C = 41.0$ MPa

7-67 $S = \dfrac{bh^2}{6}$

7-69 $S = 53.6$ in^3

7-71 S7 × 20 or W 10 × 12

7-73 S6 × 17.25 or W 8 × 10

7-75 S150 × 18.6 or W 150 × 17.9

7-77 S10 × 25.4 or W 10 × 22

7-79 S15 × 50 or W 18 × 40

7-81 S3 × 5.7 or W 4 × 13

7-83 $\sigma_{\text{comp}} = \dfrac{8P}{d^2}$

7-85 $\sigma_{\text{max}} = 12.25$ MPa

7-87 Stress varies linearly from 22.26 psi compression to 2.74 psi compression

7-89 4.92 × 9.84 in.

7-91 $\sigma_A = 2000$ psi tension $\sigma_B = 1000$ psi compression $\sigma_C = 1000$ psi compression $\sigma_D = 4000$ psi compression

7-93 43.24 in.^{3}

7-95 29,400 mm^3

7-97 $\tau = \dfrac{12Vy(h - y)}{bh^3}$

7-99 $\frac{4}{3}$

7-101 $1\frac{1}{4}$ in. spacing

7-103 $\sigma = 884$ psi $\tau = 245$ psi

7-105 250 lb/ft

7-107 $\tau_{\text{max}} = 0.530$ MPa at right support

7-109 $L = 11.55$ ft

7-111 $\sigma_T = 2147$ psi 4 ft from left end
$\sigma_C = 2684$ psi at right support
$\tau = 436$ psi at right support

7-113 (a) $\sigma = 2076$ psi
(b) $\sigma = 4200$ psi

7-115 $t = 8.5$ mm

7-117 (a)   81,000 lb
(b)   134,000 lb

7-119 $\tau_{\text{max}} = 148.6$ MPa
$\sigma_{\text{max}} = 282.1$ MPa
$\sigma_{\text{min}} = -15.1$ MPa

7-121 $\tau_{\text{max}} = 61.9$ MPa
$\sigma_{\text{max}} = 72.1$ MPa
$\sigma_{\text{min}} = -51.7$ MPa

7-123 3448 lb

CHAPTER 8

8-1 $0 \le x \le L$ $V_x = \dfrac{wL}{2} - wx$

$M_x = \dfrac{wLx}{2} - \dfrac{wx^2}{2}$

8-3 $0 \le x \le L$ $V_x = \dfrac{wL}{6} - \dfrac{wx^2}{2L}$

$M_x = \dfrac{wLx}{6} - \dfrac{wx^3}{6L}$

8-5 $0 \le x \le \dfrac{L}{2}$ $V_x = -\dfrac{P}{4}$

$M_x = -\dfrac{Px}{4}$

$\dfrac{L}{2} \le x \le L$ $V_x = -\dfrac{P}{4}$

$M_x = -\dfrac{Px}{4} + \dfrac{PL}{4}$

8-7 $0 \le x \le L$ $V_x = 0$
$M_x = M_o$

8-9 $0 \le x \le L$ $V_x = wL - wx$
$$M_x = wLx - \frac{wx^2}{2} - \frac{wL^2}{2}$$

8-11 $0 < x < L$ $V_x = -P$
$M_x = -Px$

8-13 $0 < x < 4$ $V_x = 922$
$M_x = 922x$
$4 < x < 10$ $V_x = 422$
$M_x = 922x - 500(x - 4)$
$10 < x < 14$
$V_x = -578$
$M_x = 922x - 500(x - 4) - 1000(x - 10)$
$14 < x < 18$
$V_x = -978$
$M_x = 922x - 500(x - 4) - 1000(x - 10)$
$\quad - 400(x - 14)$

8-15 $0 < x < 1.5$ $V_x = -5000$
$M_x = -5000x$
$1.5 < x < 4.5$ $V_x = 0$
$M_x = -7500$
$4.5 < x < 6.0$ $V_x = 5000$
$M_x = -7500 + 5000(x - 4.5)$

8-17 $0 < x < 5$ $V_x = 510$
$M_x = 510x$
$5 < x < 10$
$V_x = 510 - 600 - 200(x - 5)$
$M_x = 510x - 600(x - 5) - 100(x - 5)^2$
$10 < x < 12$
$V_x = 510 - 600 - 200(x - 5) + 1490$
$M_x = 510x - 600(x - 5) - 100(x - 5)^2$
$\quad + 1490(x - 10)$

8-19 $0 < x < 10$ $V_x = 1796 - 25x^2$
$$M_x = 1796x - \frac{25x^3}{3}$$
$10 \le x < 14$ $V_x = 1796 - 2500$
$M_x = 1796x - 2500(x - 6.67)$
$14 < x < 18$
$V_x = 1796 - 2500 - 1000 = -1704$
$M_x = 1796x - 2500(x - 6.67) - 1000(x - 14)$

8-21 $0 < x < 6$ $V_x = 3000$
$M_x = 3000x$

$6 < x \le 12$ $V_x = 3000 - 2000 = 1000$
$M_x = 3000x - 2000(x - 6)$
$12 \le x < 20$ $V_x = 1000 - 1000(x - 12)$
$M_x = 3000x - 2000(x - 6)$
$\quad - 500(x - 12)^2$

8-23 $0 \le x < 4$ $V_x = -1000x$
$M_x = -500x^2$
$4 < x \le 12$ $V_x = -1000x + 13,000$
$$M_x = \frac{-1000x^2}{2} + 13,000(x - 4)$$
$12 \le x < 16$
$V_x = 1000$
$M_x = -12,000(x - 6) + 13,000(x - 4)$
$16 < x < 20$
$V_x = -9000$
$M_x = -12,000(x - 6) + 13,000(x - 4)$
$\quad - 10,000(x - 16)$

8-25 $0 < x < 4$ $V_x = -2$
$M_x = -2x$
$4 < x \le 12$ $V_x = 10 - x$
$$M_x = -2x + 8(x - 4) - \frac{(x - 4)^2}{2}$$
$12 \le x < 16$
$V_x = -2$
$M_x = -2x + 8(x - 4) - 8(x - 8)$

8-27 $0 < x < 5$ $V_x = 3000$
$M_x = 3000x$
$5 < x < 10$ $V_x = 3000 - 5000 = -2000$
$M_x = 3000x - 5000(x - 5) + 5000$
$10 < x < 15$
$V_x = 3000 - 5000 - 500(x - 10)$
$M_x = 3000x - 5000(x - 5) + 5000$
$\quad - 250(x - 10)^2$
$15 < x < 20$
$V_x = 3000 - 5000 - 500(x - 10) + 7000$
$M_x = 3000x - 5000(x - 5) + 5000$
$\quad - 250(x - 10)^2 + 7000(x - 15)$

8-29 $0 < x \le 4$ $V_x = 260$
$M_x = 260x$
$4 \le x \le 14$ $V_x = 260 - 50(x - 4)$
$M_x = 260x - 25(x - 4)^2$
$14 \le x < 20$ $V_x = 260 - 500 = -240$
$M_x = 260x - 500(x - 9)$
$20 < x < 25$ $V_x = 0$
$M_x = 0$

8-31 $0 < x \leq 8$ $\qquad V_x = 2400 - 100x + \dfrac{100}{16}x^2$

$$M_x = 2400x - \dfrac{100x^2}{2} + \dfrac{100x^3}{48}$$
$$- 21{,}067$$

$8 \leq x \leq 10$ $\qquad V_x = 2400 - 400 = 2000$
$$M_x = 2400x - 400(x - 2.67)$$
$$- 21{,}067$$

8-33 $0 \leq x < 5$

$$V_x = 13.75x + \dfrac{7.5x^2}{40}$$

$$M_x = \dfrac{13.75x^2}{2} + \dfrac{7.5x^3}{120}$$

$5 < x < 15$

$$V_x = 13.75x + \dfrac{7.5x^2}{40} - 150$$

$$M_x = \dfrac{13.75x^2}{2} + \dfrac{7.5x^3}{120} - 150(x - 5)$$

$15 < x < 20$

$$V_x = 13.75x + \dfrac{7.5x^2}{40} - 150 - 200$$

$$M_x = \dfrac{13.75x^2}{2} + \dfrac{7.5x^3}{120} - 150(x - 5)$$
$$- 200(x - 15)$$

8-35 $0 < x < 4a$
$$V_x = \tfrac{6}{7}P$$
$$M_x = \tfrac{6}{7}Px - \tfrac{6}{7}Pa$$
$4a < x < 9a$
$$V_x = \tfrac{6}{7}P - 2P$$
$$M_x = \tfrac{6}{7}Px - \tfrac{6}{7}Pa - 2P(x - 4a) + 2Pa$$

8-37 $\delta_{\max} = -\dfrac{5wL^4}{384EI}$ at midspan

8-39 $\delta_{\max} = -\dfrac{PL^3}{3EI}$

8-41 $\delta_{\max} = -\dfrac{wL^4}{30EI}$

8-43 $\delta_{\max} = \dfrac{PL^3\sqrt{12}}{1728EI}$

8-45 0.6561 in.

8-47 173.6 lb

8-49 For $0 \leq x \leq L/2$, $EIy = \dfrac{wLx^3}{16} - \dfrac{wx^4}{24} - \dfrac{3wL^3x}{128}$

For $L/2 \leq x \leq L$, $EIy = \dfrac{wLx^3}{16} - \dfrac{wL}{12}(x - L/4)^3$

$$- \dfrac{11wL^3x}{(16)(24)} + \dfrac{wL^4}{(16)(48)}$$

8-51 For $0 \leq x \leq L/2$, $EIy_P = \dfrac{Px^3}{6} + \dfrac{QL^2x}{8} - \dfrac{PL^2x}{2}$

$$+ \dfrac{PL^3}{3} - \dfrac{5QL^3}{48}$$

For $L/2 \leq x \leq L$, $EIy = y_P - \dfrac{Q}{6}(x - L/2)^3$

8-53 For $0 \leq x \leq 6$ ft, $EIy = 4000x^3 - 1584 \times 10^3x$
For 6 ft $\leq x \leq 18$ ft, EIy

$$= 4000x^3 - \dfrac{500}{3}(x - 6)^4 - 1584 \times 10^3x$$

8-55 $y_A = -0.514$ in.

8-57 $\delta_v = -\dfrac{5PL^3}{48EI}$, $\delta_h = \dfrac{PL^3}{16EI}$

8-59 $M_0 = 10{,}000$ N·m, $\delta = 133$ mm to the left

8-61 $\delta = \dfrac{3wL^4}{8(3EI + kL^3)}$

8-63 $\delta = 0.0899$ in. $\qquad \searrow 51°$

8-65 $\delta = \dfrac{5wL^4}{32Eb^4}$ regardless of value of θ

8-67 $R_1 = \dfrac{3wL}{16} = R_3$, $R_2 = \dfrac{5wL}{8}$

8-69 $R_1 = \dfrac{11}{16}P$, $R_2 = \dfrac{5}{16}P$, $M_1 = \dfrac{-3}{16}PL$

8-71 $R_1 = \dfrac{17}{16}wL$, $R_2 = -\dfrac{wL}{16}$, $M_2 = \dfrac{1}{32}wL^2$

8-73 $R_1 = R_3 = \dfrac{4}{27}P$, $R_2 = \dfrac{46}{27}P$

8-75 $R_1 = \dfrac{3}{32}wL$, $R_2 = \dfrac{13}{32}wL$, $M_1 = \dfrac{-5}{192}wL^2$,

$$M_2 = \dfrac{-11}{192}wL^2$$

CHAPTER 9

9-1 $y(x) = \dfrac{1}{EI}\left[\dfrac{wL}{16}(x-0)^3 u\langle x-0\rangle - \dfrac{wx^4}{24}\right.$

$\left. + \dfrac{w(x-L/2)^4}{24} u\left\langle x - \dfrac{L}{2}\right\rangle - \dfrac{3wL^3}{128}x\right]$

9-3 $y(x) = \dfrac{1}{EI}\left[\dfrac{P}{6}x^3 - \dfrac{Q}{6}\left(x - \dfrac{L}{2}\right)^3 u\left\langle x - \dfrac{L}{2}\right\rangle\right.$

$\left. + \dfrac{L^2}{8}(Q - 4P)x + \dfrac{L^2}{48}(16P - 4Q)\right]$

9-5 $y(x) = \dfrac{1}{EI}\left[4000x^3 - \dfrac{500}{3}(x-6)^4 u\langle x-6\rangle\right.$

$\left. + \dfrac{500}{3}(x-18)^4 u\langle x-18\rangle - 1{,}584{,}000\right]$

9-7 $\delta_{\max} = \dfrac{-PL^3\sqrt{12}}{1728EI}$ at $x = L/\sqrt{12}$

9-9 $\delta_{\max} = -1.54$ mm at midpoint of beam

9-11 $\delta_{\max} = 0.371$ mm at left end

9-13 $\dfrac{PL^3}{48EI}$

9-15 $\dfrac{wL^4}{8EI}$

9-17 0.6561 in.

9-19 173.6 lb

9-21 19.4 mm

9-23 19,000 psi at $x = 60$
$\delta_{\mathrm{mid}} = 1.899$ in.; but is not $\delta_{\max}$

9-25 $\delta_{\max} = \dfrac{10^6}{3EI}$ ft at ends of beam

CHAPTER 10

10-1 114.7

10-3 57.5

10-5 46.7

10-7 75 in.

10-9 $d = 5.7$ in. (approximately)

10-11 $0.2886\sqrt{A}$ (square)

$0.2828\sqrt{A}$ (round)

10-13 $d = 30.55$ mm

10-15 $d = 5.38$ in.

10-17 13,700 lb

10-19 44.4

10-21 91.7

10-23 57.5

10-25 $\dfrac{P_p}{P_w} = 3.56$

10-27 yes

10-29 $3\frac{1}{2}$ turns

10-31 $R_i = 23.8$ mm

10-33 259 kN

10-35 **(a)** $\dfrac{L}{r} = 120$

(b) will not buckle in elastic range since $P_f/A > 20{,}000$ psi

10-37 **(a)** $L = 10.25$ m
(b) $W = 6212$ N

10-39 F.S. = 1.17

10-41 1036 lb

10-43 9000 lb

10-45 $\dfrac{L}{r} < 181$ for inelastic buckling

10-47 24,700 lb

10-49 495,000 N

10-51 221,000 N

10-53   302 lb

10-55   **(d)** $\sigma_{\text{all}} = 2170 - 17\left(\dfrac{L}{r}\right)$

10-59 $L = 14.14$ in.

10-61 $P = 24{,}700$ lb

10-63   **(a)** $R = 2.50$ in.
            (b) $R = 2.48$ in.

10-65   **(a)** $\dfrac{L}{r} = 166 > 124$, not allowed

            (b)   3560 lb

10-67   **(a)** $d = 17.05$ mm
            (b) $d = 22.2$ mm

10-69   128,000 lb

10-71   620,000 N

10-73   67,300 lb

10-75   62,600 lb

10-77   **(a)**   not allowed since $\dfrac{L}{r} = 404 > 200$

            (b)   440,000 N
            (c)   1,170,000 N

10-79   **(a)** 1.25×10^6 N

            (b) $\dfrac{L}{r} = 257 > 200$, too slender

10-81   M 6 × 33.8

10-83   213,000 lb

10-85   32,600 lb

10-87 $e \le 0.435$ in.

10-89   **(a)** $P \le 271{,}000$ N
            (b) $P \le 257{,}000$ N

10-91   **(a)** $P \le 47{,}300$
            (b) $P \le 36{,}000$ lb

10-93 $e \le 0.0059$ m

CHAPTER 11

11-1
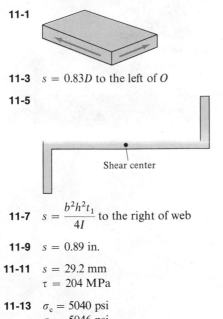

11-3 $s = 0.83D$ to the left of O

11-5

Shear center

11-7 $s = \dfrac{b^2 h^2 t_1}{4I}$ to the right of web

11-9 $s = 0.89$ in.

11-11 $s = 29.2$ mm
            $\tau = 204$ MPa

11-13 $\sigma_c = 5040$ psi
            $\sigma_t = 5046$ psi
            error $= 15.7\%, 16.7\%$

11-15 $\sigma_t = 509$ psi
            $\sigma_c = 416$ psi
            error $= 17\%, 15\%$

11-17 $b = 1.263$ in.

11-19 $R = 187.8$ mm

11-21 $\sigma_c = 6100$ psi
            $\sigma_t = 2106$ psi

11-23 $\sigma_A = 11{,}333$ psi
            $\sigma_B = -4844$ psi

11-25 $y_G = 25.558$ mm
            $\sigma_E = -56.7$ MPa
            $\sigma_A = 99.2$ MPa

11-27   In the wood:
            $\sigma_c = 2800$ psi
            $\sigma_t = 800$ psi

11-29 $\sigma_w = 1000$ psi; shear center is 0.475 in. from the bottom.

11-31   deflection of timber $= 0.52 \times$ deflection of reinforced beam

11-33 $\sigma_c = -1609$ psi
$\sigma_s = 30{,}918$ psi

11-35 $M = 1.24 \times 10^6$ in.-lb

11-37 $A_s = 2307$ mm^2

11-39 $M = 53{,}540$ N·m

CHAPTER 12

12-3 $U_t = \dfrac{1}{2E}(\sigma_x^2 + \sigma_y^2) - \dfrac{\mu}{E}(\sigma_x \sigma_y) + \dfrac{1}{2G}\tau_{xy}^2$

$U_v = \dfrac{1-2\mu}{6E}(\sigma_x^2 + \sigma_y^2 + 2\sigma_x\sigma_y)$

$U_d = \dfrac{1+\mu}{3E}(\sigma_x^2 + \sigma_y^2 - \sigma_x\sigma_y) + \dfrac{1}{2G}\tau_{xy}^2$

12-7 $U_d = 16{,}470$ Pa
$U_v = 4902$ Pa

12-9 $\sigma_{cr} = \tau_{PL}$
$\tau_{cr} = \tau_{PL}$

$\varepsilon_{cr} = \dfrac{\tau_{PL}}{2G}$

$\sigma_c = \sqrt{6}\,\tau_{PL}$

$U_{cr} = \dfrac{\tau_{PL}^2}{2G}$

12-11 For uniaxial test data in which $\sigma_{max} = \sigma_{ys}$ or σ_{PL}:

$\sigma_{max} = \dfrac{pD}{2t}$ (maximum normal stress theory)

$\sigma_{max} = \dfrac{pD}{2t}$ (maximum shear stress theory)

$\sigma_{max} = \dfrac{pD}{4t}\left(1 - \dfrac{\mu}{2}\right)$ (maximum normal strain theory)

$\sigma_{max} = \dfrac{pD}{2t}\sqrt{\dfrac{5}{4} - \mu}$ (total energy theory)

$\sigma_{max} = \dfrac{pD}{2t}\dfrac{\sqrt{3}}{2}$ (energy of distortion theory)

12-13 For torsion test data with $\tau_{max} = \tau_{PL}$:

$P = 2\pi r^2 \sqrt{\tau_{PL}^2 - \left(\dfrac{2T}{\pi r^3}\right)^2}$ (maximum shear stress theory)

$P = \sqrt{3}\pi r^2 \sqrt{\tau_{PL}^2 - \left(\dfrac{2T}{\pi r^3}\right)^2}$ (energy of distortion theory)

12-15 $r = 1.944$ in. based upon either the maximum sheer stress theory or the energy of distortion theory

CHAPTER 13

13-1 $W = \dfrac{P^2 L}{2AE}$

13-3 $\sigma_3 = 4\sigma_2 = 4\sigma_1/\sqrt{10}$

13-5 $W = 0.502$ N·m

13-7 $W = 0.513$ N·m

13-9 $W = \dfrac{T^2 L^3}{6JG}$

13-11 $W = \dfrac{P^2 L^3}{6EI}$

13-13 $\theta = \dfrac{TL^2}{2JG}$

13-15 $\delta = \dfrac{5wL^4}{384EI}$

13-17 $\delta_A = 10.9$ mm

13-19 $\delta_A = 0.513$ in.

13-21 $\delta_x = \dfrac{PL^3}{16EI}$, $\delta_y = \dfrac{5PL^3}{48EI}$

13-23 $\tau_{max} = \dfrac{TL}{\pi R^3}$

$\theta_{max} = \dfrac{TL^2}{3\pi R^4 G}$

13-25 $R_1 = R_2 = \dfrac{wL}{2}$

$M_1 = M_2 = \dfrac{wL^2}{12}$

13-27 $R_1 = \dfrac{3}{8} wL$

$R_2 = \dfrac{5}{8} wL$

$M_2 = \dfrac{-wL^2}{8}$

13-29 $R_1 = \dfrac{17}{16} wL$

$R_2 = -\dfrac{1}{16} wL$

$M_2 = \dfrac{-wL^2}{32}$

13-31 $\delta_V = 0.0366$ in.
$\delta_H = 0.00271$ in.

13-33 $\delta_H = 0.74$ mm
$\delta_V = 3.46$ mm

13-35 $\delta_H = 0.0504$ in.
$\delta_V = 0.206$ in.

CHAPTER 14

14-1   43,880 psi

14-3   70.7 MPa

14-4 11.66×10^3 psi

14-5 $P = 490$ lb

14-7   **(a)** $\sigma_{max} = 30,000$ psi
(b) $\sigma_{max} = 37,780$ psi

14-9 $\sigma_{max} = 76.81$ MPa

14-11   2681 N

14-13   97.2 hr

14-15 $P = 6201$ lb

14-17 $D = 3.308$ in.

14-19 $b = 1.953$ in.

14-21   812 mm

14-23   3 hr

14-25   fails in 11 hr

14-27 $y = 17.7$ in.
$\sigma = 9458$ psi
$\delta = 2.92$ in.

14-29   118 mm

14-31   0.087 in.

CHAPTER 15

15-1 $\varepsilon = 347.2$ μin./in.

15-3 $\Delta E_V = 0.019$ volts

15-5 $\Delta E_V = 0.025$ volts

Index

Standard SI Prefixes

NAME	SYMBOL	FACTOR
Tera	T	$1\ 000\ 000\ 000\ 000 = 10^{12}$
Giga	G	$1\ 000\ 000\ 000 = 10^{9}$
Mega	M	$1\ 000\ 000 = 10^{6}$
Kilo	k	$1\ 000 = 10^{3}$
Hecto*	h	$100 = 10^{2}$
Deka*	da	$10 = 10^{1}$
Deci*	d	$0.1 = 10^{-1}$
Centi*	c	$0.01 = 10^{-2}$
Milli	m	$0.001 = 10^{-3}$
Micro	μ	$0.000\ 001 = 10^{-6}$
Nano	n	$0.000\ 000\ 001 = 10^{-9}$
Pico	p	$0.000\ 000\ 000\ 001 = 10^{-12}$
Femto	f	$0.000\ 000\ 000\ 000\ 001 = 10^{-15}$
Atto	a	$0.000\ 000\ 000\ 000\ 000\ 001 = 10^{-18}$

* Use of these prefixes is not recommended but they are sometimes encountered.